Engineering

Mathematics

A Modern Foundation for Electronic, Electrical and Control Engineers

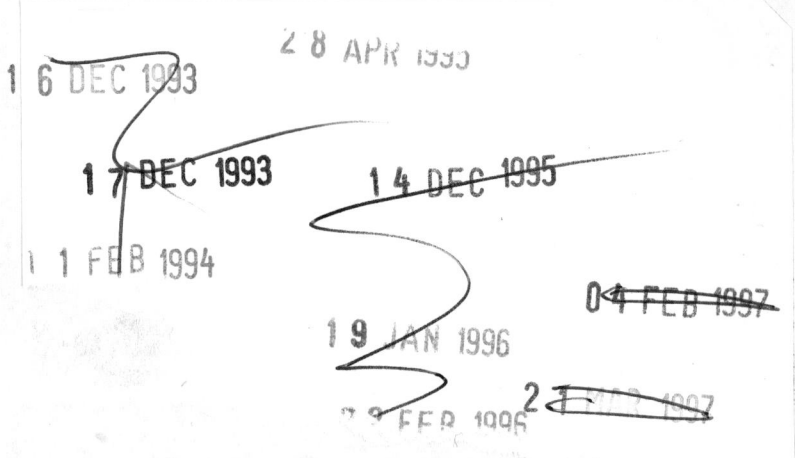

Modern Applications of Mathematics

Consulting Editors

Glyn James
Coventry University

Richard Clements
University of Bristol

Other titles in the series

Modern Engineering Mathematics   *Glyn James*
Transforms in Signals and Systems   *Peter Kraniauskas*

Engineering

Mathematics

A Modern Foundation for Electronic, Electrical and Control Engineers

Anthony Croft
Robert Davison
Martin Hargreaves

Centre for Engineering Mathematics, De Montfort University, Leicester, UK

 Addison-Wesley Publishing Company

Wokingham, England • Reading, Massachusetts • Menlo Park, California • New York
Don Mills, Ontario • Amsterdam • Bonn • Sydney • Singapore • Tokyo • Madrid
San Juan • Milan • Paris • Mexico City • Seoul • Taipei

The programs in this book have been included for their instructional
value. They have been tested with care but are not guaranteed for any
particular purpose. The publisher does not offer any warranties or
representations, nor does it accept any liabilities with respect to
the programs.

Many of the designations used by manufacturers and sellers to
distinguish their products are claimed as trademarks. Addison-Wesley
has made every attempt to supply trademark information about
manufacturers and their products mentioned in this book.

Cover designed by Designers & Partners of Oxford and
printed by The Riverside Printing Co. (Reading) Ltd.
Typeset by P & R Typesetters Ltd. (Salisbury, UK).
Printed and bound in Great Britain by
William Clowes Limited, Beccles and London

First printed 1992. Reprinted 1992.

British Library Cataloguing in Publication Data
A catalogue record for this book is available from
the British Library

Library of Congress Cataloging in Publication Data
Croft, Tony, 1957–
    Engineering mathematics: a modern foundation for electrical,
electronic, and control engineers/Anthony Croft, Robert Davison,
Martin Hargreaves.
        p.   cm.
    Includes index.
    ISBN 0-201-17557-6
    1. Engineering mathematics.   2. Electric engineering – Mathematics.
3. Electronics – Mathematics.   4. Automatic control—Mathematics.
I. Davison, Robert.   II. Hargreaves, Martin.   III. Title.
TA330.076 1992
510'.2462 – dc20   92-8335
                                                                  CIP

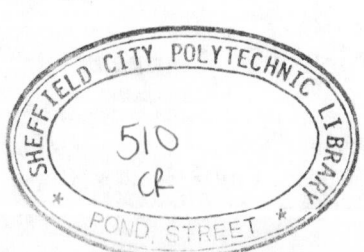

Dedications

A.C. To Jan and Thomas
R.D. To Charlie and Wikki
M.H. To the memory of my dear father, George Frederick Hargreaves

Preface

This book is designed primarily for first-year students in higher education engaged in courses on electronic, electrical and control engineering. The book starts at a level which is appropriate to readers who may not have come through the traditional educational channels. We have taken care to increase gradually the mathematical sophistication of the book in order to make it suitable not only for first-year undergraduates, but also for a wider readership, for example, HND and Combined Study degrees. The style makes it suitable for engineers who wish to engage in self-study and continuing education.

Engineers are called upon to analyse a variety of engineering systems, which can be anything from a few electronic components connected together through to a complete electrical distribution network. The analysis of these systems benefits from the intelligent application of mathematics. Indeed, many cannot be analysed without the use of mathematics. As engineering mathematics is an interdisciplinary subject it benefits from cooperation of people in both disciplines. Such cooperation has been fundamental throughout the writing of this book. One of the authors is an engineer and the other two are mathematicians. The applied nature of the book reflects this close collaboration. Every chapter includes fully worked modern engineering examples, clearly showing the relevance of the mathematics to the engineer of the 1990s. The book is structured so that the mathematical aspects can be studied without reference to the engineering examples. This makes it suitable for those interested primarily in mathematics. Throughout the text the notation used by engineers has been preferred to that used by mathematicians.

The style of the book is to develop and illustrate mathematical concepts through examples. We have tried throughout to adopt an informal approach and to describe mathematical processes using everyday language. Mathematical ideas are often developed by examples rather than by using formal proof which has been kept to a minimum. This reflects the authors' experience that engineering students learn better from practical examples, rather than from formal abstract development. Important results are highlighted for easy reference. Exercises are provided at the end of most sections; it is essential to attempt these as the only way to develop competence and understanding is through practice. A further set of miscellaneous exercises is provided at the end of each chapter. Answers to all exercises appear at the back of the book.

There is a danger that learning mathematics can become very mechanical. Some texts have a tendency to concentrate on the 'how' rather than the 'why'. This is a limited approach, leaving the reader unprepared to make the transition from contrived textbook problems to the more complex world of real engineering. This book motivates the reader by explaining why mathematics is used, thus enabling mathematics to be approached as a tool in solving engineering problems. Mathematics is the language of engineering and so it is essential to understand why it works in order to master the complex relationships present in modern engineering systems and products. This text uses engineering examples which help the reader to see the uses and benefits of the various mathematical techniques that are presented.

The most pertinent feature of the book is the wide-ranging use of engineering examples. These permeate the entire text. They range from short illustrative examples through to complete sections which can be regarded as case studies. This innovative approach has produced a book which is rare in the field of engineering mathematics; a genuine fusion of engineering and mathematics. Moreover, the examples have been carefully selected to be relevant, informative and modern. They can easily be picked out in the text as they have distinguishing titles. This dove-tailing of engineering and mathematics serves a further purpose: it helps convince the reader that the mathematical material is relevant and has not been included simply for intrinsic interest. Sterile, contrived and unrealistic examples, which are a feature of some texts, have been avoided, and the involvement of an engineer in the writing of the book has ensured that out-dated techniques have been discarded.

Traditional topics have been covered where relevant. However, the changing needs of engineers require greater emphasis to be placed on more up-to-date techniques. The growing importance of computers is reflected in the increasing use of discrete mathematics by the modern engineer. Thus we have included set theory, sequences and series, Boolean algebra, logic, difference equations and the z transform. The increased use of signal processing techniques is reflected by a greater emphasis on integral transform methods. Thus the Laplace, z and Fourier transforms have been given greater coverage than is normally found in a first-year text. Chapter 1 includes a review of some important functions and techniques that the reader is likely to have met in previous courses. This material ensures the book is self-contained and provides a convenient reference. Inevitably the choice of contents is a compromise. The topics covered were chosen after wide consultation coupled with many years teaching experience. Given the constraint of space we believe our choice is optimal.

Finally, we hope you will come to share our enthusiasm for engineering mathematics and enjoy the book.

Anthony Croft
Robert Davison
Martin Hargreaves
March 1992

Acknowledgements

We wish to thank all of the staff at Addison-Wesley who have been involved in the production of this book. In particular, a special thanks to our commissioning editor, Tim Pitts, and production editor, Susan Keany.

We also wish to thank the following colleagues who have read and commented upon various drafts of the book: Dr Eric Chowanietz, Professor Don Conway, Professor Bryan Coulbeck, Dr Martin Crane, Jane Fudge, Derek Hart, Mike Howkins, Cyril Lander, Professor Malcolm McCormick, Andrejs Ozolins, Brian Riches, Dr Harry Sharma, Dr Roy Smith, Dr David Storry, John Taylor, Don Thatcher and Bill Thompson. A particular mention must go to John Royle for his thorough appraisal of the manuscript and valuable comments.

Finally, the comments of unknown referees were most useful and are gratefully acknowledged.

Contents

Engineering functions $\quad$ 1

1.1 Introduction

The study of functions is central to engineering mathematics. Functions can be used to describe the way quantities change, for example, the variation in the voltage across an electronic component with time, the variation in position of an electric motor with time and the variation in the strength of a signal with both position and time.

In this chapter we introduce several concepts associated with functions before going on to catalogue a number of engineering functions in Section 1.4. Much of the material of Section 1.4 will already be familiar to the reader and so this section should be treated as a reference section to be dipped into whenever necessary. A number of mathematical methods are also included in Section 1.4, most of which will be familiar but they have been collected together in order to make the book complete.

When trying to understand a mathematical function it is always useful to sketch a graph in order to obtain an idea of its behaviour. The reader is encouraged to sketch such graphs whenever a new function is met. Graphics calculators are now readily available and they make this task relatively easy. If you possess such a calculator then it would be useful to make use of it whenever a new function is introduced. Software packages are also available to allow such plots to be carried out on a computer. These can be useful for plotting more complicated functions and ones that depend on more than one variable. We examine functions of more than one variable in Chapter 14.

Throughout the book we make use of the term **mathematical model**. When doing so we mean an idealization of an engineering system or a physical situation so that it can be described by mathematical equations. To reflect an engineering system very accurately, a sophisticated model, consisting of many interrelated equations may be needed. Although accurate, such a model may be cumbersome to use. Accuracy can be sacrificed in order to achieve a simple, easy-to-use model. A judgement is made as to when the right blend of accuracy and conciseness is achieved. For example, the most common mathematical model for a resistor uses Ohm's law which states that the voltage across a resistor equals the current through the resistor multiplied by the resistance value of the resistor, that is, $V = IR$. However, this model is based on a number of simplifications. It ignores any variation in current density across the cross-section of the resistor and assumes a single current value is acceptable. It also ignores the fact that if a large enough voltage is placed across the resistor then the resistor will break down. In most cases it is worth accepting these simplifications in order to obtain a concise model.

Having obtained a mathematical model, it is then used to predict the effect of changing elements or conditions within the actual system. Using the model to examine these effects is often cheaper, safer and more convenient than using the actual system.

1.2 Numbers and intervals

Numbers can be grouped into various classes, or **sets**. The **integers** are the set of numbers

$$\{\dots, -3, -2, -1, 0, 1, 2, 3, \dots\}$$

denoted by $\mathbb{Z}$. The **natural numbers** are $\{0, 1, 2, 3, \dots\}$ and this set is denoted by $\mathbb{N}$. The **positive integers**, denoted by $\mathbb{N}^+$, are given by $\{1, 2, 3, \dots\}$. Note that some numbers occur in more than one set, that is, the sets overlap.

A **rational number** has the form p/q, where p and q are integers with $q \neq 0$. For example, $5/2$, $7/118$, $-1/9$ and $3/1$ are all rational numbers. The set of rational numbers is denoted by $\mathbb{Q}$. When rational numbers are expressed as a decimal fraction they either terminate or recur infinitely.

$\frac{5}{2}$ can be expressed as 2.5 ⎱ These decimal fractions terminate,
$\frac{1}{8}$ can be expressed as 0.125 ⎰ that is, they are of finite length.

$\frac{1}{9}$ can be expressed as 0.111 111... ⎱ These are infinitely recurring
$\frac{1}{11}$ can be expressed as 0.090 909... ⎰ decimal fractions.

A number which cannot be expressed in the form p/q is called **irrational**. When written as a decimal fraction, an irrational number is infinite in length and non-recurring. The numbers π and $\sqrt{2}$ are both irrational.

It is useful to represent numbers by points on the **real line**. Figure 1.1 illustrates some rational and irrational numbers marked on the real line. Numbers which can be represented by points on the real line are known as **real numbers**. The set of real numbers is denoted by $\mathbb{R}$. This set comprises all the rational and all the irrational numbers. In Chapter 6 we shall meet complex numbers which cannot be represented as points on the real line. The real line extends indefinitely to the left and to the right so that any real number can be represented.

Sometimes we are interested in only a small section, or **interval**, of the real line. We write $[1, 3]$ to denote all the real numbers between 1 and 3 inclusive, that

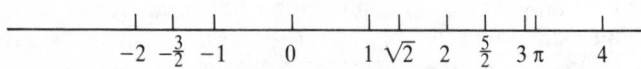

Figure 1.1   Both rational and irrational numbers are represented on the real line.

Figure 1.2 The intervals $(-6, -4)$, $[-1, 2]$, $(3, 4]$ depicted on the real line.

is, 1 and 3 are included in the interval. Thus the interval $[1, 3]$ consists of all real numbers x, such that $1 \leq x \leq 3$. The square brackets, $[\,]$, are used to denote that the end-points are included in the interval and such an interval is said to be **closed**. The interval $(1, 3)$ consists of all real numbers x, such that $1 < x < 3$. In this case the end-points are not included and the interval is said to be **open**. Parentheses, $(\,)$, denote open intervals. An interval may be open at one end and closed at the other. For example, $(1, 3]$ is open at the left and closed at the right. It consists of all real numbers x, such that $1 < x \leq 3$, and is known as a **semi-open** interval. Open and closed intervals can be represented on the real line. A closed end-point is denoted by ●; an open end-point is denoted by ○. The intervals $(-6, -4)$, $[-1, 2]$ and $(3, 4]$ are illustrated in Figure 1.2.

1.3 Basic concepts of functions

Loosely speaking, we can think of a function as a rule which, when given an input, produces a single output. If more than one output is produced, the rule is not a function. Consider the function given by the rule: 'double the input'. If 3 is the input then 6 is the output. If x is the input then $2x$ is the output, as shown in Figure 1.3.

If the doubling function has the symbol f we write:

$$f : x \rightarrow 2x$$

or more compactly,

$$f(x) = 2x$$

The last form is often written simply as $f = 2x$. If $f(x)$ is a function of x, then the value of the function when $x = 3$, for example, is written as $f(x = 3)$ or simply as $f(3)$.

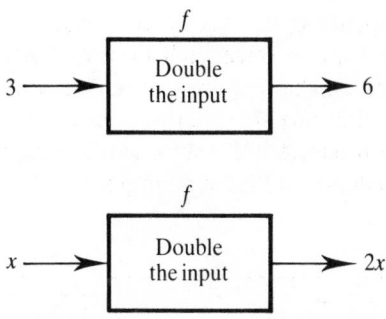

Figure 1.3 The function: 'double the input'.

Example 1.1

Given $f(x) = 2x + 1$ find:

(a) $f(3)$
(b) $f(0)$
(c) $f(-1)$
(d) $f(\alpha)$
(e) $f(2\alpha)$
(f) $f(t)$
(g) $f(t + 1)$

Solutions

(a) $f(3) = 2(3) + 1 = 7$
(b) $f(0) = 2(0) + 1 = 1$
(c) $f(-1) = 2(-1) + 1 = -1$
(d) $f(\alpha)$ is the value of $f(x)$ when x has a value of α, hence $f(\alpha) = 2\alpha + 1$
(e) $f(2\alpha) = 2(2\alpha) + 1 = 4\alpha + 1$
(f) $f(t) = 2t + 1$
(g) $f(t + 1) = 2(t + 1) + 1 = 2t + 3$ ▲

Observe from Example 1.1 that it is the rule that is important and not the letter being used. Both $f(t) = 2t + 1$ and $f(x) = 2x + 1$, instruct us to double the input and then add 1.

1.3.1 *Argument of a function*

The input to a function is often called the **argument**. In Example 1.1(d) the argument is α, while in Example 1.1(e) the argument is 2α.

Example 1.2

Given $f(x) = x/5$, write down:

(a) $f(5x)$
(b) $f(-x)$
(c) $f(x + 2)$
(d) $f(x^2)$

Solutions

(a) $f(5x) = 5x/5 = x$
(b) $f(-x) = -x/5$

(c) $f(x + 2) = (x + 2)/5$

(d) $f(x^2) = x^2/5$ ▲

Example 1.3

Given $y(t) = t^2 + t$, write down:

(a) $y(t + 2)$

(b) $y(t/2)$

Solutions

(a) $y(t + 2) = (t + 2)^2 + (t + 2) = t^2 + 5t + 6$

(b) $y(t/2) = (t/2)^2 + (t/2) = t^2/4 + t/2$ ▲

1.3.2 *Graph of a function*

A function may be represented in graphical form. The function $f(x) = 2x$ is shown in Figure 1.4. Note that the function values are plotted vertically and the x values horizontally. The horizontal axis is then called the x axis. The vertical axis is commonly referred to as the y axis, so that we often write:

$$y = f(x) = 2x$$

Since x and y can have a number of possible values, they are called **variables**: x is the **independent variable** and y is the **dependent variable**. Knowing a value of the independent variable, x, allows us to calculate the corresponding value of the dependent variable, y. To show this dependence we often write $y(x)$. The set of values that x is allowed to take is called the **domain** of the function. A domain is often an interval on the x axis. For example, if

$$f(x) = 3x + 1 \qquad -5 \leq x \leq 10 \tag{1.1}$$

the domain of the function, f, is the closed interval $[-5, 10]$. If the domain of a function is not explicitly given it is taken to be the largest set possible. For example,

$$g(x) = x^2 - 4 \tag{1.2}$$

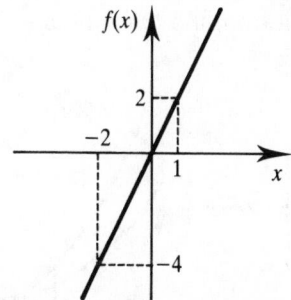

Figure 1.4 The function: $f(x) = 2x$.

has a domain of $(-\infty, \infty)$ since g is defined for every value of x and the domain has not been given otherwise. The set of values that the function takes on is called the **range**. The range of $f(x)$ in Equation (1.1) is $[-14, 31]$; the range of $g(x)$ in Equation (1.2) is $[-4, \infty)$.

Example 1.4

Consider the function, f, given by the rule: 'square the input'. This can be written as:

$$f(x) = x^2$$

The rule and the graph of f are shown in Figure 1.5. The domain of f is $(-\infty, \infty)$ and the range is $[0, \infty)$. ▲

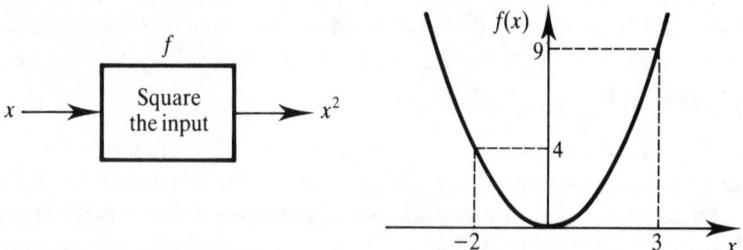

Figure 1.5   The function: 'square the input'.

Many variables of interest to engineers, for example, voltage, resistance and current, can be related by means of functions. We try to choose an appropriate letter for a particular variable; so, for example, t is used for time and P for power.

Example 1.5   Power dissipated in a resistor

The power, P, dissipated in a resistor depends on the current, I, flowing through the resistor and the resistance, R. The relationship is given by:

$$P = I^2R$$

The power dissipated in the resistor depends on the square of the current passing through it. In this case I is the independent variable and P is the dependent variable, assuming R remains constant. The function is given by the rule: 'square the input and multiply by the constant R', and the input to the function is I. The output from the function is P. This is illustrated in Figure 1.6. ▲

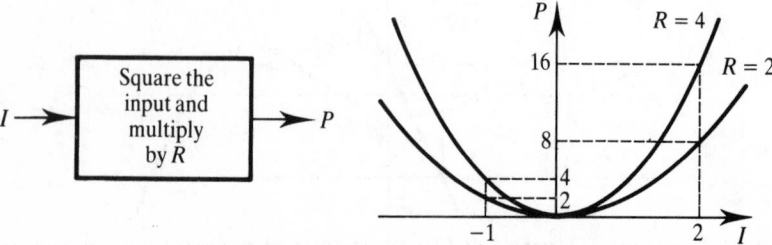

Figure 1.6   The function: $P = I^2R$.

1.3.3 *One-to-many*

Some rules relating input to output are not functions. Consider the rule: 'take plus or minus the square root of the input', that is,

$$x \rightarrow \pm\sqrt{x}$$

Now, for example, if 4 is the input, the output is $\pm\sqrt{4}$ which can be 2 or -2. Thus a single input has produced more than one output. The rule is said to be **one-to-many** meaning that one input has produced many outputs. Rules with this property are not functions. For a rule to be a function there must be a single output for any given input.

By defining a rule more specifically, it may become a function. For example, consider the rule: 'take the positive square root of the input'. This rule is a function because there is a single output for a given input. Note that the domain of this function is $[0, \infty)$ and the range is also $[0, \infty)$.

1.3.4 *Many-to-one and one-to-one functions*

Consider again the function $f(x) = x^2$ given in Example 1.4. The inputs 2 and -2 both produce the same output, 4, and the function is said to be **many-to-one**. This means that many inputs produce the same output. A many-to-one function can be recognized from its graph. If a horizontal line intersects the graph in more than one place, the function is many-to-one. Figure 1.7 illustrates a many-to-one function, $g(x)$. The inputs x_1, x_2, x_3 and x_4 all produce the same output.

A function is **one-to-one** if different inputs always produce different outputs. A horizontal line will intersect the graph of a one-to-one function in only one place. Figure 1.8 illustrates a one-to-one function, $h(x)$.

1.3.5 *Composition of functions*

Consider the function $y(x) = 2x^2$. We can think of $y(x)$ as being composed of two functions. One function is described by the rule: 'square the input' while the other function is described by the rule: 'double the input'. This is shown in Figure 1.9.

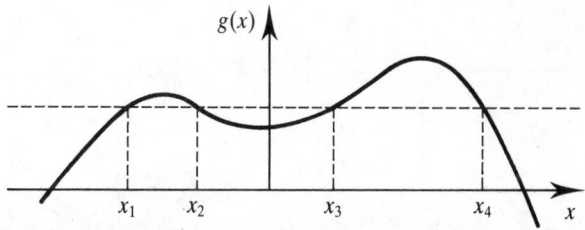

Figure 1.7 The inputs x_1, x_2, x_3 and x_4 all produce the same output, therefore $g(x)$ is a many-to-one function.

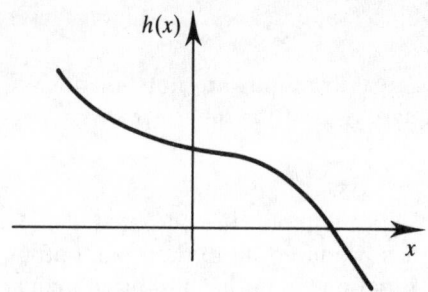

Figure 1.8 Each input produces a different output and so $h(x)$ is a one-to-one function.

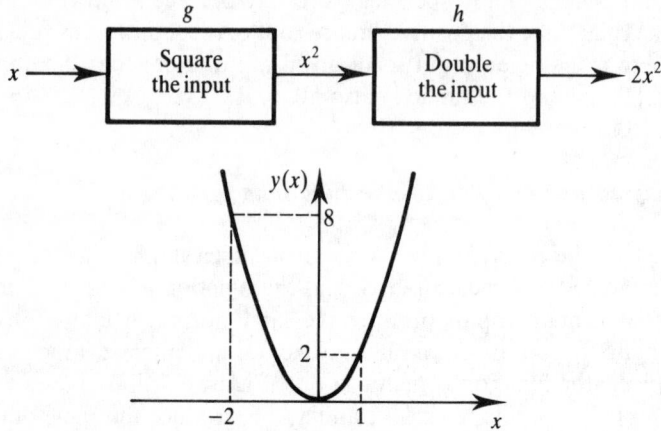

Figure 1.9 The function: $y(x) = h(g(x))$.

Mathematically, if $h(x) = 2x$ and $g(x) = x^2$ then,

$$y(x) = 2x^2 = 2(g(x)) = h(g(x))$$

The form $h(g(x))$ is known as a **composition** of the functions h and g. Note that the composition $h(g(x))$ is different from $g(h(x))$ as Example 1.6 illustrates.

Example 1.6

If $f(t) = 2t + 3$ and $g(t) = (t + 1)/2$ write expressions for the compositions

(a) $f(g(t))$

(b) $g(f(t))$

Solutions

(a) $f(g(t)) = f\left(\dfrac{t + 1}{2}\right)$

The rule describing the function f is: 'double the input and then add 3'. Hence,

$$f\left(\frac{t+1}{2}\right) = 2\left(\frac{t+1}{2}\right) + 3 = t + 4$$

So

$$f(g(t)) = t + 4$$

(b) $g(f(t)) = g(2t + 3)$

The rule for g is: 'add 1 to the input and then divide everything by 2'. So,

$$g(2t + 3) = \frac{2t + 3 + 1}{2} = t + 2$$

Hence

$$g(f(t)) = t + 2$$

Clearly $f(g(t)) \neq g(f(t))$.   ▲

1.3.6 *Inverse of a function*

Consider a function $f(x)$. It can be thought of as accepting an input x, and producing an output $f(x)$. Suppose now that this output becomes the input to the function $g(x)$, and the output from $g(x)$ is x, that is,

$$g(f(x)) = x$$

We can think of $g(x)$ as undoing the work of $f(x)$. Figure 1.10 illustrates this situation. Then $g(x)$ is the **inverse** of f, and is written as $f^{-1}(x)$.

Example 1.7

If $f(x) = 5x$ verify that the inverse of f is given by $f^{-1}(x) = x/5$.

Solution

$$f^{-1}(f(x)) = f^{-1}(5x) = \frac{5x}{5} = x$$

Hence $f^{-1}(x) = x/5$ is the inverse of $f(x) = 5x$.   ▲

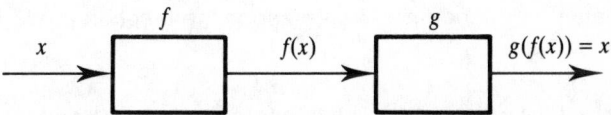

Figure 1.10   The function g is the inverse of f.

Example 1.8

If $f(x) = 2x + 1$, find $f^{-1}(x)$.

Solution

We know $f(x) = 2x + 1$ and $f^{-1}(2x + 1) = x$. The function f is given by the rule: 'double the input and then add 1' and so the inverse of $f(x)$ is given by the rule: 'subtract 1 from the input then divide by 2', that is,

$$f^{-1}(x) = \frac{x - 1}{2}$$

This inverse can be derived in a more mathematical way.

Let $y = 2x + 1$ so $f^{-1}(y) = x$. By rearrangement:

$$x = \frac{y - 1}{2}$$

and so,

$$f^{-1}(y) = \frac{y - 1}{2}$$

The inverse, f^{-1}, has been found with y as the independent variable, that is, with y as input. With an input of x we have:

$$f^{-1}(x) = \frac{x - 1}{2}$$

Once again note that the letter used for the argument of the inverse is unimportant.   ▲

Example 1.9

Given $g(x) = (x - 1)/2$ find the inverse of g.

Solution

We know $g(x) = (x - 1)/2$, and so $g^{-1}((x - 1)/2) = x$. Let $y = (x - 1)/2$ so that:

$$g^{-1}(y) = x$$

But,

$$x = 2y + 1$$

and so,

$$g^{-1}(y) = 2y + 1$$

Using the same independent variable as for the function g, we obtain:

$$g^{-1}(x) = 2x + 1$$ ▲

We note that the inverses of the functions in Examples 1.8 and 1.9 are themselves functions. They are called **inverse functions**. The inverse of $f(x) = 2x + 1$ is

$f^{-1}(x) = (x - 1)/2$, and the inverse of $g(x) = (x - 1)/2$ is $g^{-1}(x) = 2x + 1$. This illustrates the important point that if $f(x)$ and $g(x)$ are two functions and $f(x)$ is the inverse of $g(x)$, then $g(x)$ is the inverse of $f(x)$. It is important to point out that not all functions possess an inverse function. Consider $f(x) = x^2$, for $-\infty < x < \infty$.

The function, f, is given by the rule: 'square the input'. Since both a positive and negative value of x will yield the output x^2, the inverse rule is given by: 'take plus or minus the square root of the input'. As discussed earlier, this is a one-to-many rule and so is not a function. Clearly not all functions have an inverse function. In fact, only one-to-one functions have an inverse function. Suppose we restrict the domain of $f(x) = x^2$ such that $x \geq 0$. Then f is a one-to-one function and so there is an inverse function. The inverse function is $f^{-1}(x)$ given by:

$$f^{-1}(x) = +\sqrt{x}$$

Clearly,

$$f^{-1}(f(x)) = f^{-1}(x^2) = x$$

where x is the positive square root of x^2. Restricting the domain of a many-to-one function so that a one-to-one function results is a common technique of ensuring an inverse function can be found.

1.3.7 *Continuous and piecewise continuous functions*

We now introduce in an informal way, the concept of continuous and piecewise continuous functions. A more rigorous treatment follows in Chapter 7 after we have discussed limits. Figure 1.11 shows a graph of $f(x) = 1/x$. Note that there is a break, or discontinuity, in the graph at $x = 0$. The function $f(x) = 1/x$ is said to be **discontinuous** at $x = 0$.

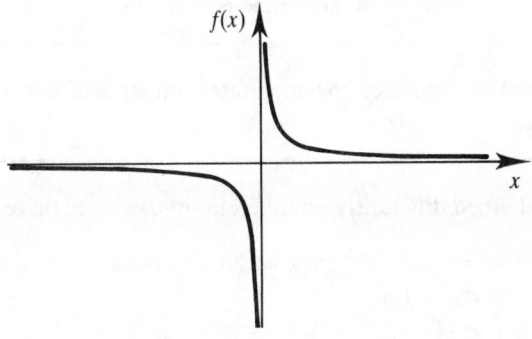

Figure 1.11 The function $f(x) = \dfrac{1}{x}$ has a discontinuity at $x = 0$.

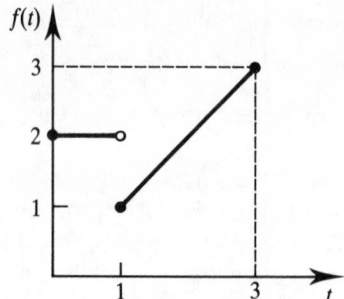

Figure 1.12 The function $f(t)$ is a piecewise continuous function with a discontinuity at $t = 1$.

■ If the graph of a function, $f(x)$, contains a break, then $f(x)$ is discontinuous.

■ A function whose graph has no breaks is a **continuous** function.

Sometimes a function is defined by different rules on different intervals of the domain. For example, consider

$$f(t) = \begin{cases} 2 & 0 \le t < 1 \\ t & 1 \le t \le 3 \end{cases}$$

The domain is $[0, 3]$ but the rule on $[0, 1)$ is different to that on $[1, 3]$. The graph of $f(t)$ is shown in Figure 1.12. Recall the convention of using ● to denote that the end-point is included and ○ to denote the end-point is excluded. Note that $f(t)$ has a discontinuity at $t = 1$. Each component, or piece, of the graph is continuous and $f(t)$ is said to be **piecewise continuous**.

■ A piecewise continuous function has a finite number of discontinuities in any given interval.

Not all functions defined differently on different intervals are piecewise continuous. For example,

$$g(t) = \begin{cases} 2 & 0 < t < 1 \\ 2t & 1 \le t < 3 \end{cases}$$

is a continuous function on the interval $(0, 3)$, as shown in Figure 1.13.

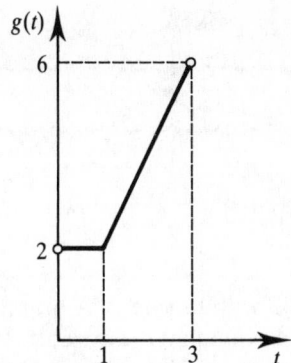

Figure 1.13   The function $g(t)$ is a continuous function on $(0, 3)$.

1.3.8 *Periodic functions*

A **periodic function** is a function which has a definite pattern which is repeated at regular intervals. More formally we say a function, $f(t)$, is periodic if:

$$f(t) = f(t + T)$$

for all values of t. The constant, T, is known as the **period** of the function.

Example 1.10   Triangular waveform

Figure 1.14 illustrates a triangular waveform. The form of the function is repeated every two seconds, that is,

$$f(t) = f(t + 2)$$

and so the function is periodic. The period is 2 seconds, that is, $T = 2$. Note that this function is continuous.   ▲

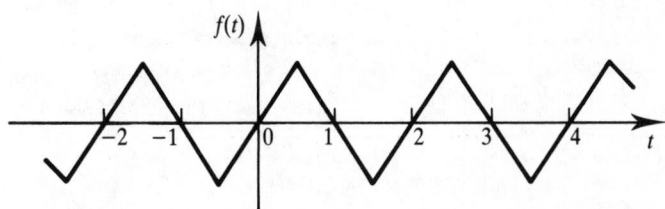

Figure 1.14   The triangular waveform is a periodic function.

Example 1.11   Square waveform

Periodic functions may be piecewise continuous. Consider the function $g(t)$ defined by

$$g(t) = \begin{cases} 1 & 0 \leq t < 1 \\ 0 & 1 \leq t < 2 \end{cases} \qquad \text{Period} = 2$$

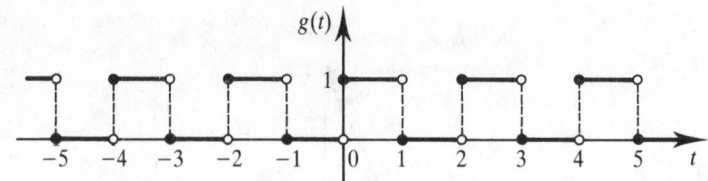

Figure 1.15 The function $g(t)$ is both piecewise continuous and periodic.

The function $g(t)$ is periodic with period 2. A graph of $g(t)$ is shown in Figure 1.15. This function is commonly referred to as a square waveform by engineers. In Figure 1.15 the open and closed end-points have been shown for mathematical correctness. Note, however, that engineers tend to omit these when sketching functions with discontinuities and usually they use a vertical line to show the discontinuity. This reflects the fact that no practical waveform can ever change its level instantaneously: even very fast rising waveforms still have a finite **risetime**. The function has discontinuities at $t = \ldots, -3, -2, -1, 0, 1, 2, 3, 4, 5, \ldots$ ▲

EXERCISES 1.1

1. Describe the rule associated with the following functions, sketch their graphs and state their domains and ranges.

   (a) $f(x) = 2x^2$

   (b) $f(x) = x^2 - 1$ $0 \leq x$

   (c) $g(t) = 3t - 4$ $0 \leq t$

   (d) $y(x) = x^3$

   (e) $f(t) = 0.5t + 2$ $-2 \leq t \leq 10$

   (f) $z(x) = 3x - 2$ $3 \leq x \leq 8$

2. If $f(x) = 5x + 4$, find:

   (a) $f(3)$

   (b) $f(-3)$

   (c) $f(\alpha)$

   (d) $f(x + 1)$

   (e) $f(3\alpha)$

   (f) $f(x^2)$

3. If $g(t) = 5t^2 - 4$, find:

   (a) $g(0)$

   (b) $g(2)$

   (c) $g(-3)$

   (d) $g(x)$

   (e) $g(2t - 1)$

4. The reactance, X_C, offered by a capacitor is given by $X_C = 1/(2\pi fC)$, where f is the frequency of the applied alternating current, and C is the capacitance of the capacitor. If $C = 10^{-6}$ F, find X_C when $f = 50$ Hz.

5. Find the inverse of the following functions:

   (a) $f(x) = x + 4$

   (b) $g(t) = 3t + 1$

   (c) $y(x) = x^3$

   (d) $h(t) = (t - 8)/3$

6. Given $f(t) = t^2 + 1$, $g(t) = 3t + 2$ and $h(t) = 1/t$, write expressions for:

   (a) $f(g(t))$

   (b) $f(h(t))$

   (c) $g(h(t))$

   (d) $h(f(t))$

   (e) $f(g(h(t)))$

7. Given $f(t) = 2t + 3$ and $g(t) = 3t$ and $h(t) = f(g(t))$ write expressions for:

   (a) $h(t)$

   (b) $f^{-1}(t)$

   (c) $g^{-1}(t)$

(d) $h^{-1}(t)$

(e) $g^{-1}(f^{-1}(t))$

What do you notice about (d) and (e)?

8. Sketch

$$f(t) = \begin{cases} t & 0 \leq t < 2 \\ 5 - 2t & 2 \leq t < 3 \end{cases}$$

Is the function piecewise continuous or continuous? State, if they exist, the position of any discontinuities.

9. The function $g(t)$ is defined by:

$$g(t) = \begin{cases} 1 & 0 \leq t \leq 1 \\ 2 - t & 1 < t < 2 \end{cases}$$

and $g(t)$ has period 2. Sketch $g(t)$ on the interval $[-1, 4]$. State any points of discontinuity.

1.4 Review of some common engineering functions and techniques

This section provides a catalogue of the more common engineering functions. The important properties and definitions are included together with some techniques. It is intended that readers will refer to this section for revision purposes and as the need arises throughout the rest of the book.

1.4.1 *Polynomial functions*

■ A **polynomial expression** has the form:

$$a_n x^n + a_{n-1} x^{n-1} + a_{n-2} x^{n-2} + \cdots + a_2 x^2 + a_1 x + a_0$$

where n is a non-negative integer, $a_n, a_{n-1}, \ldots, a_1, a_0$ are constants and x is a variable. A **polynomial function**, $P(x)$, has the form:

$$P(x) = a_n x^n + a_{n-1} x^{n-1} + a_{n-2} x^{n-2} + \cdots + a_2 x^2 + a_1 x + a_0 \tag{1.3}$$

Examples of polynomial functions include:

$$P_1(x) = 3x^2 - x + 2 \tag{1.4}$$

$$P_2(z) = 7z^4 + z^2 - 1 \tag{1.5}$$

$$P_3(t) = 3t + 9 \tag{1.6}$$

$$P_4(t) = 6 \tag{1.7}$$

where x, z and t are independent variables. It is common practice to contract the term polynomial expression to **polynomial**. By convention, a polynomial is usually written with the powers either increasing or decreasing. For example,

$$3x + 9x^2 - x^3 + 2$$

would be written either as:

$$-x^3 + 9x^2 + 3x + 2 \qquad \text{or} \qquad 2 + 3x + 9x^2 - x^3$$

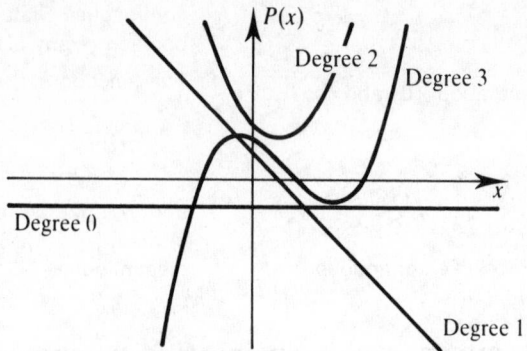

Figure 1.16   Some typical polynomials.

The **degree** of a polynomial or polynomial function is the value of the highest power. Equation (1.4) has degree 2, Equation (1.5) has degree 4, Equation (1.6) has degree 1 and Equation (1.7) has degree 0. Equation (1.3) has degree n. Polynomials with low degrees have special names. These are:

Polynomial	Degree	Name
$ax^4 + bx^3 + cx^2 + dx + e$	4	Quartic
$ax^3 + bx^2 + cx + d$	3	Cubic
$ax^2 + bx + c$	2	Quadratic
$ax + b$	1	Linear
a	0	Constant

Typical graphs of some polynomial functions are shown in Figure 1.16.

Example 1.12   Ohm's law

Ohm's law relates the current through a resistor to the voltage applied across it. The equation is:

$$V = IR$$

where V = voltage across the resistor;
   I = current through the resistor;
   R = resistance value of the resistor, which is a constant for a given temperature.

Note that the voltage is a linear polynomial function with I as the independent variable.   ▲

Example 1.13   A non-ideal voltage source

An ideal voltage source has zero internal resistance and its output voltage, V, is independent of the load applied to it, that is, V remains constant, independent of the current it supplies. Figure 1.17 shows a non-ideal voltage source. It is modelled as an ideal voltage source in series with an internal resistor with resistance R_s. The output voltage of the non-ideal voltage source is v_0 while v_R is the voltage drop across the internal resistor and i is the load current. Using Kirchhoff's voltage law,

$$V = v_R + v_0$$

and hence by Ohm's law,

$$V = iR_s + v_0$$

$$v_0 = V - iR_s$$

Note that V and R_s are constants and so the output voltage is a linear polynomial function with independent variable i. The equation gives the output voltage across the load as a function of the current through the load. The output characteristic for the non-ideal voltage source is obtained by varying the load resistor R_L and is plotted in Figure 1.18. Notice that the output voltage of the source decreases as the load current increases and is equal in value to the ideal voltage source only when there is no load current. ▲

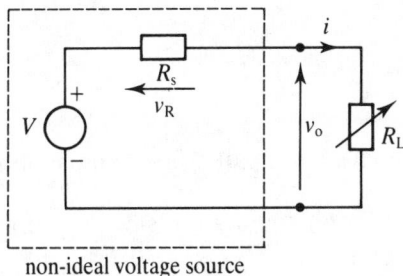

non-ideal voltage source

Figure 1.17   A non-ideal voltage source connected to a load resistor, R_L.

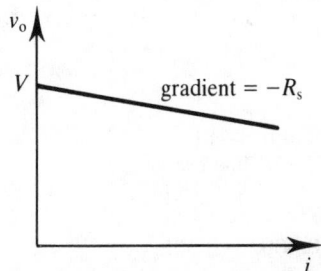

Figure 1.18   Output characteristic of a non-ideal voltage source.

Polynomial equations

■ A **polynomial equation** has the form

$$P(x) = 0 \tag{1.8}$$

where $P(x)$ is given in Equation (1.3)

The **roots** of the equation are those values of x which satisfy Equation (1.8), that is, $x = x_1$ is a root of $P(x) = 0$ if $P(x_1) = 0$. A polynomial equation of degree n has n roots.

Example 1.14

Find the roots of:

(a) $3x + 8 = 0$
(b) $x^2 + 6x - 4 = 0$
(c) $x^2 - 2x + 1 = 0$.

Solutions

(a) $3x + 8 = 0$

$$3x = -8$$

$$x = -\tfrac{8}{3}$$

The equation is linear, that is, degree 1, and so it has one root.

(b) Recall the formula for finding the roots of a quadratic equation. If

$$ax^2 + bx + c = 0$$

then

$$x = \frac{-b \pm \sqrt{b^2 - 4ac}}{2a}$$

This leads to:

$$x = \frac{-6 \pm \sqrt{36 - 4(1)(-4)}}{2}$$

$$= \frac{-6 \pm \sqrt{52}}{2} = 0.606, \ -6.606$$

The equation has degree 2 and so it has two roots.

(c) Factorizing the left-hand side yields:

$$x^2 - 2x + 1 = (x - 1)(x - 1) = 0$$

Each factor may be equated to zero, but since both factors are identical we have:

$$x - 1 = 0 \quad \text{twice}$$

that is,

$$x = 1 \quad \text{twice}$$

There is only one root, $x = 1$. This seems to contradict an earlier statement that an equation of degree n has n roots. However, in this case, $x = 1$ is known as a **repeated root**. ▲

We saw in Example 1.14(c) that a quadratic equation may factorize into a perfect square. For example, if $x^2 + 8x + 16 = 0$, then this factorizes to become $(x + 4)(x + 4) = (x + 4)^2 = 0$, a perfect square. Many equations cannot be obviously factorized. An alternative approach is to **complete the square**. For example, consider the equation $x^2 + 8x + 2 = 0$. By inspection, we can write this as:

$$[(x + 4)^2 - 16] + 2 = 0$$

that is,

$$(x + 4)^2 - 14 = 0$$

this process being known as completing the square. The solution then follows:

$$(x + 4)^2 = 14$$

$$x + 4 = \pm\sqrt{14}$$

$$x = -4 \pm \sqrt{14}$$

Example 1.15

Verify that $x = 1$ and $x = 2$ are roots of:

$$P(x) = x^4 - 2x^3 - x + 2 = 0$$

Solution

$$P(x) = x^4 - 2x^3 - x + 2$$

$$P(1) = 1 - 2 - 1 + 2 = 0$$

$$P(2) = 2^4 - 2(2^3) - 2 + 2 = 0$$

Since $P(1) = 0$ and $P(2) = 0$, then $x = 1$ and $x = 2$ are roots of the given polynomial equation and are sometimes referred to as **real roots**. Further knowledge is required to find the two remaining roots, which are known as **complex roots**. This topic is covered in Chapter 6. ▲

Example 1.16

Solve the equation

$$P(x) = x^3 + 2x^2 - 37x + 52 = 0$$

Solution

As seen in Example 1.14 a formula can be used to solve quadratic equations. For higher degree polynomial equations such simple formulae do not always exist. However, if one of the roots can be found by inspection we can proceed as follows. By inspection $P(4) = 4^3 + 2(4)^2 - 37(4) + 52 = 0$ so that $x = 4$ is a root. Hence $x - 4$ is a factor of $P(x)$. Therefore $P(x)$ can be written as:

$$P(x) = x^3 + 2x^2 - 37x + 52 = (x - 4)(x^2 + \alpha x + \beta)$$

where α and β must now be found. Expanding the right-hand side gives:

$$P(x) = x^3 + \alpha x^2 + \beta x - 4x^2 - 4\alpha x - 4\beta$$

Hence

$$x^3 + 2x^2 - 37x + 52 = x^3 + (\alpha - 4)x^2 + (\beta - 4\alpha)x - 4\beta$$

By comparing constant terms on the left- and right-hand sides we see that:

$$52 = -4\beta$$

so that

$$\beta = -13$$

By comparing coefficients of x^2 we see that:

$$2 = \alpha - 4$$

Therefore,

$$\alpha = 6$$

Hence, $P(x) = (x - 4)(x^2 + 6x - 13)$. The quadratic equation $x^2 + 6x - 13 = 0$ can be solved using the formula:

$$x = \frac{-6 \pm \sqrt{36 + 4(13)}}{2}$$

$$= \frac{-6 \pm \sqrt{88}}{2}$$

$$= 1.690, -7.690$$

We conclude that $P(x) = 0$ has roots at $x = 4$, $x = 1.690$ and $x = -7.690$. ▲

Many excellent computer software packages exist for plotting graphs and these, as well as graphics calculators, may be used to solve polynomial equations. The real roots of the equation $P(x) = 0$ are given by the values of the intercepts of the function $y = P(x)$ and the x axis, because on the x axis y is zero.

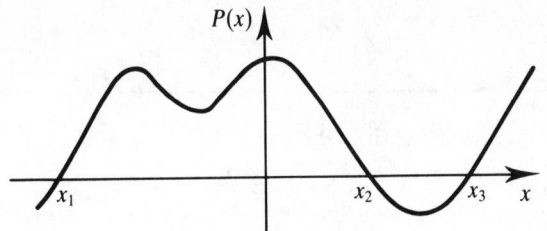

Figure 1.19 A polynomial function which cuts the x axis at points x_1, x_2 and x_3.

Figure 1.19 shows a graph of $y = P(x)$. The graph intersects the x axis at $x = x_1$, $x = x_2$ and $x = x_3$, and so the equation $P(x) = 0$ has real roots x_1, x_2 and x_3, that is, $P(x_1) = P(x_2) = P(x_3) = 0$. In Chapter 7 we examine the Newton–Raphson technique for obtaining approximate solutions of equations. Appendix I describes an alternative method for finding approximate solutions.

EXERCISES 1.2

1. State the degree of the following polynomial expressions.

   (a) $z^3 + 2z^2 - 8 + 13z$

   (b) $t^2 - 5t^5 + 2 - 8t^3$

   (c) $3w - 5w^2 + 12w^4$

   (d) $7x - x^2$

   (e) $3(2t^2 - 9t + 1)$

   (f) $2z(2z + 1)(2z - 1)$

2. Verify that the given values are roots of the following polynomial equations.

   (a) $x^2 + x - 2 = 0$ $x = -2, 1$

   (b) $2t^3 - 3t^2 - 3t + 2 = 0$ $t = -1, 0.5$

   (c) $y^3 + y^2 + y + 1 = 0$ $y = -1$

   (d) $v^4 + 4v^3 + 6v^2 + 3v = 0$ $v = 0, -1$

3. Calculate the roots of the following polynomial equations:

   (a) $3x/2 - 17 = 0$

   (b) $0.5t - 6 = 0$

   (c) $t^2 - 5t + 6 = 0$

   (d) $x^2 + x = 12$

   (e) $3x - 6 + x^2 = 0$

   (f) $t^2 = 10t - 25$

   (g) $x^3 - 6x^2 + 11x - 6 = 0$
   given $x = 1$ is a root.

   (h) $t^3 - 2t^2 - 5t + 6 = 0$
   given $t = 3$ is a root.

   (i) $v^3 - v^2 - 30v + 72 = 0$
   given $v = 4$ is a root.

4. Complete the square for the following polynomial equations and hence find their roots:

   (a) $x^2 + 2x - 8 = 0$

   (b) $x^2 - 6x - 5 = 0$

   (c) $x^2 + 4x - 6 = 0$

   (d) $x^2 - 14x - 10 = 0$

   (e) $x^2 + 5x - 49 = 0$

5. Obtain graphs of the following functions using a graphics calculator or software package.

   (a) $y = 3x^3 - x^2 + 2x + 1$ $-2 \leq x \leq 2$

   (b) $y = x^4 + \dfrac{x^3}{3} - \dfrac{5x^2}{2} + x - 1$ $-3 \leq x \leq 2$

   (c) $y = x^5 - x^2 + 2$ $-2 \leq x \leq 2$

   Hence estimate the real roots of:

   $0 = 3x^3 - x^2 + 2x + 1$ $-2 \leq x \leq 2$

   $0 = x^4 + \dfrac{x^3}{3} - \dfrac{5x^2}{2} + x - 1$ $-3 \leq x \leq 2$

   $0 = x^5 - x^2 + 2$ $-2 \leq x \leq 2$

1.4.2 Rational functions

■ A **rational function**, $R(x)$, has the form

$$R(x) = \frac{P(x)}{Q(x)}$$

where P and Q are polynomial functions; P is the **numerator** and Q is the **denominator**.

The functions,

$$R_1(x) = \frac{x + 6}{x^2 + 1} \qquad R_2(t) = \frac{t^3 - 1}{2t + 3} \qquad R_3(z) = \frac{2z^2 + z - 1}{z^2 + 3z - 2}$$

are all rational. When sketching the graph of a rational function, $y = f(x)$, it is usual to draw up a table of x and y values. Indeed this has been common practice when sketching any graph although the use of graphics calculators is now replacing this custom. It is still useful to answer questions such as:

'How does the function behave as x becomes large positively?'

'How does the function behave as x becomes large negatively?'

'What is the value of the function when $x = 0$?'

'At what values of x is the denominator zero?'

Figure 1.20 shows a graph of the function $y = (1 + 2x)/x = 1/x + 2$. As x increases, the value of y approaches 2. We write this as:

$$y \to 2 \qquad \text{as} \qquad x \to \infty$$

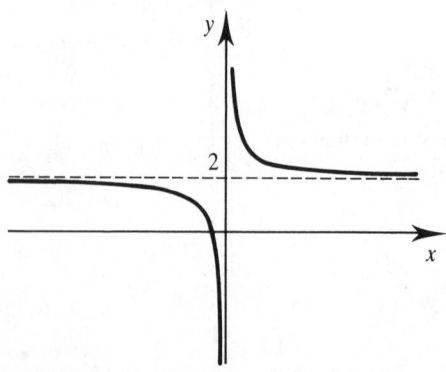

Figure 1.20   The function: $y = \dfrac{1 + 2x}{x} = \dfrac{1}{x} + 2.$

and say 'y tends to 2 as x tends to infinity'. Also from Figure 1.20, we see that:

$$y \to \pm\infty \qquad \text{as} \qquad x \to 0$$

As $x \to \infty$, the graph gets nearer and nearer to the straight line $y = 2$. We say that $y = 2$ is an **asymptote** of the graph. Similarly, $x = 0$, that is, the y axis, is an asymptote since the graph approaches the line $x = 0$ as $x \to 0$.

   If the graph of any function gets closer and closer to a straight line then that line is called an asymptote. Figure 1.21 illustrates some rational functions with their asymptotes indicated by dashed lines. In Figure 1.21(a) the asymptotes are the horizontal line $y = 3$ and the y axis, that is, $x = 0$. In Figure 1.21(b) the asymptotes are the horizontal line $y = 1$ and the vertical line $x = -2$; in Figure 1.21(c) they are $y = x + 3$ and the vertical line $x = -1$. The asymptote $y = x + 3$ being neither horizontal nor vertical, is called an **oblique asymptote**. Oblique asymptotes occur only when the degree of the numerator exceeds the degree of the denominator by one.

   We see that the vertical asymptotes occur at values of x which make the denominator zero. These values are particularly important to engineers and are known as the **poles** of the function. The function shown in Figure 1.21(a) has a pole at $x = 0$; the function shown in Figure 1.21(b) has a pole at $x = -2$ and the function shown in Figure 1.21(c) has a pole at $x = -1$.

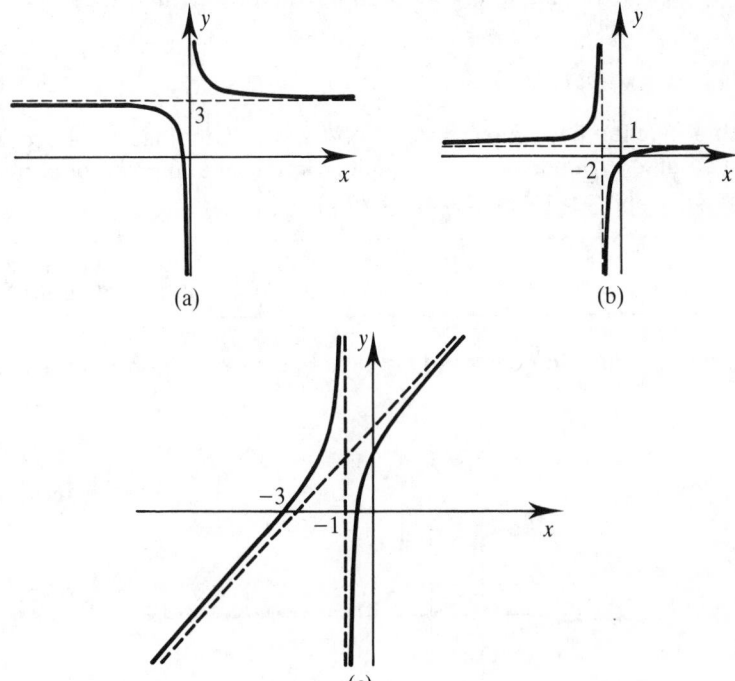

Figure 1.21   Some examples of functions with their asymptotes:

(a) $y = \dfrac{3x + 1}{x} = 3 + \dfrac{1}{x}$; (b) $y = \dfrac{x - 1}{x + 2}$; (c) $y = \dfrac{x^2 + 4x + 2}{x + 1} = x + 3 - \dfrac{1}{x + 1}$.

■ If the graph of a function, approaches a straight line, the line is known as an asymptote. Asymptotes may be horizontal, vertical or oblique.

■ Values of the independent variable where the denominator is 0 are called poles of the function.

Example 1.17

Sketch the rational function $y = x/(x^2 + x - 2)$.

Solution

For large values of x, the x^2 term in the denominator has a much greater value than the x in the numerator. Hence,

$$y \to 0 \quad \text{as} \quad x \to \infty$$

$$y \to 0 \quad \text{as} \quad x \to -\infty$$

Therefore the x axis, that is, $y = 0$, is an asymptote. Writing y as

$$y = \frac{x}{(x - 1)(x + 2)}$$

we see the function has poles at $x = 1$ and $x = -2$, that is, there are vertical asymptotes at $x = 1$ and $x = -2$. Substitution into the function of a number of values of x allows a table to be drawn up.

x	-3	-2.5	-2.1	-1.9	-1.5	-1	0	0.5	0.9	1.1	1.5	2	3
y	-0.75	-1.43	-6.77	6.55	1.20	0.50	0	-0.40	-3.10	3.55	0.86	0.50	0.30

The graph of the function can then be sketched as shown in Figure 1.22. ▲

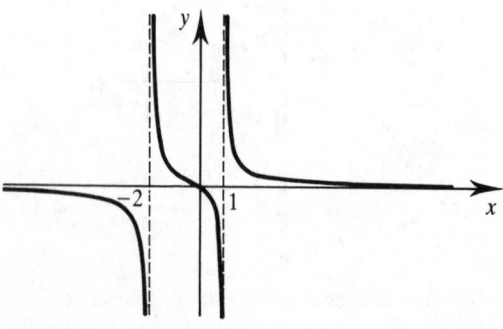

Figure 1.22   The function: $y = x/(x^2 + x - 2)$.

Example 1.18 Equivalent resistance

Consider a circuit consisting of two resistors in parallel as shown in Figure 1.23. One has a known resistance of $1\,\Omega$ and the other has a variable resistance, $R\,\Omega$. The equivalent resistance, $R_E\,\Omega$, satisfies:

$$\frac{1}{R_E} = \frac{1}{R} + \frac{1}{1} = \frac{1+R}{R}$$

Hence,

$$R_E = \frac{R}{1+R}$$

Thus the equivalent resistance is a rational function of R, with domain $R \geq 0$. The graph of this function is shown in Figure 1.24. When $R = 0$ we note that $R_E = 0$, corresponding to a short circuit. As the value of R increases, that is, $R \to \infty$, the equivalent resistance R_E approaches 1 so that $R_E = 1$ is an asymptote. ▲

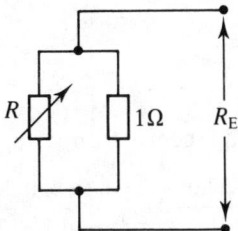

Figure 1.23 Two resistors in parallel.

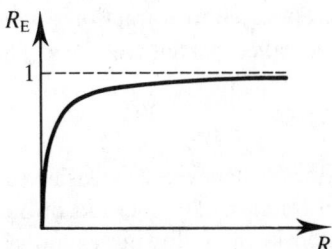

Figure 1.24 The equivalent resistance, R_E, increases as R increases.

EXERCISES 1.3

1. Sketch the following rational functions. State any asymptotes.

   (a) $f(x) = (2x + 1)/(x - 3)$

   (b) $g(s) = s/(s + 1)$

   (c) $h(z) = z/(z^2 + 1)$

   (d) $y(x) = (x + 1)/x$

   (e) $r(x) = \dfrac{2x}{(x - 1)(x - 2)}$

(f) $y(z) = \dfrac{(z-1)}{(z+1)(z-6)}$

(g) $G(s) = (3s - 1)/(s^2 - 1)$

2. Show that

$$I(x) = \frac{2x}{3} - \frac{7}{9} + \frac{23}{9(3x+2)}$$

can be expressed in the equivalent form

$$\frac{2x^2 - x + 1}{3x + 2}$$

Sketch the rational function $I(x)$ and state any asymptotes.

3. Show that

$$p(x) = 2x + \frac{1}{2} + \frac{9}{2(2x+3)}$$

can be written in the equivalent form

$$\frac{4x^2 + 7x + 6}{2x + 3}$$

Sketch the rational function $p(x)$ and state any asymptotes.

4. Show that the function

$$y(x) = x + \frac{7 - x}{x^2 + 3}$$

can be expressed in the equivalent form

$$\frac{x^3 + 2x + 7}{x^2 + 3}$$

Sketch the rational function $y(x)$ and state any asymptotes.

5. State the poles of the functions in Exercise 1.3.1 (a, b, d–g).

1.4.3 *Partial fractions*

Given a set of fractions we can add them to form a single fraction. For example,

$$\frac{1}{x} + \frac{1}{x+1} + \frac{1}{x+2} = \frac{(x+1)(x+2) + x(x+2) + x(x+1)}{x(x+1)(x+2)} = \frac{3x^2 + 6x + 2}{x^3 + 3x^2 + 2x}$$

Alternatively, if we are given a single fraction, we can break it down into the sum of easier fractions. These simple fractions, which when added together form the given fraction, are called **partial fractions**. The partial fractions of

$$\frac{3x^2 + 6x + 2}{x^3 + 3x^2 + 2x} \quad \text{are} \quad \frac{1}{x}, \frac{1}{x+1} \quad \text{and} \quad \frac{1}{x+2}$$

Partial fractions are obtained whenever it is easier to use them rather than the original more complicated fraction. The method of obtaining partial fractions of a given fraction is illustrated in the following examples.

Example 1.19

Express $(3x^2 + 6x + 2)/(x^3 + 3x^2 + 2x)$ as partial fractions.

Solution

The denominator is factorized:

$$x^3 + 3x^2 + 2x = x(x^2 + 3x + 2) = x(x + 1)(x + 2)$$

We note that the denominator has three linear factors. Each of these factors leads to a partial fraction. The factor, x, leads to a partial fraction of the form A/x, the factor $x + 1$ leads to a partial fraction of the form

$B/(x + 1)$ and the factor $x + 2$ leads to a partial fraction $C/(x + 2)$. A, B and C are constants which have yet to be found. Thus:

$$\frac{3x^2 + 6x + 2}{x^3 + 3x^2 + 2x} = \frac{3x^2 + 6x + 2}{x(x + 1)(x + 2)} = \frac{A}{x} + \frac{B}{x + 1} + \frac{C}{x + 2}$$

$$\frac{3x^2 + 6x + 2}{x(x + 1)(x + 2)} = \frac{A(x + 1)(x + 2) + Bx(x + 2) + Cx(x + 1)}{x(x + 1)(x + 2)}$$

$$3x^2 + 6x + 2 = A(x + 1)(x + 2) + Bx(x + 2) + Cx(x + 1)$$

We now need to evaluate A, B and C. There are two techniques which can be used to find these values: equating coefficients, and evaluation of the left- and right-hand sides at a specific value of x. Usually a value of x which simplifies the right-hand side is chosen. We will illustrate both techniques.

Evaluation using a specific x value

We note that the values $x = 0$, $x = -1$ and $x = -2$ all considerably simplify the right-hand side of the previous equation. We consider each value in turn.

Put $x = 0$

$$2 = A(1)(2)$$

$$A = 1$$

Put $x = -1$

$$-1 = B(-1)(1)$$

$$B = 1$$

Put $x = -2$

$$2 = C(-2)(-1)$$

$$C = 1$$

So

$$\frac{3x^2 + 6x + 2}{x^3 + 3x^2 + 2x} = \frac{1}{x} + \frac{1}{x + 1} + \frac{1}{x + 2}$$

To find A, B and C we specified a value of x. Alternatively, we could use the method of equating coefficients. This is now done.

Equating coefficients

We have:

$$3x^2 + 6x + 2 = A(x + 1)(x + 2) + Bx(x + 2) + Cx(x + 1)$$

This may be written as:

$$3x^2 + 6x + 2 = x^2(A + B + C) + x(3A + 2B + C) + 2A$$

Equating coefficients of x^2 on the left- and the right-hand sides gives:

$$3 = A + B + C$$

Equate coefficients of x,

$$6 = 3A + 2B + C$$

Equate the constant terms,

$$2 = 2A$$

These three equations in A, B and C may now be solved simultaneously to give $A = B = C = 1$ as found earlier.

Sometimes both methods are used to find a set of partial fractions. ▲

Example 1.19 illustrates the following rules for finding partial fractions:

■ (1) Factorize the denominator.
    (2) Linear factors $ax + b$ give rise to partial fractions of the form:

$$\frac{A}{ax + b}$$

Example 1.20

Noting that $x^3 + 2x^2 - 11x - 52 = (x - 4)(x^2 + 6x + 13)$ express

$$\frac{3x^2 + 11x + 14}{x^3 + 2x^2 - 11x - 52}$$

as partial fractions.

Solution

The denominator is already factorized. The quadratic factor, $x^2 + 6x + 13$, will not factorize further into two linear factors. The linear factor, $x - 4$, generates a partial fraction of the form $A/(x - 4)$. The quadratic factor, $x^2 + 6x + 13$, generates a partial fraction of the form $(Bx + C)/(x^2 + 6x + 13)$. Hence,

$$\frac{3x^2 + 11x + 14}{(x - 4)(x^2 + 6x + 13)} = \frac{A}{x - 4} + \frac{Bx + C}{x^2 + 6x + 13}$$

$$3x^2 + 11x + 14 = A(x^2 + 6x + 13) + (Bx + C)(x - 4)$$

The constants A, B and C are now found. Put $x = 4$

$$106 = A(53)$$

$$A = 2$$

Equate coefficients of x^2

$$3 = A + B$$

$$B = 1$$

Equate the constant terms

$$14 = A(13) - 4C$$

$$C = 3$$

Hence,

$$\frac{3x^2 + 11x + 14}{x^3 + 2x^2 - 11x - 52} = \frac{2}{x - 4} + \frac{x + 3}{x^2 + 6x + 13} \quad \blacktriangle$$

Example 1.20 illustrates a further rule when finding partial fractions:

■ A quadratic factor, $ax^2 + bx + c$, generates a partial fraction of the form:

$$\frac{Ax + B}{ax^2 + bx + c}$$

Example 1.21

Express:

$$\frac{2x + 5}{x^2 + 2x + 1}$$

as partial fractions.

Solution

The denominator is factorized to give $(x + 1)^2$. Here we have a case of a **repeated factor**. A repeated factor, $(x + 1)^2$, generates partial fractions of the form

$$\frac{A}{x + 1} + \frac{B}{(x + 1)^2}$$

Thus,

$$\frac{2x + 5}{x^2 + 2x + 1} = \frac{A}{x + 1} + \frac{B}{(x + 1)^2}$$

$$2x + 5 = A(x + 1) + B$$

Equating coefficients of x gives $A = 2$; equating constant terms gives $B = 3$. So,

$$\frac{2x + 5}{x^2 + 2x + 1} = \frac{2}{x + 1} + \frac{3}{(x + 1)^2} \quad \blacktriangle$$

Example 1.21 illustrates the following point:

■ A repeated factor, $(ax + b)^2$, leads to partial fractions:

$$\frac{A}{ax + b} + \frac{B}{(ax + b)^2}$$

Example 1.22

Express as partial fractions:

$$\frac{4x^3 + 10x + 4}{2x^2 + x}$$

Solution

This is an example of an **improper fraction**. The degree of the denominator must be strictly greater than the degree of the numerator for a fraction to be **proper**. If this is not the case then the fraction is improper. The partial fractions of an improper fraction are due to two sources: the factors of the denominator and the extent to which the fraction is improper.

Let n = degree of the numerator and d = degree of the denominator. For the given fraction $n - d = 1$. This is a measure of the extent to which the fraction is improper. The partial fractions will include a polynomial of degree 1, that is, $Ax + B$, in addition to the partial fractions generated by the factors of the denominator.

The factors of the denominator are x and $2x + 1$, and these generate partial fractions C/x and $D/(2x + 1)$. Hence,

$$\frac{4x^3 + 10x + 4}{2x^2 + x} = Ax + B + \frac{C}{x} + \frac{D}{2x + 1}$$

$$4x^3 + 10x + 4 = (Ax + B)x(2x + 1) + C(2x + 1) + Dx$$

The unknown constants are now evaluated.

Put $x = 0$,

$$4 = C$$

Put $x = -1/2$,

$$-3/2 = -D/2$$

$$D = 3$$

Equate coefficients of x^3,

$$4 = 2A$$

$$A = 2$$

Equate coefficients of x,

$$10 = B + 2C + D$$

$$B = -1$$

Hence,

$$\frac{4x^3 + 10x + 4}{2x^2 + x} = 2x - 1 + \frac{4}{x} + \frac{3}{2x + 1} \quad \blacktriangle$$

Example 1.22 illustrates the following point:

■ Let the degree of the numerator be n and the degree of the denominator be d. If $n \geq d$ then the fraction is improper. Improper fractions have partial fractions in addition to those generated by the factors of the denominator. These additional partial fractions take the form of a polynomial of degree $n - d$.

EXERCISES 1.4

1. Express as a single fraction:

(a) $\dfrac{1}{x + 2} + \dfrac{1}{x - 3}$

(b) $\dfrac{1}{2x} + \dfrac{3}{x + 2} - \dfrac{1}{x + 3}$

(c) $\dfrac{1}{x} + \dfrac{1}{2x^2} - \dfrac{1}{x + 4}$

(d) $\dfrac{1}{x + 2} - \dfrac{2}{(x + 2)^2} + \dfrac{1}{x - 3}$

(e) $\dfrac{2}{x^2 + 3} + \dfrac{3}{2x} - \dfrac{1}{7x - 1}$

2. Express the following as partial fractions.

(a) $\dfrac{5x}{(x + 1)(2x - 3)}$

(b) $\dfrac{3x + 2}{x^2 + 5x + 6}$

(c) $\dfrac{s + 3}{(s + 1)^2}$

(d) $\dfrac{x^3}{x^2 + 1}$

(e) $\dfrac{s^2}{s^2 + 1}$

(f) $\dfrac{s^2 - 8s - 5}{(s^2 + s + 1)(s - 4)}$

(g) $\dfrac{x^2 + x - 2}{(x - 2)^2(x + 1)}$

(h) $\dfrac{2s^3 + 3s^2 - s - 4}{s^2 + s - 1}$

(i) $\dfrac{x^3 + 4x^2 + 7x + 5}{x^2 + 3x + 2}$

1.4.4 *Exponential functions*

An **exponent** is another name for a power or index. Expressions involving exponents are called **exponential expressions**, for example, 3^4, a^b, m^n. In the exponential expression a^x, a is called the **base**; x is the exponent. Exponential expressions can be simplified and manipulated using the laws of indices (see Appendix II). These laws are summarized here.

■ $a^m a^n = a^{m+n}$ $\dfrac{a^m}{a^n} = a^{m-n}$ $a^0 = 1$ $a^{-m} = \dfrac{1}{a^m}$ $(a^m)^n = a^{mn}$

Example 1.23

Simplify:

(a) $\dfrac{a^{3x}a^{2x}}{a^{4x}}$

(b) $a^{2t}(1 - a^t) + a^{3t}$

(c) $\dfrac{(a^y)^2}{2a^y}$

(d) $\dfrac{a^{-6z}}{a^{-2z}}$

(e) $\dfrac{(2a^{3r})^2 a^{2r}}{3a^{-5r}}$

(f) $\dfrac{a^{x+y}a^y}{a^{2x}}$

(g) $\dfrac{3a^{(x/y)}a^x}{a^y}$

Solutions

(a) $\dfrac{a^{3x}a^{2x}}{a^{4x}} = \dfrac{a^{5x}}{a^{4x}} = a^x$

(b) $a^{2t}(1 - a^t) + a^{3t} = a^{2t} - a^{3t} + a^{3t} = a^{2t}$

(c) $\dfrac{(a^y)^2}{2a^y} = \dfrac{a^{2y}}{2a^y} = \dfrac{a^y}{2}$

(d) $\dfrac{a^{-6z}}{a^{-2z}} = a^{-6z-(-2z)} = a^{-4z}$

(e) $\dfrac{(2a^{3r})^2 a^{2r}}{3a^{-5r}} = \dfrac{4a^{6r}a^{2r}}{3a^{-5r}} = \dfrac{4a^{8r}}{3a^{-5r}} = \dfrac{4a^{13r}}{3}$

(f) $\dfrac{a^{x+y}a^y}{a^{2x}} = a^{x+2y-2x} = a^{2y-x}$

(g) $\dfrac{3a^{(x/y)}a^x}{a^y} = 3a^{(x/y)+x-y}$ ▲

Exponential functions

■ An **exponential function**, $f(x)$, has the form:

$$f(x) = a^x$$

where a is a positive constant called the base.

Some typical exponential functions are tabulated in Table 1.1 and are shown in Figure 1.25. Note from the graphs that these are one-to-one functions.

Table 1.1 Values of a^x for $a = 0.5$, 2 and 3.

x	0.5^x	2^x	3^x
-3	8	0.125	0.037
-2	4	0.25	0.111
-1	2	0.5	0.333
0	1	1	1
1	0.5	2	3
2	0.25	4	9
3	0.125	8	27

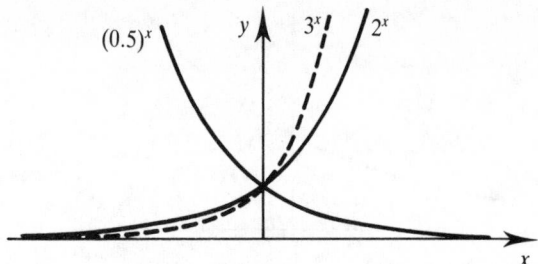

Figure 1.25 Some typical exponential functions.

An exponential function is not a polynomial function. The powers of a polynomial function are constants; the power of an exponential function, that is, the exponent, is the variable x.

■ The most widely used exponential function, commonly called **the** exponential function, is

$$f(x) = e^x$$

where e is an irrational constant (e $= 2.718\,281\,828\ldots$).

Most scientific calculators have values of e^x available. The function is tabulated in Table 1.2. The graph is shown in Figure 1.26. This particular exponential function so dominates engineering applications that whenever an engineer refers to the exponential function it almost invariably means this one. We will see later why it is so important.

As x increases positively, e^x increases very rapidly, that is, as $x \to \infty$, $e^x \to \infty$. This situation is known as **exponential growth**. As x increases negatively, e^x approaches zero, that is, as $x \to -\infty$, $e^x \to 0$. Thus $y = 0$ is an asymptote. Note that the exponential function is never negative.

Table 1.2 The values of the exponential function $f(x) = e^x$ for various values of x.

x	e^x
-3	0.050
-2	0.135
-1	0.368
0	1
1	2.718
2	7.389
3	20.086

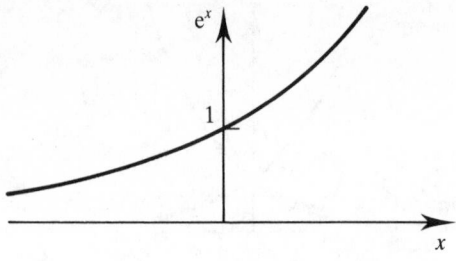

Figure 1.26   Graph of $y = e^x$ showing exponential growth.

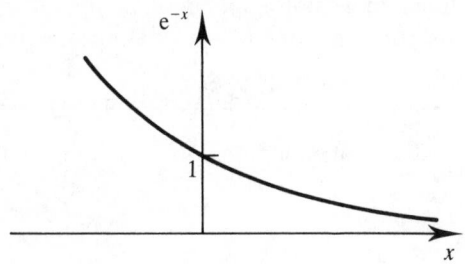

Figure 1.27   Graph of $y = e^{-x}$ showing exponential decay.

Table 1.3   The values
of the exponential
function $f(x) = e^{-x}$
for various values of x.

x	e^{-x}
-3	20.086
-2	7.389
-1	2.718
0	1
1	0.368
2	0.135
3	0.050

Figure 1.27 shows a graph of e^{-x}. As x increases positively, e^{-x} decreases to zero, that is, as $x \to \infty$, $e^{-x} \to 0$. This is known as **exponential decay**. The function is tabulated in Table 1.3.

Example 1.24   Discharge of a capacitor

Consider the circuit of Figure 1.28. Before the switch is closed, the capacitor has a voltage V across it. Suppose the switch is closed at time $t = 0$. A current then flows in the circuit and the voltage, v, across the capacitor

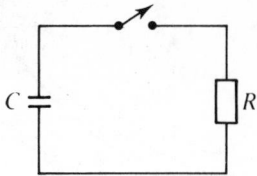

Figure 1.28   Circuit to discharge a capacitor.

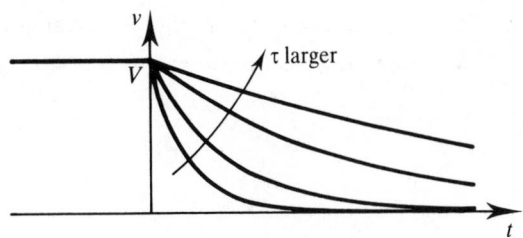

Figure 1.29   The capacitor takes longer to discharge for a larger circuit time constant, τ.

decays with time. The voltage across the capacitor is given by:

$$v = \begin{cases} V & t < 0 \\ V e^{-t/(RC)} & t \geq 0 \end{cases}$$

The quantity RC is known as the **time constant** of the circuit and is usually denoted by τ. So

$$v = \begin{cases} V & t < 0 \\ V e^{-t/\tau} & t \geq 0 \end{cases}$$

If τ is small, then the capacitor voltage decays more quickly than if τ is large. This is illustrated in Figure 1.29.   ▲

Example 1.25   The diode equation

A semiconductor diode can be modelled by the equation:

$$I = I_s(e^{qV/(kT)} - 1)$$

where: V = applied voltage (V);
$\quad$ I = diode current (A);
$\quad$ I_s = reverse saturation current (A);
$\quad$ $k = 1.38 \times 10^{-23}$ J K^{-1};
$\quad$ $q = 1.60 \times 10^{-19}$ C;
$\quad$ T = temperature (K).

This equation relates the current through the diode to the voltage across it. This is a good model for germanium diodes but only an approximate model for silicon diodes. At room temperature $q/(kT) \approx 40$ and so the equation can

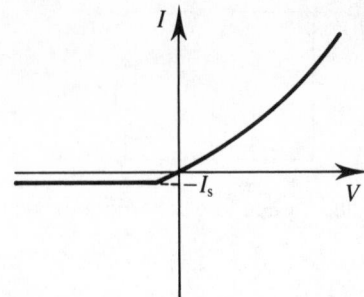

Figure 1.30   A typical diode characteristic.

be written as:

$$I = I_s(e^{40V} - 1)$$

Figure 1.30 shows a graph of *I* against *V*. Notice that for negative values of *V*, the equation may be approximated by:

$$I \approx -I_s$$

since $e^{40V} \approx 0$. The diode is said to be **reverse saturated** in this case. In reality, I_s is usually quite small for a practical device, although its size has been exaggerated in Figure 1.30. This model does not cater for the breakdown of the diode. According to the model it would be possible to apply a very large reverse voltage to a diode and yet only a small saturation current would flow. This illustrates an important point that no mathematical model covers every facet of the physical device or system it is modelling. A different model would be needed to deal with breakdown characteristics. ▲

EXERCISES 1.5

1. Sketch the following functions, using the same axes:

   $$y = e^{2x} \quad y = e^{x/2} \quad y = e^{-2x}$$

   for $-3 \leq x \leq 3$

2. Simplify,

   (a) $\dfrac{e^{2x}}{3e^{3x}}$

   (b) e^{2t-3}/e^2

   (c) $e^x(e^x + e^{2x})/e^{2x}$

   (d) $\dfrac{e^{-3}e^{-7}}{e^6e^{-2}}$

   (e) $(e^{2t})^3(e^{3t})^4/e^{10t}$

3. Sketch a graph of the function $y = 1 - e^{-x}$ for $x \geq 0$.

4. Consider the *RC* circuit of Figure 1.28. Given an initial capacitor voltage of 10 V plot the variation in capacitor voltage with time, using the same axes, for the following pairs of component values.

   (a) $R = 1\,\Omega$, $C = 1\,\mu F$

   (b) $R = 10\,\Omega$, $C = 1\,\mu F$

   (c) $R = 3.3\,\Omega$, $C = 1\,\mu F$

   (d) $R = 56\,\Omega$, $C = 0.1\,\mu F$

   Calculate the time constant, τ, in each case.

1.4.5 *Logarithm functions*

Logarithms

The equation $16 = 2^4$ may be expressed in an alternative form using **logarithms**. In logarithmic form we write:

$$\log_2 16 = 4$$

and say 'log to the base 2 of 16 equals 4'. Hence logarithms are nothing other than powers. The logarithmic form is illustrated by more examples:

$$125 = 5^3 \quad \text{so } \log_5 \quad 125 = 3$$
$$64 = 8^2 \quad \text{so } \log_8 \quad 64 = 2$$
$$16 = 4^2 \quad \text{so } \log_4 \quad 16 = 2$$
$$1000 = 10^3 \text{ so } \log_{10} 1000 = 3$$

In general,

$$\text{if } c = a^b, \text{ then } b = \log_a c$$

In practice, most logarithms use base 10 or base e. Logarithms using base e are called **natural logarithms**. $\log_{10} x$ and $\log_e x$ are usually abbreviated to $\log x$ and $\ln x$, respectively. Most scientific calculators have both logs to base 10 and logs to base e as pre-programmed functions, usually denoted as log and ln, respectively. Some calculations in communications engineering use base 2. Your calculator will probably not calculate base 2 logarithms directly. We shall see how to overcome this shortly.

Logarithmic expressions can be manipulated using the laws of logarithms. These laws are identical for any base, but it is essential when applying the laws that bases are not mixed.

■ $\log_a A + \log_a B = \log_a AB$

$\log_a A - \log_a B = \log_a\left(\dfrac{A}{B}\right)$

$n \log_a A = \log_a(A^n)$

$\log_a a = 1$

We sometimes need to change from one base to another. This can be achieved using the following rule.

■ $\log_a X = \dfrac{\log_b X}{\log_b a}$

In particular,

$$\log_2 X = \frac{\log_{10} X}{\log_{10} 2} = \frac{\log_{10} X}{0.3010}$$

Example 1.26

Simplify,

(a) $\log x + \log x^3$

(b) $3 \log x + \log x^2$

(c) $5 \ln x + \ln(1/x)$

(d) $\log(xy) + \log x - 2 \log y$

(e) $\ln(2x^3) - \ln\left(\dfrac{4}{x^2}\right) + \dfrac{1}{3} \ln 27$

Solutions

(a) Using the laws of logarithms we find:
$\log x + \log x^3 = \log x^4$

(b) $3 \log x + \log x^2 = \log x^3 + \log x^2 = \log x^5$

(c) $5 \ln x + \ln(1/x) = \ln x^5 + \ln(1/x) = \ln(x^5/x) = \ln x^4$

(d) $\log xy + \log x - 2 \log y = \log xy + \log x - \log y^2$
$= \log(xyx/y^2) = \log(x^2/y)$

(e) $\ln(2x^3) - \ln\left(\dfrac{4}{x^2}\right) + \dfrac{\ln 27}{3} = \ln\left(\dfrac{2x^3}{4/x^2}\right) + \ln 27^{1/3}$

$$= \ln\left(\frac{2x^3 x^2}{4}\right) + \ln 3$$

$$= \ln\left(\frac{x^5}{2}\right) + \ln 3 = \ln\left(\frac{3x^5}{2}\right) \quad \blacktriangle$$

Example 1.27

Find $\log_2 14$.

Solution

Using the formula for change of base we have:

$$\log_2 14 = \frac{\log_{10} 14}{\log_{10} 2} = \frac{1.146}{0.301} = 3.807 \quad \blacktriangle$$

Example 1.28 Signal ratios and decibels

The ratio between two signal levels is often of interest to engineers. For example, the output and input signals of an electronic system can be compared to see if the system has increased the level of a signal. A common case is an amplifier, where the output signal is usually much larger than the input signal. This signal ratio is often expressed in decibels (dB) given by:

$$\text{power gain (dB)} = 10 \log \frac{P_o}{P_i}$$

where P_o is the power of the output signal and P_i is the power of the input signal. The term **gain** is used because if $P_o > P_i$, then the logarithm function is positive corresponding to an increase in power. If $P_o < P_i$ then the gain is negative, corresponding to a decrease in power. A negative gain is often termed an **attenuation**.

The advantage of using decibels as a measure of gain is that if several electronic systems are connected together then it is possible to obtain the overall system gain in decibels by adding together the individual system gains. We will show this for three systems connected together, but the development is easily generalized to more systems. Let the power input to the first system be P_{i1}, and the power output from the third system be P_{o3}. Suppose the three are connected so that the power output from system 1, P_{o1}, is used as input to system 2, that is, $P_{i2} = P_{o1}$. The power output from system 2, P_{o2}, is then used as input to system 3, that is, $P_{i3} = P_{o2}$. We wish to find the overall power gain, $10 \log(P_{o3}/P_{i1})$. Now

$$\frac{P_{o3}}{P_{i1}} = \frac{P_{o3}}{P_{i3}} \frac{P_{o2}}{P_{i2}} \frac{P_{o1}}{P_{i1}}$$

because $P_{i3} = P_{o2}$ and $P_{i2} = P_{o1}$. Therefore,

$$10 \log\left(\frac{P_{o3}}{P_{i1}}\right) = 10 \log\left(\frac{P_{o3}}{P_{i3}} \frac{P_{o2}}{P_{i2}} \frac{P_{o1}}{P_{i1}}\right)$$

that is,

$$10 \log\left(\frac{P_{o3}}{P_{i1}}\right) = 10 \log\left(\frac{P_{o3}}{P_{i3}}\right) + 10 \log\left(\frac{P_{o2}}{P_{i2}}\right) + 10 \log\left(\frac{P_{o1}}{P_{i1}}\right)$$

using the laws of logarithms.

It follows that the overall power gain is equal to the sum of the individual power gains. Often engineers are more interested in voltage gain rather than power gain. The power of a signal is proportional to the square of its voltage. We define **voltage gain** (dB) by:

$$\text{voltage gain (dB)} = 10 \log\left(\frac{V_o^2}{V_i^2}\right) = 20 \log\left(\frac{V_o}{V_i}\right) \quad \blacktriangle$$

Logarithm functions

■ The **logarithm functions** are defined by:

$$f(x) = \log_a x \qquad x > 0$$

where a is a positive constant called the base.

In particular the logarithm functions, $f(x) = \log x$ and $f(x) = \ln x$ are shown in Figure 1.31 and some values are given in Table 1.4. The domain of both of these functions is $(0, \infty)$ and their ranges are $(-\infty, \infty)$. We observe from the graphs that these functions are one to one. It is important to stress that the logarithm functions, $\log_a x$, are only defined for positive values of x. The following properties should be noted.

$$\left.\begin{array}{l} \log x \to \infty \\ \ln x \to \infty \end{array}\right\} \text{ as } x \to \infty$$

$$\left.\begin{array}{l} \log x \to -\infty \\ \ln x \to -\infty \end{array}\right\} \text{ as } x \to 0$$

$$\log 1 = \ln 1 = 0$$

$$\log 10 = 1 \qquad \ln e = 1$$

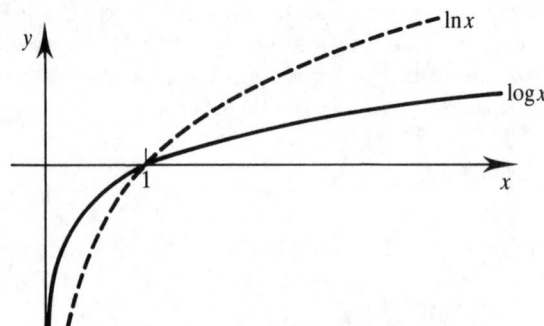

Figure 1.31   Graphs of ln x and log x.

Table 1.4   Some values for logarithm functions log x and ln x.

x	log x	ln x
0.1	-1	-2.303
0.5	-0.301	-0.693
1	0	0
2	0.301	0.693
5	0.699	1.609
10	1	2.303
50	1.699	3.912

Connection between exponential and logarithm functions

The exponential function, $f(x) = a^x$, is a one-to-one function and so an inverse function, $f^{-1}(x)$, exists. Recall

$$f^{-1}(f(x)) = x$$

So

$$f^{-1}(a^x) = x$$

Now

$$\log_a(a^x) = x \log_a a \qquad \text{using laws of logarithms}$$

$$= x \qquad \text{since } \log_a a = 1$$

Hence the inverse of $f(x) = a^x$ is $f^{-1}(x) = \log_a x$. By similar analysis the inverse of $f(x) = \log_a x$ is $f^{-1}(x) = a^x$.

■ The inverse of the exponential function, $f(x) = a^x$, is the logarithm function, that is, $f^{-1}(x) = \log_a x$.
The inverse of the logarithm function, $f(x) = \log_a x$, is the exponential function, that is, $f^{-1}(x) = a^x$.

In particular;
$\quad$ If $f(x) = e^x$, then $f^{-1}(x) = \ln x$.
$\quad$ If $f(x) = \ln x$, then $f^{-1}(x) = e^x$.
$\quad$ If $f(x) = 10^x$, then $f^{-1}(x) = \log x$.
$\quad$ If $f(x) = \log x$ then $f^{-1}(x) = 10^x$.

Example 1.29

Solve the equations:

(a) $16 = 10^x$

(b) $30 = e^x$

(c) $19 = 3e^x$

(d) $50 = 9(10^{2x})$

(e) $e^{3x} = 2e^x$

(f) $3e^{-(2x+1)} = 10$

(g) $\dfrac{17}{2 - 10^{-(x/2)}} = 60$

Solutions

(a) $\qquad 16 = 10^x$

$\qquad \log 16 = x$

$\qquad\qquad x = 1.204$

(b) $30 = e^x$

$\ln 30 = x$

$x = 3.401$

(c) $19 = 3e^x$

$e^x = \dfrac{19}{3}$

$x = \ln\left(\dfrac{19}{3}\right) = 1.846$

(d) $50 = 9(10^{2x})$

$10^{2x} = \dfrac{50}{9}$

$2x = \log\left(\dfrac{50}{9}\right)$

$x = \dfrac{\log(50/9)}{2}$

$x = 0.372$

(e) $e^{3x} = 2e^x$

$\dfrac{e^{3x}}{e^x} = 2$

$e^{2x} = 2$

$2x = \ln 2$

$x = \dfrac{\ln 2}{2} = 0.347$

(f) $3e^{-(2x+1)} = 10$

$e^{-(2x+1)} = \dfrac{10}{3}$

$-(2x + 1) = \ln\left(\dfrac{10}{3}\right)$

$2x = -\ln\left(\dfrac{10}{3}\right) - 1$

$2x = -2.204$

$x = -1.102$

(g) $\dfrac{17}{2 - 10^{-(x/2)}} = 60$

$$2 - 10^{-x/2} = \frac{17}{60}$$

$$10^{-x/2} = 2 - \frac{17}{60} = \frac{103}{60}$$

$$-\frac{x}{2} = \log\left(\frac{103}{60}\right)$$

$$x = -2\log\left(\frac{103}{60}\right) = -0.469 \quad \blacktriangle$$

Logarithmic scales for graphs

Suppose we wish to plot

$$y(x) = x^6 \qquad 1 \le x \le 10$$

This may appear a straightforward exercise but consider the variation in the x and y values. As x varies from 1 to 10, then y varies from 1 to 1 000 000, as tabulated below.

$x = 1 \qquad y = 1$
$x = 2 \qquad y = 64$
$x = 3 \qquad y = 729$
$x = 4 \qquad y = 4\,096$
$x = 5 \qquad y = 15\,625$
$x = 6 \qquad y = 46\,656$
$x = 7 \qquad y = 117\,649$
$x = 8 \qquad y = 262\,144$
$x = 9 \qquad y = 531\,441$
$x = 10 \qquad y = 1\,000\,000$

Several of these points would not be discernible on a graph and so information would be lost. This can be overcome by using a **log scale** which accommodates the large variation in y. Thus $\log y$ is plotted against x, rather than y against x. Note that in this example

$$\log y = \log x^6 = 6 \log x$$

so as x varies from 1 to 10, $\log y$ varies from 0 to 6. A plot in which one scale is logarithmic and the other is linear is known as a **log–linear** graph. Figure 1.32 shows $\log y$ plotted against x. In effect, use of the log scale has compressed a large variation into one which is much smaller and easier to observe.

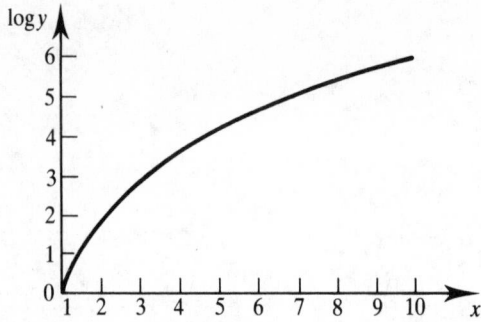

Figure 1.32 The function: $y = x^6$ plotted on a log–linear graph.

Example 1.30 Bode plot of a linear circuit

Engineers are often interested in how a circuit will respond to a sinusoidal signal. In Section 13.9 we will see that if the circuit is **linear** then, after it has settled down, the output signal is also a sinusoidal signal of the same frequency but with a different amplitude and phase (see Section 1.4.8 for details of these terms).

A **Bode plot** consists of two components:

(1) The ratio of the amplitudes of the output signal and the input signal is plotted against frequency.

(2) The phase shift between the input and output signals is plotted against frequency.

A log scale is used for the frequency in order to compress its length, for example, a typical frequency range is 0.1 Hz to 10^6 Hz which corresponds to a range of -1 to 6 on a log scale. A log scale is also used for the ratio of the signal amplitudes as this is calculated in decibels. Phase shift is plotted on a linear scale. So the signal amplitude ratio versus frequency is a log–log graph and the phase shift versus frequency is a linear–log graph. ▲

It is common to use special paper to plot graphs that require a log scale. Various types are available. The main division is between log–log paper and linear–log paper. It is also possible to choose the number of **decades** of the log scales, for instance, to plot a frequency range of 0.01 Hz to 10^4 Hz would require a $4 - (-2) = 6$ decade log scale. The advantage of using this paper is that it avoids the need to take logs of the quantities to be plotted prior to plotting them.

EXERCISES 1.6

1. Sketch

   (a) $y = \ln(2x)$

   (b) $y = 2 + \ln x$

   (c) $y = \ln(-x)$

2. Simplify

   (a) $3 \ln t - \ln t$

   (b) $6 \log t^2 + 4 \log t$

   (c) $\ln(3y^6) - 2 \ln 3 + \ln y$

(d) $\ln(6x + 4) - \ln(3x + 2)$

(e) $\dfrac{\log 9x}{2} - \log \dfrac{2}{3x}$

3. Solve

(a) $e^{3x} = 21$

(b) $10^{-2x} = 6.7$

(c) $1/(e^{-x} + 2) = 0.3$

(d) $2e^{(x/2)} - 1 = 0$

(e) $3(10^{(-4x+6)}) = 17$

(f) $(e^{x-1})^3 + e^{3x} = 500$

(g) $\sqrt{10^{2x} + 100} = 3(10^x)$

4. Calculate the voltage gain in decibels of the following amplifiers:

(a) input signal = 0.1 V, output signal = 1 V;

(b) input signal = 1 mV, output signal = 10 V;

(c) input signal = 5 mV, output signal = 8 V;

(d) input signal = 60 mV, output signal = 2 V.

5. An audio amplifier consists of two stages: a preamplifier and a main amplifier. Given the following data, calculate the voltage gain in decibels of the individual stages and the overall gain in decibels of the audio amplifier:

preamplifier: input signal = 10 mV,
output signal = 200 mV
main amplifier: input signal = 400 mV,
output signal = 3 V

6. Evaluate

(a) $\log_2 8$

(b) $\log_2 15$

(c) $\log_{16} 50$

(d) $\log_{16} 123$

(e) $\log_8 23$

(f) $\log_8 47$

1.4.6 *The trigonometric functions*

The trigonometric ratios

Consider the angle θ in the right-angled triangle ABC, as shown in Figure 1.33. We define the trigonometric ratios **sine, cosine** and **tangent** as follows:

$$\sin \theta = \frac{\text{side opposite to angle}}{\text{hypotenuse}} = \frac{BC}{AC}$$

$$\cos \theta = \frac{\text{side adjacent to angle}}{\text{hypotenuse}} = \frac{AB}{AC}$$

$$\tan \theta = \frac{\text{side opposite to angle}}{\text{side adjacent to angle}} = \frac{BC}{AB}$$

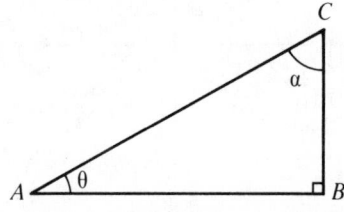

Figure 1.33   A right-angled triangle, ABC.

Note that,

$$\tan \theta = \frac{BC}{AB} = \frac{BC}{AC} \times \frac{AC}{AB} = \frac{\sin \theta}{\cos \theta}$$

Most scientific calculators have pre-programmed values of $\sin \theta$, $\cos \theta$ and $\tan \theta$. Angles can be measured in degrees or radians (see Appendix III). We will use radians unless stated otherwise. If we let $A\hat{C}B = \alpha$, then

$$\sin \alpha = \frac{AB}{AC} = \cos \theta$$

and

$$\cos \alpha = \frac{BC}{AC} = \sin \theta$$

But

$$\alpha + \theta = \frac{\pi}{2}$$

Hence,

$$\sin \theta = \cos\left(\frac{\pi}{2} - \theta\right)$$

$$\cos \theta = \sin\left(\frac{\pi}{2} - \theta\right)$$

Since θ is an angle in a right-angled triangle it cannot exceed $\pi/2$. In order to define the sine, cosine and tangent ratios for angles larger than $\pi/2$ we introduce an extended definition which is applicable to angles of any size.

Consider an arm, OP, fixed at O, which can rotate (see Figure 1.34). The angle, θ, in radians, between the arm and the positive x axis is measured

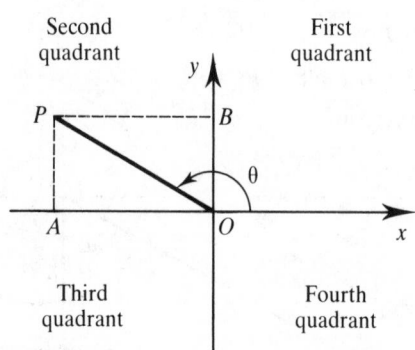

Figure 1.34   An arm, OP, fixed at O, which can rotate.

anticlockwise. The arm can be projected onto both the x and y axes. These projections are OA and OB, respectively. Whether the arm projects onto the positive or negative x and y axes depends upon which quadrant the arm is situated in. The length of the arm OP is always positive. Then,

■ $$\sin\theta = \frac{\text{projection of } OP \text{ onto } y \text{ axis}}{OP}$$

$$\cos\theta = \frac{\text{projection of } OP \text{ onto } x \text{ axis}}{OP}$$

$$\tan\theta = \frac{\text{projection of } OP \text{ onto } y \text{ axis}}{\text{projection of } OP \text{ onto } x \text{ axis}}$$

In the first quadrant, that is, $0 \leq \theta < \pi/2$, both the x and y projections are positive, so $\sin\theta$, $\cos\theta$ and $\tan\theta$ are positive. In the second quadrant, that is, $\pi/2 \leq \theta < \pi$, the x projection, OA, is negative and the y projection, OB, positive. Hence $\sin\theta$ is positive, and $\cos\theta$ and $\tan\theta$ are negative. Both the x and y projections are negative for the third quadrant and so $\sin\theta$ and $\cos\theta$ are negative while $\tan\theta$ is positive. Finally, in the fourth quadrant, the x projection is positive and the y projection is negative. Hence, $\sin\theta$ and $\tan\theta$ are negative, and $\cos\theta$ is positive (see Figure 1.35).

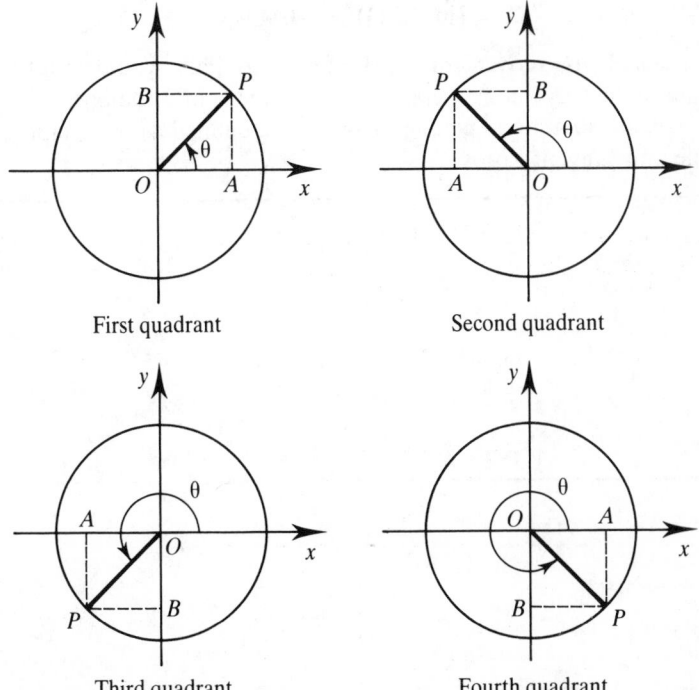

First quadrant   Second quadrant

Third quadrant   Fourth quadrant

Figure 1.35   Evaluating the trigonometric functions in each of the four quadrants.

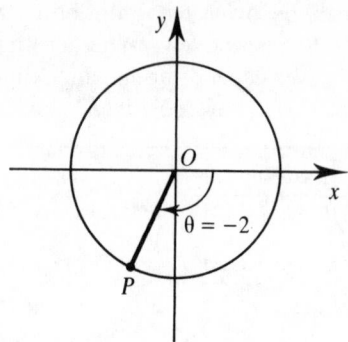

Figure 1.36   Illustration of the angle $\theta = -2$.

For angles greater than 2π, the arm OP simply rotates more than one revolution before coming to rest. Each complete revolution brings OP back to its original position. So, for example,

$$\sin(8.76) = \sin(8.76 - 2\pi) = \sin(2.477) = 0.617$$

$$\cos(14.5) = \cos(14.5 - 4\pi) = \cos(1.934) = -0.355$$

Negative angles are interpreted as a clockwise movement of the arm. Figure 1.36 illustrates an angle of -2. Note that,

$$\sin(-2) = \sin(2\pi - 2) = \sin(4.283) = -0.909$$

since an anticlockwise movement of OP of 4.283 radians would result in the arm being in the same position as a clockwise movement of 2 radians.

The **cosecant**, **secant** and **cotangent** ratios are defined as the reciprocals of the sine, cosine and tangent ratios.

■ $\cosec \theta = \dfrac{1}{\sin \theta}$

$\sec \theta = \dfrac{1}{\cos \theta}$

$\cot \theta = \dfrac{1}{\tan \theta}$

Example 1.31

Simplify:

(a) $\sin\left(\dfrac{\pi}{2} + \theta\right)$

(b) $\cos\left(\dfrac{3\pi}{2} - \theta\right)$

(c) $\tan(2\pi - \theta)$

(d) $\sin(\pi - \theta)$

where $0 \leq \theta \leq (\pi/2)$.

Solutions

(a) With reference to Figure 1.37(a) we see:

$$\sin\left(\frac{\pi}{2} + \theta\right) = OB/OP$$

$$= OB/OP'$$

$$= \cos\theta$$

(b) From Figure 1.37(b) we see:

$$\cos\left(\frac{3\pi}{2} - \theta\right) = OA/OP$$

$$= -OA'/OP'$$

$$= -\cos\left(\frac{\pi}{2} - \theta\right)$$

$$= -\sin\theta$$

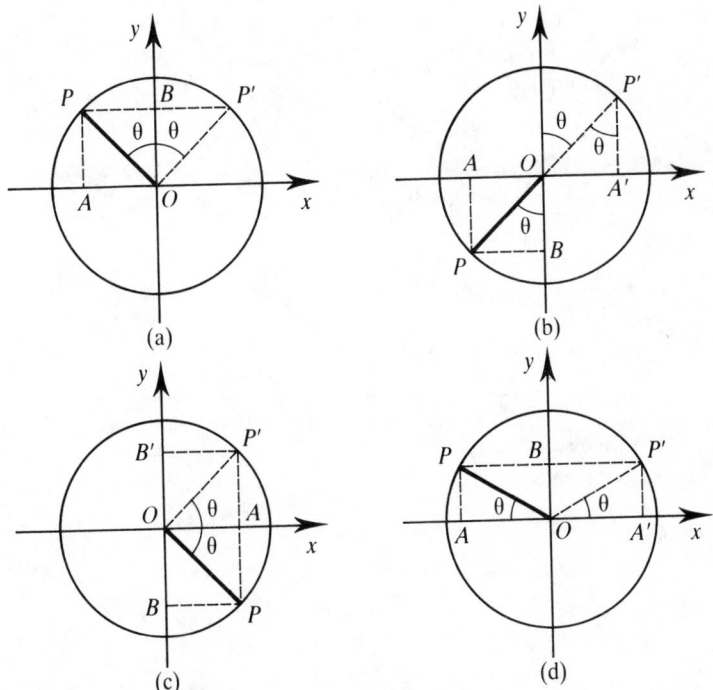

Figure 1.37   Diagrams for Example 1.31.

(c) From Figure 1.37(c) we see:

$$\tan(2\pi - \theta) = OB/OA$$

$$= -OB'/OA$$

$$= -\tan \theta$$

(d) With the aid of Figure 1.37(d) we see:

$$\sin(\pi - \theta) = OB/OP$$

$$= OB/OP'$$

$$= \sin \theta \quad \blacktriangle$$

Trigonometric identities

The formulae of Table 1.5 are commonly used and are listed for reference.

Table 1.5 Common trigonometric formulae.

$\sin(A \pm B) = \sin A \cos B \pm \sin B \cos A$
$\cos(A \pm B) = \cos A \cos B \mp \sin A \sin B$
$\tan(A \pm B) = \dfrac{\tan A \pm \tan B}{1 \mp \tan A \tan B}$
$2 \sin A \cos B = \sin(A + B) + \sin(A - B)$
$2 \cos A \cos B = \cos(A + B) + \cos(A - B)$
$2 \sin A \sin B = \cos(A - B) - \cos(A + B)$
$\sin^2 A + \cos^2 A = 1$
$1 + \cot^2 A = \text{cosec}^2 A$
$\tan^2 A + 1 = \sec^2 A$
$\cos 2A = 1 - 2 \sin^2 A = 2 \cos^2 A - 1 = \cos^2 A - \sin^2 A$
$\sin 2A = 2 \sin A \cos A$
$\sin^2 A = \dfrac{1 - \cos 2A}{2}$
$\cos^2 A = \dfrac{1 + \cos 2A}{2}$

Note: $\sin^2 A$ is the notation used for $(\sin A)^2$. Similarly $\cos^2 A$ means $(\cos A)^2$.

Example 1.32

Simplify:

(a) $\cos(\pi + \theta)$

(b) $\tan(\pi - \theta)$

(c) $\sin^3 B + \sin B \cos^2 B$

(d) $\tan A(1 + \cos 2A)$

Solutions

(a) $\cos(\pi + \theta) = \cos \pi \cos \theta - \sin \pi \sin \theta = (-1)\cos \theta - 0 \sin \theta = -\cos \theta$

(b) $\tan(\pi - \theta) = \dfrac{(\tan \pi - \tan \theta)}{(1 + \tan \pi \tan \theta)} = \dfrac{(0 - \tan \theta)}{(1 + 0 \tan \theta)} = -\tan \theta$

(c) $\sin^3 B + \sin B \cos^2 B = \sin B(\sin^2 B + \cos^2 B) = \sin B$

(d) $\tan A(1 + \cos 2A) = \dfrac{\sin A}{\cos A} 2 \cos^2 A = 2 \sin A \cos A = \sin 2A$ ▲

The sine, cosine and tangent functions

The sine, cosine and tangent functions follow directly from the trigonometric ratios. These are defined to be $f(x) = \sin x$, $f(x) = \cos x$ and $f(x) = \tan x$. Graphs can be constructed from a table of values found using a scientific calculator. They are shown in Figures 1.38, 1.39 and 1.40. Note that these functions are many-to-one.

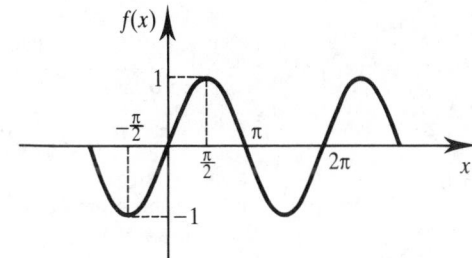

Figure 1.38 Graph of $f(x) = \sin x$.

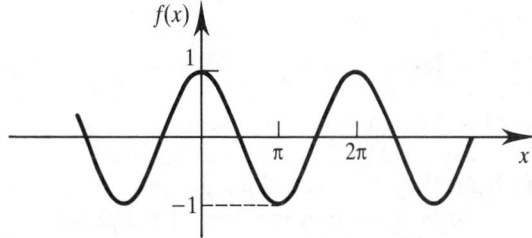

Figure 1.39 Graph of $f(x) = \cos x$.

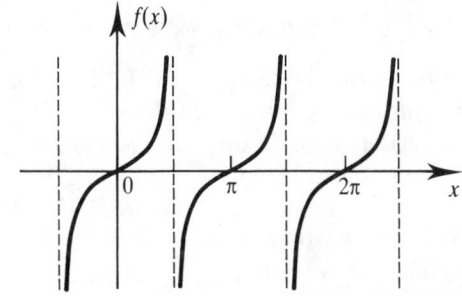

Figure 1.40 Graph of $f(x) = \tan x$.

By shifting the cosine function to the right by an amount $\pi/2$ the sine function is obtained. Similarly shifting the sine function to the left by $\pi/2$ results in the cosine function. This interchangeability between the sine and cosine functions is reflected in them being commonly referred to as **sinusoidal** functions. Notice also from the graphs two important properties of sin x and cos x:

■ $\sin x = -\sin(-x)$

$\cos x = \cos(-x)$

For example,

$$\sin \frac{\pi}{3} = -\sin\left(-\frac{\pi}{3}\right) \quad \text{and} \quad \cos \frac{\pi}{3} = \cos\left(-\frac{\pi}{3}\right).$$

EXERCISES 1.7

1. Simplify:

   (a) cot $2x$ sec $2x$

   (b) $\tan \theta \tan\left(\frac{\pi}{2} + \theta\right)$

2. Prove that

$$\frac{4}{\sin^2 2A} = 2 + \tan^2 A + \cot^2 A$$

3. Suppose u is any real number and we let

$$x = \cos u \qquad y = \sin u \qquad (1.9)$$

Then, for any u we can calculate x and y and plot the point (x, y). The resulting curve is said to be defined parametrically by Equation (1.9), and u is called a **parameter**. Show that the point (x, y) lies on the circle

$$x^2 + y^2 = 1$$

1.4.7 *Inverse trigonometric functions*

All six trigonometric functions have an inverse but we will only examine those of sin x, cos x and tan x. The inverse functions of sin x, cos x and tan x are denoted $\sin^{-1} x$, $\cos^{-1} x$ and $\tan^{-1} x$. This notation can and does cause confusion. The '-1' in $\sin^{-1} x$ is sometimes mistakenly interpreted as a power. (We write $(\sin x)^{-1}$ to denote $1/(\sin x)$.) Values of $\sin^{-1} x$, $\cos^{-1} x$ and $\tan^{-1} x$ can be found using a scientific calculator. If $y = \sin x$, then $x = \sin^{-1} y$, as in

$$\sin x = 0.7654 \qquad x = \sin^{-1}(0.7654) = 0.8717$$

Note that $y = \sin x$ is a many-to-one function. If the domain is restricted to $[-\pi/2, \pi/2]$ then the resulting function is one-to-one. This is shown in Figure 1.41. Recall from Section 1.3 that a one-to-one function has a corresponding inverse. So if the domain of $y = \sin x$ is restricted to $[-\pi/2, \pi/2]$, then an inverse function exists. A graph of $y = \sin^{-1} x$ is shown in Figure 1.42. Without the domain restriction, a one-to-many graph would result as shown in Figure 1.43. To clearly denote the inverse sine function we write:

$$y = \sin^{-1} x \qquad \frac{-\pi}{2} \leq y \leq \frac{\pi}{2}$$

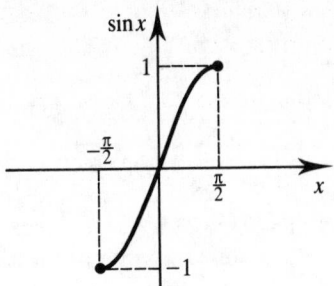

Figure 1.41 The function sin x is one-to-one if the domain is restricted to $\left[-\dfrac{\pi}{2}, \dfrac{\pi}{2}\right]$.

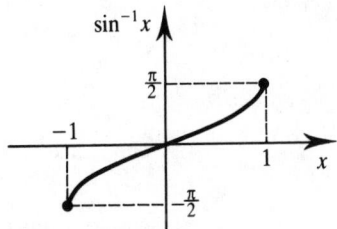

Figure 1.42 The inverse sine function, $\sin^{-1} x$.

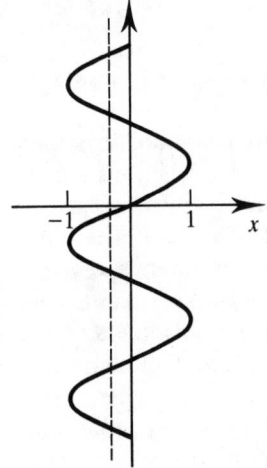

Figure 1.43 A single input produces many output values. This is not a function.

Example 1.33

Evaluate:

(a) $\sin^{-1}(0.3169)$

(b) $\sin^{-1}(-0.8061)$

Solutions

(a) $\sin^{-1}(0.3169) = 0.3225$

(b) $\sin^{-1}(-0.8061) = -0.9375$ ▲

A word of warning about inverse trigonometric functions is needed. The calculator returns a value of 0.3225 for $\sin^{-1}(0.3169)$. Note, however, that $\sin(0.3225 \pm 2n\pi) = 0.3169$, $n = 0, 1, 2, 3,\ldots$, so there are an infinite number of values of x such that $\sin x = 0.3169$. Only one of these values is returned by the calculator. This is because the domain of $y = \sin x$ is restricted to ensure it has an inverse function. To ensure the inverse functions $y = \cos^{-1} x$ and $y = \tan^{-1} x$ can be obtained, restrictions are placed on the domains of $y = \cos x$ and $y = \tan x$. By convention, $y = \cos x$ has its domain restricted to $[0, \pi]$ whereas with $y = \tan x$ the restriction is $(-\pi/2, \pi/2)$.

EXERCISES 1.8

1. Show that

$$\sin^{-1} x + \cos^{-1} x = \frac{\pi}{2}$$

2. Evaluate:

   (a) $\sin^{-1}(0.75)$

   (b) $\cos^{-1}(0.625)$

   (c) $\tan^{-1} 3$

3. Sketch the graphs of $y = \cos^{-1} x$ and $y = \tan^{-1} x$

1.4.8 *Modelling waves using* sin *x and* cos *x*

Examining the graphs of $\sin x$ and $\cos x$ reveals that they have a similar shape to waves. In fact, sine and cosine functions are often used to model waves and we will see in Chapter 13 that almost any wave can be broken down into a combination of sine and cosine functions. The main waves found in engineering are ones that vary with time and so t is often the independent variable.

The **amplitude** of a wave is the maximum displacement of the wave from its mean position. So, for example, $\sin t$ and $\cos t$ have an amplitude of 1, the amplitude of $2 \sin t$ is 2, and the amplitude of $A \sin t$ is A (see Figure 1.44). The sine and cosine functions repeat themselves at regular intervals and so are **periodic** functions. Since the waves $A \sin t$ and $A \cos t$ repeat at intervals of 2π their period, T, is 2π. Mathematically, this means

$$\sin t = \sin(t + 2\pi) \qquad \cos t = \cos(t + 2\pi)$$

or more generally,

■ $\sin t = \sin(t \pm 2n\pi)$ $n = 0, 1, 2, 3,\ldots$

   $\cos t = \cos(t \pm 2n\pi)$ $n = 0, 1, 2, 3,\ldots$

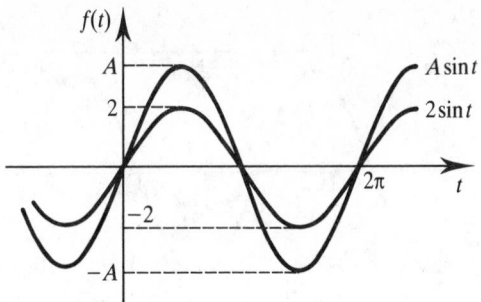

Figure 1.44 The amplitude of $f(t) = A \sin t$ is A.

These periodic functions are said to complete one **cycle** in a time interval of one period.

A more general wave is defined by $A \cos \omega t$ or $A \sin \omega t$. The symbol ω represents the **angular frequency** of the wave. It is measured in radians per second. For example, $\sin 3t$ has an angular frequency of 3 rad s^{-1}. As t increases by 1 second the angle, $3t$, increases by 3 radians. Note that $\sin t$ has an angular frequency of 1 rad s^{-1}.

The **frequency** of $A \sin \omega t$ or $A \cos \omega t$ is the number of cycles completed every second. Sin t completes one cycle in 2π seconds so in 1 second only $1/2\pi$ cycles will be completed. The frequency of $\sin t$ is thus $1/2\pi$ cycle s^{-1}. The unit 'cycles per second' is known as hertz (Hz). Since $\sin \omega t$ has an angular frequency of ω rad s^{-1}, that is, ω times faster than $\sin t$, the frequency is $\omega/2\pi$ Hz. For example, $\sin 3t$ has frequency $3/2\pi$ Hz. The symbol f is often used for frequency. So for the functions $A \sin \omega t$ and $A \cos \omega t$ we can state:

$$\text{frequency} = f = \frac{\omega}{2\pi}$$

Thus $A \sin \omega t$ and $A \cos \omega t$ may be written as $A \sin 2\pi f t$ and $A \cos 2\pi f t$. The period, T, of $A \sin \omega t$ and $A \cos \omega t$ is $2\pi/\omega = 1/f$.

A final generalization is to introduce a **phase angle**, ϕ. This allows the wave to be shifted along the time axis. It also means that either a sine function or a cosine function can be used to represent the same wave. So the general forms are:

$$A \cos(\omega t + \phi), \qquad A \sin(\omega t + \phi)$$

Figure 1.45 depicts $A \sin(\omega t + \phi)$. Note from Figure 1.45 that the actual movement of the wave along the time axis is ϕ/ω. It is easy to show this mathematically.

$$A \sin(\omega t + \phi) = A \sin \omega \left(t + \frac{\phi}{\omega} \right)$$

The quantity ϕ/ω is called the **time displacement**.

The waves met in engineering are often termed **signals** or **waveforms**. There are no rigid rules concerning the use of the words wave, signal and waveform, and often engineers use them interchangeably. We will follow this convention.

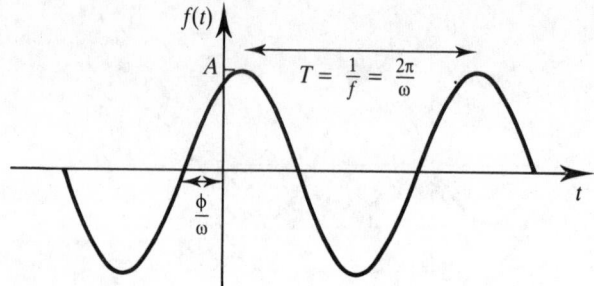

Figure 1.45 The generalized wave $A \sin(\omega t + \phi)$.

Example 1.34

State the amplitude, angular frequency and period of each of the following waves.

(a) 2 sin 3*t*

(b) $\dfrac{1}{2} \cos\left(2t + \dfrac{\pi}{6}\right)$

Solutions

(a) Amplitude, $A = 2$, angular frequency, $\omega = 3$, period, $T = 2\pi/\omega = 2\pi/3$.

(b) Amplitude, $A = 0.5$, angular frequency, $\omega = 2$,
period, $T = 2\pi/\omega = 2\pi/2 = \pi$. ▲

Example 1.35

State the amplitude, period, phase angle and time displacement of:

(a) $2 \sin(4t + 1)$ (c) $4 \cos\left(\dfrac{2t + 1}{3}\right)$

(b) $\dfrac{2 \cos(t - 0.7)}{3}$ (d) $\dfrac{3}{4} \sin\left(\dfrac{4t}{3}\right)$

Solutions

(a) Amplitude $= 2$, period $= \dfrac{2\pi}{4} = \dfrac{\pi}{2}$, phase angle $= 1$ relative to 2 sin 4*t*,
time displacement $= 0.25$.

(b) Amplitude $= \dfrac{2}{3}$, period $= 2\pi$, phase angle $= -0.7$ relative to $\left(\dfrac{2}{3}\right) \cos t$,
time displacement $= -0.7$.

(c) Amplitude $= 4$, period $= \dfrac{2\pi}{2/3} = 3\pi$, phase angle $= \dfrac{1}{3}$ relative to $4 \cos\left(\dfrac{2t}{3}\right)$,
time displacement $= 0.5$.

(d) Amplitude $= \dfrac{3}{4}$, period $= \dfrac{2\pi}{4/3} = \dfrac{3\pi}{2}$, phase angle $= 0$ relative to $\dfrac{3}{4}\sin\left(\dfrac{4t}{3}\right)$,

time displacement $= 0$. ▲

Example 1.36   Alternating current waveforms

Sine and cosine functions are often used to model alternating current (a.c.) waveforms. The equations for an a.c. current waveform are:

$$I = I_m \sin(\omega t + \phi) \qquad \text{or} \qquad I = I_m \cos(\omega t + \phi)$$

where I_m = maximum current, ω = angular frequency and ϕ = phase angle. In practice the functions can be shifted along the time axis by giving ϕ a non-zero value and so both the sine and the cosine function can be used to model any a.c. waveform; which one is used is usually a matter of convenience. ▲

Example 1.37

Write $3\sin t + 4\cos t$ in the form $a\sin(t + \phi)$.

Solution

Using the trigonometric identities of Table 1.5 we can write:

$$a\sin(t + \phi) = a(\sin t\cos\phi + \sin\phi\cos t) = (a\cos\phi)\sin t + (a\sin\phi)\cos t$$

We wish to choose a and ϕ so the expression on the right-hand side is equal to $3\sin t + 4\cos t$. We thus require

$$a\cos\phi = 3 \tag{1.10}$$
$$a\sin\phi = 4 \tag{1.11}$$

Equations (1.10) and (1.11) are solved for a and ϕ. By squaring we obtain:

$$a^2\cos^2\phi = 9$$
$$a^2\sin^2\phi = 16$$

and then by adding we get:

$$a^2(\cos^2\phi + \sin^2\phi) = 25$$
$$a^2 = 25 \qquad \text{since } \cos^2\phi + \sin^2\phi = 1$$
$$a = 5$$

Dividing Equation (1.11) by Equation (1.10), we find:

$$\frac{a\sin\phi}{a\cos\phi} = \frac{4}{3}$$
$$\tan\phi = \frac{4}{3}$$

Therefore,

$$\phi = \tan^{-1}\frac{4}{3}$$

Since both sin ϕ and cos ϕ are positive, ϕ is in the first quadrant and so $\phi = 0.927$. Therefore

$$3 \sin t + 4 \cos t = 5 \sin(t + 0.927) \quad \blacktriangle$$

Example 1.37 illustrates an important property.

■ If two waves of equal frequency, ω, are added the result is a wave of the same frequency.

Example 1.38

Express as a single cosine wave:

(a) $3 \sin 4t - 2 \cos 4t$

(b) $\dfrac{\cos 3t}{2} + \sin 3t$

Solutions

(a) Let

$$3 \sin 4t - 2 \cos 4t = a \cos(4t + \phi)$$

Then using the trigonometric identities in Table 1.5 we can write:

$$3 \sin 4t - 2 \cos 4t = a(\cos 4t \cos \phi - \sin 4t \sin \phi)$$

$$= (a \cos \phi)\cos 4t - (a \sin \phi)\sin 4t$$

Hence,

$$3 = -a \sin \phi \tag{1.12}$$

$$-2 = a \cos \phi \tag{1.13}$$

By squaring and adding we obtain:

$$9 + 4 = a^2(\sin^2 \phi + \cos^2 \phi) = a^2$$

$$a = \sqrt{13}$$

From Equations (1.12) and (1.13), both sin ϕ and cos ϕ are negative and so ϕ is in the third quadrant. Division of Equation (1.12) by Equation (1.13) yields:

$$\frac{3}{-2} = \frac{-a \sin \phi}{a \cos \phi} = -\tan \phi$$

$$\tan \phi = 1.5$$

$$\phi = 4.124$$

Hence,

$$3 \sin 4t - 2 \cos 4t = \sqrt{13} \cos(4t + 4.124)$$

(b) Let

$$\frac{\cos 3t}{2} + \sin 3t = a \cos(3t + \phi) = (a \cos \phi)\cos 3t - (a \sin \phi)\sin 3t$$

Hence

$$0.5 = a \cos \phi \tag{1.14}$$

$$1 = -a \sin \phi \tag{1.15}$$

By squaring and adding we get:

$$1.25 = a^2$$

$$a = \sqrt{1.25} = 1.12$$

Division of Equation (1.15) by Equation (1.14) yields:

$$2 = -\tan \phi$$

From Equation (1.14), $\cos \phi$ is positive; from Equation (1.15) $\sin \phi$ is negative and so ϕ is in the fourth quadrant. Hence,

$$\phi = 5.18$$

So

$$\frac{\cos 3t}{2} + \sin 3t = 1.12 \cos(3t + 5.18) \quad \blacktriangle$$

Example 1.39

If $a \cos \omega t + b \sin \omega t = R \cos(\omega t - \theta)$ show $R = \sqrt{a^2 + b^2}$ and $\tan \theta = b/a$.

Solution

Let

$$a \cos \omega t + b \sin \omega t = R \cos(\omega t - \theta)$$

Then using the trigonometric identities, we can write:

$$a \cos \omega t + b \sin \omega t = R(\cos \omega t \cos \theta + \sin \omega t \sin \theta)$$

$$= (R \cos \theta)\cos \omega t + (R \sin \theta)\sin \omega t$$

Equating the coefficients of $\cos \omega t$, and then $\sin \omega t$ gives:

$$a = R \cos \theta \tag{1.16}$$

$$b = R \sin \theta \tag{1.17}$$

Squaring Equation (1.16) and Equation (1.17) and adding gives:

$$a^2 + b^2 = R^2$$

that is,

$$R = \sqrt{a^2 + b^2}$$

Division of Equation (1.17) by Equation (1.16) gives:

$$\frac{b}{a} = \tan \theta$$

as required. ▲

From Example 1.39 we note the following result.

■ If

$$a \cos \omega t + b \sin \omega t = R \cos(\omega t - \theta)$$

then

$$R = \sqrt{a^2 + b^2} \qquad \tan \theta = \frac{b}{a} \qquad \sin \theta = \frac{b}{R} \qquad \cos \theta = \frac{a}{R}$$

EXERCISES 1.9

1. State the period of:

   (a) $2 \sin 7t$

   (b) $7 \sin(2t + 3)$

   (c) $\tan(t/2)$

   (d) $\sec 3t$

   (e) $\operatorname{cosec}(2t - 1)$

   (f) $\cot((2t/3) + 2)$

2. A sinusoidal function has an amplitude of 2/3 and a period of 2. State a possible form of the function.

3. State the phase angle and time displacement of:

   (a) $2 \sin(t + 3)$ relative to $2 \sin t$

   (b) $\sin(2t - 3)$ relative to $\sin 2t$

   (c) $\cos\left(\dfrac{t}{2} + 0.2\right)$ relative to $\cos \dfrac{t}{2}$

   (d) $\cos(2 - t)$ relative to $\cos t$

   (e) $\sin\left(\dfrac{3t + 4}{5}\right)$ relative to $\sin \dfrac{3t}{5}$

   (f) $\sin(4 - 3t)$ relative to $\sin 3t$

   (g) $\sin(2\pi t + \pi)$ relative to $\sin 2\pi t$

   (h) $3 \cos(5\pi t - 3)$ relative to $3 \cos 5\pi t$

   (i) $\sin\left(\dfrac{\pi t}{3} + 2\right)$ relative to $\sin \dfrac{\pi t}{3}$

   (j) $\cos(3\pi - t)$ relative to $\cos t$

4. Write each of the following expressions in the form (i) $A \sin(\omega t + \theta)$, (ii) $A \sin(\omega t - \theta)$, (iii) $A \cos(\omega t + \theta)$, (iv) $A \cos(\omega t - \theta)$.

   (a) $5 \sin t + 4 \cos t$

   (b) $-2 \sin 3t + 2 \cos 3t$

   (c) $4 \sin 2t - 6 \cos 2t$

   (d) $-\sin 5t - 3 \cos 5t$

1.4.9 *The hyperbolic functions*

The **hyperbolic functions** are $y(x) = \cosh x$, $y(x) = \sinh x$, $y(x) = \tanh x$, $y(x) = \operatorname{sech} x$, $y(x) = \operatorname{cosech} x$, and $y(x) = \coth x$. Cosh is a contracted form of 'hyperbolic cosine', sinh of 'hyperbolic sine' and so on. We define $\cosh x$ and $\sinh x$ by:

■ $y(x) = \cosh x = \dfrac{e^x + e^{-x}}{2}$ $y(x) = \sinh x = \dfrac{e^x - e^{-x}}{2}$

Note:

$$\cosh(-x) = \frac{e^{-x} + e^x}{2} = \cosh x$$

$$\sinh(-x) = \frac{e^{-x} - e^x}{2} = -\sinh x$$

so, for example, $\cosh 1.7 = \cosh(-1.7)$ and $\sinh(-1.7) = -\sinh 1.7$. Clearly, hyperbolic functions are nothing other than combinations of the exponential functions e^x and e^{-x}. However, these particular combinations occur so frequently in engineering it is worth introducing the $\cosh x$ and $\sinh x$ functions. The remaining hyperbolic functions are defined in terms of $\cosh x$ and $\sinh x$.

■ $y(x) = \tanh x = \dfrac{\sinh x}{\cosh x} = \dfrac{e^x - e^{-x}}{e^x + e^{-x}}$

$y(x) = \operatorname{sech} x = \dfrac{1}{\cosh x} = \dfrac{2}{e^x + e^{-x}}$

$y(x) = \operatorname{cosech} x = \dfrac{1}{\sinh x} = \dfrac{2}{e^x - e^{-x}}$

$y(x) = \coth x = \dfrac{\cosh x}{\sinh x} = \dfrac{1}{\tanh x} = \dfrac{e^x + e^{-x}}{e^x - e^{-x}}$

Values of the hyperbolic functions for various x values can be found from a scientific calculator. Usually a Hyp button followed by a Sin, Cos or Tan button is used.

Example 1.40

Evaluate:

(a) cosh 3

(b) sinh(−2)

(c) tanh 1.6

(d) sech(−2.5)

(e) coth 1

(f) cosech(−1)

Solutions

(a) cosh 3 = 10.07

(b) sinh(−2) = −3.627

(c) tanh(1.6) = 0.9217

(d) sech(−2.5) = 1/cosh(−2.5) = 0.1631

(e) coth 1 = 1/tanh 1 = 1.313

(f) cosech(−1) = 1/sinh(−1) = −0.8509 ▲

Graphs of the functions sinh x, cosh x and tanh x can be obtained using a graphics calculator.

Hyperbolic identities

Several identities involving hyperbolic functions exist. They can be verified algebraically using the definitions given, and are listed for reference. Many of the identities bear close resemblance to the trigonometric identities.

■ $\cosh^2 x - \sinh^2 x = 1$

$$1 - \tanh^2 x = \operatorname{sech}^2 x$$

$$\coth^2 x - 1 = \operatorname{cosech}^2 x$$

$$\sinh(x \pm y) = \sinh x \cosh y \pm \cosh x \sinh y$$

$$\cosh(x \pm y) = \cosh x \cosh y \pm \sinh x \sinh y$$

$$\sinh 2x = 2 \sinh x \cosh x$$

$$\cosh 2x = \cosh^2 x + \sinh^2 x$$

$$\cosh^2 x = \frac{\cosh 2x + 1}{2}$$

$$\sinh^2 x = \frac{\cosh 2x - 1}{2}$$

Note also that:

■ $e^x = \cosh x + \sinh x$ $e^{-x} = \cosh x - \sinh x$

Hence any combination of exponential terms may be expressed as a combination of cosh x and sinh x, and vice versa.

Example 1.41

Express

(a) $3e^x - 2e^{-x}$ in terms of cosh x and sinh x.

(b) $2 \sinh x + \cosh x$ in terms of e^x and e^{-x}.

Solutions

(a) $3e^x - 2e^{-x} = 3(\cosh x + \sinh x) - 2(\cosh x - \sinh x) = \cosh x + 5 \sinh x$.

(b) $2 \sinh x + \cosh x = e^x - e^{-x} + \dfrac{e^x + e^{-x}}{2} = \dfrac{3e^x - e^{-x}}{2}$. ▲

Inverse hyperbolic functions

The inverse hyperbolic functions use the familiar notation. Both $y = \sinh x$ and $y = \tanh x$ are one-to-one functions and no domain restriction is needed for an inverse to be defined. However, on $(-\infty, \infty)$, $y = \cosh x$ is a many-to-one function. If the domain is restricted to $[0, \infty)$ the resulting function is one-to-one and an inverse function can be defined. Values of $\cosh^{-1} x$, $\sinh^{-1} x$ and $\tanh^{-1} x$ can be obtained using a scientific calculator.

Example 1.42

Evaluate:

(a) $\cosh^{-1}(3.7)$

(b) $\sinh^{-1}(-2)$

(c) $\tanh^{-1}(0.5)$

Solutions

Using a calculator we get:

(a) 1.9827

(b) -1.4436

(c) 0.5493

A word of warning about the inverse of cosh x is needed. The calculator returns a value of 1.9827 for $\cosh^{-1}(3.7)$. Note, however, that $\cosh(-1.9827) = 3.7$. The value -1.9827 is not returned by the calculator; only values on $[0, \infty)$ will be returned. This is because the domain of $y = \cosh x$ is restricted to ensure it has an inverse function.

EXERCISES 1.10

1. Express $ae^x + be^{-x}$, where a and b are constants, in terms of cosh x and sinh x.

2. Express $a \cosh x + b \sinh x$, where a and b are constants, in terms of e^x and e^{-x}.

3. Show that the point $x = \cosh u$, $y = \sinh u$ lies on the curve:

$$x^2 - y^2 = 1$$

(Compare with $x = \cos u$, $y = \sin u$ lying on the circle $x^2 + y^2 = 1$, see Exercise 1.7.3).

4. Prove the identities listed earlier in this section.

1.4.10 *The modulus function*

The modulus of a positive number is simply the number itself. The modulus of a negative number is a positive number of the same magnitude. For example, the modulus of 3 is 3; the modulus of -3 is also 3. We enclose the number in vertical lines to show we are finding its modulus, thus

$$|3| = 3 \qquad |-3| = 3$$

Mathematically we define the modulus function as follows:

■ The **modulus function** is defined by:

$$f(x) = |x| = \begin{cases} x & x \geq 0 \\ -x & x < 0 \end{cases}$$

Figure 1.46 illustrates a graph of $f(x) = |x|$. The modulus of a quantity is never negative.

Consider two points on the x axis, a and b as shown in Figure 1.47. Then:

$$|a - b| = |b - a| = \text{distance from } a \text{ to } b$$

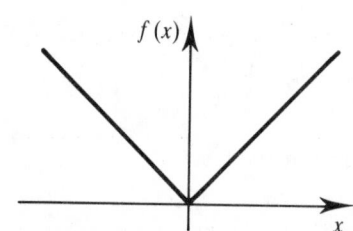

Figure 1.46 The function: $f(x) = |x|$.

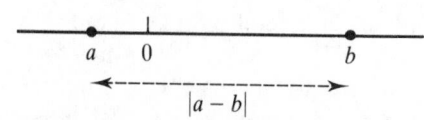

Figure 1.47 Distance from a to $b = |a - b|$.

Example 1.43

Find the distance from:

(a) $x = 2$ to $x = 9$

(b) $x = -2$ to $x = 9$

(c) $x = -2$ to $x = -9$

Solutions

(a) Distance $= |2 - 9| = |-7| = 7$

(b) Distance $= |-2 - 9| = |-11| = 11$

(c) Distance $= |-2 - (-9)| = |7| = 7$ ▲

Example 1.44

Describe the interval on the x axis defined by:

(a) $|x| < 2$

(b) $|x| \geq 3$

(c) $|x - 1| < 3$

(d) $|x + 2| > 1$

Solutions

(a) $|x| < 2$ is the same statement as $-2 < x < 2$, that is, x is numerically less than 2. Figure 1.48 illustrates this region. Note that the region is an open interval. Since the points $x = -2$ and $x = 2$ are not included, they are shown on the graph as ○.

(b) If $|x| \geq 3$ then either $x \geq 3$ or $x \leq -3$. This is shown in Figure 1.49. The required region of the x axis has two distinct parts. Since the points $x = 3$ and $x = -3$ are included in the interval of interest, they are shown on the graph as ●.

(c) $|x - 1| < 3$ is equivalent to $-3 < x - 1 < 3$, that is, $-2 < x < 4$.

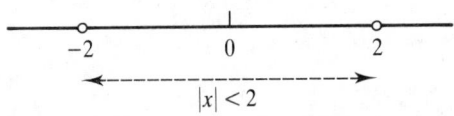

Figure 1.48   The quantity $|x| < 2$ is equivalent to $-2 < x < 2$.

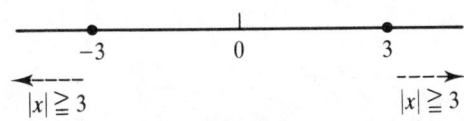

Figure 1.49   The quantity $|x| \geq 3$ is equivalent to $x \geq 3$ or $x \leq -3$.

(d) $|x + 2| > 1$ is equivalent to $x + 2 > 1$ or $x + 2 < -1$, that is, $x > -1$ or $x < -3$. ▲

The modulus function can be used to describe regions in the $x-y$ plane.

Example 1.45

Sketch the region defined by:

(a) $|x| < 2$ and $|y| < 1$
(b) $|x^2 + y^2| \leq 9$

Solutions

(a) The region is a rectangle as shown in Figure 1.50. The boundary is not part of the region as strict inequalities were used to define it. The region $|x| \leq 2, |y| \leq 1$ is the same as that in Figure 1.50 with the boundary included.

(b) $|x^2 + y^2| \leq 9$ is equivalent to $-9 \leq x^2 + y^2 \leq 9$. Note, however, that $x^2 + y^2$ is never negative and so the region is given by $0 \leq x^2 + y^2 \leq 9$.
Let $P(x, y)$ be a general point as shown in Figure 1.51. Then from Pythagoras' theorem, the distance of P to the origin is $\sqrt{x^2 + y^2}$. So,

$$(\text{distance from origin})^2 = x^2 + y^2 \leq 9$$

Then,

$$(\text{distance from origin}) \leq 3$$

This describes a disk, centre the origin, of radius 3 (see Figure 1.52). ▲

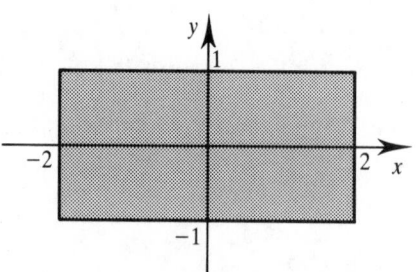

Figure 1.50   The region: $|x| < 2$ and $|y| < 1$.

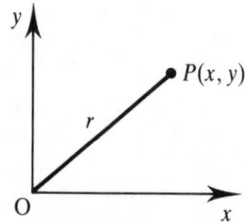

Figure 1.51   Point $P(x, y)$ is a general point.

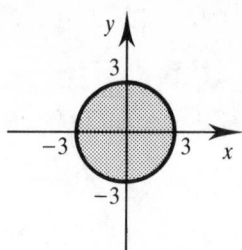

Figure 1.52   The region: $|x^2 + y^2| \leq 9$.

Example 1.46   Full wave rectifier

A fully rectified sine wave is the modulus of the sine wave. The circuit for a full wave rectifier is shown in Figure 1.53 together with the input and output waveforms. The input signal is v_{in} and the output signal is v_o. Ignoring the voltage drops across the diode gives:

$v_o = |v_{in}|$ ▲

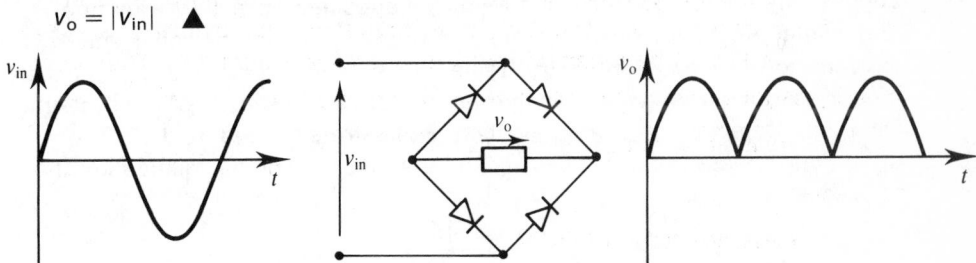

Figure 1.53   A full wave rectifier.

EXERCISES 1.11

1. Sketch the interval defined by:

   (a) $|x| > 4$

   (b) $|y - 1| \leq 3$

   (c) $|t + 6| \geq 3$

   (d) $|t^2 - 2| < 7$

   (e) $|t| < -1$

   Also, state the intervals without using the modulus signs.

2. Sketch the regions defined by

   (a) $|x| > 2, |y| < 3$

   (b) $|x + 2| < 4, |y + 1| < 3$

   (c) $|x^2 + y^2 - 1| \leq 4$

   (d) $|(x - 1)^2 + (y + 2)^2| \geq 1$

1.4.11 *The ramp function*

■ The **ramp function** is defined by,

$$f(t) = \begin{cases} ct & t \geq 0 \\ 0 & t < 0 \end{cases} \quad c \text{ constant}$$

Its graph is shown in Figure 1.54.

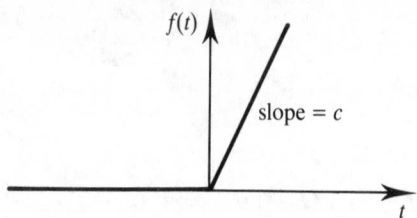

Figure 1.54   The ramp function.

Example 1.47   Telescope drive signal

In order to track the motion of the stars large telescopes are usually driven by an electric motor. The speed of this motor is controlled in order that the angular position of the telescope follows a specified trajectory with time. The whole assembly, including telescope, gears, motor, controller and sensors, forms a **position control system** or **servo-system**. This servo-system must be fed a signal corresponding to the desired trajectory of the system. Often this trajectory is a ramp function as illustrated in Figure 1.55. The drive motor is started at time $t = t_0$ and the desired angular position of the telescope is θ.   ▲

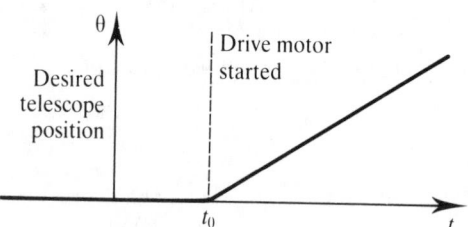

Figure 1.55   Tracking signal for a telescope.

1.4.12 *The unit step function, u(t)*

The **unit step function** is defined by:

$$u(t) = \begin{cases} 1 & t \geq 0 \\ 0 & t < 0 \end{cases}$$

Its graph is shown in Figure 1.56. Note that $u(t)$ has a discontinuity at $t = 0$. The point $(0, 1)$ is part of the function defined on $t \geq 0$. This is depicted by ●. The point $(0, 0)$ is not part of the function defined on $t < 0$. We use ○ to illustrate this.

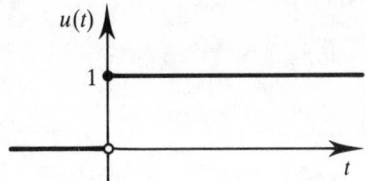

Figure 1.56 The unit step function.

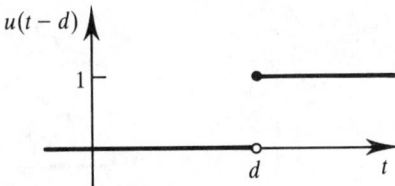

Figure 1.57 Graph of $u(t - d)$.

The position of the discontinuity may be shifted.

$$u(t - d) = \begin{cases} 1 & t \geq d \\ 0 & t < d \end{cases}$$

The graph of $u(t - d)$ is shown in Figure 1.57.

Example 1.48 RLC circuit

Consider the circuit shown in Figure 1.58. When the switch is open the voltage v, across terminals A and B is zero. If the switch is closed at $t = 0$ the voltage across A and B is V_S, where V_S is the supply voltage. Thus v can be modelled by the function:

$$v = \begin{cases} 0 & t < 0 \\ V_S & t \geq 0 \end{cases}$$

This can be written, using the unit step function, as:

$v = V_S u(t)$ ▲

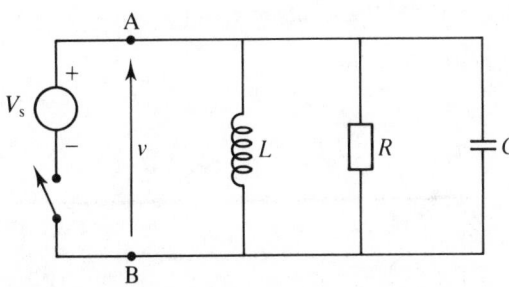

Figure 1.58 RLC circuit.

Example 1.49

Sketch the following functions:

(a) $f = u(t - 3)$

(b) $f = u(t - 1)$

(c) $f = u(t - 1) - u(t - 3)$

(d) $f = u(t - 3) - u(t - 1)$

(e) $f = e^t u(t)$

Solutions

(a) See Figure 1.59.

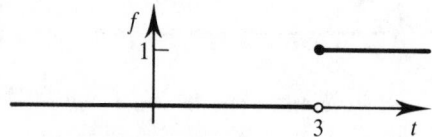

Figure 1.59   The function: $f = u(t - 3)$.

(b) See Figure 1.60.

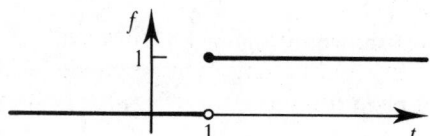

Figure 1.60   The function: $f = u(t - 1)$.

(c) See Figure 1.61.

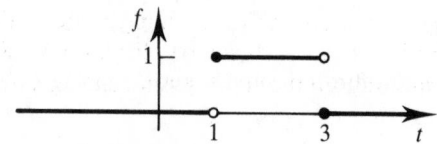

Figure 1.61   The function: $f = u(t - 1) - u(t - 3)$.

(d) See Figure 1.62.

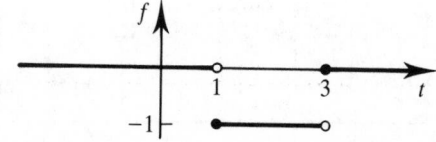

Figure 1.62   The function: $f = u(t - 3) - u(t - 1)$.

(e) See Figure 1.63. ▲

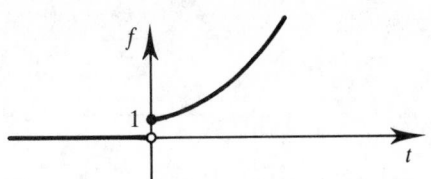

Figure 1.63   The function: $f = e^t u(t)$.

EXERCISES 1.12

Sketch the following functions:

(a) $f = u(t - 2) - u(t - 6)$

(b) $f = 3u(t - 1)$

(c) $f = 3u(t - 1) - 2u(t - 2)$

(d) $(\sin t)u(t)$

(e) $(\sin t)u(t - \pi)$

1.4.13 *The delta function or unit impulse function* $\delta(t)$

Consider the rectangle function, $R(t)$, shown in Figure 1.64. The base of the rectangle is h, the height is $1/h$ and so the area is 1. For $t > h/2$ and $t < -h/2$, the function is 0. As h decreases, the base diminishes and the height increases; the area remains constant at 1.

As h approaches 0, the base becomes infinitesimally small and the height infinitely large. The area remains at unity. The rectangle function is then called a **delta function** or **unit impulse function**. It has a value of 0 everywhere except at the origin.

$\delta(t)$ = rectangle function as h approaches 0

We write this concisely as:

$\delta(t) = R(t)$ as $h \to 0$

The position of the delta function may be changed from the origin to $t = d$. Consider a rectangle function, $R(t - d)$, shown in Figure 1.65. $R(t - d)$ is obtained by translating $R(t)$ an amount d to the right. Again, letting h approach 0 produces a delta function, this time centred on $t = d$.

$\delta(t - d) = R(t - d)$ as $h \to 0$

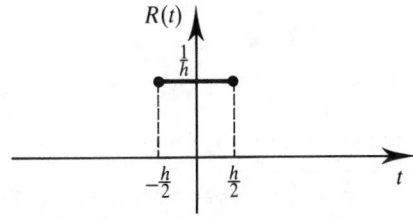

Figure 1.64   The rectangle function, $R(t)$.

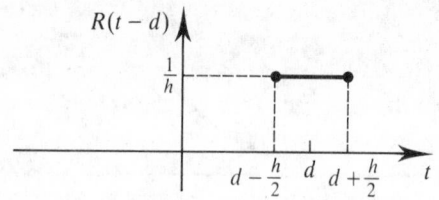

Figure 1.65 The delayed rectangle function, $R(t - d)$.

We have seen that the δ function can be regarded as bounding an area 1 between itself and the horizontal axis. More generally the area bounded by the function

$$f(t) = k\delta(t)$$

is k. We say that $k\delta(t)$ represents an impulse of strength k at the origin, and $k\delta(t - d)$ is an impulse of strength k at $t = d$. It is often useful to depict such an impulse by an arrow where the height of the arrow gives the strength of the impulse. A series of impulses is often termed an **impulse train**.

Example 1.50

A train of impulses is given by:

$$f(t) = \delta(t) + 3\delta(t - 1) + 2\delta(t - 2)$$

Depict the train graphically.

Solution

Figure 1.66 shows the representation. In Chapter 12 we shall call such a function a series of **weighted impulses** where the weights are 1, 3 and 2, respectively.   ▲

Example 1.51   Impulse response of a system

It is not possible to produce an impulse function electronically as no practical signal can have an infinite height. However, an approximation to

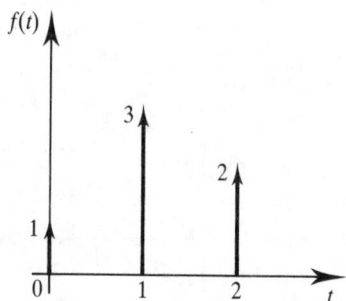

Figure 1.66   A train of impulses given by $f(t) = \delta(t) + 3\delta(t - 1) + 2\delta(t - 2)$.

an impulse function is often used, consisting of a pulse with a large voltage, V, and short duration, T. The strength of such an impulse is VT. When this pulse signal is injected into a system the output obtained is known as the **impulse response** of the system. This is a particularly useful quantity for characterizing a system and is discussed further in Chapter 11.

The approximation is valid provided the width of the pulse is an order of magnitude less than the fastest time constant in the system. If the value of T required is small in order to satisfy this constraint, then the value of V may need to be large to achieve the correct impulse strength, VT. Often this can rule out its use for many systems as the value of V is large enough to distort the system characteristics. ▲

Miscellaneous exercises

1. State the rule and sketch the graph of each of the following functions:

   (a) $f(x) = 7x - 2$

   (b) $f(t) = t^2 - 2$ $0 \le t \le 5$

   (c) $g(x) = 3e^x + 4$ $0 \le x \le 2$

   (d) $f(t) = 3 \cos t + 2$

   (e) $y(t) = (e^{2t} - 1)/2$ $t \ge 0$

   (f) $g(t) = 6 \sin(100\pi t + (\pi/2))$ $0 \le t \le 0.01$

   (g) $f(x) = x^3 + 2x + 5$ $-2 \le x \le 2$

2. Determine the inverse of each of the following functions:

   (a) $y(x) = 2x$

   (b) $f(t) = 8t - 3$

   (c) $f(x) = e^{2x}$

3. If $f(t) = \sin t$, find:

   (a) $f(\lambda)$

   (b) $f(t - \lambda)$

   (c) $f(t + \lambda)$

4. If $g(t) = \ln(t^2 + 1)$ find $g(\lambda)$ and $g(t - \lambda)$.

5. Convert the following into a single fraction:

   (a) $\dfrac{1}{x} + \dfrac{3}{x+2} + \dfrac{6}{8x+4}$

   (b) $\dfrac{1}{s} + \dfrac{2}{s^2} + \dfrac{3s+4}{8s+6}$

   (c) $\dfrac{6}{s} + \dfrac{10}{s^2} - \dfrac{s+1}{(s+2)(s+3)} + \dfrac{s-1}{(s+4)(s+3)}$

6. Reduce each of the following expressions to a single wave and in each case state the amplitude and phase angle of the resultant wave.

   (a) $2 \cos \omega t + 3 \sin \omega t$

   (b) $\cos(\omega t + (\pi/4)) + \sin \omega t$

   (c) $2 \sin(\omega t + (\pi/2)) + 4 \cos(\omega t + (\pi/4))$

   (d) $0.5 \sin(\omega t - (\pi/4)) + 1.5 \sin(\omega t + (\pi/4))$

   (e) $3 \sin \omega t + 4 \sin(\omega t + \pi)$
   $\qquad\qquad - 2 \cos(\omega t - (\pi/2))$

7. Simplify:

   (a) $\ln 3x + \ln(2/x)$

   (b) $3 \ln t + 2 \ln t^2$

   (c) $e^4 e^3 e^1$

   (d) $e^{0.5 \ln x^2}$

   (e) $(e^x + e^{-x})^2$

8. Solve

   (a) $x^2 + 2x - 35 = 0$

   (b) $x^3 + 5x^2 - 6x = 0$

   (c) $e^{t^2 - 1} = 10$

   (d) $3 \ln(2t^2 + 4t + 1) = 7.8$

   (e) $|2x - 3| = 2$

9. Sketch

   (a) $f = |-2t|$ $-3 \le t \le 3$

   (b) $f = -|2t|$ $-3 \le t \le 3$

   (c) $f = u(\sin t)$ where $u(t)$ is the unit step function

Discrete mathematics

2

2.1 Introduction

The term **discrete** is used to describe a growing number of modern branches of mathematics involving topics such as set theory, logic, Boolean algebra, difference equations and z transforms. These topics are particularly relevant to the needs of electrical and electronic engineers. **Set theory** provides us with a language for precisely specifying a great deal of mathematical work. Over the next few years this language will become particularly important as more and more emphasis is placed upon verification of software. **Boolean algebra** finds its main use in the design of digital electronic circuits. Given that an increasing proportion of electronic circuits are digital rather than analog, this is an important area of study. Digital electronic circuits confine themselves to two effective voltage levels rather than the range of voltage levels used by analog electronic circuits. These make them easier to design and manufacture as tolerances are not so critical. Digital circuits are becoming more complex each year and one of the few ways of dealing with this complexity is to use mathematics. One of the likely trends for the future is that more and more circuit designs will be proved to be correct using mathematics before they are implemented. **Difference equations** and z **transforms** are of increasing importance in fields such as digital control and digital signal processing. We shall deal with these in Chapter 12.

2.2 Set theory

■ A **set** is any collection of objects, things or states.

The objects may be numbers, letters, days of the week, or, in fact, anything under discussion. One way of describing a set is to list the whole collection of members or **elements** and enclose them in braces { }. Consider the following examples.

$A = \{1, 0\}$ the set of binary digits, one and zero

$B = \{\text{off}, \text{on}\}$ the set of possible states of a two-state system

$C = \{\text{high, low}\}$ the set of effective voltage levels in a digital electronic circuit

$D = \{0, 1, 2, 3, 4, 5, 6, 7, 8, 9\}$ the set of digits used in the decimal system

Notice that we usually use a capital letter to represent a set. To state that a particular object belongs to a particular set we use the symbol $\in$ which means 'is a member of'. So, for example, we can write:

$\text{off} \in B$ $3 \in D$

Likewise, $\notin$ means 'is not a member of' so that

$\text{low} \notin B$ $5 \notin A$

are sensible statements.

Listing members of a set is fine when there are relatively few but is useless if we are dealing with very large sets. Clearly, we could not possibly write down all the members of the set of whole numbers because there is an infinite number. To assist us special symbols have been introduced to stand for some commonly used sets. These are

---

■	$\mathbb{N}$	the set of non-negative whole numbers, 0, 1, 2, 3, ...
	$\mathbb{N}^+$	the set of positive whole numbers, 1, 2, 3, ...
	$\mathbb{Z}$	the set of whole numbers, positive, negative and zero, ... $-3, -2, -1, 0, 1, 2, 3$ ...
	$\mathbb{R}$	the set of all real numbers
	$\mathbb{R}^+$	the set of positive real numbers
	$\mathbb{R}^-$	the set of negative real numbers
	$\mathbb{Q}$	the set of rational numbers

Note that a real number is any number in the interval $(-\infty, \infty)$.

Another way of defining a set is to give a rule by which all members can be found. Consider the following notation:

$A = \{x : x \in \mathbb{R} \text{ and } x < 2\}$

This reads 'A is the set of values of x such that x is a member of the set of real numbers and x is less than 2'. Thus A corresponds to the interval $(-\infty, 2)$. Using this notation other sets can be defined.

Example 2.1

Use set notation to describe the intervals on the *x* axis given by

(a) $[0, 2]$;   (b) $[0, 2)$;   (c) $[-9, 9]$.

Solutions

(a) $\{x : x \in \mathbb{R} \text{ and } 0 \leq x \leq 2\}$
(b) $\{x : x \in \mathbb{R} \text{ and } 0 \leq x < 2\}$
(c) $\{x : x \in \mathbb{R} \text{ and } -9 \leq x \leq 9\}$ ▲

Sometimes we shall be content to use an English description of a set of objects, such as

M is the set of capacitors made by machine M

N is the set of capacitors made by machine N

Q is the set of faulty capacitors

2.2.1 *Equal sets*

Two sets are said to be equal if they contain exactly the same members. For example, the sets $\{9, 5, 2\}$ and $\{5, 9, 2\}$ are identical. The order in which we write down the members is immaterial. The sets $\{2, 2, 5, 9\}$ and $\{2, 5, 9\}$ are equal since repetition of elements is ignored.

2.2.2 *Venn diagrams*

Venn diagrams provide a graphical way of picturing sets which often aids understanding. The sets are drawn as regions, usually circles, from which various properties can be observed.

Example 2.2

Suppose we are interested in discussing the set of positive whole numbers between 1 and 10. Let $A = \{2, 3, 4, 5\}$ and $B = \{1, 3, 5, 7, 9\}$. The Venn diagram representing these sets is shown in Figure 2.1. The set containing

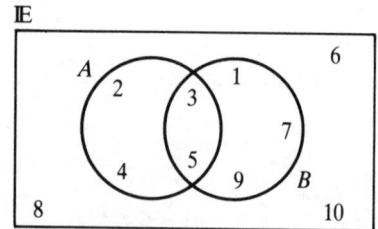

Figure 2.1   Venn diagram for Example 2.2.

all the numbers of interest is called the **universal set**, $\mathbb{E}$. $\mathbb{E}$ is represented by the rectangular region. Sets A and B are represented by the interiors of the circles and it is evident that 2, 3, 4 and 5 are members of A while 1, 3, 5, 7 and 9 are members of B. It is also clear that $6 \notin A$, $6 \notin B$, $8 \notin A$, $8 \notin B$. The elements 3 and 5 are common to both sets. ▲

■ The set containing all the members of interest is called the universal set $\mathbb{E}$.

It is useful to ask whether two or more sets have elements in common. This leads to the following definition.

2.2.3 *Intersection*

Given sets A and B, a new set which contains the elements common to both A and B is called the **intersection** of A and B, written as:

■ $A \cap B = \{x : x \in A \text{ and } x \in B\}$

In Example 2.2, we see that $A \cap B = \{3, 5\}$, that is, $3 \in A \cap B$ and $5 \in A \cap B$. If the set $A \cap B$ has no elements we say the sets A and B are **disjoint** and write $A \cap B = \varnothing$, where $\varnothing$ denotes the **empty set**.

■ A set with no elements is called an empty set and is denoted by $\varnothing$.

If $A \cap B = \varnothing$, then A and B are disjoint sets.

2.2.4 *Union*

Given two sets A and B, the set which contains all the elements of A and those of B is called the **union** of A and B, written as:

■ $A \cup B = \{x : x \in A \text{ or } x \in B \text{ or both}\}$

In Example 2.2, $A \cup B = \{1, 2, 3, 4, 5, 7, 9\}$. We note that although the elements 3 and 5 are common to both sets they are listed only once.

2.2.5 *Subsets*

If all the members of a set A are also members of a set B we say A is a **subset** of B and write $A \subset B$. We have already met a number of subsets. Convince yourself that:

$\mathbb{N} \subset \mathbb{Z}$ and $\mathbb{Z} \subset \mathbb{R}$

Example 2.3

If M represents the set of all capacitors manufactured by machine M, and M_f represents the faulty capacitors made by machine M, then clearly $M_f \subset M$.   ▲

2.2.6 *Complement*

If we are given a well-defined universal set $\mathbb{E}$ and a set A with $A \subset \mathbb{E}$, then the set of members of $\mathbb{E}$ that are not in A is called the **complement** of A and is written as $\bar{A}$. Clearly $A \cup \bar{A} = \mathbb{E}$. There are no members in the set $A \cap \bar{A}$, that is, $A \cap \bar{A} = \varnothing$.

Example 2.4

A company has a number of machines which manufacture thyristors. We consider only two machines, M and N. A small proportion made by each is faulty. Denoting the sets of faulty thyristors made by M and N, by M_f and N_f, respectively, depict this situation on a Venn diagram. Describe the sets $M_f \cup N_f$ and $\overline{M \cup N}$.

Solution

Let $\mathbb{E}$ be the universal set of all thyristors manufactured by the company. The Venn diagram is shown in Figure 2.2. Note in particular that $M \cap N = \varnothing$. There can be no thyristors in the intersection since if a thyristor is made by machine M it cannot be made by machine N and vice versa. Thus M and N are disjoint sets. Also note $M_f \subset M$, $N_f \subset N$. The set $M_f \cup N_f$ is the set of faulty thyristors manufactured by either machine M or N. The set $\overline{M \cup N}$ is the set of thyristors made by machines other than M or N.   ▲

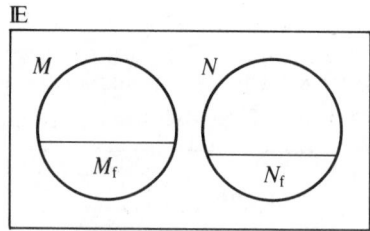

Figure 2.2   Venn diagram for Example 2.4.

We have seen how the operations $\cap$, $\cup$ can be used to define new sets. It is not difficult to show that a number of laws hold, most of which are obvious from the inspection of an appropriate Venn diagram.

2.2.7 *Laws of set algebra*

For any sets A, B, C and a universal set $\mathbb{E}$, we have the laws in Table 2.1. From these it is possible to prove the laws given in Table 2.2.

Table 2.1 The laws of set algebra.

$$
\left.\begin{array}{l}
A \cup B = B \cup A \\
A \cap B = B \cap A
\end{array}\right\} \text{Commutative laws}
$$

$$
\left.\begin{array}{l}
A \cup (B \cup C) = (A \cup B) \cup C \\
A \cap (B \cap C) = (A \cap B) \cap C
\end{array}\right\} \text{Associative laws}
$$

$$
\left.\begin{array}{l}
A \cap (B \cup C) = (A \cap B) \cup (A \cap C) \\
A \cup (B \cap C) = (A \cup B) \cap (A \cup C)
\end{array}\right\} \text{Distributive laws}
$$

$$
\left.\begin{array}{l}
A \cup \emptyset = A \\
A \cap \mathbb{E} = A
\end{array}\right\} \text{Identity laws}
$$

$$
\left.\begin{array}{l}
A \cup \overline{A} = \mathbb{E} \\
A \cap \overline{A} = \emptyset \\
\overline{\overline{A}} = A
\end{array}\right\} \text{Complement laws}
$$

Table 2.2 Laws derivable from Table 2.1.

$$
\left.\begin{array}{l}
A \cup (A \cap B) = A \\
A \cap (A \cup B) = A
\end{array}\right\} \text{Absorption laws}
$$

$$
\left.\begin{array}{l}
(A \cap B) \cup (A \cap \overline{B}) = A \\
(A \cup B) \cap (A \cup \overline{B}) = A
\end{array}\right\} \text{Minimization laws}
$$

$$
\left.\begin{array}{l}
\overline{A \cup B} = \overline{A} \cap \overline{B} \\
\overline{A \cap B} = \overline{A} \cup \overline{B}
\end{array}\right\} \text{De Morgan's laws}
$$

2.2.8 *Sets and functions*

If we are given two sets, A and B, a useful exercise is to examine relationships, given by rules, between the elements of A and the elements of B. For example, if $A = \{0, 1, 4, 9\}$ and $B = \{-3, -2, -1, 0, 1, 2, 3\}$ then each element of B is plus or minus the square root of some element of A. We can depict this as in Figure 2.3.

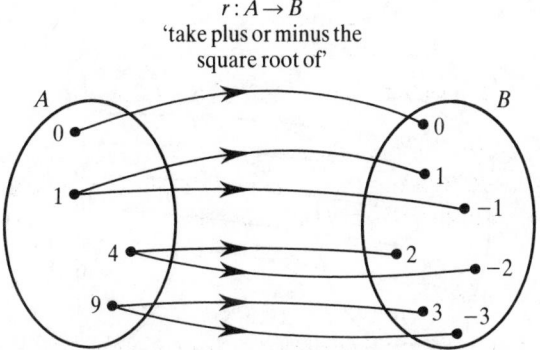

$$r : A \to B$$
'take plus or minus the
square root of'

Figure 2.3 A relation between sets A and B.

The rule, which when given an element of A, produces an element of B, is called a **relation**. If the rule of the relation is given the symbol r we write:

$$r: A \rightarrow B$$

and say 'the relation r **maps** elements of the set A to elements of the set B'. For the example above, we can write $r: 1 \rightarrow \pm 1, r: 4 \rightarrow \pm 2$, and generally $r: x \rightarrow \pm \sqrt{x}$. The set from which we choose our input is called the **domain**; the set to which we map is called the **co-domain**; the subset of the co-domain actually used is called the **range**. As we shall see this need not be the whole of the co-domain.

■ A relation r, maps elements of a set D, called the domain, to one or more elements of a set C called the co-domain. We write:

$$r: D \rightarrow C$$

Example 2.5

If $D = \{0, 1, 2, 3, 4, 5\}$ and $E = \{1, 4, 7, 10, 13, 16, 19, 21\}$ and the relation with symbol s is defined by $s: D \rightarrow E$, $s: m \rightarrow 3m + 1$, identify the domain and co-domain of s. Draw a mapping diagram to illustrate the relation. What is the range of s?

Solution

The domain of s is the set of values from which we choose our input, that is $D = \{0, 1, 2, 3, 4, 5\}$. The co-domain of s is the set to which we map, that is $E = \{1, 4, 7, 10, 13, 16, 19, 21\}$. The rule $s: m \rightarrow 3m + 1$ enables us to draw the mapping diagram. For example, $s: 3 \rightarrow 10$ and so on. The diagram is shown in Figure 2.4. The range of s is the subset of E actually used, in this case $\{1, 4, 7, 10, 13, 16\}$. We note that not all the elements of the co-domain are actually used. ▲

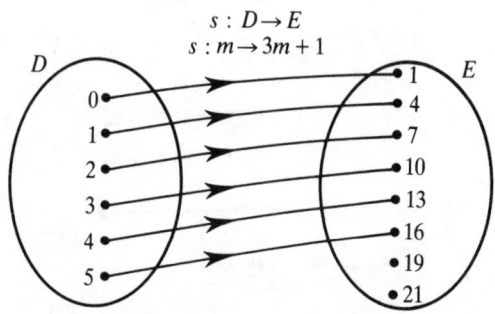

Figure 2.4   The relation s maps elements of D to E.

The notation introduced is very similar to that for functions described in Section 1.3. This is no accident. In fact, a function is a very special form of a relation. Let us recall the definition of a function:

'A function is a rule which when given an input produces a single output.'

If we study the two relations r and s, we note that when relation r received input, it could produce two outputs. On the mapping diagram this shows up as two arrows leaving some elements in A. When relation s received an input, it produced a single output. This shows up as a single arrow leaving each element in D. Hence the relation r is not a function, whereas the relation s is. This leads to the following more rigorous definition of a function.

■ A function f, is a relation which maps each element of a set D, called the domain, to a single element of a set C called the co-domain. We write:

$$f: D \to C$$

Example 2.6

If $M = \{\text{off, on}\}$, $N = \{0, 1\}$ and we define a relation r by

    $r: M \to N$

    $r: \text{off} \to 0$ $r: \text{on} \to 1$

then the relation r is a function since each element in M is mapped to a single element in N. ▲

Example 2.7

If $P = \{0, 1\}$ and $Q = \{\text{high}\}$ and we define a relation r by

    $r: P \to Q$

    $r: 1 \to \text{high}$

then r is not a function since each element in P is not mapped to an element in Q. ▲

All of the functions described in Chapter 1 have domains which are subsets of the real numbers $\mathbb{R}$. The input to each function is the particular value of the independent variable chosen from the domain and the output is the value of the dependent variable. When dealing with continuous domains the graphs we have already considered replace the mapping diagrams.

Example 2.8

Find the domain, D, of the rational function $f: D \to \mathbb{R}$ given by

$$f: x \to \frac{3x}{x - 2}$$

Solution

Since no domain is given, we choose it to be the largest set possible. This is the set of all real numbers except the value $x = 2$ at which point f is not defined. We have $D = \{x: x \in \mathbb{R}, x \neq 2\}$.   ▲

EXERCISES 2.1

1. If $A = \{1, 3, 5, 7, 9, 11\}$ and $B = \{3, 4, 5, 6\}$ find $A \cap B$ and $A \cup B$.

2. Write out all the members of the following sets:

    (a) $A = \{x: x \in \mathbb{N} \text{ and } x < 10\}$

    (b) $B = \{x: x \in \mathbb{R} \text{ and } 0 \leq x \leq 10 \text{ and } x \text{ is divisible by } 3\}$.

3. Use Venn diagrams to illustrate the following for general sets C and D:

    (a) $C \cap D$, (b) $C \cup D$, (c) $C \cap \bar{D}$, (d) $\overline{C \cup D}$, (e) $\overline{C \cap D}$.

4. By drawing Venn diagrams verify De Morgan's laws:

    $$\overline{A \cap B} = \bar{A} \cup \bar{B} \quad \text{and} \quad \overline{A \cup B} = \bar{A} \cap \bar{B}$$

5. For sets $A = \{0, 1, 2\}$ and $B = \{3, 4\}$, draw a mapping diagram to illustrate the following relations. Determine which relations are functions. For those that are not functions, give reasons for your decision.

    (a) $r: A \rightarrow B$, $r: 0 \rightarrow 3$, $r: 1 \rightarrow 4$, $r: 2 \rightarrow 4$

    (b) $s: A \rightarrow B$, $s: 0 \rightarrow 3$, $s: 0 \rightarrow 4$, $s: 1 \rightarrow 3$, $s: 2 \rightarrow 3$

    (c) $t: A \rightarrow B$, $t: 0 \rightarrow 3$, $t: 1 \rightarrow 4$

6. If $A = \{1, 3, 5, 7\}$ and $B = \{1, 2, 3, 4\}$, draw a mapping diagram to illustrate the relation $r: A \rightarrow B$, where r is the relation 'is bigger than'. Is r a function?

7. Specify the range of the following functions using set notation. Sketch a graph of each function clearly indicating the domain and range.

    (a) $f(x) = 3x + 1 \quad x \in \mathbb{R}$

    (b) $g(t) = t^2 + 2 \quad t \in \mathbb{R}$

    (c) $y(x) = x - 3 \quad x \in \mathbb{R}, 2 \leq x \leq 10$

    (d) $g(x) = x^4 - 3 \quad x \in \mathbb{R}$

    (e) $f(t) = t^2 - 3t + 4 \quad t \in \mathbb{R}, 15 \leq t$

    (f) $f(t) = t^3 - 3t^2 + 2t + 1 \quad t \in \mathbb{R}, -3 \leq t \leq 8$

    (g) $f(t) = \sin 100\pi t \quad t \in \mathbb{R}$

    (h) $f(t) = 3 \sin 400\pi t \quad t \in \mathbb{R}, 0 \leq t \leq \frac{1}{800}$

    (i) $f(x) = e^{3x} \quad x \in \mathbb{R}, x \geq 0$

2.3 Logic

In Section 2.4 we will examine Boolean algebra. This concerns itself with the manipulation of logic statements and so is suitable for analysing digital logic circuits. In this section we introduce the basic concepts of logic by means of logic gates as these form the usual starting point for engineers studying this topic.

2.3.1 *The OR gate*

The OR gate is an electronic device which receives two inputs each in the form of a binary digit, that is 0 or 1, and produces a binary digit as output, depending

$$A \quad \rightarrow \quad F = A + B$$
$$B$$

Figure 2.5 Symbol for an OR gate.

Table 2.3 The truth table for an OR gate with inputs *A* and *B*.

A	*B*	*F = A + B*
1	1	1
1	0	1
0	1	1
0	0	0

upon the values of the two inputs. It is represented by the symbol shown in Figure 2.5.

A and *B* are the two inputs, and *F* is the single output. As high (1) or low (0) voltages are applied to *A* and *B* various possible outputs are achieved, these being defined by means of a **truth table** as shown in Table 2.3. So, for example, if a low (0) voltage is applied to *A* and a high (1) voltage is applied to *B*, the output is a high (1) voltage at *F*. We note that a '1' appears in the right-hand column of the truth table whenever *A* or *B* takes the value 1, hence the name OR gate. We use the symbol + to represent OR. Because it connects the variables *A* and *B*, OR is known as a **logical connective**. We shall meet other logical connectives shortly. This connective is also known as a **disjunction**, so that *A* + *B* is said to be the disjunction of *A* and *B*.

2.3.2 *The AND gate*

It is possible to construct another electronic device called an AND gate which works in a similar way except that the output only takes the value 1 when both inputs are 1. The symbol for this gate is shown in Figure 2.6 and the complete truth table is shown in Table 2.4. The logical connective AND is given the symbol · and is known as a **conjunction** so that *A* · *B* is said to be the conjunction of *A* and *B*.

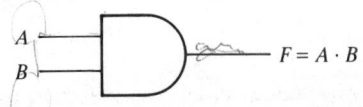

Figure 2.6 Symbol for an AND gate.

Table 2.4 The truth table for an AND gate with inputs A and B.

A	B	$F = A \cdot B$
1	1	1
1	0	0
0	1	0
0	0	0

2.3.3 *The inverter or NOT gate*

The inverter is a device with one input and one output and has the symbol shown in Figure 2.7. It has a truth table defined by Table 2.5. If the input is A, then the output is represented by the symbol $\bar{A}$, known as the complement of A.

$$A \qquad \triangleright\!\!\!o \qquad F = \bar{A}$$

Figure 2.7 Symbol for an inverter.

Table 2.5
The truth table for a NOT gate.

A	$F = \bar{A}$
1	0
0	1

2.3.4 *The NOR gate*

This gate is logically equivalent to a NOT gate in series with an OR gate as shown in Figure 2.8. It is represented by the symbol shown in Figure 2.9 and has its truth table defined in Table 2.6.

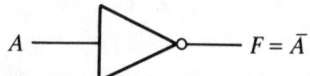

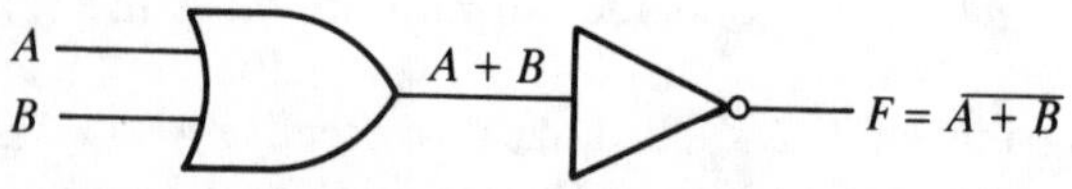

$$A \quad B \qquad A + B \qquad F = \overline{A + B}$$

Figure 2.8 A NOT gate in series with an OR gate.

$$A \quad B \qquad F = \overline{A + B}$$

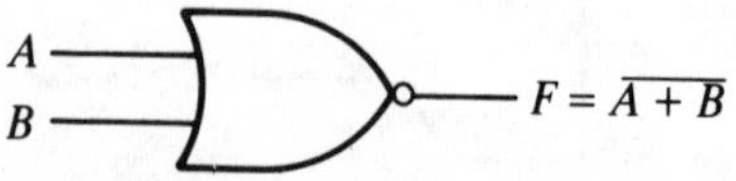

Figure 2.9 Symbol for a NOR gate.

Table 2.6   The truth table
for a NOR gate.

A	B	$F = \overline{A + B}$
1	1	0
1	0	0
0	1	0
0	0	1

2.3.5 *The NAND gate*

This gate is logically equivalent to a NOT gate in series with an AND gate as shown in Figure 2.10. It is represented by the symbol shown in Figure 2.11 and has the truth table defined by Table 2.7.

Although we have only examined gates with two inputs it is possible for a gate to have more than two. For example, the Boolean expression for a four-input NAND gate would be $F = \overline{A \cdot B \cdot C \cdot D}$ while that of a four-input OR gate would be $F = A + B + C + D$, where A, B, C and D are the inputs, and F is the output. Logic gates form the building blocks for more complicated digital electronic circuits.

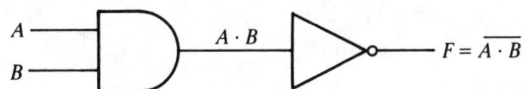

Figure 2.10   A NOT gate in series with an AND gate.

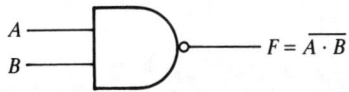

Figure 2.11   Symbol for a NAND gate.

Table 2.7   The truth
table for a NAND gate.

A	B	$F = \overline{A \cdot B}$
1	1	0
1	0	1
0	1	1
0	0	1

2.4 Boolean algebra

Suppose A and B are binary digits, that is, 1 or 0. These, together with the logical connectives $+$ and $\cdot$ and also the complement NOT, form what is known as a Boolean algebra. The quantities A and B are known as **Boolean variables**. Expressions such as $A + B$, $A \cdot B$ and $\bar{A}$ are known as **Boolean expressions**. More complex Boolean expressions can be built up using more Boolean variables together with combinations of $+$, $\cdot$ and NOT; for example, we can draw up a truth table for expressions such as $(A \cdot B) + (C \cdot \bar{D})$.

We shall restrict our attention to the logic gates described in the last section although the techniques of Boolean algebra are more widely applicable. A Boolean variable can only take the values 0 or 1. For our purposes a **Boolean algebra** is a set of Boolean variables with the two operations $\cdot$ and $+$, together with the operation of taking the complement, for which certain laws hold.

2.4.1 *Laws of Boolean algebra*

For any Boolean variables A, B, C, we have the laws in Table 2.8. From these it is possible to prove the laws given in Table 2.9. You will notice that these laws are analogous to those of set algebra if we interpret $+$ as $\cup$, $\cdot$ as $\cap$, 1 as the

Table 2.8 Laws of Boolean algebra.

$$\left.\begin{array}{l} A + B = B + A \\ A \cdot B = B \cdot A \end{array}\right\} \text{Commutative laws}$$

$$\left.\begin{array}{l} A + (B + C) = (A + B) + C \\ A \cdot (B \cdot C) = (A \cdot B) \cdot C \end{array}\right\} \text{Associative laws}$$

$$\left.\begin{array}{l} A \cdot (B + C) = (A \cdot B) + (A \cdot C) \\ A + (B \cdot C) = (A + B) \cdot (A + C) \end{array}\right\} \text{Distributive laws}$$

$$\left.\begin{array}{l} A + 0 = A \\ A \cdot 1 = A \end{array}\right\} \text{Identity laws}$$

$$\left.\begin{array}{l} A + \bar{A} = 1 \\ A \cdot \bar{A} = 0 \\ \bar{\bar{A}} = A \end{array}\right\} \text{Complement laws}$$

Table 2.9 Laws derived from the laws of Table 2.8.

$$\left.\begin{array}{l} A + (A \cdot B) = A \\ A \cdot (A + B) = A \end{array}\right\} \text{Absorption laws}$$

$$\left.\begin{array}{l} (A \cdot B) + (A \cdot \bar{B}) = A \\ (A + B) \cdot (A + \bar{B}) = A \end{array}\right\} \text{Minimization laws}$$

$$\left.\begin{array}{l} \overline{A + B} = \bar{A} \cdot \bar{B} \\ \overline{A \cdot B} = \bar{A} + \bar{B} \end{array}\right\} \text{De Morgan's laws}$$

$$A + 1 = 1$$
$$A \cdot 0 = 0$$

universal set $\mathbb{E}$, and 0 as the empty set $\emptyset$. In ordinary algebra, multiplication takes precedence over addition. In Boolean algebra $\cdot$ takes precedence over $+$. So, for example, we can write the first absorption law without brackets, that is

$$A + A \cdot B = A$$

Similarly, the first minimization law becomes:

$$A \cdot B + A \cdot \bar{B} = A$$

We shall follow this rule of precedence from now on.

Example 2.9

Find the truth table for the Boolean expression $A + B \cdot \bar{C}$.

Solution

We construct the table by noting that A, B and C are Boolean variables, that is, they can take the values 0 or 1. The first stage in the process is to form all possible combinations of A, B and C, as shown in Table 2.10. Then we complete the table by forming $\bar{C}$, then $B \cdot \bar{C}$ and finally $A + B \cdot \bar{C}$, using the truth tables defined earlier. So, for example, whenever $C = 1$, $\bar{C} = 0$. The complete process is shown in Table 2.11. Work through the table to ensure you understand how it was constructed. ▲

Table 2.10
The possible combinations for three variables, A, B and C.

A	B	C
1	1	1
1	1	0
1	0	1
1	0	0
0	1	1
0	1	0
0	0	1
0	0	0

Table 2.11   The truth table for $A + B \cdot \bar{C}$.

A	B	C	$\bar{C}$	$B \cdot \bar{C}$	$A + B \cdot \bar{C}$
1	1	1	0	0	1
1	1	0	1	1	1
1	0	1	0	0	1
1	0	0	1	0	1
0	1	1	0	0	0
0	1	0	1	1	1
0	0	1	0	0	0
0	0	0	1	0	0

2.4.2 *Logical equivalence*

We know from the distributive laws of Boolean algebra that

$$A + (B \cdot \bar{C}) = (A + B) \cdot (A + \bar{C})$$

Let us construct the truth table for the right-hand side of this expression.

Table 2.12   Truth table for $(A + B) \cdot (A + \bar{C})$.

A	*B*	*C*	$\bar{C}$	*A* + *B*	$A + \bar{C}$	$(A + B) \cdot (A + \bar{C})$
1	1	1	0	1	1	1
1	1	0	1	1	1	1
1	0	1	0	1	1	1
1	0	0	1	1	1	1
0	1	1	0	1	0	0
0	1	0	1	1	1	1
0	0	1	0	0	0	0
0	0	0	1	0	1	0

If we now observe the final column of Table 2.12 we see it is the same as that of Table 2.11. We say that $A + (B \cdot \bar{C})$ is **logically equivalent** to $(A + B) \cdot (A + \bar{C})$. Figures 2.12 and 2.13 show the two ways in which these logically equivalent circuits could be constructed using OR gates, AND gates and inverters. Clearly different electronic circuits can be constructed to perform the same logical task. We shall shortly explore a way of simplifying circuits to reduce the number of components required.

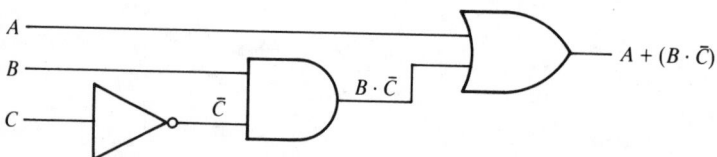

Figure 2.12   Circuit to implement $A + (B \cdot \bar{C})$.

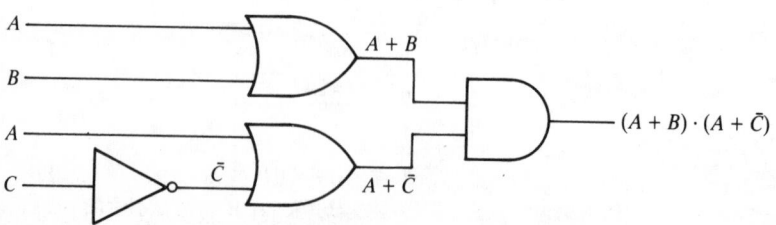

Figure 2.13   Circuit to implement $(A + B) \cdot (A + \bar{C})$.

Example 2.10

Find the Boolean expression and truth table for the electronic circuit shown in Figure 2.14.

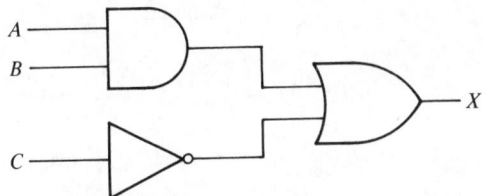

Figure 2.14   Circuit for Example 2.10.

Solution

By labelling intermediate points in the circuit we see that $X = A \cdot B + \bar{C}$. In order to obtain the truth table we form all possible combinations of A, B and C, followed by $A \cdot B$, $\bar{C}$ and finally $X = A \cdot B + \bar{C}$. The complete calculation is shown in Table 2.13.   ▲

Table 2.13   The truth table for Figure 2.14.

A	B	C	$A \cdot B$	$\bar{C}$	$X = A \cdot B + \bar{C}$
1	1	1	1	0	1
1	1	0	1	1	1
1	0	1	0	0	0
1	0	0	0	1	1
0	1	1	0	0	0
0	1	0	0	1	1
0	0	1	0	0	0
0	0	0	0	1	1

What we would now like to be able to do is carry out the reverse process, that is, start with a truth table and find an appropriate Boolean expression so that the required electronic device can be constructed.

Example 2.11

Given inputs A, B and C, find a Boolean expression for X as given by the truth table in Table 2.14.

Solution

To find an equivalent Boolean expression the procedure is as follows. Look down the rows of the truth table and select those with a right-hand side equal to 1. In this example, there are five such rows; 1, 2, 4, 6 and 8. Each of these rows gives rise to a term in the required Boolean expression. Each

Table 2.14 The truth table for a system with inputs A, B and C and an output X.

A	B	C	X
1	1	1	1
1	1	0	1
1	0	1	0
1	0	0	1
0	1	1	0
0	1	0	1
0	0	1	0
0	0	0	1

term is constructed so that it has a value 1 for the input values of that row. For example, for the input values of row 1, that is 1, 1, 1, we find $A \cdot B \cdot C$ has the value 1, whereas for the input values of row 2, that is 1, 1, 0, we find $A \cdot B \cdot \bar{C}$ has the value 1. Carrying out this process for the other rows we find that the required expression is

$$X = A \cdot B \cdot C + A \cdot B \cdot \bar{C} + A \cdot \bar{B} \cdot \bar{C} + \bar{A} \cdot B \cdot \bar{C} + \bar{A} \cdot \bar{B} \cdot \bar{C} \tag{2.1}$$

that is, a disjunction of terms, each term corresponding to one of the selected rows. This important expression is known as a **disjunctive normal form** (d.n.f.). We note that the truth table is the same as that of Example 2.10 which had Boolean expression $(A \cdot B) + \bar{C}$. The d.n.f. we have just calculated, while correct, is not the simplest. ▲

More generally, to find the required d.n.f. from a truth table we write down an expression of the form

$$(\quad) + (\quad) + \cdots + (\quad)$$

where each term has the value 1 for the input values of that row. We could now construct an electronic circuit corresponding to Equation (2.1) using a number of AND and OR gates together with inverters, and it would do the required job in the sense that the desired truth table would be achieved.

However, we know from Example 2.10 that the much simpler expression $(A \cdot B) + \bar{C}$ has the same truth table and if a circuit were to be built corresponding to this it would require fewer components. Clearly what we need is a technique for finding the simplest expression which does the desired job since the d.n.f. is not in general the simplest. It is not obvious what is meant by 'simplest expression'. In what follows we shall be concerned with finding the simplest d.n.f. It is nevertheless possible that a logically equivalent statement exists which would give a simpler circuit. Simplification can be achieved using the laws of Boolean algebra as we shall see in Examples 2.13 and 2.14.

Example 2.12 The exclusive OR gate

We have already looked at the OR gate in Section 2.3. The full name for this type of OR gate is the **inclusive** OR gate. It is so called because it gives an output of 1 when either or both inputs are 1. The **exclusive** OR gate only gives an output of 1 when either but not both inputs are 1. The truth table for this gate is given in Table 2.15 and its symbol is shown in Figure 2.15. Using the truth table, the d.n.f. for the gate is:

$$F = A \cdot \bar{B} + \bar{A} \cdot B$$

The exclusive OR often arises in the design of digital logic circuits. In fact, it is possible to buy integrated circuits that contain exclusive OR gates as basic units. ▲

Table 2.15 The truth table for an exclusive OR gate.

A	B	F
1	1	0
1	0	1
0	1	1
0	0	0

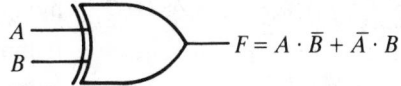

Figure 2.15 Symbol for an exclusive OR gate.

Example 2.13

Use the laws of Boolean algebra given in Table 2.8 to simplify the following expressions.

(a) $A \cdot B + A \cdot \bar{B}$

(b) $A + \bar{A} \cdot \bar{B}$

(c) $A + \bar{A} \cdot \bar{B} \cdot C$

(d) $A \cdot B \cdot C + A \cdot \bar{B} \cdot C + A \cdot B \cdot \bar{C} + A \cdot \bar{B} \cdot \bar{C} + \bar{A} \cdot \bar{B} \cdot C$

Solutions

(a) Using the distributive law we can write

$$A \cdot B + A \cdot \bar{B} = A \cdot (B + \bar{B})$$

Using the complement law, $B + \bar{B} = 1$ and hence

$$A \cdot B + A \cdot \bar{B} = A \cdot 1$$

$$= A \qquad \text{using the identity law}$$

Hence $A \cdot B + A \cdot \bar{B}$ simplifies to A. Note that this is the first minimization law given in Table 2.9.

(b) $\quad A + \bar{A} \cdot \bar{B} = (A + \bar{A}) \cdot (A + \bar{B}) \qquad$ by the distributive law

$$= 1 \cdot (A + \bar{B}) \qquad\qquad \text{by the complement law}$$

$$= A + \bar{B} \qquad\qquad\quad \text{using the identity law}$$

(c) Note that $A + \bar{A} \cdot \bar{B} \cdot C$ can be written as $A + (\bar{A} \cdot \bar{B}) \cdot C$ using the associative laws. Then

$$A + (\bar{A} \cdot \bar{B}) \cdot C = (A + \bar{A} \cdot \bar{B}) \cdot (A + C) \qquad \text{by the distributive law}$$

$$= (A + \bar{B}) \cdot (A + C) \qquad\quad \text{using part (b)}$$

$$= A + \bar{B} \cdot C \qquad\qquad\quad\; \text{by the distributive law}$$

(d) $\quad A \cdot B \cdot C + A \cdot \bar{B} \cdot C + A \cdot B \cdot \bar{C} + A \cdot \bar{B} \cdot \bar{C} + \bar{A} \cdot \bar{B} \cdot C$

can be rearranged using the commutative law to give:

$$A \cdot B \cdot C + A \cdot B \cdot \bar{C} + A \cdot \bar{B} \cdot C + A \cdot \bar{B} \cdot \bar{C} + \bar{A} \cdot \bar{B} \cdot C$$

This equals:

$$A \cdot B \cdot (C + \bar{C}) + A \cdot \bar{B} \cdot (C + \bar{C}) + \bar{A} \cdot \bar{B} \cdot C \qquad \text{by the distributive law}$$

$$= A \cdot B \cdot 1 + A \cdot \bar{B} \cdot 1 + \bar{A} \cdot \bar{B} \cdot C \qquad\qquad \text{using the complement law}$$

$$= A \cdot B + A \cdot \bar{B} + \bar{A} \cdot \bar{B} \cdot C \qquad\qquad\quad \text{using the identity law}$$

$$= A \cdot (B + \bar{B}) + \bar{A} \cdot \bar{B} \cdot C \qquad\qquad\quad\;\; \text{using the distributive law}$$

$$= A + \bar{A} \cdot \bar{B} \cdot C \qquad\qquad\qquad\qquad \text{using the complement and}$$
$$\qquad\qquad\qquad\qquad\qquad\qquad\qquad\qquad\; \text{identity laws}$$

Using the result of part (c) this can be further simplified to $A + \bar{B} \cdot C$ ▲

Example 2.14 Design of a binary full-adder circuit

The binary adder circuit is a common type of digital logic circuit. For example, the accumulator of a microprocessor is essentially a binary adder. The term **full-adder** is used to describe a circuit which can add together two binary digits and also add the carry-out digit from a previous stage. The outputs from the full-adder consist of the sum value and the carry-out value. By connecting together a series of full-adders it is possible to add together two binary words. (A binary word is a group of binary digits, such as 0111 1010.) For example, adding together two 4-bit binary words would require four full-adder circuits. This is shown in Figure 2.16.

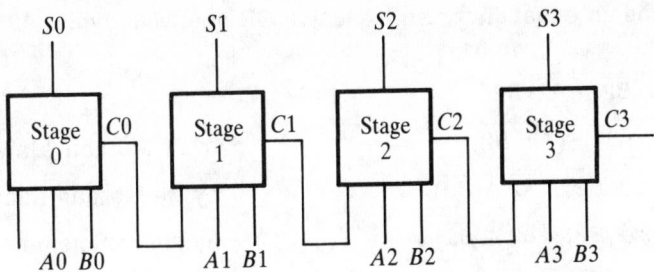

Figure 2.16 Four full-adders connected to allow two 4-bit binary words to be added.

The inputs $A0$–$A3$ and $B0$–$B3$ hold the two binary words that are to be added. The outputs $S0$–$S3$ hold the result of carrying out the addition. The lines $C0$–$C3$ hold the carry-out values from each of the stages. Sometimes there will be a carry-in to stage 0 as a result of a previous calculation. Let us consider the design of stage 2 in more detail. The design of the other stages will be identical. First of all a truth table for the circuit is derived. This is shown in Table 2.16.

Notice that there are three inputs to the circuit, $C1$, $A2$ and $B2$. There are also two outputs from the circuit, $S2$ and $C2$. Writing expressions for the outputs in d.n.f. yields:

$$S2 = \overline{C1} \cdot \overline{A2} \cdot B2 + \overline{C1} \cdot A2 \cdot \overline{B2} + C1 \cdot \overline{A2} \cdot \overline{B2} + C1 \cdot A2 \cdot B2$$

$$C2 = \overline{C1} \cdot A2 \cdot B2 + C1 \cdot \overline{A2} \cdot B2 + C1 \cdot A2 \cdot \overline{B2} + C1 \cdot A2 \cdot B2$$

It is important to reduce these expressions to as simple a form as possible in order to minimize the number of electronic gates needed to implement the expressions. So, starting with $S2$

$$S2 = \overline{C1} \cdot \overline{A2} \cdot B2 + \overline{C1} \cdot A2 \cdot \overline{B2} + C1 \cdot \overline{A2} \cdot \overline{B2} + C1 \cdot A2 \cdot B2$$

$$= C1 \cdot (A2 \cdot B2 + \overline{A2} \cdot \overline{B2}) + \overline{C1} \cdot (A2 \cdot \overline{B2} + \overline{A2} \cdot B2) \qquad \text{by the}$$

distributive law   (2.2)

Let

$$X = A2 \cdot \overline{B2} + \overline{A2} \cdot B2 \qquad\qquad (2.3)$$

Table 2.16 Truth table for a full-adder.

C1	A2	B2	S2	C2
0	0	0	0	0
0	0	1	1	0
0	1	0	1	0
0	1	1	0	1
1	0	0	1	0
1	0	1	0	1
1	1	0	0	1
1	1	1	1	1

Notice this is an equation for an exclusive OR gate with inputs $A2$ and $B2$. Using Equation (2.3) we have:

$$\overline{X} = \overline{A2 \cdot \overline{B2} + \overline{A2} \cdot B2}$$

$$= (\overline{A2 \cdot \overline{B2}}) \cdot (\overline{\overline{A2} \cdot B2}) \qquad \text{by De Morgan's law}$$

$$= (\overline{A2} + \overline{\overline{B2}}) \cdot (\overline{\overline{A2}} + \overline{B2}) \qquad \text{by De Morgan's law}$$

$$= (\overline{A2} + B2) \cdot (A2 + \overline{B2}) \qquad \text{by the complement law}$$

$$= \overline{A2} \cdot A2 + \overline{A2} \cdot \overline{B2} + B2 \cdot A2 + B2 \cdot \overline{B2} \qquad \text{by the distributive law}$$

$$= 0 + \overline{A2} \cdot \overline{B2} + B2 \cdot A2 + 0 \qquad \text{by the complement law}$$

$$= \overline{A2} \cdot \overline{B2} + B2 \cdot A2 \qquad \text{by the identity law}$$

$$= A2 \cdot B2 + \overline{A2} \cdot \overline{B2} \qquad \text{by the commutative law}$$

It is now possible to write Equation (2.2) as:

$$S2 = C1 \cdot \overline{X} + \overline{C1} \cdot X$$

This is the expression for an exclusive OR gate with inputs $C1$ and X. It is therefore possible to obtain $S2$ with two exclusive OR gates which are usually available as a basic building block on an integrated circuit. The circuit for $S2$ is shown in Figure 2.17. Turning to $C2$, we have

$$C2 = \overline{C1} \cdot A2 \cdot B2 + C1 \cdot \overline{A2} \cdot B2 + C1 \cdot A2 \cdot \overline{B2} + C1 \cdot A2 \cdot B2$$

$$= A2 \cdot B2 \cdot (\overline{C1} + C1) + C1 \cdot (\overline{A2} \cdot B2 + A2 \cdot \overline{B2}) \qquad \text{by the distributive law}$$

$$= A2 \cdot B2 + C1 \cdot X \qquad \text{by the complement law}$$

since $\overline{C1} + C1 = 1$, where X is given by Equation (2.3). The output, X, has already been generated to produce $S2$ but can be used again provided the exclusive OR gate can stand feeding two inputs. Assuming this is so then the final circuit for the full-adder is shown in Figure 2.18. ▲

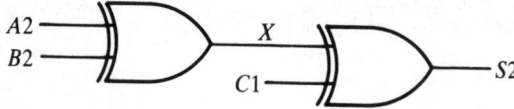

Figure 2.17 Circuit to implement $S2 = C1 \cdot \overline{X} + \overline{C1} \cdot X$, where $X = A2 \cdot \overline{B2} + \overline{A2} \cdot B2$.

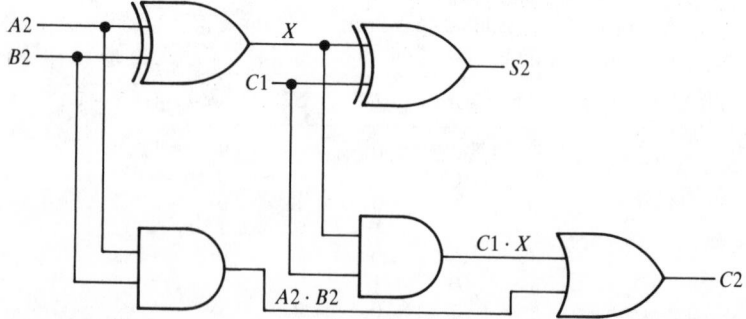

Figure 2.18 Circuit to implement the stage 2 full-adder.

EXERCISES 2.2

1. Construct a truth table showing $\overline{A \cdot B}$ and $\overline{A} + \overline{B}$ in order to verify the logical equivalence expressed in De Morgan's law $\overline{A \cdot B} = \overline{A} + \overline{B}$. Carry out a similar exercise to verify $\overline{A + B} = \overline{A} \cdot \overline{B}$.

2. Let $B = 1$ and then $B = 0$ in the absorption laws, and use the identity laws to obtain (a) $A + A = A$ and (b) $A \cdot A = A$. Verify your results using truth tables.

3. Derive Boolean expressions and truth tables for the circuits shown in Figure 2.19.

4. Simplify the following Boolean expressions using Boolean algebra.

   (a) $A \cdot B + A \cdot \overline{B} + B \cdot C + A \cdot B \cdot C$

   (b) $A \cdot (C + A) + C \cdot B + D + C + B \cdot \overline{C} + C \cdot A$

   (c) $A \cdot B \cdot C \cdot D + A \cdot B \cdot \overline{C} + A \cdot B \cdot C \cdot \overline{D} + A \cdot \overline{B} \cdot C \cdot D + A \cdot \overline{B} \cdot \overline{C} \cdot D$

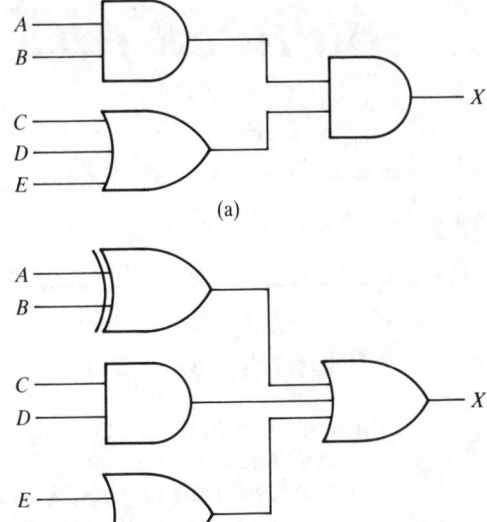

(a)

(b)

Figure 2.19 Circuits for Exercise 2.2.3.

Miscellaneous exercises

1. List all the elements of the following sets:

   (a) $S = \{n : n \in \mathbb{Z}, 5 \leq n^2 \leq 50\}$

   (b) $S = \{m : m \in \mathbb{N}, 5 \leq m^2 \leq 50\}$

   (c) $S = \{m : m \in \mathbb{N}, m^2 + 2m - 15 = 0\}$

2. If $A = \{n : n \in \mathbb{Z}, -10 \leq n \leq 20\}$, $B = \{m : m \in \mathbb{N}, m > 15\}$ list the members of $A \cap B$ and write down an expression for $A \cup B$.

3. Specify the range of the following functions using set notation. Sketch a graph of each function clearly indicating the domain and the range.

   (a) $f(t) = e^{3t}$ $t \in \mathbb{R}, -2 \leq t \leq 5$

   (b) $f(x) = \log_{10} x$ $x \in \mathbb{R}, 1 \leq x \leq 100$

   (c) $f(x) = 3|x| + 2$ $x \in \mathbb{R}$

   (d) $f(x) = x/(x + 2)$ $x \in \mathbb{R}, x \neq -2$

   (e) $f(t) = 6 \sin(400\pi t + 2\pi/3)$ $t \in \mathbb{R}$

   (f) $f(t) = 3 \sin(100\pi t)$ $t \in \mathbb{R}, 0 \leq t \leq \frac{1}{200}$

   (g) $f(x) = \cosh x$ $x \in \mathbb{R}$

   (h) $f(x) = \sinh x$ $x \in \mathbb{R}, x \geq 2$

4. Reduce the following expressions using Boolean algebra:

   (a) $\overline{A + C} + \overline{A \cdot B} \cdot (B + C)$

   (b) $(\overline{A} + B) \cdot B \cdot C + (A + \overline{C}) \cdot (B + A \cdot C)$

   (c) $\overline{(A + B)} \cdot (\overline{A} + \overline{C})$

Sequences and series 3

3.1 Introduction

Much of the material in this chapter is of a fundamental nature and is applicable to many different areas of engineering. For example, if continuous signals or waveforms, such as those described in Chapter 1, are sampled at periodic intervals we obtain a sequence of measured values. Sequences also arise when we attempt to obtain approximate solutions of equations which model physical phenomena. Such approximations are necessary if a solution is to be obtained using a digital computer. For many problems of practical interest to engineers a computer solution is the only possibility. The z transform is an example of an infinite series which is particularly important in the field of digital signal processing. Signal processing is concerned with modifying signals in order to improve them in some way. For example, the signals received from space satellites have to undergo extensive processing in order to counteract the effects of noise, and to filter out unwanted frequencies before they can provide, say, acceptable visual images. Digital signal processing is signal processing carried out using a computer. So, skill in manipulating sequences and series is crucial. Later chapters will develop these concepts and show examples of their use in solving real engineering problems.

3.2 Sequences

■ A **sequence** is a set of numbers or terms, not necessarily distinct, written down in a definite order.

For example,

$$1, 3, 5, 7, 9 \quad \text{and} \quad 1, \tfrac{1}{2}, \tfrac{1}{4}, \tfrac{1}{8}, \tfrac{1}{16}$$

are both sequences. These sequences have a finite number of terms but we shall frequently deal with ones involving an infinite number of terms. The notation '...' will signify this situation. Thus,

$$2, 4, 6, 8, \ldots$$

and

$$1, -1, 1, -1, 1, -1, \ldots$$

are both infinite sequences. In general situations we shall write a sequence as:

$$x[1], x[2], x[3], \ldots$$

or more compactly,

$$x[k] \qquad k = 1, 2, 3, \ldots$$

An alternative notation is

$$x_1, x_2, x_3, \ldots$$

The former notation is usually used in signal processing where the terms in the sequence represent the values of the signal. The latter notation arises in the numerical solution of equations. Hence both forms will be required. Often $x[1]$ will be the first term of the sequence although this is not always the case. The sequence:

$$\ldots, x[-3], x[-2], x[-1], x[0], x[1], x[2], x[3], \ldots$$

is usually written as:

$$x[k] \qquad k = \ldots, -3, -2, -1, 0, 1, 2, 3, \ldots$$

A complete sequence, as opposed to a specific term of a sequence, is often written using braces, for example

$$\{x[k]\} = x[1], x[2], \ldots$$

although it is common to write $x[k]$ when there is no confusion, and this is the convention we shall adopt in this book.

A sequence can also be regarded as a function whose domain is a subset of the set of integers. For example, the function defined by:

$$x: \mathbb{N} \to \mathbb{R} \qquad x: k \to \frac{3k}{2}$$

is the sequence

$$x[0] = 0 \qquad x[1] = \tfrac{3}{2} \qquad x[2] = 3 \qquad x[3] = \tfrac{9}{2}, \ldots$$

The values in the range of the function are the terms of the sequence. The independent variable is k. Functions of this sort differ from those described in Chapter 1 because the independent variable is not selected from a continuous interval but rather is **discrete**. It is, nevertheless, possible to represent $x[k]$ graphically as illustrated in Examples 3.1–3.3, but instead of a piecewise continuous curve, we now have a collection of isolated points.

Example 3.1

Graph the sequences given by:

(a)

$$x[k] = \begin{cases} 0 & k < 0 \\ 1 & k \geq 0 \end{cases} \quad k = \ldots, -3, -2, -1, 0, 1, 2, \ldots, \text{that is } k \in \mathbb{Z}$$

(b)

$$x[k] = \begin{cases} 1 & k \text{ even} \\ -1 & k \text{ odd} \end{cases} \quad k \in \mathbb{Z}$$

Solutions

(a) The graphs are obtained by plotting the terms of the sequence against k (see Figure 3.1). This sequence is known as the **unit step sequence**. We shall denote this by $u[k]$.

(b) The sequence $x[k]$ is shown in Figure 3.2. ▲

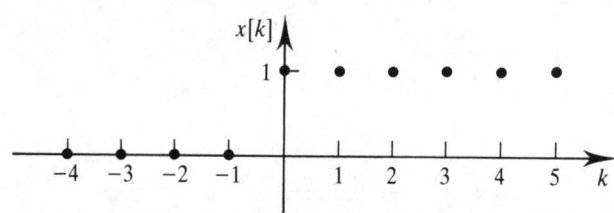

Figure 3.1   The unit step sequence.

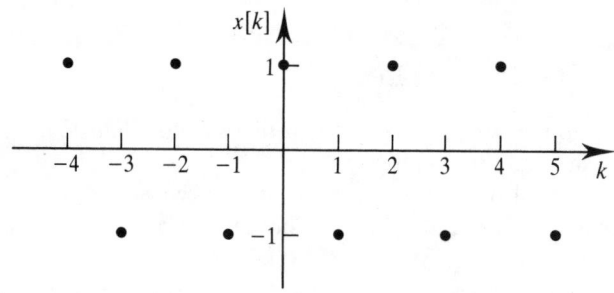

Figure 3.2   The sequence $x[k] = \begin{cases} 1 & k \text{ even} \\ -1 & k \text{ odd}. \end{cases}$

Example 3.2

Graph the sequence defined by:

$$x[k] = \begin{cases} 1 & k = 0 \\ 0 & k \neq 0 \end{cases} \quad k \in \mathbb{Z}$$

Solution

Figure 3.3 shows the graph of this sequence which is commonly called the **Kronecker delta sequence**. ▲

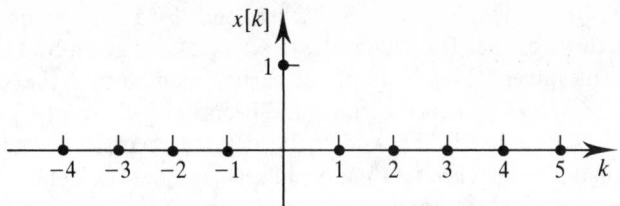

Figure 3.3 The Kronecker delta sequence.

Example 3.3

The sequence $x[k]$ is obtained by measuring or **sampling** the continuous function $f(t) = \sin t$, $t \in \mathbb{R}$, at $t = -2\pi$, $-3\pi/2$, $-\pi$, $-\pi/2$, 0, $\pi/2$, π, $3\pi/2$ and 2π. Write down the terms of this sequence and show them on a graph.

Solution

The function $f(t) = \sin t$, for $-2\pi \leq t \leq 2\pi$ is shown in Figure 3.4. We sample the continuous function at the required points. The sample values are shown as ●. From the graph we see that:

$$x[k] = 0, 1, 0, -1, 0, 1, 0, -1, 0 \qquad k = -4, -3, \ldots, 3, 4$$

The graph of $x[k]$ is shown in Figure 3.5. ▲

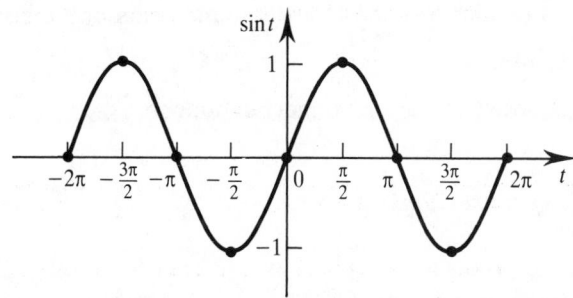

Figure 3.4 The function $f(t) = \sin t$ with sampled points shown.

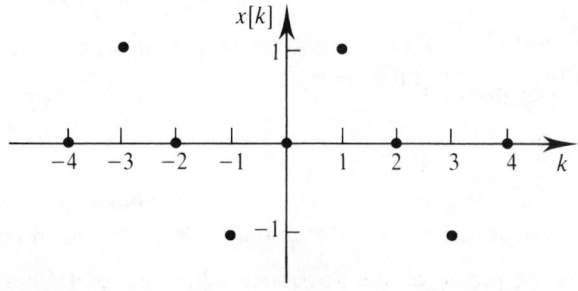

Figure 3.5 Sequence formed from sampling $f(t) = \sin t$.

Sometimes it is possible to describe a sequence by a rule giving the kth term. For example, the sequence for which $x[k] = 2^k$, $k \in \mathbb{N}$, is given by 1, 2, 4, 8 .... On occasions, a rule gives $x[k]$ in terms of earlier members of the sequence. For example, the previous sequence could have been defined by $x[k] = 2x[k-1]$, $x[0] = 1$. The sequence is then said to be defined **recursively** and the defining formula is called a **recurrence relation** or **difference equation**. Difference equations are particularly important in digital signal processing and are dealt with in Chapter 12.

Example 3.4

Write down the terms $x[k]$ for $k = 0, \ldots, 7$ of the sequence defined recursively as:

$$x[k] = x[k-2] + x[k-1]$$

where $x[0] = 1$ and $x[1] = 1$.

Solution

The values of $x[0]$ and $x[1]$ are given. We find:

$$x[2] = x[0] + x[1] = 2$$
$$x[3] = x[1] + x[2] = 3$$

Continuing in this fashion we find the first eight terms of the sequence are:

1, 1, 2, 3, 5, 8, 13, 21

This sequence is known as the Fibonacci sequence. ▲

3.2.1 *Arithmetic progressions*

An arithmetic progression is a sequence where each term is found by adding a fixed quantity, called the **common difference**, to the previous term.

Example 3.5

Write down the first five terms of the arithmetic progression where the first term is 1 and the common difference is 3.

Solution

The second term is found by adding the common difference, 3, to the first term, 1, and so the second term is 4. Continuing in this way we can construct the sequence:

1, 4, 7, 10, 13, ... ▲

A more general arithmetic progression has first term a and common difference d, that is

$a, a + d, a + 2d, a + 3d, \ldots$

It is easy to see that the kth term is:

$a + (k - 1)d$

All arithmetic progressions can be written recursively as $x[k] = x[k - 1] + d$.

■ Arithmetic progression: $a, a + d, a + 2d, + \ldots$

$a = $ first term, $d = $ common difference, kth term $= a + (k - 1)d$

Example 3.6

Find the 10th and 20th terms of the arithmetic progression with a first term 5 and common difference -4.

Solution

Here $a = 5$ and $d = -4$. The kth term is $5 - 4(k - 1)$. Therefore the 10th term is $5 - 4(9) = -31$ and the 20th term is $5 - 4(19) = -71$. ▲

3.2.2 Geometric progressions

A geometric progression is a sequence where each term is found by multiplying the previous term by a fixed quantity called the **common ratio**.

Example 3.7

Write down the geometric progression whose first term is 1 and whose common ratio is $\frac{1}{2}$.

Solution

The second term is found by multiplying the first by the common ratio, $\frac{1}{2}$, that is $\frac{1}{2} \times 1 = \frac{1}{2}$. Continuing in this way we obtain the sequence:

$1, \frac{1}{2}, \frac{1}{4}, \frac{1}{8}, \ldots$ ▲

A general geometric progression has first term a and common ratio r and can therefore be written as:

$a, ar, ar^2, ar^3, \ldots$

and it is easy to see that the kth term is ar^{k-1}. All geometric progressions can be written recursively as $x[k] = rx[k - 1]$.

■ Geometric progression:

$a, ar, ar^2, \ldots$

a = first term, r = common ratio, kth term = ar^{k-1}

3.2.3 *More general sequences*

We have already met a number of infinite sequences. For example,

(1) $x[k] = 2, 4, 6, 8, \ldots$

(2) $x[k] = 1, \frac{1}{2}, \frac{1}{4}, \ldots$

In case (1) the terms of the sequence go on increasing without bound. We say the sequence is **unbounded**. On the other hand, in case (2) it is clear that successive terms get smaller and smaller and as $k \to \infty$, $x[k] \to 0$. The notion of getting closer and closer to a fixed value is very important in mathematics and gives rise to the concept of a **limit**. In case (2) we say 'the limit of $x[k]$ as k tends to infinity is 0' and we write this concisely as:

$$\lim_{k \to \infty} x[k] = 0$$

We say that the sequence converges to 0, and because its terms do not increase without bound we say it is **bounded**.

More formally, we say that a sequence $x[k]$ **converges** to a limit l if, by proceeding far enough along the sequence, all subsequent terms can be made to lie as close to l as we wish. Whenever a sequence is not convergent it is said to be **divergent**.

It is possible to have sequences which are bounded but nevertheless do not converge to a limit. The sequence

$x[k] = -1, 1, -1, 1, -1, 1, \ldots$

clearly fails to have a limit as $k \to \infty$ although it is bounded, that is, its values all lie within a given range. This particular sequence is said to **oscillate**.

It is possible to evaluate the limit of a sequence, when such a limit exists, from knowledge of its general term. To be able to do this we can make use of certain rules, the proofs of which are beyond the scope of this book, but which we now state:

■ If $x[k]$ and $y[k]$ are two sequences such that $\lim_{k \to \infty} x[k] = l_1$, and $\lim_{k \to \infty} y[k] = l_2$, where l_1 and l_2 are finite, then:

(1) The sequence given by $x[k] \pm y[k]$ has limit $l_1 \pm l_2$.

(2) The sequence given by $cx[k]$, where c is a constant, has limit, cl_1.

(3) The sequence $x[k]y[k]$ has limit $l_1 l_2$.

(4) The sequence $x[k]/y[k]$ has limit l_1/l_2 provided $l_2 \neq 0$.

Furthermore, we can always assume that

$$\lim_{k \to \infty} \frac{1}{k^m} = 0 \qquad \text{for any constant } m > 0$$

Example 3.8

Find, if possible, the limit of each of the following sequences, $x[k]$.

(a) $x[k] = 1/k$ $k \in \mathbb{N}^+$

(b) $x[k] = 5$ $k \in \mathbb{N}^+$

(c) $x[k] = 3 + 1/k$ $k \in \mathbb{N}^+$

(d) $x[k] = 1/(k + 1)$ $k \in \mathbb{N}^+$

(e) $x[k] = k^2$ $k \in \mathbb{N}^+$

Solutions

(a) The sequence $x[k]$ is given by:

$$1, \tfrac{1}{2}, \tfrac{1}{3}, \tfrac{1}{4}, \ldots$$

Successive terms get smaller and smaller, and as $k \to \infty$, $x[k] \to 0$. By proceeding far enough along the sequence we can get as close to the limit 0 as we wish. Hence $\lim_{k \to \infty} x[k] = \lim_{k \to \infty} 1/k = 0$.

(b) The sequence $x[k]$ is given by $5, 5, 5, 5, \ldots$. This sequence has limit 5.

(c) The sequence $3, 3, 3, 3, \ldots$ has limit 3. The sequence $1, \tfrac{1}{2}, \tfrac{1}{3}, \ldots$ has limit 0. Therefore, using Rule 1 we have:

$$\lim_{k \to \infty} 3 + \frac{1}{k} = 3 + 0 = 3$$

The terms of the sequence $x[k] = 3 + 1/k$ are given by $4, 3\tfrac{1}{2}, 3\tfrac{1}{3}, \ldots$, and by proceeding far enough along we can make all subsequent terms lie as close to the limit 3 as we wish.

(d) The sequence $x[k] = 1/(k + 1)$, $k \in \mathbb{N}^+$, is given by:

$$\tfrac{1}{2}, \tfrac{1}{3}, \tfrac{1}{4}, \ldots$$

and has limit 0.

(e) The sequence $x[k] = k^2$, $k \in \mathbb{N}^+$ is given by $1, 4, 9, 16, \ldots$, and increases without bound. This sequence has no limit – it is divergent. ▲

Example 3.9

Given a sequence with general term $x[k] = (k - 1)/(k + 1)$, find $\lim_{k \to \infty} x[k]$.

Solution

It is meaningless to simply write $k = \infty$ to obtain $\lim_{k \to \infty} x[k] = (\infty - 1)/(\infty + 1)$, since such a quantity is undefined. What we should do is try to rewrite $x[k]$ in a form in which we can sensibly let $k \to \infty$. In this case, we proceed by expressing $(k - 1)/(k + 1)$ in partial fractions as follows:

$$\frac{k - 1}{k + 1} = A + \frac{B}{k + 1}$$

Therefore,

$$k - 1 = A(k + 1) + B$$

Equating coefficients of k yields: $1 = A$.
    Letting $k = -1$ gives $-2 = B$. Hence,

$$\frac{k - 1}{k + 1} = 1 - \frac{2}{k + 1}$$

Then, as $k \to \infty$ the term $2/(k + 1) \to 0$ so that,

$$\lim_{k \to \infty} x[k] = \lim_{k \to \infty} \left(1 - \frac{2}{k + 1} \right)$$

$$= \lim_{k \to \infty} (1) - \lim_{k \to \infty} \left(\frac{2}{k + 1} \right) \qquad \text{by Rule 1}$$

$$= 1 - 0$$

$$= 1$$

Alternatively, by dividing both numerator and denominator by k, we could write:

$$\frac{k - 1}{k + 1} = \frac{1 - (1/k)}{1 + (1/k)}$$

Then, as $k \to \infty$, $1/k \to 0$ so that,

$$\lim_{k \to \infty} \left(\frac{1 - (1/k)}{1 + (1/k)} \right) = \frac{\lim_{k \to \infty} (1 - (1/k))}{\lim_{k \to \infty} (1 + (1/k))} \qquad \text{by Rule 4}$$

$$= \frac{1}{1}$$

$$= 1 \text{ as before} \quad \blacktriangle$$

Example 3.10

Given a sequence with general term:

$$x[k] = \frac{3k^2 - 5k + 6}{k^2 + 2k + 1}$$

find $\lim_{k \to \infty} x[k]$.

Solution

Dividing the numerator and denominator by k^2 introduces terms which tend to zero as $k \to \infty$, that is

$$\frac{3k^2 - 5k + 6}{k^2 + 2k + 1} = \frac{3 - (5/k) + (6/k^2)}{1 + (2/k) + (1/k^2)}$$

Then as $k \to \infty$, we find:

$$\lim_{k \to \infty} x[k] = \tfrac{3}{1} = 3 \quad \blacktriangle$$

Example 3.11

Examine the behaviour of $k^2/(3k + 1)$ as $k \to \infty$.

Solution

$$\frac{k^2}{3k + 1} = \frac{k}{3 + (1/k)}$$

As $k \to \infty$, $1/k \to 0$ so that the denominator approaches 3. On the other hand as $k \to \infty$ the numerator tends to infinity so that this sequence diverges to infinity. $\blacktriangle$

EXERCISES 3.1

1. Graph the sequences given by:

   (a) $x[k] = k \qquad k \in \mathbb{N}$

   (b) $x[k] = \begin{cases} 3 & k = 2 \\ 0 & \text{otherwise} \end{cases} \qquad k \in \mathbb{N}$

   (c) $x[k] = e^{-k} \qquad k \in \mathbb{N}$

2. The sequence $x[k]$ is obtained by sampling $f(t) = \cos(t + 2)$, $t \in \mathbb{R}$. The sampling begins at $t = 0$ and thereafter at $t = 1, 2, 3, \ldots$. Write down the first six terms of the sequence.

3. A sequence, $x[k]$, is defined by:

   $$x[k] = \frac{k^2}{2} + k \qquad k \in \mathbb{N}$$

   State the first five terms of the sequence.

4. Write down the first five terms, and plot graphs, of the sequences given recursively by:

   (a) $x[k] = \dfrac{x[k-1]}{2} \qquad x[0] = 1$

   (b) $x[k] = 3x[k-1] - 2x[k-2]$
   $\qquad x[0] = 2 \qquad x[1] = 1$

5. A recurrence relation is defined by:

$$x[n + 1] = x[n] + 10 \quad x[0] = 1 \quad n \in \mathbb{N}$$

Find $x[1]$, $x[2]$, $x[3]$ and $x[4]$.

6. A sequence is defined by means of the recurrence relation:

$$x[n + 1] = x[n] + n^2 \quad x[0] = 1 \quad n \in \mathbb{N}$$

Write down the first five terms.

7. Consider the difference equation

$$x[n + 2] - x[n + 1] = 3x[n] \quad n \in \mathbb{N}$$

If $x[0] = 1$ and $x[1] = 2$, find the terms $x[2]$, $x[3]$, ..., $x[6]$.

8. Write down the 10th and 19th terms of the arithmetic progressions:

(a) $8, 11, 14, \ldots$

(b) $8, 5, 2, \ldots$

9. An arithmetic progression is given by:

$$b, \frac{2b}{3}, \frac{b}{3}, 0, \ldots$$

(a) State the 6th term.

(b) State the kth term.

(c) If the 20th term has a value of 15, find b.

10. Write down the 5th and 10th terms of the geometric progression $8, 4, 2, \ldots$.

11. Find the 10th and 20th terms of the geometric progression with first term 3 and common ratio 2.

12. A geometric progression is given by:

$$a, ar, ar^2, ar^3, \ldots$$

If $|(k + 1)\text{th term}| > |k\text{th term}|$ and $(k + 1)\text{th term} \times k\text{th term} < 0$, which of the following, if any, must be true?

(a) $r > 1$; (b) $a > 1$; (c) $r < -1$; (d) a is negative; (e) $-1 < r < 1$.

13. A geometric progression has first term $a = 1$. The 9th term exceeds the 5th term by 240. Find possible values for the 8th term.

14. If $x[k] = (3k + 2)/k$ find $\lim_{k \to \infty} x[k]$.

15. Find $\lim_{k \to \infty} (3k + 2)/(k^2 + 7)$.

16. Find the limits as k tends to infinity, if they exist, of the following sequences:

(a) $x[k] = k^3$

(b) $x[k] = (2k + 3)/(4k + 2)$

(c) $x[k] = (k^2 + k)/(k^2 + k + 1)$

17. Find $\lim_{k \to \infty} \left(\dfrac{6k + 7}{3k - 2} \right)^4$.

18. Find $\lim_{k \to \infty} x[k]$, if it exists, when

(a) $x[k] = (-1)^k$

(b) $x[k] = 2 - \dfrac{k}{10}$

(c) $x[k] = (1/3)^k$

(d) $x[k] = \dfrac{3k^3 - 2k^2 + 4}{5k^3 + 2k^2 + 4}$

(e) $x[k] = (1/5)^{2k}$

3.3 Series

Whenever the terms of a sequence are added together we obtain what is known as a **series**. For example, if we add the terms of the sequence, $1, \frac{1}{2}, \frac{1}{4}, \frac{1}{8}$, we obtain the series, S, where

$$S = 1 + \tfrac{1}{2} + \tfrac{1}{4} + \tfrac{1}{8}$$

This series ends after the 4th term and is said to be a **finite series**. Other series we shall meet continue indefinitely and are said to be **infinite series**.

Given an arbitrary sequence $x[k]$, we introduce the **sigma** notation

$$S_n = \sum_{k=1}^{n} x[k]$$

to mean the sum $x[1] + x[2] + \cdots + x[n]$, the first and last values of k being shown below and above the Greek letter Σ, which is pronounced 'sigma'. If the first term of the sequence is $x[0]$ rather than $x[1]$ we would write $\Sigma_{k=0}^{n} x[k]$.

3.3.1 *Sum of a finite arithmetic series*

An arithmetic series is the sum of an arithmetic progression. Consider the sum

$$S = 1 + 2 + 3 + 4 + 5$$

Clearly this sums to 15. When there are many more terms it is necessary to find a more efficient way of adding them up. The equation for S can be written in two ways:

$$S = 1 + 2 + 3 + 4 + 5$$

and

$$S = 5 + 4 + 3 + 2 + 1$$

If we add these two equations together we get:

$$2S = 6 + 6 + 6 + 6 + 6$$

There are 5 terms so that

$$2S = 5 \times 6 = 30$$

that is

$$S = 15$$

Now a general arithmetic series with k terms can be written as:

$$S_k = a + (a + d) + (a + 2d) + \cdots + (a + (k - 1)d)$$

but rewriting this back-to-front, we have:

$$S_k = (a + (k - 1)d) + (a + (k - 2)d) + \cdots + (a + d) + a$$

Adding together the first term in each series produces $2a + (k - 1)d$. Adding the second terms together produces $2a + (k - 1)d$. Indeed adding together the ith terms yields $2a + (k - 1)d$. Hence,

$$2S_k = \underbrace{(2a + (k - 1)d) + (2a + (k - 1)d) + \cdots + (2a + (k - 1)d)}_{k \text{ times}}$$

that is

$$2S_k = k(2a + (k - 1)d)$$

so that,

$$S_k = \frac{k}{2}(2a + (k - 1)d)$$

This formula tells us the sum to k terms of the arithmetic series with first term a and common difference d.

■ Sum of an arithmetic series: $S_k = \dfrac{k}{2}(2a + (k-1)d)$

Example 3.12

Find the sum of the arithmetic series containing 30 terms, with first term 1 and common difference 4.

Solution

We wish to find S_k

$$S_k = \underbrace{1 + 5 + 9 + \cdots}_{30\ \text{terms}}$$

Using $S_k = \dfrac{k}{2}(2a + (k-1)d)$ we find $S_{30} = \dfrac{30}{2}(2 + 29 \times 4) = 1770.$ ▲

Example 3.13

Find the sum of the arithmetic series with first term 1, common difference 3 and with last term 100.

Solution

We already know that the kth term of an arithmetic progression is given by $a + (k-1)d$. In this case the last term is 100. We can use this fact to find the number of terms. Thus,

$$100 = 1 + 3(k-1)$$

that is

$$3(k-1) = 99$$
$$k - 1 = 33$$
$$k = 34$$

That is, there are 34 terms in this series. Therefore the sum, S_{34}, is given by

$$S_{34} = \tfrac{34}{2}\{2(1) + (33)(3)\}$$
$$= 17(101)$$
$$= 1717 \quad ▲$$

3.3.2 *Sum of a finite geometric series*

A geometric series is the sum of the terms of a geometric progression. If we sum the geometric progression, $1, \frac{1}{2}, \frac{1}{4}, \frac{1}{8}, \frac{1}{16}$ we find:

$$S = 1 + \tfrac{1}{2} + \tfrac{1}{4} + \tfrac{1}{8} + \tfrac{1}{16} \tag{3.1}$$

If there had been a large number of terms it would have been impractical to add them all directly. However, let us multiply Equation (3.1) by the common ratio, $\frac{1}{2}$:

$$\tfrac{1}{2}S = \tfrac{1}{2} + \tfrac{1}{4} + \tfrac{1}{8} + \tfrac{1}{16} + \tfrac{1}{32} \tag{3.2}$$

so that, subtracting Equation (3.2) from Equation (3.1) we find:

$$S - \tfrac{1}{2}S = 1 - \tfrac{1}{32}$$

since most terms cancel out. Therefore $\frac{1}{2}S = \frac{31}{32}$ and so $S = \frac{31}{16} = 1\frac{15}{16}$.

We can apply this approach more generally: when we have a geometric progression with first term a and common ratio r, the sum to k terms is:

$$S_k = a + ar + ar^2 + ar^3 + \cdots + ar^{k-1}$$

Multiplying by r gives

$$rS_k = ar + ar^2 + ar^3 + \cdots + ar^{k-1} + ar^k$$

Subtraction gives $S_k - rS_k = a - ar^k$, so that:

$$S_k = \frac{a(1 - r^k)}{1 - r} \qquad \text{provided } r \neq 1$$

This formula gives the sum to k terms of the geometric series with first term a and common ratio r.

■ Sum of a geometric series: $S_k = \dfrac{a(1 - r^k)}{1 - r} \qquad r \neq 1$

3.3.3 *Sum of an infinite series*

When dealing with infinite series the situation is more complicated. Nevertheless, it is frequently the case that the answer to many problems can be expressed as an infinite series. In certain circumstances, the sum of a series tends to a finite answer as more and more terms are included and we say the series has **converged**. To illustrate this idea, consider the graphical interpretation of the series $1 + \frac{1}{2} + \frac{1}{4} + \frac{1}{8} + \cdots$, as given in Figure 3.6.

Start at 0 and move a length 1: total distance moved $= 1$
Move further, a length $\frac{1}{2}$: total distance moved $= 1\frac{1}{2}$
Move further, a length $\frac{1}{4}$: total distance moved $= 1\frac{3}{4}$

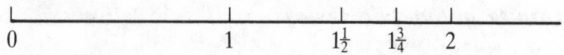

Figure 3.6 Graphical interpretation of the series, $1 + \frac{1}{2} + \frac{1}{4} + \frac{1}{8} + \cdots$.

At each stage the extra distance moved is half the distance remaining up to the point $x = 2$. It is obvious that the total distance we move cannot exceed 2 although we can get as close to 2 as we like by adding on more and more terms. We say that the series $1 + \frac{1}{2} + \frac{1}{4} + \cdots$ converges to 2. The sequence of total distances moved, given previously,

$$1, 1\frac{1}{2}, 1\frac{3}{4}, 1\frac{7}{8}, \ldots$$

is called the **sequence of partial sums** of the series.

For any given infinite series $\sum_{k=1}^{\infty} x[k]$, we can form the sequence of partial sums,

$$S_1 = \sum_{k=1}^{1} x[k] = x[1]$$

$$S_2 = \sum_{k=1}^{2} x[k] = x[1] + x[2]$$

$$S_3 = \sum_{k=1}^{3} x[k] = x[1] + x[2] + x[3]$$

$$\vdots$$

If the sequence $\{S_n\}$ converges to a limit S, we say that the infinite series has a sum S or that it has converged to S. Clearly not all infinite series will converge. For example, consider the series,

$$1 + 2 + 3 + 4 + 5 + \cdots$$

The sequence of partial sums is 1, 3, 6, 10, 15, ... This sequence diverges to infinity and so the series $1 + 2 + 3 + 4 + 5 + \cdots$ is divergent. It is necessary to establish tests or convergence criteria to help us decide whether a given series converges or diverges.

3.3.4 *Divergence test*

If the terms, $x[k]$, of the series do not tend to zero as $k \to \infty$, then the series must diverge. In order to converge, the terms $x[k]$ must eventually approach zero. However, a word of warning is necessary. The fact that $x[k] \to 0$ as $k \to \infty$ is not sufficient to guarantee that the series $\sum_{k=1}^{\infty} x[k]$ will converge. In fact, the **harmonic series** is one such case. This series is given by:

$$\sum_{k=1}^{\infty} x[k] = 1 + \frac{1}{2} + \frac{1}{3} + \frac{1}{4} + \cdots$$

Clearly $x[k] = 1/k$, $k \in \mathbb{N}^+$, and $x[k] \to 0$ as $k \to \infty$. However, it is possible to show that the series diverges.

3.3.5 *Comparison test*

Suppose we have a series $\Sigma_{k=1}^{\infty} x[k]$ which is such that $x[k] \geq 0$ for all k, and which is already known to converge. If we have another series $\Sigma_{k=1}^{\infty} y[k]$, for which $y[k] \geq 0$ for all k, and furthermore, $y[k] \leq x[k]$ for all k, then $\Sigma_{k=1}^{\infty} y[k]$ will also converge. Loosely speaking, $y[k]$ has smaller terms than $x[k]$. Note that this test only applies to series with non-negative terms.

Conversely if we know a series $\Sigma_{k=1}^{\infty} x[k]$, with $x[k] \geq 0$, diverges and if $y[k] \geq x[k]$ for all k, then the series $\Sigma_{k=1}^{\infty} y[k]$ must diverge also.

Example 3.14

Show that the series $\Sigma_{k=1}^{\infty} y[k]$ diverges where $y[k] = 1/\sqrt{k}$.

Solution

The harmonic series $\Sigma_{k=1}^{\infty} x[k] = \Sigma_{k=1}^{\infty}(1/k)$ diverges. We note that this is a sum of positive terms, that is, $x[k] > 0$ for all k. Also, since $\sqrt{k} \leq k$ for all k it follows that:

$$\frac{1}{\sqrt{k}} \geq \frac{1}{k} \qquad \text{for all } k$$

Hence $y[k] \geq x[k]$ for all k, and so, by the comparison test $\Sigma_{k=1}^{\infty} y[k] = \Sigma_{k=1}^{\infty}(1/\sqrt{k})$ diverges also. ▲

Example 3.15

The series $\Sigma_{k=1}^{\infty}(1/k^2)$ is known to converge. What can you deduce about $\Sigma_{k=1}^{\infty}(1/k^3)$?

Solution

Both series are sums of positive terms. Furthermore since $k^3 \geq k^2$ for all k, it follows that $(1/k^3) \leq (1/k^2)$ for all k. Now we are told that $\Sigma_{k=1}^{\infty}(1/k^2)$ converges and hence by the comparison test $\Sigma_{k=1}^{\infty}(1/k^3)$ must converge also. ▲

More generally it can be shown that $\Sigma_{k=1}^{\infty}(1/k^m)$ converges for $m > 1$ but diverges if $m \leq 1$.

3.3.6 *Ratio test*

Suppose we have a series $\sum_{k=1}^{\infty} x[k]$ with $x[k] > 0$ for all k. If also $x[k+1]/x[k]$ tends to a limit l as $k \to \infty$, then

(1) If $l > 1$, $\sum_{k=1}^{\infty} x[k]$ is divergent.
(2) If $l < 1$, $\sum_{k=1}^{\infty} x[k]$ is convergent.

If $l = 1$ no conclusion can be drawn. Note that this test applies to series with positive terms.

Example 3.16

Use the ratio test to assess whether $\sum_{k=1}^{\infty}(1/k!)$ converges or diverges.

Solution

Here we have $x[k] = 1/k!$ and $x[k+1] = 1/(k+1)!$. Clearly $x[k] > 0$ for all k. We consider the quantity:

$$\frac{x[k+1]}{x[k]} = \frac{k!}{(k+1)!} = \frac{k(k-1)(k-2)\cdots(3)(2)(1)}{(k+1)(k)(k-1)(k-2)\cdots(3)(2)(1)} = \frac{1}{k+1}$$

As $k \to \infty$, $1/(k+1) \to 0$ and so, by the ratio test $\sum_{k=1}^{\infty} x[k] = \sum_{k=1}^{\infty} 1/k!$ is convergent. ▲

Example 3.17

Find the outcome of applying the ratio test to the series:

$$\tfrac{1}{2} + \tfrac{2}{3} + \tfrac{3}{4} + \tfrac{4}{5} + \cdots$$

Solution

Writing $x[1] = 1/2$, $x[2] = 2/3$ we see that $x[k] = k/(k+1)$ and $x[k+1] = (k+1)/(k+2)$. Then

$$\frac{x[k+1]}{x[k]} = \frac{k+1}{k+2} \div \frac{k}{k+1}$$

$$= \frac{k+1}{k+2} \times \frac{k+1}{k}$$

$$= \frac{k^2 + 2k + 1}{k^2 + 2k}$$

Now

$$\lim_{k \to \infty} \frac{k^2 + 2k + 1}{k^2 + 2k} = 1$$

and so no conclusion can be drawn from the ratio test (see Exercise 3.2.14). We can see from the divergence test that the series diverges because $x[k]$ does not tend to zero. ▲

3.3.7 *Alternating series*

The comparison test applies to series with non-negative terms and the ratio test applies to series with positive terms. Let us now consider series in which there are alternately positive and negative terms. Such series are called **alternating series**. For example, consider the series $\sum_{k=1}^{\infty} x[k]$ where $x[k] = (-1)^{k+1}/k^2$. Writing out the first few terms, we see:

$$\sum_{k=1}^{\infty} x[k] = 1 - \tfrac{1}{4} + \tfrac{1}{9} - \tfrac{1}{16} + \cdots$$

Clearly the terms alternate in sign. Related to this series is the series $\sum_{k=1}^{\infty} |x[k]|$, obtained by considering the absolute value of each of the terms, that is,

$$\sum_{k=1}^{\infty} |x[k]| = 1 + \tfrac{1}{4} + \tfrac{1}{9} + \tfrac{1}{16} + \cdots$$

It can be shown that if $\sum_{k=1}^{\infty} x[k]$ is an alternating series such that $\sum_{k=1}^{\infty} |x[k]|$ converges, then $\sum_{k=1}^{\infty} x[k]$ converges also. The converse of this statement is not true. Just because the alternating series $\sum_{k=1}^{\infty} x[k]$ converges, we cannot assume that $\sum_{k=1}^{\infty} |x[k]|$ will converge. If $\sum_{k=1}^{\infty} |x[k]|$ converges the series $\sum_{k=1}^{\infty} x[k]$ is said to be **absolutely convergent**. If $\sum_{k=1}^{\infty} x[k]$ converges but $\sum_{k=1}^{\infty} |x[k]|$ diverges the series $\sum_{k=1}^{\infty} x[k]$ is said to be **conditionally convergent**. In many respects alternating series are easier to test for convergence.

3.3.8 *Leibniz' test*

If $\sum_{k=1}^{\infty} x[k]$ is an alternating series for which $x[k] \to 0$ as $k \to \infty$, and for which $|x[k+1]| < |x[k]|$ for all k then it can be shown that the alternating series is convergent. In other words, if the terms of an alternating series get smaller and smaller in absolute value and tend to zero, the series is convergent. This test is therefore particularly straightforward to apply.

Example 3.18

Show that the alternating harmonic series, $1 - \tfrac{1}{2} + \tfrac{1}{3} - \tfrac{1}{4} + \cdots$, converges.

Solution

We note that $|x[k]| = 1/k$ and $|x[k+1]| = 1/(k+1)$. Clearly $|x[k+1]| < |x[k]|$ for all k and the terms of the series tend to zero. Hence the alternating harmonic series converges. We noted earlier that the harmonic series, $1 + \tfrac{1}{2} + \tfrac{1}{3} + \cdots$, diverges. Therefore, the alternating harmonic series is conditionally convergent. ▲

Example 3.19

Determine whether or not the series $\sum_{k=1}^{\infty} x[k] = \sum_{k=1}^{\infty} (-1)^k 3^k / k!$ is absolutely convergent, conditionally convergent, or divergent.

Solution

Consider $|x[k]| = 3^k / k!$. Then $|x[k+1]| = 3^{k+1} / (k+1)!$. Therefore,

$$\left| \frac{x[k+1]}{x[k]} \right| = \left| \frac{3^{k+1}k!}{(k+1)!3^k} \right| = \frac{3}{k+1} \to 0 \qquad \text{as } k \to \infty$$

Therefore, by the ratio test, $\sum_{k=1}^{\infty} |x[k]|$ converges. Hence $\sum_{k=1}^{\infty} x[k]$ converges also, and so the given series is absolutely convergent. ▲

Other tests exist which are more powerful and apply to wider classes of series but these are beyond the scope of this book.

3.3.9 *Sum of an infinite geometric series*

In the case of an infinite geometric series, it is possible to derive a simple formula for its sum when convergence takes place. We have already seen that the sum to k terms is given by

$$S_k = \frac{a(1 - r^k)}{1 - r} \qquad r \neq 1$$

What we must do is allow k to become large so that more and more terms are included in the sum. Provided that $-1 < r < 1$, then $r^k \to 0$ as $k \to \infty$. Then $S_k \to a/(1 - r)$. When this happens we write:

$$S_\infty = \frac{a}{1 - r}$$

where S_∞ is known as the 'sum to infinity'. If $r > 1$ or $r < -1$, r^k fails to approach a finite limit as $k \to \infty$ and the geometric series diverges.

■ Sum of an infinite geometric series: $S_\infty = \dfrac{a}{1 - r}$ $\qquad -1 < r < 1$

Example 3.20

Find the sum to k terms of the following series and deduce their sums to infinity:

(a) $1 + \frac{1}{3} + \frac{1}{9} + \frac{1}{27} + \cdots$

(b) $12 + 6 + 3 + 1\frac{1}{2} + \cdots$

Solutions

(a) This is a geometric series with first term 1 and common ratio 1/3.
Therefore,

$$S_k = \frac{a(1 - r^k)}{1 - r} = \frac{1(1 - (1/3)^k)}{(2/3)} = \frac{3}{2}\left(1 - \left(\frac{1}{3}\right)^k\right)$$

As $k \to \infty$, $(1/3)^k \to 0$ so that $S_\infty = 3/2$.

(b) This is a geometric series with first term 12 and common ratio $\frac{1}{2}$.
Therefore,

$$S_k = 24(1 - (1/2)^k)$$

As $k \to \infty$, $(1/2)^k \to 0$ so that $S_\infty = 24$. This could, of course, have been
obtained directly from the formula for the sum to infinity. ▲

EXERCISES 3.2

1. An arithmetic series has a first term of 4 and its 30th term is 1000. Find the sum to 30 terms.

2. Find the sum to 20 terms of the arithmetic series with first term a, and common difference d, given by:

   (a) $a = 4$, $d = 3$

   (b) $a = 4$, $d = -3$.

3. If the sum to ten terms of an arithmetic series is 100 and its common difference, d, is -3, find its first term.

4. The sum to 20 terms of an arithmetic series is identical to the sum to 22 terms. If the common difference is -2, find the first term.

5. Find the sum to five terms of the geometric series with first term 1 and common ratio 1/3. Find the sum to infinity.

6. Find the sum of the first 20 terms of the geometric series with first term 3 and common ratio 1.5.

7. Find the sum of the arithmetic series with first term 2, common difference 2, and last term 50.

8. The sum to infinity of a geometric series is 4 times the first term. Find the common ratio.

9. The sum to infinity of a geometric series is twice the sum of the first two terms. Find possible values of the common ratio.

10. Express the alternating harmonic series $1 - \frac{1}{2} + \frac{1}{3} - \frac{1}{4} + \cdots$ in sigma notation.

11. Write down the first six terms of the series $\Sigma_{k=0}^{\infty} z^{-k}$.

12. By comparison with the convergent series $\Sigma_{k=1}^{\infty} 1/k^2$ show that,

    $$\frac{3}{1.2} + \frac{3}{2.3} + \frac{3}{3.4} + \cdots$$

    converges.

13. Use the ratio test to ascertain whether,

    (a) $\Sigma_{k=0}^{\infty}(2k + 1)/2^k$

    (b) $\Sigma_{k=0}^{\infty} 2^k/(2k + 1)$

    converge or diverge.

14. Apply the divergence test to the series $\frac{1}{2} + \frac{2}{3} + \frac{3}{4} + \cdots$. What can you deduce about this series?

15. Test the following alternating series for convergence. For those that converge, state whether the convergence is absolute or conditional.

    (a) $\Sigma_{k=1}^{\infty}(-1)^{k+1}/3^k$

    (b) $\Sigma_{k=1}^{\infty}(-1)^k/(k^2 + 2k + 3)$

    (c) $\Sigma_{k=1}^{\infty}(-1)^{k+1}k/(4k + 1)$

    (d) $\Sigma_{k=1}^{\infty}(-1)^{k+1}/(2k - 1)$

16. Determine whether or not the series $\Sigma_{k=0}^{\infty} 5^k/(k + 1)!$ converges.

17. Explain why $\Sigma_{k=1}^{\infty} x[k]$ is the same as $\Sigma_{n=1}^{\infty} x[n]$. Further, explain why both can be written as $\Sigma_{k=0}^{\infty} x[k + 1]$.

3.4 The binomial theorem

It is straightforward to show that the expression $(a + b)^2$ can be written as $a^2 + 2ab + b^2$. It is slightly more complicated to expand the expression $(a + b)^3$ to $a^3 + 3a^2b + 3ab^2 + b^3$. However, it is often necessary to expand quantities such as $(a + b)^6$ or $(a + b)^{10}$ say, and then the algebra becomes extremely lengthy. A simple technique for expanding expressions of the form $(a + b)^n$, where n is a positive integer, is given by Pascal's triangle.

Pascal's triangle is the triangle of numbers shown in Figure 3.7, where it is observed that every entry is obtained by adding the two entries on either side in the preceding row, always starting and finishing a row with a '1'. You will note that the third row down, 1 2 1, gives the coefficients in the expansion of $(a + b)^2 = 1a^2 + 2ab + 1b^2$, while the fourth row, 1 3 3 1, gives the coefficients in the expansion of $(a + b)^3 = 1a^3 + 3a^2b + 3ab^2 + 1b^3$. Furthermore, the terms in these expansions are composed of decreasing powers of a and increasing powers of b. When we come to expand the quantity $(a + b)^4$ the row beginning '1 4' in the triangle will provide us with the necessary coefficients in the expansion and we must simply take care to put in place the appropriate powers of a and b. Thus $(a + b)^4 = 1a^4 + 4a^3b + 6a^2b^2 + 4ab^3 + 1b^4$.

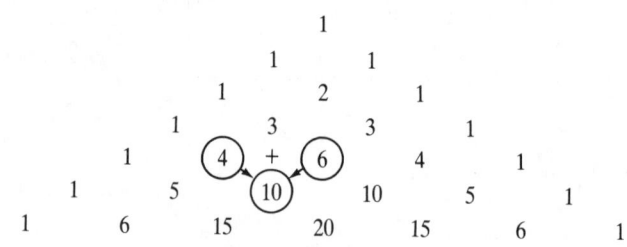

Figure 3.7 Pascal's triangle.

Example 3.21

Use Pascal's triangle to expand $(a + b)^6$.

Solution

We look to the row commencing '1 6', that is 1 6 15 20 15 6 1, because $a + b$ is raised to the power 6. This row provides the necessary coefficients. Thus,

$$(a + b)^6 = a^6 + 6a^5b + 15a^4b^2 + 20a^3b^3 + 15a^2b^4 + 6ab^5 + b^6$$

Example 3.22

Expand $(1 + x)^7$ using Pascal's triangle.

Solution

Forming the row commencing '1 7' we select the coefficients

1 7 21 35 35 21 7 1

In this example, $a = 1$ and $b = x$ so that,

$$(1 + x)^7 = 1 + 7x + 21x^2 + 35x^3 + 35x^4 + 21x^5 + 7x^6 + x^7 \quad \blacktriangle$$

When it is necessary to expand the quantity $(a + b)^n$ for large n, it is clearly inappropriate to use Pascal's triangle since an extremely large triangle would have to be constructed. However, it is frequently the case that in such situations only the first few terms in the expansion are required. This is where the **binomial theorem** is useful.

■ The binomial theorem states that when $n \in \mathbb{N}^+$,

$$(a + b)^n = a^n + na^{n-1}b + \frac{n(n-1)}{2!}a^{n-2}b^2 + \frac{n(n-1)(n-2)}{3!}a^{n-3}b^3$$
$$+ \cdots + b^n$$

It is also frequently quoted for the case when $a = 1$ and $b = x$, so that

---

■ $\quad (1 + x)^n = 1 + nx + \frac{n(n-1)}{2!}x^2 + \frac{n(n-1)(n-2)}{3!}x^3 + \cdots + x^n \qquad (3.3)$

Example 3.23

Expand $(1 + x)^{10}$ up to the terms in x^3.

Solution

We could use Pascal's triangle to answer this question and look to the row commencing '1   10' but to find this row considerable calculations would be needed. We shall use the binomial theorem in the form of Equation (3.3). Taking $n = 10$, we find

$$(1 + x)^{10} = 1 + 10x + \frac{10(9)}{2!}x^2 + \frac{(10)(9)(8)}{3!}x^3 + \cdots$$
$$= 1 + 10x + 45x^2 + 120x^3 + \cdots$$

so that, up to and including x^3, the expansion is

$$1 + 10x + 45x^2 + 120x^3 \quad \blacktriangle$$

We have assumed in the foregoing discussion that n is a positive integer in which case the expansion given by Equation (3.3) will eventually terminate. In Example 3.23 this would occur when we reached the term in x^{10}. It can be shown, however, that when n is not a positive integer the function $(1 + x)^n$ and the expansion given by:

$$(1 + x)^n = 1 + nx + \frac{n(n-1)}{2!}x^2 + \frac{n(n-1)(n-2)}{3!}x^3 + \cdots \qquad (3.4)$$

have the same value provided $-1 < x < 1$. However, when n is not a positive integer the series does not terminate and we must deal with an infinite series. This series converges when $-1 < x < 1$ and the expansion is then said to be valid. When x lies outside this interval the infinite series diverges and so bears no relation to the value of $(1 + x)^n$. The expansion is then said to be invalid.

■ The binomial theorem:

$$(1 + x)^n = 1 + nx + \frac{n(n-1)}{2!}x^2 + \frac{n(n-1)(n-2)}{3!}x^3 + \cdots \qquad -1 < x < 1$$

Example 3.24

Use the binomial theorem to expand $1/(1 + x)$ in ascending powers of x up to and including the term in x^3.

Solution

$1/(1 + x)$ can be written as $(1 + x)^{-1}$. Using the binomial theorem given by Equation (3.4) with $n = -1$, we find

$$(1 + x)^{-1} = 1 + (-1)x + \frac{(-1)(-2)}{2!}x^2 + \frac{(-1)(-2)(-3)}{3!}x^3 + \cdots$$

$$= 1 - x + x^2 - x^3 + \cdots$$

provided $-1 < x < 1$. Consequently, if in future applications we come across the series $1 - x + x^2 - x^3 + \cdots$, we shall be able to write it in the form $(1 + x)^{-1}$. This **closed form** avoids the use of an infinite series and so is easier to handle. We shall make use of this technique in Chapter 12 when we meet the z transform. ▲

Let us consider another example.

Example 3.25

Obtain a quadratic approximation to $(1 - 2x)^{1/2}$ using the binomial theorem.

Solution

Using Equation (3.4) with x replaced by $-2x$ and $n = \frac{1}{2}$ we have:

$$(1 - 2x)^{1/2} = 1 + (\tfrac{1}{2})(-2x) + \frac{(1/2)(-1/2)}{2!}(-2x)^2 + \cdots$$

$$= 1 - x - \tfrac{1}{2}x^2 + \cdots$$

provided that $-1 < -2x < 1$, that is $-\frac{1}{2} < x < \frac{1}{2}$. A quadratic approximation is therefore $1 - x - \frac{1}{2}x^2$. ▲

EXERCISES 3.3

1. Use Pascal's triangle to expand $(a + b)^8$.

2. Use Pascal's triangle to expand $(2x + 3y)^4$.

3. Expand $(a - 2b)^5$.

4. Use the binomial theorem to find the expansion of $(3 - 2x)^6$ up to and including the term in x^3.

5. Obtain the first four terms in the expansion of $(1 + \frac{1}{2}x)^{10}$.

6. Obtain the first five terms in the expansion of $(1 + 2x)^{1/2}$. State the range of values of x for which the expansion is valid. Choose a value of x within the range of validity and compute values of your expansion for comparison with the true function values.

7. Expand $(1 + \frac{1}{2}x)^{-4}$ in ascending powers of x up to the term in x^4, stating the range of values of x for which the expansion is valid.

8. Expand $(1 + \frac{1}{x})^{-1/2}$ in descending powers up to the fourth term.

9. (a) Expand $(1 + x^2)^4$.

   (b) Expand $(1 + 1/x^2)^4$.

10. A function, $f(x)$, is given by:

$$f(x) = \left(1 + \frac{1}{x}\right)^{1/2}$$

(a) Obtain the first four terms in the expansion of $f(x)$ in descending powers of x. State the range of values of x for which the expansion is valid.

(b) By writing $f(x)$ in the form:

$$f(x) = x^{-1/2}(1 + x)^{1/2}$$

obtain the first four terms in the expansion of $f(x)$ in ascending powers of x. State the range of values of x for which the expansion is valid.

11. The function, $g(x)$, is defined by:

$$g(x) = \frac{1}{(1 + x^2)^4}$$

(a) Obtain the first four terms in the expansion of $g(x)$ in ascending powers of x. State the range of values of x for which the expansion is valid.

(b) By rewriting $g(x)$ in an appropriate form, obtain the first four terms in the expansion of $g(x)$ in descending powers of x. State the range of values of x for which the expansion is valid.

3.5 Power series

A particularly important class of series are known as **power series** and these are infinite series involving integer powers of the variable x. For example,

$$1 + x + x^2 + x^3 + \cdots$$

and

$$1 + x + \frac{x^2}{2} + \frac{x^3}{3!} + \cdots$$

are both power series. Note that a power series can be regarded as an infinite polynomial. Many common functions can be expressed in terms of a power series, for example

$$\sin x = x - \frac{x^3}{3!} + \frac{x^5}{5!} - \cdots \qquad x \text{ in radians}$$

which converges for any value of x. For example,

$$\sin(0.5) = 0.5 - \frac{(0.5)^3}{6} + \frac{(0.5)^5}{120} - \cdots$$

Taking just the first three terms, we find:

$$\sin(0.5) \approx 0.5 - 0.020\,833\,3 + 0.000\,260\,4 = 0.479\,427\,1$$

as compared with the true value, $\sin 0.5 = 0.479\,425\,5$.

More generally, a power series is only meaningful if the series converges for the particular value of x chosen. We define an important quantity known as the **radius of convergence**, R, as the largest value for which an x chosen in the interval $-R < x < R$, causes the series to converge.

The open interval $(-R, R)$ is known as the **interval of convergence**. Tests for convergence of a power series are carried out using techniques such as those described in Section 3.3. Further consideration will be given to power series in Chapter 9, but for future reference we give some common expansions now.

■ $\sin x = x - \dfrac{x^3}{3!} + \dfrac{x^5}{5!} - \cdots \qquad x \text{ in radians}$

   $\cos x = 1 - \dfrac{x^2}{2!} + \dfrac{x^4}{4!} - \cdots \qquad x \text{ in radians}$

   $e^x = 1 + x + \dfrac{x^2}{2!} + \dfrac{x^3}{3!} + \dfrac{x^4}{4!} + \cdots$

all of which converge for any value of x.

Each of these series converges rapidly when x is small, and so can be used to obtain useful approximations. In particular, we note that:

■ If x is small and measured in radians

   $\sin x \approx x \qquad \text{and} \qquad \cos x \approx 1 - \dfrac{x^2}{2!}$

These formulae are known as the **small-angle approximations**.

EXERCISES 3.4

1. The power series expansion of e^x is given by:

$$e^x = 1 + x + \frac{x^2}{2!} + \frac{x^3}{3!} + \cdots$$

and is valid for any x. Take four terms of the series when $x = 0, 0.1, 0.5$ and 1, to compare the sum to four terms with the value of e^x obtained from your calculator. Comment upon the result.

2. Using the power series expansion for $\cos x$

   (a) Write down the power series expansion for $\cos 2x$

   (b) Write down the power series expansion for $\cos(x/2)$.

   By considering the power series expansion for $\cos(-x)$ show that $\cos x = \cos(-x)$.

3. By considering the power series expansion of e^x find $\Sigma_{k=0}^{\infty} 1/k!$.

4. Obtain a cubic approximation to $e^x \sin x$.

5. (a) State the power series expansion for e^{-x}.

   (b) By using your solution to (a) and the expansion for e^x, deduce the power series expansions of $\cosh x$ and $\sinh x$.

Miscellaneous exercises

1. Write down and graph the first five terms of the sequences $x[k]$ defined by

   (a) $x[k] = (-1)^k \qquad k \in \mathbb{N}$
   (b) $x[k] = (-1)^k/(2k + 1) \qquad k \in \mathbb{N}$

2. Find expressions for the kth terms of the sequences whose first five terms are

   (a) $1, 9, 17, 25, 33$
   (b) $-1, 1, -1, 1, -1$

3. For the Fibonacci sequence $1, 1, 2, 3, 5, 8, \ldots$ show that:

   $$\lim_{k \to \infty} \frac{x[k+1]}{x[k]} = \frac{1}{2}(1 + \sqrt{5})$$

   Hint: write $x[k + 1] = x[k] + x[k - 1]$, form $x[k + 1]/x[k]$ and take limits.

4. Use the binomial theorem to expand $(1 + x + x^2)^5$ as far as the term in x^3.

5. Use the binomial theorem to expand $1/(3 + x)^3$ in ascending powers of x as far as the term in x^3.

6. The power series expansion for $\ln(1 + x)$ is given by:

   $$\ln(1 + x) = x - \frac{x^2}{2} + \frac{x^3}{3} - \frac{x^4}{4} + \cdots$$

   and is valid for $-1 < x \le 1$. Take a number of values of x in this interval and obtain an approximate value of $\ln(1 + x)$ by means of this series. Compare your answers with the values obtained from your calculator.

7. By multiplying both numerator and denominator of $(\sqrt{k + 1} - \sqrt{k})/2$ by $\sqrt{k + 1} + \sqrt{k}$ find

   $$\lim_{k \to \infty} \frac{\sqrt{k + 1} - \sqrt{k}}{2}$$

8. Find $\lim_{k \to \infty} \left(\dfrac{3k - 1}{2k + 7}\right)^3$.

9. Find $\lim_{k \to \infty} \dfrac{2k^5 - 3k^2}{7k^7 + 2k}$.

10. Write down the first eight terms of the series $\Sigma_{k=1}^{n} k$. By noting that this is an arithmetic series show that

    $$\sum_{k=1}^{n} k = \frac{n(n + 1)}{2}.$$

11. Write down the first six terms of the sequence defined by the recurrence relation:

$$x[n + 3] = x[n + 2] - 2x[n]$$
$$x[0] = 0 \quad x[1] = 2 \quad x[2] = 3$$

12. Find the limit, if it exists, as $k \to \infty$ of the geometric progression:

$$a, ar, ar^2, \ldots, ar^{k-1}, \ldots$$

when

(a) $-1 < r < 1$

(b) $r > 1$

(c) $r < -1$

(d) $r = 1$

(e) $r = -1$

13. An arithmetic series has a first term of 4 and the 10th term is 0.

(a) Find S_{20}.

(b) If $S_n = 0$, find n.

14. A geometric series has:

$$S_3 = \tfrac{37}{8} \qquad S_6 = \tfrac{3367}{512}$$

Find the first term and the common ratio.

Vectors

4

4.1 Introduction

Certain physical quantities are fully described by a single number: for example, the mass of a stone, the speed of a car. Such quantities are called **scalars**. On the other hand, some quantities are not fully described until a direction is specified in addition to the number. For example, a velocity of 30 metres per second due east is different from a velocity of 30 metres per second due north. These quantities are called **vectors** and it is important to distinguish them from scalars. There are many engineering applications in which vector and scalar quantities play important roles. For example, speed, potential, work and energy are scalar quantities, while velocity, electric and magnetic forces, the position of a robot and the state–space representation of a system can all be described by vectors. A variety of mathematical techniques have been developed to enable useful calculations to be carried out using vectors and in this chapter these will be discussed.

4.2 Vectors and scalars: basic concepts

Scalars are the simplest quantities with which to deal; the specification of a single number is all that is required. Vectors also have a direction and it is useful to consider a graphical representation. Thus the line segment AB of length 4 in Figure 4.1 can represent a vector in the direction shown by the arrow on AB. This vector is denoted by $\overrightarrow{AB}$. Note that $\overrightarrow{AB} \neq \overrightarrow{BA}$. The vector $\overrightarrow{AB}$ is directed from A to B, but $\overrightarrow{BA}$ is directed from B to A.

An alternative notation is frequently used: we denote $\overrightarrow{AB}$ by **a**. This bold notation is commonly used in textbooks but the notation $\underline{a}$ is preferable for handwritten work. We now need to refer to the diagram to appreciate the intended direction. The length of the line segment represents the **magnitude**, or **modulus**, of the vector and we use the notation $|\overrightarrow{AB}|$, $|\mathbf{a}|$ or simply a to denote this. Note that whereas **a** is a vector, $|\mathbf{a}|$ is a scalar.

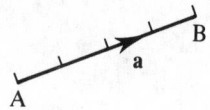

Figure 4.1   A vector, $\overrightarrow{AB}$, of length 4.

4.2.1 *Negative vectors*

The vector $-\mathbf{a}$ is a vector in the opposite direction to, but with the same magnitude as, $\mathbf{a}$. Geometrically it will be $\overrightarrow{BA}$. Thus $-\mathbf{a}$ is the same as $\overrightarrow{BA}$.

4.2.2 *Equal vectors*

Two vectors are equal if they have the same magnitude and direction. In Figure 4.2 vectors $\overrightarrow{CD}$ and $\overrightarrow{AB}$ are equal even though their locations differ. This is a useful property of vectors: a vector can be translated, maintaining its length and direction without changing the vector itself. There are exceptions to this property. For example, we shall soon meet position vectors which are used to locate specific fixed points in space. They clearly cannot be moved around freely. Nevertheless most of the vectors we shall meet can be translated freely, and as such are often called **free vectors**.

4.2.3 *Vector addition*

It is frequently useful to add two or more vectors together and the addition of vectors is defined by the **triangle law**. Referring to Figure 4.3, if we wish to add

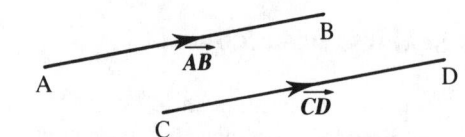

Figure 4.2   Two equal vectors.

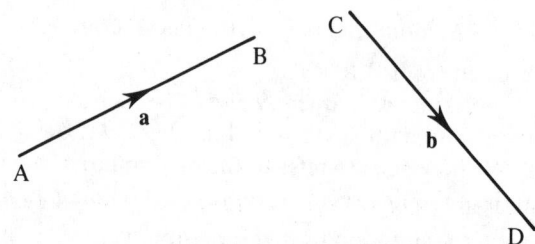

Figure 4.3   Two vectors, $\overrightarrow{AB}$ and $\overrightarrow{CD}$.

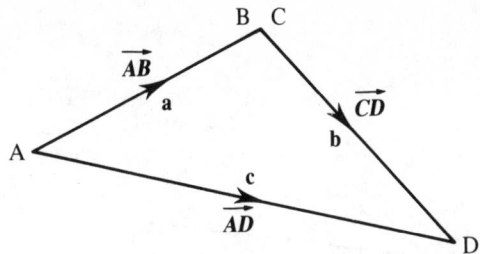

Figure 4.4   Addition of the two vectors of Figure 4.3 using the triangle law.

$\overrightarrow{AB}$ to $\overrightarrow{CD}$, $\overrightarrow{CD}$ is translated until C and B coincide. As mentioned earlier, this translation does not change the vector $\overrightarrow{CD}$. Then the sum, $\overrightarrow{AB} + \overrightarrow{CD}$, is defined by the vector represented by the third side of the completed triangle, that is $\overrightarrow{AD}$ (Figure 4.4). Thus,

$$\overrightarrow{AB} + \overrightarrow{CD} = \overrightarrow{AD}$$

Similarly, if $\overrightarrow{AB}$ is denoted by **a**, $\overrightarrow{CD}$ is denoted by **b**, and $\overrightarrow{AD}$ is denoted by **c**, then we have

$$\mathbf{c} = \mathbf{a} + \mathbf{b}$$

We note that to add **a** and **b** a triangle is formed using **a** and **b** as two of the sides in such a way that the head of one vector touches the tail of the other as shown in Figure 4.4. The sum **a** + **b** is then represented by the third side.

It is possible to prove the following rules:

■ $\mathbf{a} + \mathbf{b} = \mathbf{b} + \mathbf{a}$ vector addition is commutative

   $\mathbf{a} + (\mathbf{b} + \mathbf{c}) = (\mathbf{a} + \mathbf{b}) + \mathbf{c}$ vector addition is associative

---

To see why it is appropriate to add vectors using the triangle law consider Examples 4.1 and 4.2.

Example 4.1 Routing of an automated vehicle

An unmanned vehicle moves on tracks around a factory floor carrying electrical components from stores at A to workers at C as illustrated in Figure 4.5. The vehicle may arrive at C either directly, or via point B. The movement from A to B can be represented by a vector $\overrightarrow{AB}$ known as a **displacement vector**, whose magnitude is the distance between points A and B. Similarly, movement from B to C is represented by $\overrightarrow{BC}$, and movement directly from A to C is represented by $\overrightarrow{AC}$. Clearly, since A, B and C are fixed points, these displacement vectors are fixed too. Since the head of the

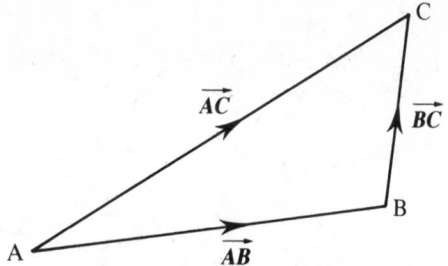

Figure 4.5 Routing of an automated vehicle from stores at A to workers at C.

vector $\overrightarrow{AB}$ touches the tail of the vector $\overrightarrow{BC}$ we are ready to use the triangle law of vector addition to find the combined effect of the two displacements:

$$\overrightarrow{AB} + \overrightarrow{BC} = \overrightarrow{AC}$$

This means that the combined effect of displacements $\overrightarrow{AB}$ and $\overrightarrow{BC}$ is the displacement $\overrightarrow{AC}$. We say that $\overrightarrow{AC}$ is the **resultant** of $\overrightarrow{AB}$ and $\overrightarrow{BC}$. ▲

■ In considering motion from point A to point B, the vector $\overrightarrow{AB}$ is called a displacement vector.

Example 4.2 Resultant of two forces acting on a body

A force $\mathbf{F}_1$ of 2 N acts vertically downwards, and a force $\mathbf{F}_2$ of 3 N acts horizontally to the right, upon the body shown in Figure 4.6. Translating $\mathbf{F}_1$ until its tail touches the head of $\mathbf{F}_2$ we can apply the triangle law to find the combined effect of the two forces. This is a single force $\mathbf{R}$ known as the resultant of $\mathbf{F}_2$ and $\mathbf{F}_1$. We write:

$$\mathbf{R} = \mathbf{F}_2 + \mathbf{F}_1$$

and say that $\mathbf{R}$ is the vector sum of $\mathbf{F}_2$ and $\mathbf{F}_1$. The resultant force, $\mathbf{R}$, acts at an angle of θ to the vertical, where $\tan \theta = 3/2$, and has a magnitude given by Pythagoras' theorem as $\sqrt{2^2 + 3^2} = \sqrt{13}$ N.

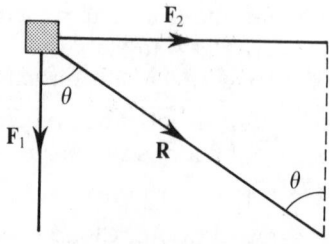

Figure 4.6 Resultant force, $\mathbf{R}$, produced by a vertical force, $\mathbf{F}_1$, and a horizontal force, $\mathbf{F}_2$.

Example 4.3

Vectors **p**, **q**, **r** and **s** form the sides of the square shown in Figure 4.7. Express:

(a) **p** in terms of **r**

(b) **s** in terms of **q**

(c) diagonal $\overrightarrow{BD}$ in terms of **q** and **r**.

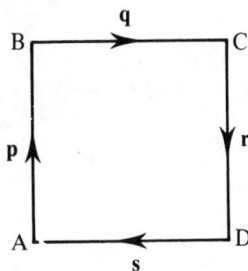

Figure 4.7   The vectors **p**, **q**, **r** and **s** form the sides of a square.

Solutions

(a) The vector **p** has the same length as **r** but has the opposite direction. Therefore **p** = −**r**.

(b) Vector **s** has the same length as **q** but has the opposite direction. Therefore **s** = −**q**.

(c) The head of $\overrightarrow{BC}$ coincides with the tail of $\overrightarrow{CD}$. Therefore, by the triangle law of addition, the third side of triangle BCD represents the sum of $\overrightarrow{BC}$ and $\overrightarrow{CD}$, that is

$$\overrightarrow{BD} = \overrightarrow{BC} + \overrightarrow{CD}$$

$$= \mathbf{q} + \mathbf{r} \quad \blacktriangle$$

4.2.4 *Vector subtraction*

Subtraction of one vector from another is performed by adding the corresponding negative vector, that is if we seek **a** − **b**, we form **a** + (−**b**).

Example 4.4

Consider the rectangle illustrated in Figure 4.8 with **a** and **b** as shown. Express in terms of A, B, C or D the vectors **a** − **b** and **b** − **a**.

Solution

We have:

$$\mathbf{b} = \overrightarrow{AD}$$

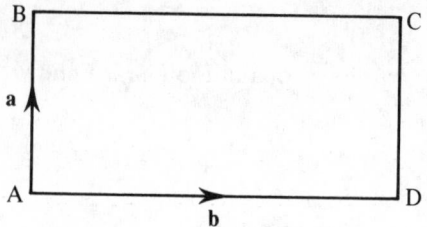

Figure 4.8 The rectangle ABCD.

hence

$$-\mathbf{b} = \overrightarrow{DA}$$

Then

$$\mathbf{a} - \mathbf{b} = \mathbf{a} + (-\mathbf{b})$$
$$= \overrightarrow{AB} + \overrightarrow{DA}$$
$$= \overrightarrow{DA} + \overrightarrow{AB} \qquad \text{by commutativity}$$

$\overrightarrow{DA}$ and $\overrightarrow{AB}$ are shown in Figure 4.9. Their sum is given by the triangle law, that is

$$\mathbf{a} - \mathbf{b} = \overrightarrow{DA} + \overrightarrow{AB} = \overrightarrow{DB}$$

Similarly,

$$\mathbf{a} = \overrightarrow{AB}$$

hence

$$-\mathbf{a} = \overrightarrow{BA}$$

Then

$$\mathbf{b} - \mathbf{a} = \mathbf{b} + (-\mathbf{a})$$
$$= \overrightarrow{AD} + \overrightarrow{BA}$$
$$= \overrightarrow{BA} + \overrightarrow{AD}$$

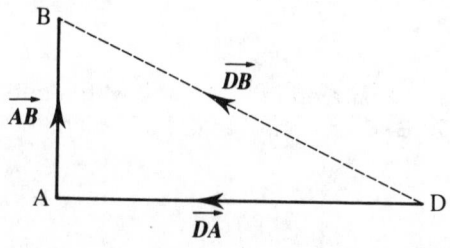

Figure 4.9 The vectors $\overrightarrow{AB}$ and $\overrightarrow{DA}$.

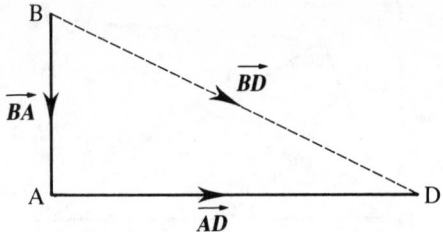

Figure 4.10 The vectors $\vec{BA}$ and $\vec{AD}$.

$\vec{BA}$ and $\vec{AD}$ are shown in Figure 4.10. Again their sum is given by the triangle law, that is

$$\mathbf{b} - \mathbf{a} = \vec{BA} + \vec{AD} = \vec{BD} \quad \blacktriangle$$

Example 4.5

Referring to Figure 4.11, if $\mathbf{r}_1$ and $\mathbf{r}_2$ are as shown, find the vector $\mathbf{b} = \vec{QP}$ represented by the third side of the triangle OPQ.

Solution

From Figure 4.11 we note from the triangle rule that:

$$\vec{QP} = \vec{QO} + \vec{OP} = \vec{OP} + \vec{QO} \text{ by commutativity.}$$

But $\vec{QO} = -\mathbf{r}_2$ and $\vec{QP} = \mathbf{b}$, and so

$$\mathbf{b} = \mathbf{r}_1 - \mathbf{r}_2 \quad \blacktriangle$$

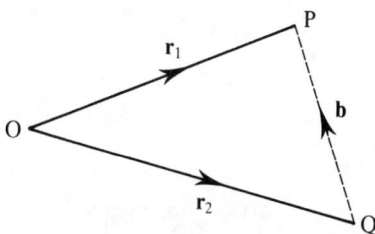

Figure 4.11 The two vectors $\mathbf{r}_1$ and $\mathbf{r}_2$.

Vectors do not necessarily lie in a two-dimensional plane. Three-dimensional vectors are commonly used as is illustrated in the following example.

Example 4.6

OPQR is the tetrahedron shown in Figure 4.12. Let $\vec{OP} = \mathbf{p}$, $\vec{OQ} = \mathbf{q}$ and $\vec{OR} = \mathbf{r}$. Express $\vec{PQ}$, $\vec{QR}$ and $\vec{RP}$ in terms of $\mathbf{p}$, $\mathbf{q}$ and $\mathbf{r}$.

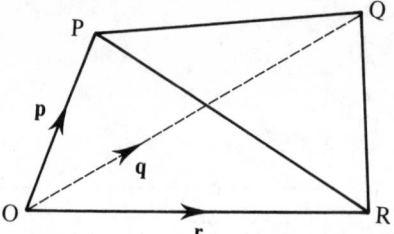

Figure 4.12   The tetrahedron OPQR.

Solution

Consider the triangle OPQ shown in Figure 4.13. We note that $\vec{OQ}$ represents the third side of the triangle formed when **p** and $\vec{PQ}$ are placed head to tail. Using the triangle law we find:

$$\vec{OP} + \vec{PQ} = \vec{OQ}$$

Therefore,

$$\vec{PQ} = \vec{OQ} - \vec{OP}$$

$$= \mathbf{q} - \mathbf{p}$$

Similarly, $\vec{QR} = \mathbf{r} - \mathbf{q}$ and $\vec{RP} = \mathbf{p} - \mathbf{r}$. ▲

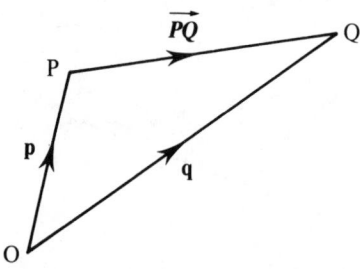

Figure 4.13   The triangle OPQ.

4.2.5 *Multiplication of a vector by a scalar*

If k is any positive scalar and **a** is a vector then $k\mathbf{a}$ is a vector in the same direction as **a** but k times as long. If k is any negative scalar, $k\mathbf{a}$ is a vector in the opposite direction to **a**, and k times as long. By way of example, study the vectors in Figure 4.14. Clearly 2**a** is twice as long as **a** but has the same direction. The vector $\frac{1}{2}\mathbf{b}$ is half as long as **b** but has the same direction as **b**. It is possible to prove the following rules.

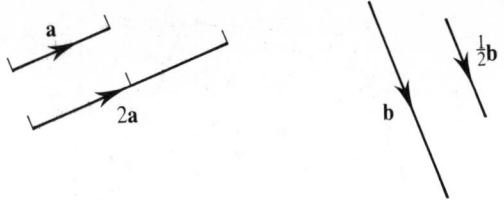

Figure 4.14 Scalar multiplication of a vector.

■ For any scalars k and l, and any vectors **a** and **b**:

$k(\mathbf{a} + \mathbf{b}) = k\mathbf{a} + k\mathbf{b}$

$(k + l)\mathbf{a} = k\mathbf{a} + l\mathbf{a}$

$k(l\mathbf{a}) = (kl)\mathbf{a}$

The vector $k\mathbf{a}$ is said to be a **scalar multiple** of **a**.

Example 4.7 Magnetic vector quantities

The magnetic field intensity, **H**, is a vector quantity with units of ampere per metre (A m^{-1}). The magnetic flux density, **B**, is also a vector quantity with units of weber per square metre (Wb m^{-2}) or tesla (T). The vector **B** can be calculated by multiplying **H** by a scalar quantity μ, that is

$\mathbf{B} = \mu\mathbf{H}$

μ is known as the permeability of a material and has units of weber per ampere per metre (Wb A^{-1} m^{-1}). Confirm for yourself that the units match on both sides of the equation. ▲

Example 4.8

In the triangle ABC, M is the midpoint of AB. Let $\overrightarrow{AB}$ be denoted by **a**, and $\overrightarrow{BC}$ by **b**. Express $\overrightarrow{AC}$, $\overrightarrow{CA}$ and $\overrightarrow{CM}$ in terms of **a** and **b**.

Solution

The situation is sketched in Figure 4.15. Using the triangle rule for addition we find

$\overrightarrow{AB} + \overrightarrow{BC} = \overrightarrow{AC}$

Therefore,

$\overrightarrow{AC} = \mathbf{a} + \mathbf{b}$

It follows that $\overrightarrow{CA} = -\overrightarrow{AC} = -(\mathbf{a} + \mathbf{b})$.

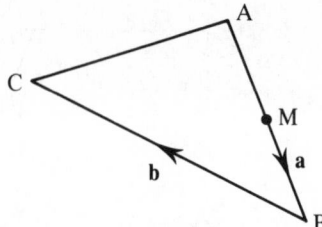

Figure 4.15   The triangle ABC, with midpoint M of side AB.

Again by the triangle rule applied to triangle CMB we find:

$$\vec{CM} = \vec{CB} + \vec{BM}$$

Now $\vec{BM} = \tfrac{1}{2}\vec{BA} = -\tfrac{1}{2}\mathbf{a}$ and so,

$$\vec{CM} = -\mathbf{b} + (-\tfrac{1}{2}\mathbf{a})$$

$$= -(\mathbf{b} + \tfrac{1}{2}\mathbf{a}) \quad \blacktriangle$$

4.2.6 *Unit vectors*

Vectors which have length 1 are called **unit vectors**. If **a** has length 3, for example, then a unit vector in the direction of **a** is clearly $\tfrac{1}{3}\mathbf{a}$. More generally we denote the unit vector in the direction **a** by **â**. Recall that the length or modulus of **a** is $|\mathbf{a}|$ and so we can write

---

■ $\hat{\mathbf{a}} = \dfrac{\mathbf{a}}{|\mathbf{a}|}$

Note that $|\mathbf{a}|$ and hence $1/|\mathbf{a}|$ are scalars.

4.2.7 *Orthogonal vectors*

If the angle between two vectors **a** and **b** is 90°, that is **a** and **b** are perpendicular, then **a** and **b** are said to be **orthogonal**.

EXERCISES 4.1

Throughout these exercises, use diagrams to illustrate your answers.

1. For the arbitrary points A, B, C, D and E, find a single vector which is equivalent to

    (a) $\vec{DC} + \vec{CB}$

    (b) $\vec{CE} + \vec{DC}$

2. Figure 4.16 shows a cube. Let $\mathbf{p} = \vec{AB}$, $\mathbf{q} = \vec{AD}$ and $\mathbf{r} = \vec{AE}$. Express the vectors representing $\vec{BD}$, $\vec{AC}$ and $\vec{AG}$ in terms of **p**, **q** and **r**.

3. In the triangle ABC, M is the midpoint of BC, and N is the midpoint of AC. Show that $\vec{NM} = \tfrac{1}{2}\vec{AB}$.

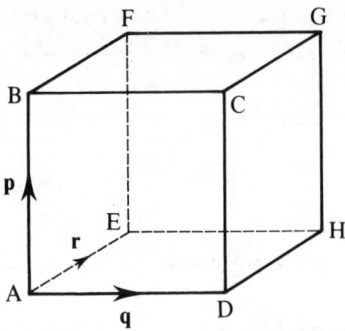

Figure 4.16 Figure for Exercise 4.1, Question (2).

4.3 Cartesian components

Consider the $x-y$ plane in Figure 4.17. The general point P has coordinates (x, y). We can join the origin to P by a vector $\overrightarrow{OP}$, which is called the position vector of P, which we often denote by **r**. The modulus of **r** is $|\mathbf{r}| = r$, and is the length of $\overrightarrow{OP}$. It is possible to express **r** in terms of the numbers of x and y. If we denote a unit vector along the x axis by **i**, and a unit vector along the y axis by **j** (we usually omit the ^ here), then it is clear from the definition of scalar multiplication that $\overrightarrow{OM} = x\mathbf{i}$, and $\overrightarrow{MP} = y\mathbf{j}$. It follows from the triangle law of addition that:

■ $\mathbf{r} = \overrightarrow{OP} = \overrightarrow{OM} + \overrightarrow{MP} = x\mathbf{i} + y\mathbf{j}$

Clearly the vectors **i** and **j** are orthogonal. The numbers x and y are the **i** and **j** **components** of **r**. Furthermore, using Pythagoras' theorem we can deduce that

■ $r = \sqrt{x^2 + y^2}$

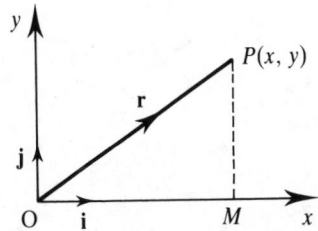

Figure 4.17 The $x-y$ plane with point P.

Alternative notations which are sometimes useful are:

$$\blacksquare \quad \mathbf{r} = \vec{OP} = \begin{pmatrix} x \\ y \end{pmatrix}$$

and

$$\mathbf{r} = \vec{OP} = (x, y)$$

When written in these forms $\begin{pmatrix} x \\ y \end{pmatrix}$ is called a **column vector** and (x, y) is called a **row vector**. To avoid confusion with the coordinates (x, y) we shall not use row vectors here but they will be needed in Chapter 14. We will also use the column vector notation for more general vectors, thus,

$$a\mathbf{i} + b\mathbf{j} = \begin{pmatrix} a \\ b \end{pmatrix}$$

We said earlier that a vector can be translated, maintaining its length and direction without changing the vector itself. While this is true generally, position vectors form an important exception. Position vectors are constrained to their specific position and must always remain tied to the origin.

Example 4.9

If A is the point with coordinates $(5, 4)$ and B is the point with coordinates $(-3, 2)$ find the position vectors of A and B, and the vector $\vec{AB}$.
Further, find $|\vec{AB}|$.

Solution

The position vector of A is $5\mathbf{i} + 4\mathbf{j} = \begin{pmatrix} 5 \\ 4 \end{pmatrix}$, which we shall denote by **a**. The position vector of B is $-3\mathbf{i} + 2\mathbf{j} = \begin{pmatrix} -3 \\ 2 \end{pmatrix}$, which we shall denote by **b**.
Application of the triangle law to triangle OAB (Figure 4.18) gives:

$$\vec{OA} + \vec{AB} = \vec{OB}$$

that is

$$\mathbf{a} + \vec{AB} = \mathbf{b}$$

Therefore,

$$\vec{AB} = \mathbf{b} - \mathbf{a}$$
$$= (-3\mathbf{i} + 2\mathbf{j}) - (5\mathbf{i} + 4\mathbf{j})$$
$$= -8\mathbf{i} - 2\mathbf{j}$$

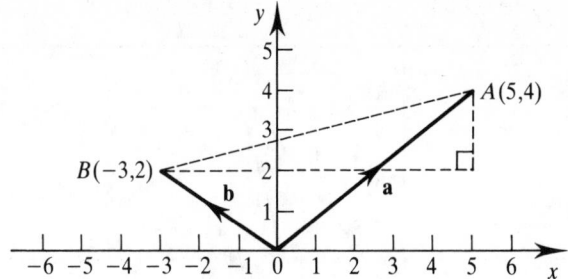

Figure 4.18   Points A and B in the x–y plane.

Alternatively, in terms of column vectors

$$\mathbf{b} - \mathbf{a} = \begin{pmatrix} -3 \\ 2 \end{pmatrix} - \begin{pmatrix} 5 \\ 4 \end{pmatrix}$$

$$= \begin{pmatrix} -8 \\ -2 \end{pmatrix}$$

We note that subtraction (and likewise addition) of column vectors is carried out component by component. To find $|\overrightarrow{AB}|$ we must obtain the length of the vector $\overrightarrow{AB}$. Referring to Figure 4.18, we note that this quantity is the length of the hypotenuse of a right-angled triangle with perpendicular sides 8 and 2. That is $|\overrightarrow{AB}| = \sqrt{8^2 + 2^2} = \sqrt{68} = 8.25.$ ▲

   More generally we have the following result:

■ Given vectors $\mathbf{a} = \overrightarrow{OA} = a_1\mathbf{i} + a_2\mathbf{j}$ and $\mathbf{b} = \overrightarrow{OB} = b_1\mathbf{i} + b_2\mathbf{j}$ (Figure 4.19), then
$$\overrightarrow{AB} = \mathbf{b} - \mathbf{a} = (b_1 - a_1)\mathbf{i} + (b_2 - a_2)\mathbf{j}$$
and
$$|\overrightarrow{AB}| = |\mathbf{b} - \mathbf{a}| = |(b_1 - a_1)\mathbf{i} + (b_2 - a_2)\mathbf{j}|$$
$$= \sqrt{(b_1 - a_1)^2 + (b_2 - a_2)^2}$$

Example 4.10

If $\mathbf{a} = \begin{pmatrix} 7 \\ 3 \end{pmatrix}$ and $\mathbf{b} = \begin{pmatrix} -2 \\ 5 \end{pmatrix}$,

(a) find $\mathbf{a} + \mathbf{b}$, $\mathbf{a} - \mathbf{b}$, $\mathbf{b} + \mathbf{a}$ and $\mathbf{b} - \mathbf{a}$, commenting upon the results
(b) find $2\mathbf{a} - 3\mathbf{b}$
(c) find $|\mathbf{a} - \mathbf{b}|$.

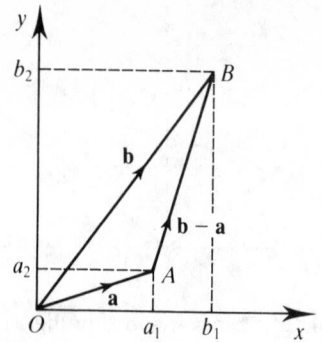

Figure 4.19 The quantity $|\mathbf{b}-\mathbf{a}| = \sqrt{(b_1 - a_1)^2 + (b_2 - a_2)^2}$.

Solutions

(a) $\mathbf{a} + \mathbf{b} = \begin{pmatrix} 7 \\ 3 \end{pmatrix} + \begin{pmatrix} -2 \\ 5 \end{pmatrix} = \begin{pmatrix} 5 \\ 8 \end{pmatrix}$

$\mathbf{a} - \mathbf{b} = \begin{pmatrix} 7 \\ 3 \end{pmatrix} - \begin{pmatrix} -2 \\ 5 \end{pmatrix} = \begin{pmatrix} 9 \\ -2 \end{pmatrix}$

$\mathbf{b} + \mathbf{a} = \begin{pmatrix} -2 \\ 5 \end{pmatrix} + \begin{pmatrix} 7 \\ 3 \end{pmatrix} = \begin{pmatrix} 5 \\ 8 \end{pmatrix}$

$\mathbf{b} - \mathbf{a} = \begin{pmatrix} -2 \\ 5 \end{pmatrix} - \begin{pmatrix} 7 \\ 3 \end{pmatrix} = \begin{pmatrix} -9 \\ 2 \end{pmatrix}$

We note that addition is commutative whereas subtraction is not.

(b) $2\mathbf{a} - 3\mathbf{b} = 2\begin{pmatrix} 7 \\ 3 \end{pmatrix} - 3\begin{pmatrix} -2 \\ 5 \end{pmatrix} = \begin{pmatrix} 14 \\ 6 \end{pmatrix} - \begin{pmatrix} -6 \\ 15 \end{pmatrix} = \begin{pmatrix} 20 \\ -9 \end{pmatrix}$

(c) $|\mathbf{a} - \mathbf{b}| = |9\mathbf{i} - 2\mathbf{j}| = \sqrt{9^2 + (-2)^2} = \sqrt{85}$ ▲

The previous development readily generalizes to the three-dimensional case. Taking Cartesian axes x, y and z, any point in three-dimensional space can be represented by giving x, y and z coordinates (Figure 4.20). Denoting unit vectors along these axes by $\mathbf{i}$, $\mathbf{j}$ and $\mathbf{k}$, respectively, we can write the vector from O to $P(x, y, z)$ as:

■ $\overrightarrow{OP} = \mathbf{r} = x\mathbf{i} + y\mathbf{j} + z\mathbf{k}$

$$= \begin{pmatrix} x \\ y \\ z \end{pmatrix}$$

The vectors $\mathbf{i}$, $\mathbf{j}$ and $\mathbf{k}$ are orthogonal.

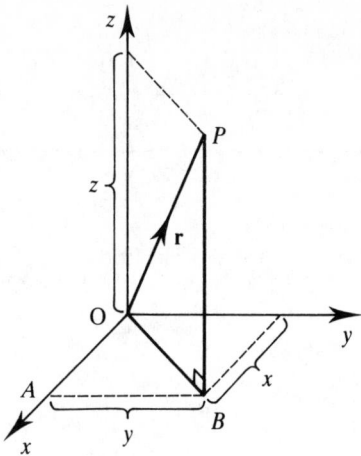

Figure 4.20   The quantity $|\mathbf{r}| = \sqrt{x^2 + y^2 + z^2}$.

Example 4.11

If $\mathbf{r} = x\mathbf{i} + y\mathbf{j} + z\mathbf{k}$ show that the modulus of $\mathbf{r}$ is $r = \sqrt{x^2 + y^2 + z^2}$.

Solution

Recalling Figure 4.20 we first calculate the length of OB. Now OAB is a right-angled triangle with perpendicular sides OA $= x$ and AB $= y$. Therefore by Pythagoras' theorem OB has length $\sqrt{x^2 + y^2}$. Then, applying Pythagoras' theorem to right-angled triangle OBP which has perpendicular sides OB and BP $= z$, we find:

$$|\mathbf{r}| = OP = \sqrt{OB^2 + BP^2}$$
$$= \sqrt{(x^2 + y^2) + z^2}$$
$$= \sqrt{x^2 + y^2 + z^2}$$

as required. ▲

■   If $\mathbf{r} = x\mathbf{i} + y\mathbf{j} + z\mathbf{k}$

then $|\mathbf{r}| = \sqrt{x^2 + y^2 + z^2}$

In three dimensions we have the following general result:

■   Given vectors $\mathbf{a} = \overrightarrow{OA} = a_1\mathbf{i} + a_2\mathbf{j} + a_3\mathbf{k}$ and $\mathbf{b} = \overrightarrow{OB} = b_1\mathbf{i} + b_2\mathbf{j} + b_3\mathbf{k}$, then

$$\overrightarrow{AB} = \mathbf{b} - \mathbf{a} = (b_1 - a_1)\mathbf{i} + (b_2 - a_2)\mathbf{j} + (b_3 - a_3)\mathbf{k}$$

and

$$|\vec{AB}| = |\mathbf{b} - \mathbf{a}| = |(b_1 - a_1)\mathbf{i} + (b_2 - a_2)\mathbf{j} + (b_3 - a_3)\mathbf{k}|$$
$$= \sqrt{(b_1 - a_1)^2 + (b_2 - a_2)^2 + (b_3 - a_3)^2}$$

Example 4.12

If $\mathbf{a} = 3\mathbf{i} - 2\mathbf{j} + \mathbf{k}$ and $\mathbf{b} = -2\mathbf{i} + \mathbf{j} - 5\mathbf{k}$, find:

(a) $|\mathbf{a}|$ (d) $\hat{\mathbf{b}}$

(b) $\hat{\mathbf{a}}$ (e) $\mathbf{b} - \mathbf{a}$

(c) $|\mathbf{b}|$ (f) $|\mathbf{b} - \mathbf{a}|$

Solutions

(a) $|\mathbf{a}| = \sqrt{3^2 + (-2)^2 + 1^2} = \sqrt{14}$

(b) $\hat{\mathbf{a}} = \dfrac{\mathbf{a}}{|\mathbf{a}|} = \dfrac{1}{\sqrt{14}}(3\mathbf{i} - 2\mathbf{j} + \mathbf{k}) = \dfrac{3}{\sqrt{14}}\mathbf{i} - \dfrac{2}{\sqrt{14}}\mathbf{j} + \dfrac{1}{\sqrt{14}}\mathbf{k}$

(c) $|\mathbf{b}| = \sqrt{(-2)^2 + 1^2 + (-5)^2} = \sqrt{30}$

(d) $\hat{\mathbf{b}} = \dfrac{\mathbf{b}}{|\mathbf{b}|} = \dfrac{1}{\sqrt{30}}(-2\mathbf{i} + \mathbf{j} - 5\mathbf{k}) = \dfrac{-2}{\sqrt{30}}\mathbf{i} + \dfrac{1}{\sqrt{30}}\mathbf{j} - \dfrac{5}{\sqrt{30}}\mathbf{k}$

(e) $\mathbf{b} - \mathbf{a} = -5\mathbf{i} + 3\mathbf{j} - 6\mathbf{k}$

(f) $|\mathbf{b} - \mathbf{a}| = \sqrt{(-5)^2 + 3^2 + (-6)^2} = \sqrt{70}$ ▲

4.3.1 *The zero vector*

A vector, all the components of which are zero, is called a **zero vector** and is denoted by **0** to distinguish it from the scalar 0. Clearly the zero vector has a length of 0; it is unusual in that it has arbitrary direction.

Example 4.13 Robot positions

Position vectors provide a useful means of determining the position of a robot. There are many different types of robot but a common type uses a series of rigid links connected together by flexible joints. Usually the mechanism is anchored at one point. A typical example is illustrated in Figure 4.21.

The anchor point is X and the tip of the robot is situated at point Y. The final link is sometimes called the hand of the robot. The hand often has rotating and gripping facilities and its size relative to the rest of the robot is usually quite small. Each of the robot links can be represented by a vector (see Figure 4.21). The vector **d** corresponds to the hand. A common

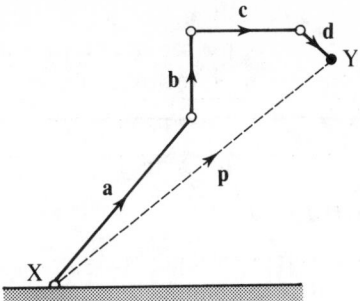

Figure 4.21 A typical robot configuration with vectors representing the robot links.

requirement in robotics is to be able to calculate the position of the tip of the hand to ensure it does not collide with other objects. This can be achieved by defining a set of Cartesian coordinates with origin at the anchor point of the robot, X. Each of the link vectors can then be represented in terms of these coordinates. For example, in the case of the robot in Figure 4.21:

$$\mathbf{a} = a_1\mathbf{i} + a_2\mathbf{j} + a_3\mathbf{k}$$

$$\mathbf{b} = b_1\mathbf{i} + b_2\mathbf{j} + b_3\mathbf{k}$$

$$\mathbf{c} = c_1\mathbf{i} + c_2\mathbf{j} + c_3\mathbf{k}$$

$$\mathbf{d} = d_1\mathbf{i} + d_2\mathbf{j} + d_3\mathbf{k}$$

The position of the tip of the hand can be calculated by adding together these vectors. So,

$$\mathbf{p} = \mathbf{a} + \mathbf{b} + \mathbf{c} + \mathbf{d}$$

$$= (a_1 + b_1 + c_1 + d_1)\mathbf{i} + (a_2 + b_2 + c_2 + d_2)\mathbf{j} + (a_3 + b_3 + c_3 + d_3)\mathbf{k} \quad \blacktriangle$$

4.3.2 *Linear combinations, dependence and independence*

Suppose we have two vectors **a** and **b**. If we form arbitrary scalar multiples of these, that is $k_1\mathbf{a}$ and $k_2\mathbf{b}$, and add these together, we obtain a new vector **c** where $\mathbf{c} = k_1\mathbf{a} + k_2\mathbf{b}$. The vector **c** is said to be a **linear combination** of **a** and **b**. Note that scalar multiplication and addition of vectors are the only operations allowed when forming a linear combination. Vector **c** is said to **depend linearly** on **a** and **b**. Of course we could also write:

$$\mathbf{a} = \frac{1}{k_1}\mathbf{c} - \frac{k_2}{k_1}\mathbf{b} \qquad \text{provided } k_1 \neq 0$$

so that **a** depends linearly on **c** and **b**. Provided $k_2 \neq 0$, then

$$\mathbf{b} = \frac{1}{k_2}\mathbf{c} - \frac{k_1}{k_2}\mathbf{a}$$

so that **b** depends linearly on **c** and **a**. The set of vectors $\{a, b, c\}$ is said to be **linearly dependent** and any one of the set can be written as a linear combination of the other two. In general, we have the following definition:

■ A set of n vectors $\{a_1, a_2, \ldots, a_n\}$ is linearly dependent if the expression

$$k_1 a_1 + k_2 a_2 + \cdots + k_n a_n = 0$$

can be satisfied by finding scalar constants $k_1, k_2, \ldots, k_n$, not all of which are zero. If the only way we can make the combination zero is by choosing all the k_is to be zero, then the given set of vectors is said to be **linearly independent**.

Example 4.14

Show that the vectors **i** and **j** are linearly independent.

Solution

We form the expression $k_1 \mathbf{i} + k_2 \mathbf{j} = 0$ and try to choose k_1 and k_2 so that the equation is satisfied. Using column vectors we have:

$$k_1 \begin{pmatrix} 1 \\ 0 \end{pmatrix} + k_2 \begin{pmatrix} 0 \\ 1 \end{pmatrix} = \begin{pmatrix} 0 \\ 0 \end{pmatrix}$$

that is

$$\begin{pmatrix} k_1 \\ 0 \end{pmatrix} + \begin{pmatrix} 0 \\ k_2 \end{pmatrix} = \begin{pmatrix} k_1 \\ k_2 \end{pmatrix} = \begin{pmatrix} 0 \\ 0 \end{pmatrix}$$

The only way we can satisfy the equation is by choosing $k_1 = 0$ and $k_2 = 0$ and hence we conclude that the vectors **i** and **j** are linearly independent. Geometrically, we note that since they are perpendicular, no scalar multiple of **i** can give **j** and vice versa. ▲

Example 4.15

The vectors

$$\begin{pmatrix} 1 \\ 2 \\ 3 \end{pmatrix} \quad \begin{pmatrix} 5 \\ 1 \\ 9 \end{pmatrix} \quad \begin{pmatrix} 13 \\ -1 \\ 21 \end{pmatrix}$$

are linearly dependent because, for example,

$$3 \begin{pmatrix} 5 \\ 1 \\ 9 \end{pmatrix} - 2 \begin{pmatrix} 1 \\ 2 \\ 3 \end{pmatrix} - 1 \begin{pmatrix} 13 \\ -1 \\ 21 \end{pmatrix} = \begin{pmatrix} 0 \\ 0 \\ 0 \end{pmatrix} \quad ▲$$

EXERCISES 4.2

1. P and Q lie in the x–y plane. Find $\overrightarrow{PQ}$, where P is the point with coordinates $(5, 1)$ and Q is the point with coordinates $(-1, 4)$. Find $|\overrightarrow{PQ}|$.

2. A and B lie in the x–y plane. If A is the point $(3, 4)$ and B is the point $(1, -5)$ write down the vectors $\overrightarrow{OA}$, $\overrightarrow{OB}$ and $\overrightarrow{AB}$. Find a unit vector in the direction of $\overrightarrow{AB}$.

3. If $\mathbf{a} = 4\mathbf{i} - \mathbf{j} + 3\mathbf{k}$ and $\mathbf{b} = -2\mathbf{i} + 2\mathbf{j} - \mathbf{k}$, find unit vectors in the directions of $\mathbf{a}$, $\mathbf{b}$ and $\mathbf{b} - \mathbf{a}$.

4. If $\mathbf{a} = 5\mathbf{i} - 2\mathbf{j}$, $\mathbf{b} = 3\mathbf{i} - 7\mathbf{j}$ and $\mathbf{c} = -3\mathbf{i} + 4\mathbf{j}$, express, in terms of $\mathbf{i}$ and $\mathbf{j}$,

$$\mathbf{a} + \mathbf{b} \qquad \mathbf{a} + \mathbf{c} \qquad \mathbf{c} - \mathbf{b} \qquad 3\mathbf{c} - 4\mathbf{b}$$

Draw diagrams to illustrate your results. Repeat the calculations using column vector notation.

5. Write down a unit vector which is parallel to the line $y = 7x - 3$.

6. Find $\overrightarrow{PQ}$ where P is the point in three-dimensional space with coordinates $(4, 1, 3)$ and Q is the point with coordinates $(1, 2, 4)$. Find the distance between P and Q. Further, find the position vector of the point dividing PQ in the ratio $1:3$.

7. If P, Q and R have coordinates $(3, 2, 1)$, $(2, 1, 2)$ and $(1, 3, 3)$, respectively, use vectors to determine which pair of points are closest to each other.

8. Consider the robot of Example 4.13. The link vectors have the following values:

$$\mathbf{a} = 12\mathbf{i} + 18\mathbf{j} + \mathbf{k}$$

$$\mathbf{b} = 6\mathbf{i} - 3\mathbf{j} + 8\mathbf{k}$$

$$\mathbf{c} = 3\mathbf{i} + 2\mathbf{j} - 4\mathbf{k}$$

$$\mathbf{d} = 0.5\mathbf{i} - 0.2\mathbf{j} + 0.6\mathbf{k}$$

Calculate the length of each of the links and the position vector of the tip of the robot.

9. Show that the vectors $\mathbf{a} = \mathbf{i} + \mathbf{j}$ and $\mathbf{b} = -\mathbf{i} + \mathbf{j}$ are linearly independent.

10. Show that the vectors $\mathbf{i}$, $\mathbf{j}$ and $\mathbf{k}$ are linearly independent.

4.4 The scalar product

Given any two vectors $\mathbf{a}$ and $\mathbf{b}$, there are two ways in which we can define their product. These are known as the scalar product and the vector product. As the names suggest, the result of finding a scalar product is a scalar whereas the result of finding a vector product is a vector. The **scalar product** of $\mathbf{a}$ and $\mathbf{b}$ is written as:

$\mathbf{a} \cdot \mathbf{b}$

This notation gives rise to the alternative name **dot product**. It is defined by the formula

---

■ $\mathbf{a} \cdot \mathbf{b} = |\mathbf{a}||\mathbf{b}| \cos \theta$

---

where θ is the angle between the two vectors as shown in Figure 4.22.

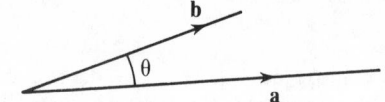

Figure 4.22 Two vectors **a** and **b** separated by angle θ.

From the definition of the scalar product, it is possible to show that the following rules hold.

■ $\mathbf{a} \cdot \mathbf{b} = \mathbf{b} \cdot \mathbf{a}$ the scalar product is commutative

     $k(\mathbf{a} \cdot \mathbf{b}) = (k\mathbf{a} \cdot \mathbf{b})$ where k is a scalar

     $(\mathbf{a} + \mathbf{b}) \cdot \mathbf{c} = (\mathbf{a} \cdot \mathbf{c}) + (\mathbf{b} \cdot \mathbf{c})$ the distributive rule

---

It is important at this stage to realize that notation is very important in vector work. You should not use a $\times$ to denote the scalar product because this is the symbol we shall use for the vector product.

Example 4.16

If **a** and **b** are parallel vectors, show that $\mathbf{a} \cdot \mathbf{b} = |\mathbf{a}||\mathbf{b}|$. If **a** and **b** are orthogonal show that their scalar product is zero.

Solution

If **a** and **b** are parallel then the angle between them is zero. Therefore $\mathbf{a} \cdot \mathbf{b} = |\mathbf{a}||\mathbf{b}| \cos 0° = |\mathbf{a}||\mathbf{b}|$. If **a** and **b** are orthogonal, then the angle between them is $90°$ and $\mathbf{a} \cdot \mathbf{b} = |\mathbf{a}||\mathbf{b}| \cos 90° = 0$.

Similarly we can show that if **a** and **b** are two non-zero vectors for which $\mathbf{a} \cdot \mathbf{b} = 0$, then **a** and **b** must be orthogonal. ▲

■   If **a** and **b** are parallel vectors, $\mathbf{a} \cdot \mathbf{b} = |\mathbf{a}||\mathbf{b}|$.

     If **a** and **b** are orthogonal vectors, $\mathbf{a} \cdot \mathbf{b} = 0$

An immediate consequence of the previous result is the following useful set of formulae:

■ $\mathbf{i} \cdot \mathbf{i} = \mathbf{j} \cdot \mathbf{j} = \mathbf{k} \cdot \mathbf{k} = 1$

     $\mathbf{i} \cdot \mathbf{j} = \mathbf{j} \cdot \mathbf{k} = \mathbf{k} \cdot \mathbf{i} = 0$

Example 4.17

If $\mathbf{a} = a_1\mathbf{i} + a_2\mathbf{j} + a_3\mathbf{k}$ and $\mathbf{b} = b_1\mathbf{i} + b_2\mathbf{j} + b_3\mathbf{k}$ show that $\mathbf{a} \cdot \mathbf{b} = a_1 b_1 + a_2 b_2 + a_3 b_3$.

Solution

We have

$$\mathbf{a} \cdot \mathbf{b} = (a_1\mathbf{i} + a_2\mathbf{j} + a_3\mathbf{k}) \cdot (b_1\mathbf{i} + b_2\mathbf{j} + b_3\mathbf{k})$$

$$= a_1\mathbf{i} \cdot (b_1\mathbf{i} + b_2\mathbf{j} + b_3\mathbf{k}) + a_2\mathbf{j} \cdot (b_1\mathbf{i} + b_2\mathbf{j} + b_3\mathbf{k})$$

$$+ a_3\mathbf{k} \cdot (b_1\mathbf{i} + b_2\mathbf{j} + b_3\mathbf{k})$$

$$= a_1b_1\mathbf{i} \cdot \mathbf{i} + a_1b_2\mathbf{i} \cdot \mathbf{j} + a_1b_3\mathbf{i} \cdot \mathbf{k} + a_2b_1\mathbf{j} \cdot \mathbf{i} + a_2b_2\mathbf{j} \cdot \mathbf{j} + a_2b_3\mathbf{j} \cdot \mathbf{k}$$

$$+ a_3b_1\mathbf{k} \cdot \mathbf{i} + a_3b_2\mathbf{k} \cdot \mathbf{j} + a_3b_3\mathbf{k} \cdot \mathbf{k}$$

$$= a_1b_1 + a_2b_2 + a_3b_3$$

as required. Thus, given two vectors in component form their scalar product is the sum of the products of corresponding components. ▲

The result developed in Example 4.17 is important and should be memorized:

■ If $\mathbf{a} = a_1\mathbf{i} + a_2\mathbf{j} + a_3\mathbf{k}$ and $\mathbf{b} = b_1\mathbf{i} + b_2\mathbf{j} + b_3\mathbf{k}$, then

$$\mathbf{a} \cdot \mathbf{b} = a_1b_1 + a_2b_2 + a_3b_3$$

Example 4.18

If $\mathbf{a} = a_1\mathbf{i} + a_2\mathbf{j} + a_3\mathbf{k}$, find:

(a) $\mathbf{a} \cdot \mathbf{a}$

(b) $|\mathbf{a}|^2$.

Solutions

(a) Using the previous result we find:

$$\mathbf{a} \cdot \mathbf{a} = (a_1\mathbf{i} + a_2\mathbf{j} + a_3\mathbf{k}) \cdot (a_1\mathbf{i} + a_2\mathbf{j} + a_3\mathbf{k})$$

$$= a_1^2 + a_2^2 + a_3^2$$

(b) From Example 4.11 we know that the modulus of $\mathbf{r} = x\mathbf{i} + y\mathbf{j} + z\mathbf{k}$ is $\sqrt{x^2 + y^2 + z^2}$ and therefore,

$$|\mathbf{a}| = \sqrt{a_1^2 + a_2^2 + a_3^2}$$

so that,

$$|\mathbf{a}|^2 = a_1^2 + a_2^2 + a_3^2 \quad ▲$$

We note the general result that:

■ $a \cdot a = |a|^2$

Example 4.19

If $a = 3i + j - k$ and $b = 2i + j + 2k$ find $a \cdot b$ and the angle between a and b.

Solution

We have

$$a \cdot b = (3)(2) + (1)(1) + (-1)(2)$$
$$= 6 + 1 - 2$$
$$= 5$$

Furthermore, from the definition of the scalar product $a \cdot b = |a||b| \cos \theta$. Now,

$$|a| = \sqrt{9 + 1 + 1} = \sqrt{11} \qquad \text{and} \qquad |b| = \sqrt{4 + 1 + 4} = 3$$

Therefore,

$$\cos \theta = \frac{a \cdot b}{|a||b|} = \frac{5}{3\sqrt{11}}$$

from which we deduce that $\theta = 59.8°$ or 1.04 radians.   ▲

Example 4.20 Movement of a charged particle in an electric field

Figure 4.23 shows two charged plates situated in a vacuum. Plate A has an excess of positive charge, while plate B has an excess of negative charge.

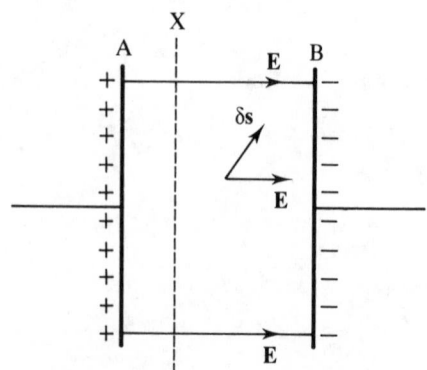

Figure 4.23   Two charged plates situated in a vacuum.

Such an arrangement gives rise to an electric field. An electric field is an example of a **vector field**. Vector fields are dealt with more generally in Chapter 14. Mathematically, a vector field is a region of space with a vector defined at each point. The concept is useful for analysing physical problems in which the direction as well as the strength of a quantity is important.

In Figure 4.23 the electric field vector **E**, in the region of space between the charged plates has a direction perpendicular to the plates pointing from A to B. The magnitude of the electric field vector is constant in this region if end effects are ignored. If a charged particle is required to move against an electric field, then work must be done to achieve this. For example, to transport a positively charged particle from the surface of plate B to the surface of plate A would require work to be done. This would lead to an increase in potential of the charged particle.

If **s** represents the displacement and V the potential it is conventional to write δ**s** to represent a very small change in displacement, and δV to represent a very small change in potential.

If a unit charge is moved a small displacement δ**s** in an electric field (Figure 4.23) then the change in potential δV is given by:

$$\delta V = -\mathbf{E} \cdot \delta \mathbf{s} \tag{4.1}$$

This is an example of the use of a scalar product. Notice that although **E** and δ**s** are vector quantities the change in potential, δV, is a scalar.

Consider again the charged plates of Figure 4.23. If a unit charge is moved a small displacement along the plane X, then δ**s** is perpendicular to **E**. So,

$$\delta V = -\mathbf{E} \cdot \delta \mathbf{s} = -|\mathbf{E}||\delta \mathbf{s}| \cos \theta$$

With $\theta = 90°$, we find $\delta V = 0$. The surface X is known as an **equipotential surface** because movement of a charged particle along this surface does not give rise to a change in its potential. Movement of a charge in a direction parallel to the electric field gives rise to the maximum change in potential, as for this case $\theta = 0°$. ▲

EXERCISES 4.3

1. If $\mathbf{a} = 3\mathbf{i} - 7\mathbf{j}$ and $\mathbf{b} = 2\mathbf{i} + 4\mathbf{j}$ find $\mathbf{a} \cdot \mathbf{b}$, $\mathbf{b} \cdot \mathbf{a}$, $\mathbf{a} \cdot \mathbf{a}$ and $\mathbf{b} \cdot \mathbf{b}$.

2. Evaluate $(-13\mathbf{i} - 5\mathbf{j}) \cdot (-3\mathbf{i} + 4\mathbf{j})$.

3. Find the angle between the vectors $4\mathbf{i} - 2\mathbf{j}$ and $3\mathbf{i} - 3\mathbf{j}$.

4. If $\mathbf{a} = 7\mathbf{i} + 8\mathbf{j}$ and $\mathbf{b} = 5\mathbf{i}$ find $\mathbf{a} \cdot \hat{\mathbf{b}}$.

5. If $\mathbf{r}_1 = \begin{pmatrix} 3 \\ 1 \\ 2 \end{pmatrix}$ and $\mathbf{r}_2 = \begin{pmatrix} 5 \\ 1 \\ 0 \end{pmatrix}$ find $\mathbf{r}_1 \cdot \mathbf{r}_1$, $\mathbf{r}_1 \cdot \mathbf{r}_2$ and $\mathbf{r}_2 \cdot \mathbf{r}_2$.

6. Given that $\mathbf{p} = 2\mathbf{q}$ simplify $\mathbf{p} \cdot \mathbf{q}$, $(\mathbf{p} + 5\mathbf{q}) \cdot \mathbf{q}$, and $(\mathbf{q} - \mathbf{p}) \cdot \mathbf{p}$.

7. Find the modulus of $\mathbf{a} = \mathbf{i} - \mathbf{j} - \mathbf{k}$, the modulus of $\mathbf{b} = 2\mathbf{i} + \mathbf{j} + 2\mathbf{k}$ and the scalar product $\mathbf{a} \cdot \mathbf{b}$. Deduce the angle between $\mathbf{a}$ and $\mathbf{b}$.

8. If $\mathbf{a} = 2\mathbf{i} + 2\mathbf{j} - \mathbf{k}$ and $\mathbf{b} = 3\mathbf{i} - 6\mathbf{j} + 2\mathbf{k}$, find $|\mathbf{a}|$, $|\mathbf{b}|$, $\mathbf{a} \cdot \mathbf{b}$ and the angle between $\mathbf{a}$ and $\mathbf{b}$.

9. Use a vector method to show that the diagonals of the rhombus shown in Figure 4.24 intersect at 90°.

10. Use the scalar product to find a two-dimensional vector $\mathbf{a} = a_1\mathbf{i} + a_2\mathbf{j}$ perpendicular to the vector $4\mathbf{i} - 2\mathbf{j}$.

11. If $\mathbf{a} = 3\mathbf{i} - 2\mathbf{j}$, $\mathbf{b} = 7\mathbf{i} + 5\mathbf{j}$ and $\mathbf{c} = 9\mathbf{i} - \mathbf{j}$, show that $\mathbf{a} \cdot (\mathbf{b} - \mathbf{c}) = (\mathbf{a} \cdot \mathbf{b}) - (\mathbf{a} \cdot \mathbf{c})$.

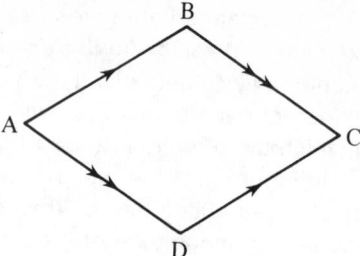

Figure 4.24   The rhombus ABCD.

4.5 The vector product

The result of finding the vector product of $\mathbf{a}$ and $\mathbf{b}$ is a vector of length $|\mathbf{a}||\mathbf{b}|\sin\theta$, where θ is the angle between $\mathbf{a}$ and $\mathbf{b}$. The direction of this vector is such that it is perpendicular to $\mathbf{a}$ and to $\mathbf{b}$, and so it is perpendicular to the plane containing $\mathbf{a}$ and $\mathbf{b}$ (Figure 4.25). There are, however, two possible directions for this vector, but it is conventional to choose the one associated with the application of the right-handed screw rule. Imagine turning a right-handed screw in the sense from $\mathbf{a}$ towards $\mathbf{b}$ as shown. A right-handed screw is one which, when turned clockwise, enters the material into which it is being screwed. The direction in which the screw advances is the direction of the required vector product. The symbol we shall use to denote the vector product is $\times$. Formally, we write,

---

■ $\mathbf{a} \times \mathbf{b} = |\mathbf{a}||\mathbf{b}|\sin\theta\,\hat{\mathbf{e}}$

---

where $\hat{\mathbf{e}}$ is the unit vector required to define the appropriate direction, that is $\hat{\mathbf{e}}$ is a unit vector perpendicular to $\mathbf{a}$ and to $\mathbf{b}$ in a sense defined by the right-handed screw rule. To evaluate $\mathbf{b} \times \mathbf{a}$ we must imagine turning the screw from the direction of $\mathbf{b}$ towards that of $\mathbf{a}$. The screw will advance as shown in Figure 4.26.

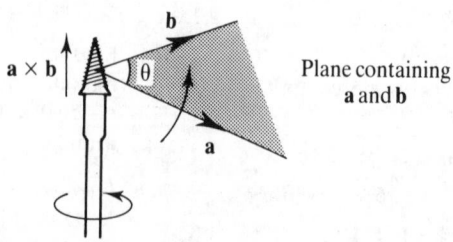

Figure 4.25 $\mathbf{a} \times \mathbf{b}$ is perpendicular to the plane containing $\mathbf{a}$ and $\mathbf{b}$. The right-handed screw rule allows the direction of $\mathbf{a} \times \mathbf{b}$ to be found.

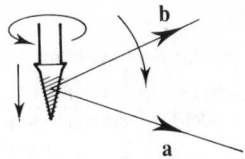

Figure 4.26   Right-handed screw rule allows the direction of **b** × **a** to be found.

We notice immediately that $\mathbf{a} \times \mathbf{b} \neq \mathbf{b} \times \mathbf{a}$ since their directions are different. From the definition of the vector product, it is possible to show that the following rules hold:

■ $\mathbf{a} \times \mathbf{b} = -(\mathbf{b} \times \mathbf{a})$ the vector product is not commutative

  $\mathbf{a} \times (\mathbf{b} + \mathbf{c}) = (\mathbf{a} \times \mathbf{b}) + (\mathbf{a} \times \mathbf{c})$ the distributive rule

  $k(\mathbf{a} \times \mathbf{b}) = (k\mathbf{a}) \times \mathbf{b} = \mathbf{a} \times (k\mathbf{b})$ where k is a scalar

---

Example 4.21

If **a** and **b** are parallel show that $\mathbf{a} \times \mathbf{b} = \mathbf{0}$.

Solution

If **a** and **b** are parallel then the angle between **a** and **b** is zero. Therefore, $\mathbf{a} \times \mathbf{b} = |\mathbf{a}||\mathbf{b}| \sin 0\,\hat{\mathbf{e}} = \mathbf{0}$. Note that the answer is still a vector, and that we denote the zero vector $0\mathbf{i} + 0\mathbf{j} + 0\mathbf{k}$ by **0**, to distinguish it from the scalar 0. In particular, we note that

    $\mathbf{i} \times \mathbf{i} = \mathbf{j} \times \mathbf{j} = \mathbf{k} \times \mathbf{k} = \mathbf{0}$ ▲

■ If **a** and **b** are parallel, then

   $\mathbf{a} \times \mathbf{b} = \mathbf{0}$

In particular:

■ $\mathbf{i} \times \mathbf{i} = \mathbf{j} \times \mathbf{j} = \mathbf{k} \times \mathbf{k} = \mathbf{0}$

Example 4.22

Show that $\mathbf{i} \times \mathbf{j} = \mathbf{k}$ and find expressions for $\mathbf{j} \times \mathbf{k}$ and $\mathbf{k} \times \mathbf{i}$.

Solution

We note that $\mathbf{i}$ and $\mathbf{j}$ are perpendicular so that $|\mathbf{i} \times \mathbf{j}| = |\mathbf{i}||\mathbf{j}| \sin 90° = 1$. Furthermore, the unit vector perpendicular to $\mathbf{i}$ and to $\mathbf{j}$ in the sense defined by the right-handed screw rule is $\mathbf{k}$. Therefore, $\mathbf{i} \times \mathbf{j} = \mathbf{k}$ as required. Similarly you should be able to show that $\mathbf{j} \times \mathbf{k} = \mathbf{i}$ and $\mathbf{k} \times \mathbf{i} = \mathbf{j}$. ▲

■ $\mathbf{i} \times \mathbf{j} = \mathbf{k}$ $\qquad$ $\mathbf{j} \times \mathbf{k} = \mathbf{i}$ $\qquad$ $\mathbf{k} \times \mathbf{i} = \mathbf{j}$

$\mathbf{j} \times \mathbf{i} = -\mathbf{k}$ $\qquad$ $\mathbf{k} \times \mathbf{j} = -\mathbf{i}$ $\qquad$ $\mathbf{i} \times \mathbf{k} = -\mathbf{j}$

Example 4.23

Simplify $(\mathbf{a} \times \mathbf{b}) - (\mathbf{b} \times \mathbf{a})$.

Solution

Use the result $\mathbf{a} \times \mathbf{b} = -(\mathbf{b} \times \mathbf{a})$ to obtain:

$$(\mathbf{a} \times \mathbf{b}) - (\mathbf{b} \times \mathbf{a}) = (\mathbf{a} \times \mathbf{b}) - (-(\mathbf{a} \times \mathbf{b}))$$
$$= (\mathbf{a} \times \mathbf{b}) + (\mathbf{a} \times \mathbf{b})$$
$$= 2(\mathbf{a} \times \mathbf{b}) \quad ▲$$

Example 4.24

(a) If $\mathbf{a} = a_1\mathbf{i} + a_2\mathbf{j} + a_3\mathbf{k}$ and $\mathbf{b} = b_1\mathbf{i} + b_2\mathbf{j} + b_3\mathbf{k}$, show that:

$$\mathbf{a} \times \mathbf{b} = (a_2 b_3 - a_3 b_2)\mathbf{i} - (a_1 b_3 - a_3 b_1)\mathbf{j} + (a_1 b_2 - a_2 b_1)\mathbf{k}$$

(b) If $\mathbf{a} = 2\mathbf{i} + \mathbf{j} + 3\mathbf{k}$ and $\mathbf{b} = 3\mathbf{i} + 2\mathbf{j} + \mathbf{k}$ find $\mathbf{a} \times \mathbf{b}$

Solutions

(a) $\mathbf{a} \times \mathbf{b} = (a_1\mathbf{i} + a_2\mathbf{j} + a_3\mathbf{k}) \times (b_1\mathbf{i} + b_2\mathbf{j} + b_3\mathbf{k})$

$\qquad = a_1\mathbf{i} \times (b_1\mathbf{i} + b_2\mathbf{j} + b_3\mathbf{k})$

$\qquad\quad + a_2\mathbf{j} \times (b_1\mathbf{i} + b_2\mathbf{j} + b_3\mathbf{k})$

$\qquad\quad + a_3\mathbf{k} \times (b_1\mathbf{i} + b_2\mathbf{j} + b_3\mathbf{k})$

$\qquad = a_1 b_1(\mathbf{i} \times \mathbf{i}) + a_1 b_2(\mathbf{i} \times \mathbf{j}) + a_1 b_3(\mathbf{i} \times \mathbf{k})$

$\qquad\quad + a_2 b_1(\mathbf{j} \times \mathbf{i}) + a_2 b_2(\mathbf{j} \times \mathbf{j}) + a_2 b_3(\mathbf{j} \times \mathbf{k})$

$\qquad\quad + a_3 b_1(\mathbf{k} \times \mathbf{i}) + a_3 b_2(\mathbf{k} \times \mathbf{j}) + a_3 b_3(\mathbf{k} \times \mathbf{k})$

Using the results of Examples 4.21 and 4.22, we find that the expression for $\mathbf{a} \times \mathbf{b}$ simplifies to:

$$\mathbf{a} \times \mathbf{b} = (a_2 b_3 - a_3 b_2)\mathbf{i} - (a_1 b_3 - a_3 b_1)\mathbf{j} + (a_1 b_2 - a_2 b_1)\mathbf{k}$$

(b) Using the result of part (a) we have

$$\mathbf{a} \times \mathbf{b} = ((1)(1) - (3)(2))\mathbf{i} - ((2)(1) - (3)(3))\mathbf{j} + ((2)(2) - (1)(3))\mathbf{k}$$

$$= -5\mathbf{i} + 7\mathbf{j} + \mathbf{k}$$

Verify for yourself that $\mathbf{b} \times \mathbf{a} = 5\mathbf{i} - 7\mathbf{j} - \mathbf{k}$. ▲

Readers already familiar with the manipulation of determinants will realize that $\mathbf{a} \times \mathbf{b}$ can conveniently be written as:

$$\mathbf{a} \times \mathbf{b} = \begin{vmatrix} \mathbf{i} & \mathbf{j} & \mathbf{k} \\ a_1 & a_2 & a_3 \\ b_1 & b_2 & b_3 \end{vmatrix}$$

We shall deal with this approach in Chapter 5.

■ If $\mathbf{a} = a_1\mathbf{i} + a_2\mathbf{j} + a_3\mathbf{k}$ and $\mathbf{b} = b_1\mathbf{i} + b_2\mathbf{j} + b_3\mathbf{k}$, then

$$\mathbf{a} \times \mathbf{b} = (a_2 b_3 - a_3 b_2)\mathbf{i} - (a_1 b_3 - a_3 b_1)\mathbf{j} + (a_1 b_2 - a_2 b_1)\mathbf{k}$$

$$= \begin{vmatrix} \mathbf{i} & \mathbf{j} & \mathbf{k} \\ a_1 & a_2 & a_3 \\ b_1 & b_2 & b_3 \end{vmatrix}$$

Example 4.25 Force on a charged particle moving in a magnetic field

It is possible to model the effect of magnetism by means of a vector field. The magnetic field vector has the symbol **B**. Consider Figure 4.27. If a particle with charge q, moves with velocity, **v**, in a magnetic field, **B**, it experiences a force due to the field. The equation for this force is:

$$\mathbf{F} = q\mathbf{v} \times \mathbf{B}$$

where $\mathbf{v}$ = velocity of the particle (m s^{-1}), $\mathbf{B}$ = magnetic field (T), q = quantity of charge (C), and $\mathbf{F}$ = force exerted on the charged particle (N).

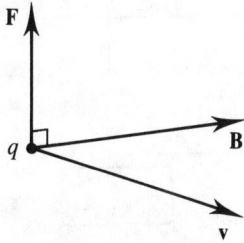

Figure 4.27 Force, **F**, exerted on a particle with charge, q, when moving with velocity, **v**, in a magnetic field, **B**.

Notice that the force exerted is a vector quantity and is obtained by evaluating a vector product. The direction of the force is perpendicular to the velocity and magnetic field vectors, and is determined using the right-handed screw rule.

This is a very useful formula as it can be utilized to calculate the forces exerted on conductors in an electric motor. It can also be used to analyse electricity generators in which the motion of conductors in a magnetic field leads to the movement of charge within the conductors, thus generating electricity. ▲

Example 4.26 The Hall effect in a semiconductor

A frequent requirement in the semiconductor industry is to be able to measure the density of holes in a p-type semiconductor and the density of electrons in an n-type semiconductor. This can be achieved by using the **Hall effect**. We will consider the case of a p-type semiconductor but the derivation for an n-type semiconductor is similar.

Consider the piece of semiconductor shown in Figure 4.28. A d.c. voltage, V, is applied to the ends of the semiconductor. This gives rise to a flow of current composed mainly of holes as they are the majority carriers for a p-type semiconductor. This current can be represented by a vector pointing in the x direction and denoted by **I**. A magnetic field, **B**, is applied to the semiconductor in the y direction. The moving holes experience a force, $\mathbf{F_B}$, per unit volume, caused by the magnetic field given by:

$$\mathbf{F_B} = \frac{1}{A}\mathbf{I} \times \mathbf{B}$$

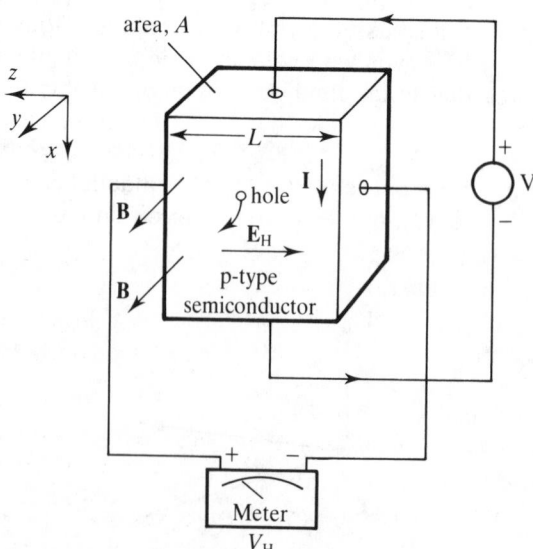

Figure 4.28   Hall effect in a p-type semiconductor.

where A is the cross-sectional area of the semiconductor. This causes the holes to drift in the z direction and so causes an excess of positive charge to appear on one side of the semiconductor. This gives rise to a voltage known as the Hall voltage, V_H. As this excess charge builds up it creates an electric field, E_H, in the negative z direction, which in turn exerts an opposing force on the holes. This force is given by $F_E = qp_0E_H$ where q = elementary charge = 1.60×10^{-19} C, and p_0 = density of holes (holes per metre^3). Equilibrium is reached when the two forces are equal in magnitude, that is $|F_B| = |F_E|$. Now,

$$|F_E| = qp_0|E_H| \qquad |F_B| = \frac{|I \times B|}{A} = \frac{IB}{A}$$

In equilibrium the magnitude of the electric field, $|E_H|$, is constant and so we can write $|E_H| = V_H/L$, where L is the width of the semiconductor. Hence,

$$\frac{IB}{A} = qp_0\frac{V_H}{L}$$

so that

$$p_0 = \frac{BIL}{V_HqA}$$

So, by measuring the value of the Hall voltage, it is possible to calculate the density of the holes, p_0, in the semiconductor. ▲

EXERCISES 4.4

1. If $a = i - 2j + 3k$ and $b = 2i - j - k$, find:

   (a) $a \times b$

   (b) $b \times a$

2. If $a = i - 2j$ and $b = 5i + 5k$ find $a \times b$.

3. If $a = i + j - k$, $b = i - j$ and $c = 2i + k$ find:

   (a) $(a \times b) \times c$

   (b) $a \times (b \times c)$

4. If $p = 6i + 7j - 2k$ and $q = 3i - j + 4k$ find $|p|$, $|q|$ and $|p \times q|$. Deduce the sine of the angle between p and q.

5. For arbitrary vectors p and q simplify:

   (a) $(p + q) \times p$

   (b) $(p + q) \times (p - q)$

6. If $c = i + j$ and $d = 2i + k$, find a unit vector perpendicular to both c and d. Further, find the sine of the angle between c and d.

7. A, B, C are the points with coordinates $(1, 2, 3)$, $(3, 2, 1)$ and $(-1, 1, 0)$, respectively. Find a unit vector perpendicular to the plane containing A, B and C.

8. If $a = 7i - 2j - 5k$ and $b = 5i + j + 3k$, find a vector perpendicular to a and b.

9. If $a = 7i - j + k$, $b = 3i - j - 2k$ and $c = 9i + j - 3k$, show that:

   $$a \times (b + c) = (a \times b) + (a \times c)$$

4.6 Vectors of *n*-dimensions

The examples we have discussed have all concerned two and three-dimensional vectors. Our understanding has been helped by the fact that two-dimensional vectors can be drawn in the plane of the paper and three-dimensional vectors can be visualized in the three-dimensional space in which we live. However, there are some situations when the generalization to higher dimensions is appropriate, but no convenient geometrical interpretation is available. Nevertheless, many of the concepts we have discussed are still applicable. For example, we can introduce the four-dimensional vectors:

$$\mathbf{a} = \begin{pmatrix} 3 \\ 1 \\ 2 \\ 4 \end{pmatrix} \quad \text{and} \quad \mathbf{b} = \begin{pmatrix} 1 \\ 0 \\ 3 \\ 1 \end{pmatrix}$$

It is natural to define the magnitude, or **norm**, of $\mathbf{a}$ as $\sqrt{3^2 + 1^2 + 2^2 + 4^2} = \sqrt{30}$ and the scalar product of $\mathbf{a}$ and $\mathbf{b}$ as $\mathbf{a} \cdot \mathbf{b} = (3)(1) + (1)(0) + (2)(3) + (4)(1) = 13$. An *n*-dimensional vector will have *n* components. Operations such as addition, subtraction and scalar multiplication are defined in an obvious way.

It is also possible to define a set of variables as a vector. This turns out to be a useful way of modelling a physical system. The system is described by means of a vector which consists of an ordered set of variables sufficient to describe the state of the system. Such a vector is called a **state vector**. This concept is explored in more detail in Chapter 10.

Example 4.27 Mesh current vector

A circuit has a set of mesh currents $\{I_1, I_2, I_3, I_4\}$ from which we can form a current vector

$$\mathbf{I} = \begin{pmatrix} I_1 \\ I_2 \\ I_3 \\ I_4 \end{pmatrix}$$

No geometrical interpretation is possible but nevertheless this vector provides a useful mathematical way of handling the mesh currents. We shall see how vectors such as these can be manipulated in Section 5.11. ▲

EXERCISES 4.5

1. If

$$a = \begin{pmatrix} 1 \\ 1 \\ 0 \\ 1 \\ 1 \end{pmatrix} \quad \text{and} \quad b = \begin{pmatrix} 3 \\ 2 \\ 1 \\ 0 \\ 1 \end{pmatrix}$$

   find the norm of **a**, the norm of **b** and **a** · **b**. Further, find the norm of **a** − **b**.

2. Two non-zero vectors are mutually orthogonal if their scalar product is zero. Determine which of the following are mutually orthogonal.

$$a = \begin{pmatrix} 1 \\ 2 \\ 4 \\ -1 \end{pmatrix} \quad b = \begin{pmatrix} 2 \\ 1 \\ 0 \\ 0 \end{pmatrix}$$

$$c = \begin{pmatrix} 3 \\ 0 \\ 1 \\ 0 \end{pmatrix} \quad d = \begin{pmatrix} 0 \\ 0 \\ 7 \\ 2 \end{pmatrix}$$

$$e = \begin{pmatrix} 3 \\ 0 \\ -2 \\ -5 \end{pmatrix}$$

Miscellaneous exercises

1. Find **a** · **b** and **a** × **b** when

   (a) **a** = 7**i** − **j** + **k**, **b** = 3**i** + 2**j** + 5**k**
   (b) **a** = 6**i** − 6**j** − 6**k**, **b** = **i** − **j** − **k**.

2. For the triangle ABC, express as simply as possible the vector $\overrightarrow{AB} + \overrightarrow{BC} + \overrightarrow{CA}$.

3. If **a** = 7**i** − **j** + 2**k** and **b** = 8**i** + **j** + **k**, find |**a**|, |**b**| and **a** · **b**. Deduce the cosine of the angle between **a** and **b**.

4. If **a** = 6**i** − **j** + 2**k** and **b** = 3**i** − **j** + 3**k**, find |**a**|, |**b**|, |**a** × **b**|. Deduce the sine of the angle between **a** and **b**.

5. If **a** = 7**i** + 9**j** − 3**k** and **b** = 2**i** − 4**j**, find **â**, **b̂**, $\widehat{a \times b}$.

6. By combining the scalar and vector products other types of products can be defined. The **triple scalar product** for three vectors is defined as (**a** × **b**) · **c** which is a scalar. If **a** = 3**i** − **j** + 2**k**, **b** = 2**i** − 2**j** − **k**, **c** = 3**i** + **j**, find **a** × **b** and (**a** × **b**) · **c**. Show that (**a** × **b**) · **c** = **a** · (**b** × **c**).

7. The **triple vector product** is defined by (**a** × **b**) × **c**. Find the triple vector product of the vectors given in Question 6. Also find **a** · **c**, **b** · **c** and verify that:

$$(a \cdot c)b - (b \cdot c)a = (a \times b) \times c$$

   Further, find **a** × (**b** × **c**) and confirm that **a** × (**b** × **c**) ≠ (**a** × **b**) × **c**.

8. Show that the vectors **p** = 3**i** − 2**j** + **k**, **q** = 2**i** + **j** − 4**k** and **r** = **i** − 3**j** + 5**k** form the three sides of a right-angled triangle.

9. Find a unit vector parallel to the line $y = 7x − 3$. Find a unit vector parallel to $y = 2x + 7$. Use the scalar product to find the angle between these two lines.

10. An electric charge q which moves with a velocity **v** produces a magnetic field **B** given by:

$$B = \frac{\mu q}{4\pi} \frac{v \times r}{|r|^2} \quad \text{where } \mu = \text{constant.}$$

   Find **B** if **r** = 3**i** + **j** − 2**k** and **v** = **i** − 2**j** + 3**k**.

11. In the triangle ABC, denote $\overrightarrow{AB}$ by **c**, $\overrightarrow{AC}$ by **b** and $\overrightarrow{CB}$ by **a**. Use the scalar product to prove the cosine rule:

$$a^2 = b^2 + c^2 − 2bc \cos A.$$

Matrix algebra

5

5.1 Introduction

Matrices provide a means of storing large quantities of information in such a way that each piece can be easily identified and manipulated. They permit the solution of large systems of linear equations to be carried out in a logical and formal way so that computer implementation follows naturally. Applications of matrices extend over many areas of engineering including electrical network analysis and robotics.

An example of an extremely large electrical network is the National Grid in Britain. The equations governing this network are expressed in matrix form for analysis by computer because solutions are required at regular intervals throughout the day and night in order to make decisions such as whether or not a power station should be connected to, or removed from, the grid.

To obtain the trajectory of a robot it is necessary to perform matrix calculations to find the speed at which various motors within the robot should operate. This is a complicated problem as it is necessary to ensure that a robot reaches its required destination and does not collide with another object during its movement.

5.2 Basic definitions

■ A **matrix** is a rectangular array or block of numbers.

For example,

$$A = \begin{pmatrix} 1 & 2 & 3 \\ 4 & 5 & 6 \\ -1 & 2 & 4 \end{pmatrix} \quad B = \begin{pmatrix} 1 & 2 & -2 \\ 3 & 4 & 0.5 \end{pmatrix} \quad C = (1 \quad -1 \quad 1)$$

are all matrices. Note that we usually use a capital letter to denote a matrix, and enclose the array of numbers in brackets. To describe the size of a matrix we quote its number of rows and columns in that order so, for example, an $r \times s$ matrix has r rows and s columns. We say the matrix has **order** $r \times s$.

■ An $r \times s$ matrix has r rows and s columns.

Example 5.1

Describe the sizes of the matrices A, B and C on p. 154, and give examples of matrices of order 3×1, 3×2 and 4×2.

Solutions

A has order 3×3, B has order 2×3 and C has order 1×3.

$$\begin{pmatrix} -1 \\ -2 \\ 3 \end{pmatrix} \text{ is a } 3 \times 1 \text{ matrix} \qquad \begin{pmatrix} 1 & 2 \\ 3 & 4 \\ 5 & 6 \end{pmatrix} \text{ is a } 3 \times 2 \text{ matrix}$$

and

$$\begin{pmatrix} -1 & -1 \\ -1 & 2 \\ 2 & -0.5 \\ 1 & 0 \end{pmatrix} \text{ is a } 4 \times 2 \text{ matrix} \quad ▲$$

More generally, if the matrix A has m rows and n columns we can write:

$$A = \begin{pmatrix} a_{11} & a_{12} & \cdots & a_{1n} \\ a_{21} & a_{22} & \cdots & a_{2n} \\ \vdots & \vdots & \ddots & \vdots \\ a_{m1} & a_{m2} & \cdots & a_{mn} \end{pmatrix}$$

where a_{ij} represents the number or **element** in the ith row and jth column. A matrix with a single column can also be regarded as a column vector.

The operations of addition, subtraction and multiplication are defined upon matrices and these are explained in Section 5.3.

5.3 Addition, subtraction and multiplication

5.3.1 *Matrix addition and subtraction*

Two matrices can be added (or subtracted) if they have the same shape and size, that is, the same order. Their sum (or difference) is found by adding (or subtracting) corresponding elements as the following example shows.

Example 5.2

If

$$A = \begin{pmatrix} 1 & 5 & -2 \\ 3 & 1 & 1 \end{pmatrix} \quad \text{and} \quad B = \begin{pmatrix} 1 & 2 & 0 \\ -1 & 1 & 4 \end{pmatrix}$$

find $A + B$ and $A - B$.

Solutions

$$A + B = \begin{pmatrix} 1 & 5 & -2 \\ 3 & 1 & 1 \end{pmatrix} + \begin{pmatrix} 1 & 2 & 0 \\ -1 & 1 & 4 \end{pmatrix} = \begin{pmatrix} 2 & 7 & -2 \\ 2 & 2 & 5 \end{pmatrix}$$

$$A - B = \begin{pmatrix} 1 & 5 & -2 \\ 3 & 1 & 1 \end{pmatrix} - \begin{pmatrix} 1 & 2 & 0 \\ -1 & 1 & 4 \end{pmatrix} = \begin{pmatrix} 0 & 3 & -2 \\ 4 & 0 & -3 \end{pmatrix} \quad \blacktriangle$$

On the other hand, the matrices $\begin{pmatrix} 1 & 2 \\ 0 & 1 \end{pmatrix}$ and $\begin{pmatrix} 3 \\ 1 \end{pmatrix}$ cannot be added or subtracted because they have different orders.

Example 5.3

If $C = \begin{pmatrix} a & b \\ c & d \end{pmatrix}$ and $D = \begin{pmatrix} e & f \\ g & h \end{pmatrix}$ show that $C + D = D + C$.

Solution

$$C + D = \begin{pmatrix} a & b \\ c & d \end{pmatrix} + \begin{pmatrix} e & f \\ g & h \end{pmatrix} = \begin{pmatrix} a + e & b + f \\ c + g & d + h \end{pmatrix}$$

$$D + C = \begin{pmatrix} e & f \\ g & h \end{pmatrix} + \begin{pmatrix} a & b \\ c & d \end{pmatrix} = \begin{pmatrix} e + a & f + b \\ g + c & h + d \end{pmatrix}$$

Now $a + e$ is exactly the same as $e + a$ because addition of numbers is **commutative**. The same observation can be made of $b + f$, $c + g$ and $d + h$. Hence $C + D = D + C$. The addition of these matrices is therefore commutative. This may seem an obvious statement but we shall shortly meet matrix multiplication which is not commutative, so in general commutativity should not be simply assumed. ▲

The result obtained in Example 5.3 is true more generally:

■ Matrix addition is commutative, that is,

$A + B = B + A$

It is also easy to show that:

■ Matrix addition is **associative**, that is,

$$A + (B + C) = (A + B) + C$$

5.3.2 *Scalar multiplication*

Given any matrix A, we can multiply it by a number, that is, a scalar, to form a new matrix of the same order as A. This multiplication is performed by multiplying every element of A by the number.

Example 5.4

If

$$A = \begin{pmatrix} 1 & 3 \\ -2 & 1 \\ 0 & 1 \end{pmatrix}$$

find $2A$, $-3A$ and $\frac{1}{2}A$

Solutions

$$2A = 2\begin{pmatrix} 1 & 3 \\ -2 & 1 \\ 0 & 1 \end{pmatrix} = \begin{pmatrix} 2 & 6 \\ -4 & 2 \\ 0 & 2 \end{pmatrix}$$

$$-3A = -3\begin{pmatrix} 1 & 3 \\ -2 & 1 \\ 0 & 1 \end{pmatrix} = \begin{pmatrix} -3 & -9 \\ 6 & -3 \\ 0 & -3 \end{pmatrix}$$

and

$$\frac{1}{2}A = \frac{1}{2}\begin{pmatrix} 1 & 3 \\ -2 & 1 \\ 0 & 1 \end{pmatrix} = \begin{pmatrix} \frac{1}{2} & \frac{3}{2} \\ -1 & \frac{1}{2} \\ 0 & \frac{1}{2} \end{pmatrix} \quad \blacktriangle$$

In general we have:

■ If $A = \begin{pmatrix} a_{11} & a_{12} & \cdots & a_{1n} \\ a_{21} & a_{22} & \cdots & a_{2n} \\ \vdots & \vdots & \ddots & \vdots \\ a_{m1} & a_{m2} & \cdots & a_{mn} \end{pmatrix}$ then $kA = \begin{pmatrix} ka_{11} & ka_{12} & \cdots & ka_{1n} \\ ka_{21} & ka_{22} & \cdots & ka_{2n} \\ \vdots & \vdots & \ddots & \vdots \\ ka_{m1} & ka_{m2} & \cdots & ka_{mn} \end{pmatrix}$

5.3.3 *Matrix multiplication*

Matrix multiplication is defined in a special way which at first seems strange but is in fact very useful. If A is a $p \times q$ matrix and B is an $r \times s$ matrix we can form the product AB only if $q = r$, that is, only if the number of columns in A is the same as the number of rows in B. The product is then a $p \times s$ matrix C, that is,

$$C = AB \quad \text{where} \quad \begin{array}{l} A \text{ is } p \times q \\ B \text{ is } q \times s \\ C \text{ is } p \times s \end{array}$$

To understand how the product is actually formed consider the following examples.

Example 5.5

If

$$A = \begin{pmatrix} a_{11} & a_{12} \\ a_{21} & a_{22} \end{pmatrix} \quad \text{and} \quad B = \begin{pmatrix} b_{11} & b_{12} \\ b_{21} & b_{22} \end{pmatrix}$$

find AB.

Solution

Clearly A and B both have order 2×2, so the number of columns in A is the same as the number of rows in B and we can proceed. The product, C, will also have order 2×2 and is defined by:

$$C = \begin{pmatrix} a_{11} & a_{12} \\ a_{21} & a_{22} \end{pmatrix}\begin{pmatrix} b_{11} & b_{12} \\ b_{21} & b_{22} \end{pmatrix} = \begin{pmatrix} a_{11}b_{11} + a_{12}b_{21} & a_{11}b_{12} + a_{12}b_{22} \\ a_{21}b_{11} + a_{22}b_{21} & a_{21}b_{12} + a_{22}b_{22} \end{pmatrix}$$

$$= \begin{pmatrix} c_{11} & c_{12} \\ c_{21} & c_{22} \end{pmatrix}$$

To obtain the element c_{11} in C you can imagine taking the first row of A and pairing its elements with those in the first column of B. The paired elements are then multiplied together and added to form c_{11}:

$$c_{11} = (a_{11} \quad a_{12})\begin{pmatrix} b_{11} \\ b_{21} \end{pmatrix} = a_{11} \times b_{11} + a_{12} \times b_{21} = a_{11}b_{11} + a_{12}b_{21}$$

Similarly, to find the c_{21} element of C take the second row of A and pair its elements with the first column of B. The paired elements are then multiplied and added to form c_{21}:

$$c_{21} = (a_{21} \quad a_{22})\begin{pmatrix} b_{11} \\ b_{21} \end{pmatrix} = a_{21} \times b_{11} + a_{22} \times b_{21} = a_{21}b_{11} + a_{22}b_{21}$$

Similarly $c_{12} = a_{11}b_{12} + a_{12}b_{22}$ and $c_{22} = a_{21}b_{12} + a_{22}b_{22}$. ▲

■ If A is a $p \times q$ matrix and B is a $q \times s$ matrix, then the product $C = AB$ will be a $p \times s$ matrix. To find c_{ij} we take the ith row of A and pair its elements with the jth column of B. The paired elements are multiplied together and added to form c_{ij}.

Example 5.6

If $B = \begin{pmatrix} 1 & 2 & 3 \\ 4 & 5 & 6 \end{pmatrix}$ and $C = \begin{pmatrix} 1 \\ 2 \\ 4 \end{pmatrix}$ find BC.

Solution

B has order 2×3 and C has order 3×1 so clearly the product BC exists and will have order 2×1. BC is formed as follows:

$$BC = \begin{pmatrix} 1 & 2 & 3 \\ 4 & 5 & 6 \end{pmatrix} \begin{pmatrix} 1 \\ 2 \\ 4 \end{pmatrix} = \begin{pmatrix} 1 \times 1 + 2 \times 2 + 3 \times 4 \\ 4 \times 1 + 5 \times 2 + 6 \times 4 \end{pmatrix} = \begin{pmatrix} 17 \\ 38 \end{pmatrix}$$

Note the order of the product, 2×1, can be determined at the start by considering the orders of B and C. ▲

Example 5.7

Find AB where

$$A = \begin{pmatrix} 1 & 2 \\ 3 & 4 \\ -1 & 0 \end{pmatrix} \quad \text{and} \quad B = \begin{pmatrix} -1 \\ 2 \\ -1 \end{pmatrix}$$

Solution

A and B have orders 3×2 and 3×1, respectively, and consequently the product, AB, cannot be formed. ▲

Example 5.8

Given

$$A = \begin{pmatrix} 1 & 1 & 1 \\ 2 & 1 & 0 \\ 3 & -1 & 2 \end{pmatrix} \quad \text{and} \quad B = \begin{pmatrix} 0 & 3 & 1 \\ 4 & -1 & 0 \\ 2 & 2 & 1 \end{pmatrix}$$

find, if possible, AB and BA, and comment upon the result.

Solutions

A and B both have order 3×3 and the products AB and BA can both be formed. Both will have order 3×3.

$$AB = \begin{pmatrix} 1 & 1 & 1 \\ 2 & 1 & 0 \\ 3 & -1 & 2 \end{pmatrix} \begin{pmatrix} 0 & 3 & 1 \\ 4 & -1 & 0 \\ 2 & 2 & 1 \end{pmatrix} = \begin{pmatrix} 6 & 4 & 2 \\ 4 & 5 & 2 \\ 0 & 14 & 5 \end{pmatrix}$$

$$BA = \begin{pmatrix} 0 & 3 & 1 \\ 4 & -1 & 0 \\ 2 & 2 & 1 \end{pmatrix} \begin{pmatrix} 1 & 1 & 1 \\ 2 & 1 & 0 \\ 3 & -1 & 2 \end{pmatrix} = \begin{pmatrix} 9 & 2 & 2 \\ 2 & 3 & 4 \\ 9 & 3 & 4 \end{pmatrix}$$

Clearly AB and BA are not the same. Matrix multiplication is not usually commutative and we must pay particular attention to this detail when we are working with matrices. ▲

■ In general $AB \neq BA$ and so matrix multiplication is not commutative.

In the product AB we say that B has been **premultiplied** by A, or alternatively A has been **postmultiplied** by B.

Example 5.9

Given

$$A = \begin{pmatrix} 1 & -1 & 2 \\ 3 & 0 & 2 \\ -1 & 3 & 5 \end{pmatrix} \quad B = \begin{pmatrix} -1 & 2 & 9 \\ 1 & 0 & 0 \\ 3 & -2 & 1 \end{pmatrix} \quad C = \begin{pmatrix} 4 & 1 & 5 \\ 2 & 3 & 1 \\ 0 & 1 & 5 \end{pmatrix}$$

find BC, $A(BC)$, AB and $(AB)C$, commenting upon the result.

Solutions

$$BC = \begin{pmatrix} -1 & 2 & 9 \\ 1 & 0 & 0 \\ 3 & -2 & 1 \end{pmatrix} \begin{pmatrix} 4 & 1 & 5 \\ 2 & 3 & 1 \\ 0 & 1 & 5 \end{pmatrix} = \begin{pmatrix} 0 & 14 & 42 \\ 4 & 1 & 5 \\ 8 & -2 & 18 \end{pmatrix}$$

$$A(BC) = \begin{pmatrix} 1 & -1 & 2 \\ 3 & 0 & 2 \\ -1 & 3 & 5 \end{pmatrix} \begin{pmatrix} 0 & 14 & 42 \\ 4 & 1 & 5 \\ 8 & -2 & 18 \end{pmatrix} = \begin{pmatrix} 12 & 9 & 73 \\ 16 & 38 & 162 \\ 52 & -21 & 63 \end{pmatrix}$$

$$AB = \begin{pmatrix} 1 & -1 & 2 \\ 3 & 0 & 2 \\ -1 & 3 & 5 \end{pmatrix}\begin{pmatrix} -1 & 2 & 9 \\ 1 & 0 & 0 \\ 3 & -2 & 1 \end{pmatrix} = \begin{pmatrix} 4 & -2 & 11 \\ 3 & 2 & 29 \\ 19 & -12 & -4 \end{pmatrix}$$

$$(AB)C = \begin{pmatrix} 4 & -2 & 11 \\ 3 & 2 & 29 \\ 19 & -12 & -4 \end{pmatrix}\begin{pmatrix} 4 & 1 & 5 \\ 2 & 3 & 1 \\ 0 & 1 & 5 \end{pmatrix} = \begin{pmatrix} 12 & 9 & 73 \\ 16 & 38 & 162 \\ 52 & -21 & 63 \end{pmatrix}$$

We note that $A(BC) = (AB)C$ so that in this case matrix multiplication is associative. ▲

The result obtained in Example 5.9 is also true in general:

■ Matrix multiplication is associative:

$$(AB)C = A(BC)$$

EXERCISES 5.1

1. Evaluate

(a) $(1 \quad 4)\begin{pmatrix} -1 \\ 4 \end{pmatrix}$

(b) $(3 \quad 7)\begin{pmatrix} 3 \\ -4 \end{pmatrix}$

(c) $\begin{pmatrix} 5 & 2 \\ 1 & 3 \end{pmatrix}\begin{pmatrix} 9 \\ 1 \end{pmatrix}$

(d) $\begin{pmatrix} 1 & 4 \\ -1 & 3 \end{pmatrix}\begin{pmatrix} 5 & 2 \\ -3 & 4 \end{pmatrix}$

(e) $\begin{pmatrix} 5 & 1 \\ 29 & 6 \end{pmatrix}\begin{pmatrix} 6 & -1 \\ -29 & 5 \end{pmatrix}$

2. If $A = \begin{pmatrix} 1 & 1 \\ 3 & 4 \end{pmatrix}$, $B = \begin{pmatrix} 2 \\ 1 \end{pmatrix}$, $C = \begin{pmatrix} -7 & 1 \\ 0 & 4 \end{pmatrix}$,

$D = (3 \quad 2 \quad 1)$ and $E = \begin{pmatrix} 2 & 3 & 4 \\ 1 & 2 & -1 \end{pmatrix}$

find, if possible,

(a) $A + D$, $C - A$ and $D - E$

(b) AB, BA, CA, AC, DA, DB, BD, EB, BE and AE

(c) $7C$, $-3D$ and kE, where k is a scalar.

3. Plot the points A, B, C with position vectors given by:

$$\mathbf{v}_1 = \begin{pmatrix} 1 \\ 0 \end{pmatrix} \qquad \mathbf{v}_2 = \begin{pmatrix} 2 \\ 0 \end{pmatrix} \qquad \mathbf{v}_3 = \begin{pmatrix} 2 \\ 3 \end{pmatrix}$$

respectively. Treating these vectors as matrices of order 2×1 find the products $M\mathbf{v}_1$, $M\mathbf{v}_2$, $M\mathbf{v}_3$ when

(a) $M = \begin{pmatrix} 1 & 0 \\ 0 & -1 \end{pmatrix}$

(b) $M = \begin{pmatrix} 0 & 1 \\ 1 & 0 \end{pmatrix}$

(c) $M = \begin{pmatrix} 0 & -1 \\ 1 & 0 \end{pmatrix}$

In each case draw a diagram to illustrate the effect upon the vectors of multiplication by the matrix.

4. Find AB and BA where

$$A = \begin{pmatrix} 1 & 3 & 2 \\ -1 & 0 & 4 \\ 5 & 1 & -1 \end{pmatrix}$$

$$B = \begin{pmatrix} 5 & 2 & 1 \\ 0 & 3 & 4 \\ 1 & 3 & 5 \end{pmatrix}$$

5. Given that A^2 means the product of a matrix A with itself, find A^2 when

$$A = \begin{pmatrix} 4 & 2 \\ 1 & 3 \end{pmatrix}. \text{ Find } A^3.$$

6. If $A = \begin{pmatrix} 1 & 3 \\ -2 & 4 \end{pmatrix}$, $B = \begin{pmatrix} 2 & 1 \\ -4 & 5 \end{pmatrix}$ find

AB, BA, $A + B$ and $(A + B)^2$. Show that

$$(A + B)^2 = A^2 + AB + BA + B^2$$

Why is $(A + B)^2$ not equal to $A^2 + 2AB + B^2$?

7. Find, if possible,

(a) $\begin{pmatrix} 1 & 0 & 0 & 0 \\ 0 & 0 & -1 & 0 \\ 0 & 1 & 0 & 0 \\ 0 & 0 & 0 & 1 \end{pmatrix}\begin{pmatrix} 1 \\ 1 \\ 2 \\ 1 \end{pmatrix}$

(b) $\begin{pmatrix} 0 & 0 & 1 & 0 \\ 0 & 1 & 0 & 0 \\ -1 & 0 & 0 & 0 \\ 0 & 0 & 0 & 1 \end{pmatrix}\begin{pmatrix} 2 \\ 5 \\ 5 \\ 1 \end{pmatrix}$

8. Find $\begin{pmatrix} 1 & 3 & 6 \\ 2 & -5 & 7 \end{pmatrix}\begin{pmatrix} 1 & 2 \\ 3 & -5 \\ 6 & 7 \end{pmatrix}$.

9. Given the vector $\mathbf{v} = \begin{pmatrix} 1 \\ 2 \\ 3 \end{pmatrix}$ calculate the vectors obtained when $\mathbf{v}$ is multiplied by the following matrices.

(a) $\begin{pmatrix} 6 & 2 & 9 \\ 1 & 3 & 2 \\ -1 & 2 & -3 \end{pmatrix}$

(b) $\begin{pmatrix} -1 & 0 & 3 \\ 7 & 1 & 9 \\ 1 & 3 & 4 \end{pmatrix}$

(c) $\begin{pmatrix} 1 & 3 & 1 \\ 9 & 2 & 6 \\ 2 & 8 & 0 \end{pmatrix}$

(d) $\begin{pmatrix} 3 & 1 & 2 \\ 6 & 5 & 4 \end{pmatrix}$

(e) $\begin{pmatrix} 6 & 8 & 3 \\ 9 & 6 & 4 \\ 5 & 3 & 9 \\ 2 & 5 & 2 \end{pmatrix}$

5.4 Robot coordinate frames

In Chapter 4 we saw that vectors provide a useful tool for the analysis of the position of robots. By assigning a vector to each of the links the position vector corresponding to the tip of the robot can be calculated (Figure 5.1). In practice the inverse problem is more likely: calculate the link vectors to achieve a particular position vector. Usually a desired position for the tip of the robot is known and link positions to achieve this are required. The problem is made more complicated because the position of a link depends upon the movements of all the joints

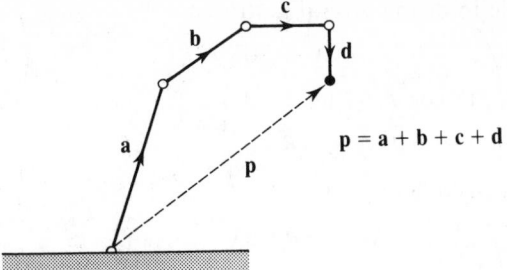

Figure 5.1 A robot with links represented by vectors **a, b, c** and **d**.

between it and the anchor point. The solution of this problem can be quite complicated, especially when the robot has several links. One way forward is to define the position of a link by its own local set of coordinates. This is usually termed a **coordinate frame** because it provides a frame of reference for the link. Matrix operations can then be used to relate the coordinate frames, thus allowing a link position to be defined with respect to a convenient coordinate frame. A common requirement is to be able to relate the link positions to a **world coordinate frame**. If a robot is being used in conjunction with other machines then the world coordinate frame may have an origin some distance away from the robot. The advantage of defining link coordinate frames is that the position of a link is easily defined within its own coordinate frame and the movement of coordinate frames relative to each other can be expressed by means of matrix equations.

5.4.1 *Translation and rotation of vectors*

An introduction to the mathematics involved in analysing the movement of robots can be obtained by examining the way in which vectors can be translated and rotated using matrix operations.

Consider the point P with position vector given by:

$$\mathbf{r} = \begin{pmatrix} x \\ y \\ z \end{pmatrix}$$

In order to translate and rotate this vector it is useful to introduce an augmented vector **V** given by

$$\mathbf{V} = \begin{pmatrix} x \\ y \\ z \\ 1 \end{pmatrix} \tag{5.1}$$

It is then possible to define several matrices:

$$\text{Rot}(x, \theta) = \begin{pmatrix} 1 & 0 & 0 & 0 \\ 0 & \cos\theta & -\sin\theta & 0 \\ 0 & \sin\theta & \cos\theta & 0 \\ 0 & 0 & 0 & 1 \end{pmatrix} \tag{5.2}$$

$$\text{Rot}(y, \theta) = \begin{pmatrix} \cos\theta & 0 & \sin\theta & 0 \\ 0 & 1 & 0 & 0 \\ -\sin\theta & 0 & \cos\theta & 0 \\ 0 & 0 & 0 & 1 \end{pmatrix} \tag{5.3}$$

$$\text{Rot}(z, \theta) = \begin{pmatrix} \cos\theta & -\sin\theta & 0 & 0 \\ \sin\theta & \cos\theta & 0 & 0 \\ 0 & 0 & 1 & 0 \\ 0 & 0 & 0 & 1 \end{pmatrix} \tag{5.4}$$

$$\text{Trans}(a, b, c) = \begin{pmatrix} 1 & 0 & 0 & a \\ 0 & 1 & 0 & b \\ 0 & 0 & 1 & c \\ 0 & 0 & 0 & 1 \end{pmatrix} \tag{5.5}$$

Matrices (5.2)–(5.4) allow vectors to be rotated by an angle θ around axes x, y and z, respectively. For example, the product $\text{Rot}(x, \theta)\mathbf{V}$ has the effect of rotating $\mathbf{r}$ through an angle θ about the x axis. Matrix (5.5) allows a vector to be translated a units in the x direction, b units in the y direction and c units in the z direction.

It is possible to combine these matrices to calculate the effect of several operations on a vector. In doing so, it is important to maintain the correct order of operations as matrix multiplication is non-commutative.

For example, the position of a vector that has first been translated and then rotated about the x axis can be defined by:

$$\mathbf{V}_{\text{new}} = \text{Rot}(x, \theta)\,\text{Trans}(a, b, c)\,\mathbf{V}_{\text{old}}$$

A few examples will help to clarify these ideas.

Example 5.10

Rotate the vector

$$\mathbf{r} = \begin{pmatrix} 1 \\ 1 \\ 2 \end{pmatrix}$$

through 90° about the x axis.

Solution

$$\mathbf{r}_{old} = \begin{pmatrix} 1 \\ 1 \\ 2 \end{pmatrix} \qquad \mathbf{V}_{old} = \begin{pmatrix} 1 \\ 1 \\ 2 \\ 1 \end{pmatrix}$$

$$\text{Rot}(x, 90°) = \begin{pmatrix} 1 & 0 & 0 & 0 \\ 0 & \cos 90° & -\sin 90° & 0 \\ 0 & \sin 90° & \cos 90° & 0 \\ 0 & 0 & 0 & 1 \end{pmatrix}$$

$$= \begin{pmatrix} 1 & 0 & 0 & 0 \\ 0 & 0 & -1 & 0 \\ 0 & 1 & 0 & 0 \\ 0 & 0 & 0 & 1 \end{pmatrix}$$

$$\mathbf{V}_{new} = \begin{pmatrix} 1 & 0 & 0 & 0 \\ 0 & 0 & -1 & 0 \\ 0 & 1 & 0 & 0 \\ 0 & 0 & 0 & 1 \end{pmatrix}\begin{pmatrix} 1 \\ 1 \\ 2 \\ 1 \end{pmatrix}$$

$$= \begin{pmatrix} 1 \\ -2 \\ 1 \\ 1 \end{pmatrix}$$

So,

$$\mathbf{r}_{new} = \begin{pmatrix} 1 \\ -2 \\ 1 \end{pmatrix} \quad \blacktriangle$$

Example 5.11

Translate the vector

$$\mathbf{r} = \begin{pmatrix} 1 \\ 3 \\ 2 \end{pmatrix}$$

by

$$\begin{pmatrix} 1 \\ 2 \\ 3 \end{pmatrix}$$

and then rotate by 90° about the y axis.

Solution

$$\mathbf{r}_{\text{old}} = \begin{pmatrix} 1 \\ 3 \\ 2 \end{pmatrix} \qquad \mathbf{V}_{\text{old}} = \begin{pmatrix} 1 \\ 3 \\ 2 \\ 1 \end{pmatrix}$$

To translate $\mathbf{r}_{\text{old}}$, we form

$$\text{Trans}(1, 2, 3) = \begin{pmatrix} 1 & 0 & 0 & 1 \\ 0 & 1 & 0 & 2 \\ 0 & 0 & 1 & 3 \\ 0 & 0 & 0 & 1 \end{pmatrix}$$

Then

$$\text{Trans}(1, 2, 3)\mathbf{V}_{\text{old}} = \begin{pmatrix} 1 & 0 & 0 & 1 \\ 0 & 1 & 0 & 2 \\ 0 & 0 & 1 & 3 \\ 0 & 0 & 0 & 1 \end{pmatrix}\begin{pmatrix} 1 \\ 3 \\ 2 \\ 1 \end{pmatrix} = \begin{pmatrix} 2 \\ 5 \\ 5 \\ 1 \end{pmatrix}$$

To rotate by 90° about the y axis we require

$$\text{Rot}(y, 90°) = \begin{pmatrix} \cos 90° & 0 & \sin 90° & 0 \\ 0 & 1 & 0 & 0 \\ -\sin 90° & 0 & \cos 90° & 0 \\ 0 & 0 & 0 & 1 \end{pmatrix}$$

$$= \begin{pmatrix} 0 & 0 & 1 & 0 \\ 0 & 1 & 0 & 0 \\ -1 & 0 & 0 & 0 \\ 0 & 0 & 0 & 1 \end{pmatrix}$$

The vector $\begin{pmatrix} 2 \\ 5 \\ 5 \\ 1 \end{pmatrix}$ is premultiplied by this matrix to give

$$V_{new} = Rot(y, 90°)\,Trans(1, 2, 3)\,V_{old}$$

$$= \begin{pmatrix} 0 & 0 & 1 & 0 \\ 0 & 1 & 0 & 0 \\ -1 & 0 & 0 & 0 \\ 0 & 0 & 0 & 1 \end{pmatrix} \begin{pmatrix} 2 \\ 5 \\ 5 \\ 1 \end{pmatrix} = \begin{pmatrix} 5 \\ 5 \\ -2 \\ 1 \end{pmatrix}$$

Hence $r_{new} = \begin{pmatrix} 5 \\ 5 \\ -2 \end{pmatrix}$. ▲

5.5 Some special matrices

5.5.1 *Square matrices*

A matrix which has the same number of rows as columns is called a **square** matrix. Thus

$$\begin{pmatrix} 1 & 2 & 3 \\ -1 & 0 & 1 \\ 3 & 2 & 1 \end{pmatrix}$$ is a square matrix, while $\begin{pmatrix} -1 & 3 & 0 \\ 2 & 4 & 1 \end{pmatrix}$ is not.

5.5.2 *Diagonal matrices*

Some square matrices have elements which are zero everywhere except on the leading diagonal (top-left to bottom-right). Such matrices are said to be **diagonal**. Thus

$$\begin{pmatrix} 1 & 0 & 0 \\ 0 & 2 & 0 \\ 0 & 0 & -1 \end{pmatrix} \quad \begin{pmatrix} 1 & 0 \\ 0 & b \end{pmatrix} \quad \begin{pmatrix} 3 & 0 & 0 & 0 \\ 0 & 2 & 0 & 0 \\ 0 & 0 & 1 & 0 \\ 0 & 0 & 0 & 0 \end{pmatrix}$$

are all diagonal matrices, whereas

$$\begin{pmatrix} 1 & 2 & 4 \\ 0 & 1 & 0 \\ 3 & 0 & 1 \end{pmatrix}$$

is not.

5.5.3 *Identity matrices*

Diagonal matrices which have only 1s on their leading diagonals, for example,

$$\begin{pmatrix} 1 & 0 \\ 0 & 1 \end{pmatrix} \quad \text{and} \quad \begin{pmatrix} 1 & 0 & 0 \\ 0 & 1 & 0 \\ 0 & 0 & 1 \end{pmatrix}$$

are called **identity** matrices and are denoted by the letter I.

Example 5.12

Find IA where $I = \begin{pmatrix} 1 & 0 \\ 0 & 1 \end{pmatrix}$ and $A = \begin{pmatrix} 2 & 4 & 4 \\ 3 & -1 & 0 \end{pmatrix}$ and comment upon the result.

Solution

$$IA = \begin{pmatrix} 1 & 0 \\ 0 & 1 \end{pmatrix}\begin{pmatrix} 2 & 4 & 4 \\ 3 & -1 & 0 \end{pmatrix} = \begin{pmatrix} 2 & 4 & 4 \\ 3 & -1 & 0 \end{pmatrix}$$

The effect of premultiplying A by I has been to leave A unaltered. The product is identical to the original matrix A, and this is why I is called an identity matrix. ▲

In general, if A is an arbitrary matrix and I is an identity matrix of the appropriate size, then

$$IA = A$$

■ If A is a square matrix then $IA = AI = A$

5.5.4 *The transpose of a matrix*

If A is an arbitrary $m \times n$ matrix, a related matrix is the **transpose** of A, written A^T, found by interchanging the rows and columns of A. Thus the first row of A becomes the first column of A^T and so on. A^T is an $n \times m$ matrix.

Example 5.13

If $A = \begin{pmatrix} 1 & -1 \\ 2 & 4 \end{pmatrix}$ find A^T.

Solution

$$A^T = \begin{pmatrix} 1 & 2 \\ -1 & 4 \end{pmatrix} \quad \blacktriangle$$

Example 5.14

If $A = \begin{pmatrix} 4 & 2 & 6 \\ 1 & 8 & 7 \end{pmatrix}$ find A^T and evaluate AA^T.

Solutions

$$A^T = \begin{pmatrix} 4 & 1 \\ 2 & 8 \\ 6 & 7 \end{pmatrix}$$

$$AA^T = \begin{pmatrix} 4 & 2 & 6 \\ 1 & 8 & 7 \end{pmatrix} \begin{pmatrix} 4 & 1 \\ 2 & 8 \\ 6 & 7 \end{pmatrix} = \begin{pmatrix} 56 & 62 \\ 62 & 114 \end{pmatrix} \quad \blacktriangle$$

5.5.5 *Symmetric matrices*

If a square matrix A and its transpose A^T are identical, then A is said to be a **symmetric** matrix.

Example 5.15

If $A = \begin{pmatrix} 5 & -4 & 2 \\ -4 & 6 & 9 \\ 2 & 9 & 13 \end{pmatrix}$ find A^T.

Solution

$$A^T = \begin{pmatrix} 5 & -4 & 2 \\ -4 & 6 & 9 \\ 2 & 9 & 13 \end{pmatrix}$$

which is clearly equal to A. Hence A is a symmetric matrix. Note that a symmetric matrix is symmetrical about its leading diagonal. $\blacktriangle$

5.5.6 *Skew symmetric matrices*

If a square matrix A is such that $A^T = -A$ then A is said to be **skew symmetric**.

Example 5.16

If $A = \begin{pmatrix} 0 & 5 \\ -5 & 0 \end{pmatrix}$, find A^T and deduce that A is skew symmetric.

Solution

We have $A^T = \begin{pmatrix} 0 & -5 \\ 5 & 0 \end{pmatrix}$ which is clearly equal to $-A$. Hence A is skew symmetric. ▲

EXERCISES 5.2

1. Treating the column vector $\mathbf{x} = \begin{pmatrix} x \\ y \\ z \end{pmatrix}$ as a 3 × 1 matrix, find $I\mathbf{x}$ where I is the 3 × 3 identity matrix.

2. If $A = \begin{pmatrix} a & b \\ c & d \end{pmatrix}$ show that AA^T is a symmetric matrix.

3. If $A = \begin{pmatrix} 2 & 1 & 3 \\ 4 & 2 & 1 \\ -1 & 3 & 2 \end{pmatrix}$ and $B = \begin{pmatrix} 1 & -7 & 0 \\ 0 & 2 & 5 \\ 3 & 4 & 5 \end{pmatrix}$ find A^T, B^T, AB and $(AB)^T$. Deduce that $(AB)^T = B^T A^T$.

4. Determine the type of matrix obtained when two diagonal matrices are multiplied together.

5. If $A = \begin{pmatrix} a & b \\ c & d \end{pmatrix}$ is skew symmetric, show that $a = d = 0$, that is, the diagonal elements are zero.

5.6 The inverse of a 2 × 2 matrix

When we are dealing with ordinary numbers it is often necessary to carry out the operation of division. Thus, for example, if we know that $3x = 4$, then clearly $x = 4/3$. If we are given matrices A and C and know that

$$AB = C$$

how do we find B? It might be tempting to write:

$$B = \frac{C}{A}$$

Unfortunately, this would be entirely wrong since division of matrices is not defined. However, given expressions like $AB = C$ it is often necessary to be able to find the appropriate expression for B. This is where we need to introduce the concept of an inverse matrix.

If A is a square matrix and we can find another matrix B with the property that

$$AB = BA = I$$

then B is said to be the **inverse** of A and is written A^{-1}, that is,

$$AA^{-1} = A^{-1}A = I$$

If B is the inverse of A, then A is also the inverse of B. Note that A^{-1} does not mean a reciprocal; there is no such thing as matrix division. A^{-1} is the notation we use for the inverse of A.

■ $AA^{-1} = A^{-1}A = I$

Since A is a square matrix, A^{-1} is also square and of the same order, so that the products AA^{-1} and $A^{-1}A$ can be formed. The term 'inverse' cannot be applied to a matrix which is not square.

Example 5.17

If $A = \begin{pmatrix} 2 & 1 \\ 3 & 2 \end{pmatrix}$ show that the matrix $\begin{pmatrix} 2 & -1 \\ -3 & 2 \end{pmatrix}$ is the inverse of A.

Solution

Forming the products

$$\begin{pmatrix} 2 & 1 \\ 3 & 2 \end{pmatrix}\begin{pmatrix} 2 & -1 \\ -3 & 2 \end{pmatrix} = \begin{pmatrix} 1 & 0 \\ 0 & 1 \end{pmatrix}$$

$$\begin{pmatrix} 2 & -1 \\ -3 & 2 \end{pmatrix}\begin{pmatrix} 2 & 1 \\ 3 & 2 \end{pmatrix} = \begin{pmatrix} 1 & 0 \\ 0 & 1 \end{pmatrix}$$

we see that $\begin{pmatrix} 2 & -1 \\ -3 & 2 \end{pmatrix}$ is the inverse of A.   ▲

5.6.1 *Finding the inverse of a matrix*

For 2 × 2 matrices a simple formula exists to find the inverse of

$$A = \begin{pmatrix} a & b \\ c & d \end{pmatrix}$$

This formula states

$$\blacksquare \quad A^{-1} = \frac{1}{ad - bc} \begin{pmatrix} d & -b \\ -c & a \end{pmatrix}$$

Example 5.18

If $A = \begin{pmatrix} 3 & 5 \\ 1 & 2 \end{pmatrix}$ find A^{-1}.

Solution

Clearly $ad - bc = 6 - 5 = 1$, so that

$$A^{-1} = \frac{1}{1} \begin{pmatrix} 2 & -5 \\ -1 & 3 \end{pmatrix} = \begin{pmatrix} 2 & -5 \\ -1 & 3 \end{pmatrix}$$

The solution should always be checked by forming AA^{-1}. ▲

Example 5.19

If $A = \begin{pmatrix} 1 & 5 \\ 2 & 4 \end{pmatrix}$ find A^{-1}.

Solution

Here we have $ad - bc = 4 - 10 = -6$. Therefore,

$$A^{-1} = \frac{1}{-6} \begin{pmatrix} 4 & -5 \\ -2 & 1 \end{pmatrix} = \begin{pmatrix} -\frac{2}{3} & \frac{5}{6} \\ \frac{1}{3} & -\frac{1}{6} \end{pmatrix} \quad ▲$$

Example 5.20

If $A = \begin{pmatrix} 2 & 4 \\ 1 & 2 \end{pmatrix}$ find A^{-1}.

Solution

This time, $ad - bc = 4 - 4 = 0$, so when we come to form $1/(ad - bc)$ we find $1/0$ which is not defined. We cannot form the inverse of A in this case; it does not exist. ▲

Clearly not all square matrices have inverses. The quantity $ad - bc$ is obviously the important determining factor since only if $ad - bc \neq 0$ can we find A^{-1}. This quantity is therefore given a special name: the **determinant** of A, denoted by $|A|$, or det A. Given any 2×2 matrix A, its determinant, $|A|$, is the scalar $ad - bc$. This is easily remembered as

[product of $\searrow$ diagonal] − [product of $\swarrow$ diagonal]

If A is the matrix $\begin{pmatrix} a & b \\ c & d \end{pmatrix}$, we write its determinant as $\begin{vmatrix} a & b \\ c & d \end{vmatrix}$. Note that the straight lines $||$ indicate that we are discussing the determinant, which is a scalar, rather than the matrix itself. If the matrix A is such that $|A| = 0$, then it has no inverse and is said to be **singular**. If $|A| \neq 0$ then A^{-1} exists and A is said to be **non-singular**.

■ A singular matrix A has $|A| = 0$

A non-singular matrix A has $|A| \neq 0$

Example 5.21

If $A = \begin{pmatrix} 1 & 2 \\ 5 & 0 \end{pmatrix}$ and $B = \begin{pmatrix} -1 & 2 \\ -3 & 1 \end{pmatrix}$ find $|A|$, $|B|$ and $|AB|$.

Solutions

$$|A| = \begin{vmatrix} 1 & 2 \\ 5 & 0 \end{vmatrix} = (1)(0) - (2)(5) = -10$$

$$|B| = \begin{vmatrix} -1 & 2 \\ -3 & 1 \end{vmatrix} = (-1)(1) - (2)(-3) = 5$$

$$AB = \begin{pmatrix} 1 & 2 \\ 5 & 0 \end{pmatrix}\begin{pmatrix} -1 & 2 \\ -3 & 1 \end{pmatrix} = \begin{pmatrix} -7 & 4 \\ -5 & 10 \end{pmatrix}$$

$$|AB| = (-7)(10) - (4)(-5) = -50$$

We note that $|A||B| = |AB|$. ▲

The result obtained in Example 5.21 is true more generally:

■ If A and B are square matrices of the same order $|A|\,|B| = |AB|$.

5.6.2 *Orthogonal matrices*

A non-singular square matrix A such that $A^T = A^{-1}$ is said to be **orthogonal**. Consequently, if A is orthogonal $AA^T = A^T A = I$.

Example 5.22

Find the inverse of $A = \begin{pmatrix} 0 & -1 \\ 1 & 0 \end{pmatrix}$. Deduce that A is an orthogonal matrix.

Solution

From the formula for the inverse of a 2 × 2 matrix we find:

$$A^{-1} = \frac{1}{1} \begin{pmatrix} 0 & 1 \\ -1 & 0 \end{pmatrix} = \begin{pmatrix} 0 & 1 \\ -1 & 0 \end{pmatrix}$$

This is clearly equal to the transpose of A. Hence A is an orthogonal matrix. ▲

To find the inverses of larger matrices we shall need to study determinants further. This is done in Section 5.7.

EXERCISES 5.3

1. If $A = \begin{pmatrix} 5 & 6 \\ -4 & 8 \end{pmatrix}$ find A^{-1}.

2. If $A = \begin{pmatrix} 1 & 1 \\ 0 & 3 \end{pmatrix}$ and $B = \begin{pmatrix} 2 & 1 \\ -1 & 3 \end{pmatrix}$ find AB, $(AB)^{-1}$, B^{-1}, A^{-1} and $B^{-1}A^{-1}$. Deduce that $(AB)^{-1} = B^{-1}A^{-1}$.

3. Given that the matrix

$$M = \begin{pmatrix} \cos \omega t & -\sin \omega t & 0 \\ \sin \omega t & \cos \omega t & 0 \\ 0 & 0 & 1 \end{pmatrix}$$

is orthogonal, find M^{-1}.

4. If $A = \begin{pmatrix} 3 & 0 \\ -1 & 4 \end{pmatrix}$ and $B = \begin{pmatrix} 7 & 8 \\ 4 & 3 \end{pmatrix}$ find $|AB|$, $|BA|$.

5. If $A = \begin{pmatrix} a & b \\ c & d \end{pmatrix}$, $B = \begin{pmatrix} e & f \\ g & h \end{pmatrix}$ find AB, $|A|$, $|B|$, $|AB|$. Verify that $|AB| = |A||B|$.

6. If $A = \begin{pmatrix} 1 & 2 \\ 3 & 4 \end{pmatrix}$ find A^{-1}. Find values of the constants a and b such that $A + aA^{-1} = bI$.

5.7 Determinants

■ If $A = \begin{pmatrix} a_{11} & a_{12} & a_{13} \\ a_{21} & a_{22} & a_{23} \\ a_{31} & a_{32} & a_{33} \end{pmatrix}$, the value of its determinant, $|A|$, is given by

$$|A| = a_{11} \begin{vmatrix} a_{22} & a_{23} \\ a_{32} & a_{33} \end{vmatrix} - a_{12} \begin{vmatrix} a_{21} & a_{23} \\ a_{31} & a_{33} \end{vmatrix} + a_{13} \begin{vmatrix} a_{21} & a_{22} \\ a_{31} & a_{32} \end{vmatrix}$$

If we choose an element of A, a_{ij} say, and cross out its row and column and form the determinant of the four remaining elements, this determinant is known as the **minor** of the element a_{ij}.

A moment's study will therefore reveal that the determinant of A is given by:

$$|A| = (a_{11} \times \text{its minor}) - (a_{12} \times \text{its minor}) + (a_{13} \times \text{its minor})$$

This method of evaluating a determinant is known as **expansion along the first row.**

Example 5.23

Find the determinant of the matrix

$$A = \begin{pmatrix} 1 & 2 & 1 \\ -1 & 3 & 4 \\ 5 & 1 & 2 \end{pmatrix}$$

Solution

The determinant of A, written as

$$\begin{vmatrix} 1 & 2 & 1 \\ -1 & 3 & 4 \\ 5 & 1 & 2 \end{vmatrix}$$

is found by expanding along its first row;

$$|A| = 1 \begin{vmatrix} 3 & 4 \\ 1 & 2 \end{vmatrix} - 2 \begin{vmatrix} -1 & 4 \\ 5 & 2 \end{vmatrix} + 1 \begin{vmatrix} -1 & 3 \\ 5 & 1 \end{vmatrix}$$

$$= 1(2) - 2(-22) + 1(-16)$$

$$= 2 + 44 - 16$$

$$= 30 \quad \blacktriangle$$

Example 5.24

Find the minors of the elements 1 and 4 in the matrix

$$B = \begin{pmatrix} 7 & 2 & 3 \\ 1 & 0 & 3 \\ 0 & 4 & 2 \end{pmatrix}$$

Solutions

To find the minor of 1 delete its row and column to form the determinant

$\begin{vmatrix} 2 & 3 \\ 4 & 2 \end{vmatrix}$. The required minor is therefore $4 - 12 = -8$.

Similarly, the minor of 4 is $\begin{vmatrix} 7 & 3 \\ 1 & 3 \end{vmatrix} = 21 - 3 = 18.$ $\blacktriangle$

In addition to finding the minor of each element in a matrix, it is often useful to find a related quantity – the **cofactor** of each element. The cofactor is found by imposing on the minor a positive or negative sign depending upon its position, that is, a **place sign**, according to the following rule.

$$
\begin{array}{ccc}
+ & - & + \\
- & + & - \\
+ & - & +
\end{array}
$$

Example 5.25

If

$$
A = \begin{pmatrix} 3 & 2 & 7 \\ 9 & 1 & 0 \\ 3 & -1 & 2 \end{pmatrix}
$$

find the cofactors of 9 and 7.

Solutions

The minor of 9 is $\begin{vmatrix} 2 & 7 \\ -1 & 2 \end{vmatrix} = 4 - (-7) = 11$, but since its place sign is negative, the required cofactor is -11.

The minor of 7 is $\begin{vmatrix} 9 & 1 \\ 3 & -1 \end{vmatrix} = -9 - 3 = -12$. Its place sign is positive, so that the required cofactor is simply -12. ▲

Determinants can also be used to evaluate the vector product of two vectors. If $\mathbf{a} = a_1\mathbf{i} + a_2\mathbf{j} + a_3\mathbf{k}$ and $\mathbf{b} = b_1\mathbf{i} + b_2\mathbf{j} + b_3\mathbf{k}$, we showed in Section 4.5 that $\mathbf{a} \times \mathbf{b}$ is the vector defined by

$$
\mathbf{a} \times \mathbf{b} = (a_2b_3 - a_3b_2)\mathbf{i} + (a_3b_1 - a_1b_3)\mathbf{j} + (a_1b_2 - a_2b_1)\mathbf{k}
$$

If we consider the expansion of the determinant given by:

$$
\begin{vmatrix} \mathbf{i} & \mathbf{j} & \mathbf{k} \\ a_1 & a_2 & a_3 \\ b_1 & b_2 & b_3 \end{vmatrix}
$$

we find the same result. This definition is therefore a convenient mechanism for evaluating a vector product.

■ If $\mathbf{a} = a_1\mathbf{i} + a_2\mathbf{j} + a_3\mathbf{k}$ and $\mathbf{b} = b_1\mathbf{i} + b_2\mathbf{j} + b_3\mathbf{k}$, then

$$
\mathbf{a} \times \mathbf{b} = \begin{vmatrix} \mathbf{i} & \mathbf{j} & \mathbf{k} \\ a_1 & a_2 & a_3 \\ b_1 & b_2 & b_3 \end{vmatrix}
$$

Example 5.26

If $\mathbf{a} = 3\mathbf{i} + \mathbf{j} - 2\mathbf{k}$ and $\mathbf{b} = 4\mathbf{i} + 5\mathbf{k}$ find $\mathbf{a} \times \mathbf{b}$.

Solution

We have

$$\mathbf{a} \times \mathbf{b} = \begin{vmatrix} \mathbf{i} & \mathbf{j} & \mathbf{k} \\ 3 & 1 & -2 \\ 4 & 0 & 5 \end{vmatrix}$$

$$= 5\mathbf{i} - 23\mathbf{j} - 4\mathbf{k} \quad \blacktriangle$$

5.7.1 *Cramer's rule*

A useful application of determinants is to the solution of simultaneous equations. Consider the case of three simultaneous equations in three unknowns:

$$a_{11}x + a_{12}y + a_{13}z = b_1$$

$$a_{21}x + a_{22}y + a_{23}z = b_2$$

$$a_{31}x + a_{32}y + a_{33}z = b_3$$

Cramer's rule states that x, y and z are given by the following ratios of determinants.

$$x = \frac{\begin{vmatrix} b_1 & a_{12} & a_{13} \\ b_2 & a_{22} & a_{23} \\ b_3 & a_{32} & a_{33} \end{vmatrix}}{\begin{vmatrix} a_{11} & a_{12} & a_{13} \\ a_{21} & a_{22} & a_{23} \\ a_{31} & a_{32} & a_{33} \end{vmatrix}} \quad y = \frac{\begin{vmatrix} a_{11} & b_1 & a_{13} \\ a_{21} & b_2 & a_{23} \\ a_{31} & b_3 & a_{33} \end{vmatrix}}{\begin{vmatrix} a_{11} & a_{12} & a_{13} \\ a_{21} & a_{22} & a_{23} \\ a_{31} & a_{32} & a_{33} \end{vmatrix}} \quad z = \frac{\begin{vmatrix} a_{11} & a_{12} & b_1 \\ a_{21} & a_{22} & b_2 \\ a_{31} & a_{32} & b_3 \end{vmatrix}}{\begin{vmatrix} a_{11} & a_{12} & a_{13} \\ a_{21} & a_{22} & a_{23} \\ a_{31} & a_{32} & a_{33} \end{vmatrix}}$$

Note that in all cases the determinant in the denominator is identical and its elements are the coefficients on the left-hand side of the simultaneous equations. When this determinant is zero, Cramer's method will clearly fail.

Example 5.27

Solve

$$3x + 2y - z = 4$$

$$2x - y + 2z = 10$$

$$x - 3y - 4z = 5$$

Solution

We find

$$x = \dfrac{\begin{vmatrix} 4 & 2 & -1 \\ 10 & -1 & 2 \\ 5 & -3 & -4 \end{vmatrix}}{\begin{vmatrix} 3 & 2 & -1 \\ 2 & -1 & 2 \\ 1 & -3 & -4 \end{vmatrix}} = \dfrac{165}{55} = 3$$

Verify for yourself that $y = -2$ and $z = 1$. ▲

EXERCISES 5.4

1. Find $\begin{vmatrix} 4 & 6 \\ 2 & 8 \end{vmatrix}$, $\begin{vmatrix} 1 & 3 & 4 \\ 2 & 1 & 0 \\ 3 & 5 & -1 \end{vmatrix}$ and

$\begin{vmatrix} 6 & 7 & 2 \\ 1 & 4 & 3 \\ -1 & 1 & 4 \end{vmatrix}$.

2. Find $\begin{vmatrix} \cos \omega t & \sin \omega t \\ -\sin \omega t & \cos \omega t \end{vmatrix}$.

3. Evaluate $\begin{vmatrix} 5 & 0 & 0 \\ 6 & 3 & 2 \\ 4 & 5 & 7 \end{vmatrix}$ and $\begin{vmatrix} 9 & 0 & 0 \\ 0 & 7 & 0 \\ 0 & 0 & 8 \end{vmatrix}$.

4. If $A = \begin{pmatrix} 2 & -1 & 7 \\ 0 & 8 & 4 \\ 3 & 6 & 4 \end{pmatrix}$, find $|A|$ and $|A^T|$.

Comment upon your result.

5. Use Cramer's rule to solve
   (a) $2x - 3y + z = 0$
   $5x + 4y + z = 10$
   $2x - 2y - z = -1$
   (b) $3x + y = -1$
   $2x - y + z = -1$
   $5x + 5y - 7z = -16$

6. Given
   $$A = \begin{pmatrix} 3 & 7 & 6 \\ -2 & 1 & 0 \\ 4 & 2 & -5 \end{pmatrix}$$
   (a) find $|A|$
   (b) find the cofactors of the elements of row 2, that is, $-2, 1, 0$.
   (c) calculate
   $-2 \times$ (cofactor of -2)
   $+ 1 \times$ (cofactor of 1)
   $+ 0 \times$ (cofactor of 0).
   What do you deduce?

7. If $\mathbf{a} = 7\mathbf{i} + 11\mathbf{j} - 2\mathbf{k}$ and $\mathbf{b} = 6\mathbf{i} - 3\mathbf{j} + \mathbf{k}$ find $\mathbf{a} \times \mathbf{b}$.

5.8 The inverse of a 3 × 3 matrix

One application of determinants is to find the inverse of a 3×3 matrix. Given a 3×3 matrix, A, its inverse is found as follows.

(1) Find the transpose of A, by interchanging the rows and columns of A.

(2) Replace each element of A^T by its cofactor, that is, by its minor together with its associated place sign. The resulting matrix is known as the **adjoint** of A, denoted $adj(A)$.

(3) Finally, the inverse of A is given by:

■ $A^{-1} = \dfrac{adj(A)}{|A|}$

Example 5.28

Find the inverse of:

$$A = \begin{pmatrix} 1 & -2 & 0 \\ 3 & 1 & 5 \\ -1 & 2 & 3 \end{pmatrix}$$

Solution

$$A^T = \begin{pmatrix} 1 & 3 & -1 \\ -2 & 1 & 2 \\ 0 & 5 & 3 \end{pmatrix}$$

Replacing each element of A^T by its cofactor, we find:

$$adj(A) = \begin{pmatrix} -7 & 6 & -10 \\ -14 & 3 & -5 \\ 7 & 0 & 7 \end{pmatrix}$$

The determinant of A is given by,

$$|A| = 1 \begin{vmatrix} 1 & 5 \\ 2 & 3 \end{vmatrix} - (-2) \begin{vmatrix} 3 & 5 \\ -1 & 3 \end{vmatrix} + 0 \begin{vmatrix} 3 & 1 \\ -1 & 2 \end{vmatrix}$$

$$= (1)(-7) + (2)(14)$$

$$= 21$$

Therefore,

$$A^{-1} = \frac{adj(A)}{|A|} = \frac{1}{21} \begin{pmatrix} -7 & 6 & -10 \\ -14 & 3 & -5 \\ 7 & 0 & 7 \end{pmatrix}$$

Note, this solution should be checked by forming AA^{-1} to give I. ▲

It is clear that should $|A| = 0$ then no inverse will exist since then the quantity $1/|A|$ is undefined. Recall that such a matrix is said to be singular.

■ For any square matrix A, the following statements are equivalent:

$|A| = 0$

A is singular

A has no inverse

EXERCISES 5.5

1. Find $\text{adj}(A)$, $|A|$ and, if it exists, A^{-1}, if

(a)

$$A = \begin{pmatrix} 2 & -3 & 1 \\ 5 & 4 & 1 \\ 2 & -2 & -1 \end{pmatrix}$$

(b)

$$A = \begin{pmatrix} 3 & 1 & 0 \\ 2 & -1 & 1 \\ 5 & 5 & -7 \end{pmatrix}$$

(c)

$$A = \begin{pmatrix} 2 & -1 & 4 \\ 5 & -2 & 9 \\ 3 & 2 & -1 \end{pmatrix}$$

2. If $P = \begin{pmatrix} 10 & -5 & -4 \\ -5 & 10 & -3 \\ -4 & -3 & 8 \end{pmatrix}$,

find $\text{adj}(P)$ and $|P|$. Deduce P^{-1}.

5.9 Application to the solution of simultaneous equations

The matrix techniques we have developed allow the solution of simultaneous equations to be found in a systematic way.

Example 5.29

Use a matrix method to solve the simultaneous equations

$$2x + 4y = 14$$
$$x - 3y = -8 \tag{5.6}$$

Solution

We first note that the system of equations can be written in matrix form as follows:

$$\begin{pmatrix} 2 & 4 \\ 1 & -3 \end{pmatrix} \begin{pmatrix} x \\ y \end{pmatrix} = \begin{pmatrix} 14 \\ -8 \end{pmatrix} \tag{5.7}$$

To understand this expression it is necessary that matrix multiplication has been fully mastered, for, by multiplying out the left-hand side, we find

$$\begin{pmatrix} 2 & 4 \\ 1 & -3 \end{pmatrix}\begin{pmatrix} x \\ y \end{pmatrix} = \begin{pmatrix} 2x + 4y \\ 1x - 3y \end{pmatrix}$$

and the form (5.7) follows immediately.

We can write Equation (5.7) as

$$AX = B \tag{5.8}$$

where A is the matrix $\begin{pmatrix} 2 & 4 \\ 1 & -3 \end{pmatrix}$, X is the matrix $\begin{pmatrix} x \\ y \end{pmatrix}$ and B is the matrix $\begin{pmatrix} 14 \\ -8 \end{pmatrix}$.

In order to find $X = \begin{pmatrix} x \\ y \end{pmatrix}$ it is now necessary to make X the subject of the equation $AX = B$. We can premultiply Equation (5.8) by A^{-1}, the inverse of A, provided such an inverse exists, to give:

$$A^{-1}AX = A^{-1}B$$

Then, noting that $A^{-1}A = I$, we find:

$$IX = A^{-1}B$$

that is, $X = A^{-1}B$

using the properties of the identity matrix. We have now made X the subject of the equation as required and we see that to find X we must premultiply the right-hand side of Equation (5.8) by the inverse of A.

In this case

$$A^{-1} = \frac{1}{-10}\begin{pmatrix} -3 & -4 \\ -1 & 2 \end{pmatrix}$$

$$= \begin{pmatrix} 3/10 & 2/5 \\ 1/10 & -1/5 \end{pmatrix}$$

and

$$A^{-1}B = \begin{pmatrix} 3/10 & 2/5 \\ 1/10 & -1/5 \end{pmatrix}\begin{pmatrix} 14 \\ -8 \end{pmatrix}$$

$$= \begin{pmatrix} 1 \\ 3 \end{pmatrix}$$

that is, $X = \begin{pmatrix} x \\ y \end{pmatrix} = \begin{pmatrix} 1 \\ 3 \end{pmatrix}$, so that $x = 1$ and $y = 3$ is the required solution. ▲

■ If $AX = B$ then $X = A^{-1}B$ provided A^{-1} exists.

This technique can be applied to three equations in three unknowns in an analogous way.

Example 5.30

Express the following equations in the form $AX = B$ and hence solve them:

$$3x + 2y - z = 4$$
$$2x - y + 2z = 10$$
$$x - 3y - 4z = 5$$

Solution

Using the rules of matrix multiplication, we find:

$$\begin{pmatrix} 3 & 2 & -1 \\ 2 & -1 & 2 \\ 1 & -3 & -4 \end{pmatrix} \begin{pmatrix} x \\ y \\ z \end{pmatrix} = \begin{pmatrix} 4 \\ 10 \\ 5 \end{pmatrix}$$

which is in the form $AX = B$. The matrix A is called the **coefficient matrix** and is simply the coefficients of x, y and z in the equations. As before,

$$AX = B$$
$$A^{-1}AX = A^{-1}B$$
$$IX = X = A^{-1}B$$

We must therefore find the inverse of A in order to solve the equations. To invert A we use the adjoint. If

$$A = \begin{pmatrix} 3 & 2 & -1 \\ 2 & -1 & 2 \\ 1 & -3 & -4 \end{pmatrix}$$

then

$$A^T = \begin{pmatrix} 3 & 2 & 1 \\ 2 & -1 & -3 \\ -1 & 2 & -4 \end{pmatrix}$$

and you should verify that adj(A) is given by:

$$\text{adj}(A) = \begin{pmatrix} 10 & 11 & 3 \\ 10 & -11 & -8 \\ -5 & 11 & -7 \end{pmatrix}$$

The determinant of A is found by expanding along the first row

$$|A| = 3\begin{vmatrix} -1 & 2 \\ -3 & -4 \end{vmatrix} - 2\begin{vmatrix} 2 & 2 \\ 1 & -4 \end{vmatrix} - 1\begin{vmatrix} 2 & -1 \\ 1 & -3 \end{vmatrix}$$

$$= (3)(10) - (2)(-10) - (1)(-5)$$

$$= 30 + 20 + 5$$

$$= 55$$

Therefore,

$$A^{-1} = \frac{\text{adj}(A)}{|A|} = \frac{1}{55}\begin{pmatrix} 10 & 11 & 3 \\ 10 & -11 & -8 \\ -5 & 11 & -7 \end{pmatrix}$$

Finally, the solution X is given by:

$$X = \begin{pmatrix} x \\ y \\ z \end{pmatrix} = A^{-1}B = \frac{1}{55}\begin{pmatrix} 10 & 11 & 3 \\ 10 & -11 & -8 \\ -5 & 11 & -7 \end{pmatrix}\begin{pmatrix} 4 \\ 10 \\ 5 \end{pmatrix}$$

$$= \begin{pmatrix} 3 \\ -2 \\ 1 \end{pmatrix}$$

that is, the solution is $x = 3$, $y = -2$ and $z = 1$. ▲

EXERCISES 5.6

Solve the following equations $AX = B$ by finding A^{-1}, if it exists.

(a) $\begin{pmatrix} 6 & 3 \\ 5 & 2 \end{pmatrix}\begin{pmatrix} x \\ y \end{pmatrix} = \begin{pmatrix} 12 \\ 9 \end{pmatrix}$

(b) $\begin{pmatrix} 4 & 4 \\ 1 & 3 \end{pmatrix}\begin{pmatrix} x \\ y \end{pmatrix} = \begin{pmatrix} 20 \\ 11 \end{pmatrix}$

(c) $\begin{pmatrix} 2 & -1 \\ 3 & 2 \end{pmatrix}\begin{pmatrix} x \\ y \end{pmatrix} = \begin{pmatrix} -4 \\ 1 \end{pmatrix}$

(d) $\begin{pmatrix} 4 & 1 & 3 \\ 2 & -1 & 4 \\ 0 & 1 & 5 \end{pmatrix}\begin{pmatrix} x \\ y \\ z \end{pmatrix} = \begin{pmatrix} 20 \\ 20 \\ 20 \end{pmatrix}$

(e) $\begin{pmatrix} 4 & 1 & 3 \\ 2 & -1 & 4 \\ 0 & 1 & 5 \end{pmatrix}\begin{pmatrix} x \\ y \\ z \end{pmatrix} = \begin{pmatrix} 15 \\ 12 \\ 17 \end{pmatrix}$

(f) $\begin{pmatrix} 4 & 1 & 3 \\ 2 & -1 & 4 \\ 0 & 1 & 5 \end{pmatrix}\begin{pmatrix} x \\ y \\ z \end{pmatrix} = \begin{pmatrix} 0 \\ 0 \\ 0 \end{pmatrix}$

5.10 Gaussian elimination

An alternative technique for the solution of simultaneous equations is that of Gaussian elimination which we introduce by means of the following trivial example.

Example 5.31

Use Gaussian elimination to solve

$$2x + 3y = 1$$
$$x + y = 3$$

Solution

First consider the equations with a step pattern imposed as follows:

$$2x + 3y = 1$$
$$x + y = 3$$

Our aim will be to perform various operations on these equations to remove or eliminate all the values underneath the step. You will probably remember from your early work on simultaneous equations that in order to eliminate a variable from an equation, that equation can be multiplied by any suitable number and then added to or subtracted from another equation. In this example we can eliminate the x term from below the step by multiplying the second equation by 2 and subtracting the first equation. Since the first equation is entirely above the step we shall leave it as it stands. This whole process will be written as follows:

$$
\begin{array}{lll}
R_1 & 2x + 3y = 1 & (1)\\
R_2 \rightarrow 2R_2 - R_1 & 0x - y = 5 & (2)
\end{array}
\qquad (5.9)
$$

where the symbol R_1 means that equation 1 is unaltered, and $R_2 \rightarrow 2R_2 - R_1$ means that equation 2 has been replaced by 2 × equation 2 − equation 1. All this may seem to be overcomplicating a simple problem but a moment's study of Equation (5.9) will reveal why this 'stepped' form is useful. Because the value under the step is zero we can read off y from the last line, that is, $-y = 5$, so that

$$y = -5$$

Knowing y we can then move up to the first equation and substitute for y to find x.

$$2x + 3(-5) = 1$$
$$x = 8$$

This last stage is known as **back substitution**. ▲

Before we consider another example, let us note some important points.

(1) It is necessary to write down the operations used as indicated previously. This aids checking and provides a record of the working used.

(2) The operations allowed to eliminate unwanted variables are:

    (a) any equation can be multiplied by any non-zero constant;

    (b) any equation can be added to or subtracted from any other equation;

    (c) equations can be interchanged.

It is often convenient to use matrices to carry out this method, in which case the operations allowed are referred to as **row operations**. The advantage of using matrices is that it is then unnecessary to write down x, y (and later z) each time. To do this, we first form the **augmented matrix**:

$$\begin{pmatrix} 2 & 3 & 1 \\ 1 & 1 & 3 \end{pmatrix}$$

so called because the coefficient matrix $\begin{pmatrix} 2 & 3 \\ 1 & 1 \end{pmatrix}$ is augmented by the right-hand

side matrix $\begin{pmatrix} 1 \\ 3 \end{pmatrix}$. It is to be understood that this notation means $2x + 3y = 1$, and

so on, so that we no longer write down x and y. Each row of the augmented matrix corresponds to one equation. The aim, as before, is to carry out row operations on the stepped form:

$$\begin{pmatrix} 2 & 3 & 1 \\ 1 & 1 & 3 \end{pmatrix}$$

in order to obtain values of zero under the step. Clearly to achieve the required form, the row operations we performed earlier are required, that is,

$$\begin{matrix} R_1 \\ R_2 \to 2R_2 - R_1 \end{matrix} \qquad \begin{pmatrix} 2 & 3 & 1 \\ 0 & -1 & 5 \end{pmatrix}$$

The last line means $0x - 1y = 5$, that is, $y = -5$, and finally back substitution yields x, as before.

This technique has other advantages in that it allows us to observe other forms of behaviour. We shall see that some equations have a unique solution, some have no solutions, while others have an infinite number.

Example 5.32

Use Gaussian elimination to solve:

$$2x + 3y = 4$$

$$4x + 6y = 7$$

Solution

In augmented matrix form we have:

$$\left(\begin{array}{cc|c} 2 & 3 & 4 \\ 4 & 6 & 7 \end{array}\right)$$

We proceed to eliminate entries under the step:

$$\begin{array}{c} R_1 \\ R_2 \to R_2 - 2R_1 \end{array} \quad \left(\begin{array}{cc|c} 2 & 3 & 4 \\ 0 & 0 & -1 \end{array}\right)$$

Study of the last line seems to imply that $0x + 0y = -1$ which is clearly nonsense. When this happens the equations have no solutions and we say that the simultaneous equations are **inconsistent**. ▲

Example 5.33

Use Gaussian elimination to solve:

$$x + y = 0$$

$$2x + 2y = 0$$

Solution

In augmented matrix form we have:

$$\left(\begin{array}{cc|c} 1 & 1 & 0 \\ 2 & 2 & 0 \end{array}\right)$$

Eliminating entries under the step we find:

$$\begin{array}{c} R_1 \\ R_2 \to R_2 - 2R_1 \end{array} \quad \left(\begin{array}{cc|c} 1 & 1 & 0 \\ 0 & 0 & 0 \end{array}\right)$$

This last line implies that $0x + 0y = 0$. This is not an inconsistency, but we are now observing a third type of behaviour. Whenever this happens we need to introduce what are called **free variables**. The first row starts off with a non-zero x. There is now no row which starts off with a non-zero y. We therefore say y is free and choose it to be anything we please, that is,

$$y = \lambda \qquad \lambda \text{ is our free choice}$$

Then back substitution occurs as before.

$$x + \lambda = 0$$

$$x = -\lambda$$

The solution is therefore $x = -\lambda$, $y = \lambda$, where λ is any number. There are thus an infinite number of solutions, for example,

$$x = -1 \qquad y = 1$$

or

$$x = \frac{1}{2} \qquad y = -\frac{1}{2}$$

and so on. ▲

Observation of the coefficient matrices in the last two examples shows that they have a determinant of zero. Whenever this happens we shall find the equations are either inconsistent or have an infinite number of solutions. We shall next consider the generalization of this method to three equations in three unknowns.

Example 5.34

Solve by Gaussian elimination:

$$x - 4y - 2z = 21$$
$$2x + y + 2z = 3$$
$$3x + 2y - z = -2$$

Solution

We first form the augmented matrix and add the stepped pattern as indicated.

$$\begin{pmatrix} 1 & -4 & -2 & 21 \\ 2 & 1 & 2 & 3 \\ 3 & 2 & -1 & -2 \end{pmatrix}$$

The aim is to eliminate all numbers underneath the steps by carrying out appropriate row operations. This should be carried out by eliminating unwanted numbers in the first column first. We find:

$$\begin{matrix} R_1 \\ R_2 \rightarrow R_2 - 2R_1 \\ R_3 \rightarrow R_3 - 3R_1 \end{matrix} \begin{pmatrix} 1 & -4 & -2 & 21 \\ 0 & 9 & 6 & -39 \\ 0 & 14 & 5 & -65 \end{pmatrix}$$

We have combined the elimination of unwanted numbers in the first column into one stage. We now remove unwanted numbers in the second column:

$$\begin{matrix} R_1 \\ R_2 \\ R_3 \rightarrow R_3 - \frac{14}{9}R_2 \end{matrix} \begin{pmatrix} 1 & -4 & -2 & 21 \\ 0 & 9 & 6 & -39 \\ 0 & 0 & -\frac{39}{9} & -\frac{13}{3} \end{pmatrix}$$

and the elimination is complete. Although $R_3 \rightarrow R_3 + \frac{14}{4}R_1$ would eliminate the 14, it would re-introduce a non-zero term into the first column. It is therefore essential to use the second row and not the first to eliminate this element. We can now read off z since the last equation states $0x + 0y - \frac{13}{3}z = -13/3$, that is, $z = 1$. Back substitution of $z = 1$ in the second equation gives $y = -5$ and finally, substitution of z and y into the first equation gives $x = 3$. ▲

Example 5.35

Solve the following equations by Gaussian elimination:

$$x - y + z = 3$$
$$x + 5y - 5z = 2$$
$$2x + y - z = 1$$

Solution

Forming the augmented matrix, we find:

$$\begin{pmatrix} 1 & -1 & 1 & 3 \\ 1 & 5 & -5 & 2 \\ 2 & 1 & -1 & 1 \end{pmatrix}$$

Then, as before, we aim to eliminate all non-zero entries under the step. Starting with those in the first column, we find:

$$\begin{matrix} R_1 \\ R_2 \to R_2 - R_1 \\ R_3 \to R_3 - 2R_1 \end{matrix} \quad \begin{pmatrix} 1 & -1 & 1 & 3 \\ 0 & 6 & -6 & -1 \\ 0 & 3 & -3 & -5 \end{pmatrix}$$

Then,

$$\begin{matrix} R_1 \\ R_2 \\ R_3 \to 2R_3 - R_2 \end{matrix} \quad \begin{pmatrix} 1 & -1 & 1 & 3 \\ 0 & 6 & -6 & -1 \\ 0 & 0 & 0 & -9 \end{pmatrix}$$

This last line implies that $0x + 0y + 0z = -9$ which is clearly inconsistent. We conclude that there are no solutions. ▲

You will see from Examples 5.34 and 5.35 that not only have all entries under the step been reduced to zero, but also each successive row contains more leading zeros than the previous one. We say the system has been reduced to **echelon** form. More generally the system has been reduced to echelon form if for $i < j$ the number of leading zeros in row j is larger than the number in row i. Consider Example 5.36.

Example 5.36

Solve the following equations by Gaussian elimination:

$$2x - y + z = 2$$
$$-2x + y + z = 4$$
$$6x - 3y - 2z = -9$$

Solution

Forming the augmented matrix, we have

$$\begin{pmatrix} 2 & -1 & 1 & 2 \\ -2 & 1 & 1 & 4 \\ 6 & -3 & -2 & -9 \end{pmatrix}$$

Eliminating the unwanted values in the first column, we find:

$$\begin{array}{c} R_1 \\ R_2 \to R_2 + R_1 \\ R_3 \to R_3 - 3R_1 \end{array} \quad \begin{pmatrix} 2 & -1 & 1 & 2 \\ 0 & 0 & 2 & 6 \\ 0 & 0 & -5 & -15 \end{pmatrix}$$

Entries under the step are now zero. To reduce the matrix to echelon form we must ensure each successive row has more leading zeros than the row before. We continue:

$$\begin{array}{c} R_1 \\ R_2 \\ R_3 \to 2R_3 + 5R_2 \end{array} \quad \begin{pmatrix} 2 & -1 & 1 & 2 \\ 0 & 0 & 2 & 6 \\ 0 & 0 & 0 & 0 \end{pmatrix}$$

which is now in echelon form.

In this form there is a row which starts off with a non-zero x value, that is, the first row, there is a row which starts off with a non-zero z value but no row which starts off with a non-zero y value. Therefore, we choose y to be the free variable, $y = \lambda$ say. From the second row we have $z = 3$ and from the first $2x - y + z = 2$, so that $2x = \lambda - 1$, that is, $x = (\lambda - 1)/2$. ▲

5.10.1 *Finding the inverse matrix using row operations*

A similar technique can be used to find the inverse of a square matrix A where this exists. Suppose we are given the matrix A and wish to find its inverse B. Then we know:

$$AB = I$$

that is,

$$\begin{pmatrix} a_{11} & a_{12} & a_{13} \\ a_{21} & a_{22} & a_{23} \\ a_{31} & a_{32} & a_{33} \end{pmatrix} \begin{pmatrix} b_{11} & b_{12} & b_{13} \\ b_{21} & b_{22} & b_{23} \\ b_{31} & b_{32} & b_{33} \end{pmatrix} = \begin{pmatrix} 1 & 0 & 0 \\ 0 & 1 & 0 \\ 0 & 0 & 1 \end{pmatrix}$$

We form the augmented matrix

$$\begin{pmatrix} a_{11} & a_{12} & a_{13} & 1 & 0 & 0 \\ a_{21} & a_{22} & a_{23} & 0 & 1 & 0 \\ a_{31} & a_{32} & a_{33} & 0 & 0 & 1 \end{pmatrix}$$

Now carry out row operations on this matrix in such a way that the left-hand side is reduced to a 3×3 identity matrix. The matrix which then remains on the right-hand side is the required inverse.

Example 5.37

Find the inverse of

$$A = \begin{pmatrix} -1 & 8 & -2 \\ -6 & 49 & -10 \\ -4 & 34 & -5 \end{pmatrix}$$

by row reduction to the identity.

Solution

We form the augmented matrix

$$\begin{pmatrix} -1 & 8 & -2 & 1 & 0 & 0 \\ -6 & 49 & -10 & 0 & 1 & 0 \\ -4 & 34 & -5 & 0 & 0 & 1 \end{pmatrix}$$

We now carry out row operations on the whole matrix to reduce the left-hand side to an identity matrix. This means we must eliminate all the elements off the diagonal. Work through the following calculation yourself:

$$\begin{matrix} R_1 \\ R_2 \rightarrow R_2 - 6R_1 \\ R_3 \rightarrow R_3 - 4R_1 \end{matrix} \begin{pmatrix} -1 & 8 & -2 & 1 & 0 & 0 \\ 0 & 1 & 2 & -6 & 1 & 0 \\ 0 & 2 & 3 & -4 & 0 & 1 \end{pmatrix}$$

This has removed all the off-diagonal entries in column 1. To remove those in column 2:

$$\begin{matrix} R_1 \rightarrow R_1 - 8R_2 \\ R_2 \\ R_3 \rightarrow R_3 - 2R_2 \end{matrix} \begin{pmatrix} -1 & 0 & -18 & 49 & -8 & 0 \\ 0 & 1 & 2 & -6 & 1 & 0 \\ 0 & 0 & -1 & 8 & -2 & 1 \end{pmatrix}$$

To remove those in column 3:

$$\begin{matrix} R_1 \rightarrow R_1 - 18R_3 \\ R_2 \rightarrow R_2 + 2R_3 \\ R_3 \end{matrix} \begin{pmatrix} -1 & 0 & 0 & -95 & 28 & -18 \\ 0 & 1 & 0 & 10 & -3 & 2 \\ 0 & 0 & -1 & 8 & -2 & 1 \end{pmatrix}$$

We must now adjust the '−1' entries to obtain the identity matrix:

$$\begin{matrix} R_1 \to -R_1 \\ R_2 \\ R_3 \to -R_3 \end{matrix} \begin{pmatrix} 1 & 0 & 0 & 95 & -28 & 18 \\ 0 & 1 & 0 & 10 & -3 & 2 \\ 0 & 0 & 1 & -8 & 2 & -1 \end{pmatrix}$$

Finally, the required inverse is the matrix remaining on the right-hand side:

$$\begin{pmatrix} 95 & -28 & 18 \\ 10 & -3 & 2 \\ -8 & 2 & -1 \end{pmatrix}$$

You should check this result by evaluating AA^{-1}. ▲

EXERCISES 5.7

1. Solve the following equations by Gaussian elimination:

(a) $2x - 3y = 32$

$3x + 7y = -21$

(b) $2x + y - 3z = -5$

$x - y + 2z = 12$

$7x - 2y + 3z = 37$

(c) $x + y - z = 1$

$3x - y + 5z = 3$

$7x + 2y + 3z = 7$

(d) $2x + y - z = -9$

$3x - 2y + 4z = 5$

$-2x - y + 7z = 33$

(e) $4x + 7y + 8z = 2$

$5x + 8y + 13z = 0$

$3x + 5y + 7z = 1$

2. Use Gaussian elimination to solve

$x + y + z = 7$

$x - y + 2z = 9$

$2x + y - z = 1$

3. Find the inverses of the following matrices using the technique of Example 5.37:

(a) $\begin{pmatrix} 4 & 1 \\ 3 & 2 \end{pmatrix}$

(b) $\begin{pmatrix} 4 & 2 & 1 \\ 0 & 3 & 4 \\ -1 & 1 & 3 \end{pmatrix}$

(c) $\begin{pmatrix} 1 & 0 & 3 \\ 2 & 1 & 5 \\ -7 & 2 & 1 \end{pmatrix}$

5.11 Analysis of electrical networks

Matrix algebra is very useful for the analysis of certain types of electrical network. For such networks it is possible to produce a mathematical model consisting

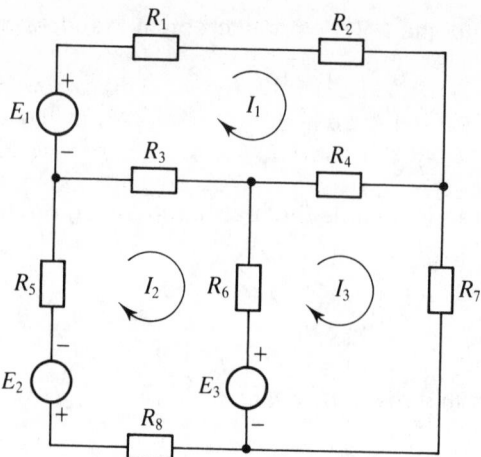

Figure 5.2 An electrical network with mesh currents shown.

of simultaneous equations which can be solved using the techniques just described. We will consider the case when the network consists of resistors and voltage sources. The technique is similar for other types of network.

In order to develop this approach, it is necessary to develop a systematic method for writing the circuit equations. The method adopted depends on what the unknown variables are. A common problem is that the voltage sources and the resistor values are known and it is desired to know the current values in each part of the network. This can be formulated as a matrix equation. Given

$$V = RI'$$

where

V = voltage vector for the network

I' = current vector for the network

R = matrix of resistor values

the problem is to calculate I' when V and R are known. I' is used to avoid confusion with the identity matrix.

Any size of electrical network can be analysed using this approach. We will limit the discussion to the case where I' has three components, for simplicity. The extension to larger networks is straightforward. Consider the electrical network of Figure 5.2. Mesh currents have been drawn for each of the loops in the circuit. A **mesh** is defined as a loop that cannot contain a smaller closed current path. For convenience, each mesh current is drawn in a clockwise direction even though it may turn out to be in the opposite direction when the calculations have been performed. The net current in each branch of the circuit can be obtained by combining the mesh currents. These are termed the **branch currents**. The concept of a mesh current may appear slightly abstract but it does provide a convenient mechanism for analysing electrical networks. We will examine an approach that avoids the use of mesh currents later in this section.

The next stage is to make use of Kirchhoff's voltage law for each of the meshes in the network. This states that the algebraic sum of the voltages around any closed loop in an electrical network is zero. Therefore the sum of the voltage rises must equal the sum of voltage drops. When applying Kirchhoff's voltage law it is important to use the correct sign for a voltage source depending on whether or not it is 'aiding' a mesh current.

For mesh 1

$$E_1 = I_1 R_1 + I_1 R_2 + (I_1 - I_3)R_4 + (I_1 - I_2)R_3$$
$$E_1 = I_1(R_1 + R_2 + R_4 + R_3) + I_2(-R_3) + I_3(-R_4)$$

For mesh 2

$$-E_2 - E_3 = I_2 R_5 + (I_2 - I_1)R_3 + (I_2 - I_3)R_6 + I_2 R_8$$
$$-E_2 - E_3 = I_1(-R_3) + I_2(R_5 + R_3 + R_6 + R_8) + I_3(-R_6)$$

For mesh 3

$$E_3 = (I_3 - I_2)R_6 + (I_3 - I_1)R_4 + I_3 R_7$$
$$E_3 = I_1(-R_4) + I_2(-R_6) + I_3(R_6 + R_4 + R_7)$$

These equations can be written in matrix form as:

$$
\begin{pmatrix} E_1 \\ -E_2 - E_3 \\ E_3 \end{pmatrix} = \begin{pmatrix} R_1 + R_2 + R_4 + R_3 & -R_3 & -R_4 \\ -R_3 & R_5 + R_3 + R_6 + R_8 & -R_6 \\ -R_4 & -R_6 & R_6 + R_4 + R_7 \end{pmatrix} \begin{pmatrix} I_1 \\ I_2 \\ I_3 \end{pmatrix}
$$

Example 5.38

Consider the electrical network of Figure 5.3. It has the same structure as that of Figure 5.2 but with actual values for the voltage sources and resistors. Branch currents as well as mesh currents have been shown. Calculate the mesh currents and hence the branch currents for the network.

Solution

We have already obtained the equations for this network. Substituting actual values for the resistors and voltage sources gives:

$$
\begin{pmatrix} 3 \\ -2 - 4 \\ 4 \end{pmatrix} = \begin{pmatrix} 10 & -3 & -1 \\ -3 & 14 & -2 \\ -1 & -2 & 6 \end{pmatrix} \begin{pmatrix} I_1 \\ I_2 \\ I_3 \end{pmatrix}
$$

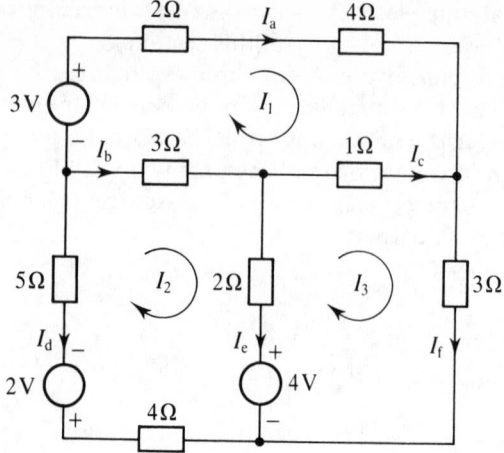

Figure 5.3   The electrical network of Figure 5.2 with values for the source voltages and resistors added.

This is now in the form $V = RI'$. We shall solve these equations by Gaussian elimination. Forming the augmented matrix, we have:

$$\begin{pmatrix} 10 & -3 & -1 & 3 \\ -3 & 14 & -2 & -6 \\ -1 & -2 & 6 & 4 \end{pmatrix}$$

Then

$$\begin{array}{c} R_1 \\ R_2 \rightarrow 10R_2 + 3R_1 \\ R_3 \rightarrow 10R_3 + R_1 \end{array} \begin{pmatrix} 10 & -3 & -1 & 3 \\ 0 & 131 & -23 & -51 \\ 0 & -23 & 59 & 43 \end{pmatrix}$$

Then

$$\begin{array}{c} R_1 \\ R_2 \\ R_3 \rightarrow 131R_3 + 23R_2 \end{array} \begin{pmatrix} 10 & -3 & -1 & 3 \\ 0 & 131 & -23 & -51 \\ 0 & 0 & 7200 & 4460 \end{pmatrix}$$

Hence,

$$I_3 = \frac{4460}{7200} = 0.619 \text{ A}$$

Similarly,

$$I_2 = \frac{-51 + 23(0.619)}{131} = -0.281 \text{ A}$$

Finally,

$$I_1 = \frac{3 + 0.619 + 3(-0.281)}{10} = 0.278 \text{ A}$$

The branch currents are then

$I_a = I_1 = 278 \text{ mA}$

$I_b = I_2 - I_1 = -281 - 278 = -559 \text{ mA}$

$I_c = I_3 - I_1 = 619 - 278 = 341 \text{ mA}$

$I_d = -I_2 = 281 \text{ mA}$

$I_e = I_2 - I_3 = -281 - 619 = -900 \text{ mA}$

$I_f = I_3 = 619 \text{ mA}$ ▲

An alternative approach to analysing an electrical network is to use the **node voltage method**. For our purposes the nodes of an electrical network can be thought of as the 'islands' of equal potential that lie between electrical components and sources. The procedure is as follows:

(1) Pick a reference node. In order to simplify the equations this is usually chosen to be the node which is common to the largest number of voltage sources and/or the largest number of branches.

(2) Assign a node voltage variable to all of the other nodes. If two nodes are separated solely by a voltage source then only one of the nodes need be assigned a voltage variable. The node voltages are all measured with respect to the reference node.

(3) At each node, write Kirchhoff's current law in terms of the node voltages. Note that once the node voltages have been calculated it is easy to obtain the branch currents.

We will again examine the network of Figure 5.2, but this time use the node voltage method. The network is shown in Figure 5.4 with node voltages assigned and branch currents labelled. The reference node is indicated by using the earth symbol. Writing Kirchhoff's current law for each node, we obtain:

node a

$I_a = I_a$

$$\frac{V_b + E_1 - V_a}{R_1} = \frac{V_a - V_d}{R_2}$$

$V_b R_2 + E_1 R_2 - V_a R_2 = V_a R_1 - V_d R_1$

$V_a(R_1 + R_2) - V_b R_2 - V_d R_1 = E_1 R_2$

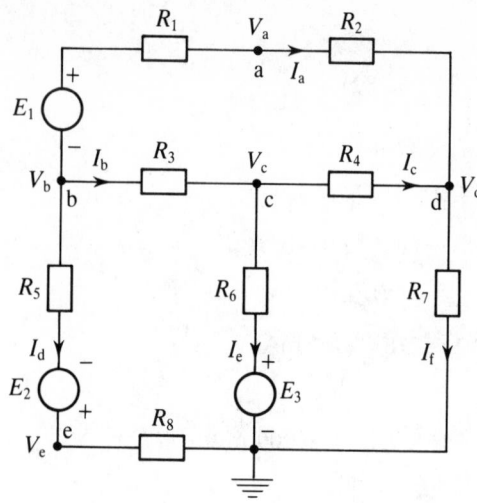

Figure 5.4   The network of Figure 5.2 with node voltages labelled.

node b

$$I_a + I_b + I_d = 0$$

$$\frac{V_b + E_1 - V_a}{R_1} + \frac{V_b - V_c}{R_3} + \frac{V_b + E_2 - V_e}{R_5} = 0$$

Rearrangement yields:

$$V_b R_3 R_5 + E_1 R_3 R_5 - V_a R_3 R_5 + V_b R_1 R_5 - V_c R_1 R_5 + V_b R_1 R_3$$
$$+ E_2 R_1 R_3 - V_e R_1 R_3 = 0$$

that is,

$$V_a R_3 R_5 - V_b(R_1 R_3 + R_1 R_5 + R_3 R_5) + V_c R_1 R_5 + V_e R_1 R_3$$
$$= E_1 R_3 R_5 + E_2 R_1 R_3$$

node c

$$I_b = I_c + I_e$$

$$\frac{V_b - V_c}{R_3} = \frac{V_c - V_d}{R_4} + \frac{V_c - E_3}{R_6}$$

so that

$$V_b R_4 R_6 - V_c R_4 R_6 = V_c R_3 R_6 - V_d R_3 R_6 + V_c R_3 R_4 - E_3 R_3 R_4$$

that is,

$$V_b R_4 R_6 - V_c(R_4 R_6 + R_3 R_6 + R_3 R_4) + V_d R_3 R_6 = -E_3 R_3 R_4$$

node d

$$I_a + I_c = I_f$$

$$\frac{V_a - V_d}{R_2} + \frac{V_c - V_d}{R_4} = \frac{V_d}{R_7}$$

$$V_a R_4 R_7 - V_d R_4 R_7 + V_c R_2 R_7 - V_d R_2 R_7 = V_d R_2 R_4$$

$$V_a R_4 R_7 + V_c R_2 R_7 - V_d (R_4 R_7 + R_2 R_7 + R_2 R_4) = 0$$

node e

$$I_d = I_d$$

$$\frac{V_b + E_2 - V_e}{R_5} = \frac{V_e}{R_8}$$

$$V_b R_8 - V_e (R_5 + R_8) = -E_2 R_8$$

These equations can be written in matrix form $AV = B$, where A is the matrix

$$\begin{pmatrix} R_1 + R_2 & -R_2 & 0 & -R_1 & 0 \\ R_3 R_5 & -R_1 R_3 - R_1 R_5 - R_3 R_5 & R_1 R_5 & 0 & R_1 R_3 \\ 0 & R_4 R_6 & -R_4 R_6 - R_3 R_6 - R_3 R_4 & R_3 R_6 & 0 \\ R_4 R_7 & 0 & R_2 R_7 & -R_4 R_7 - R_2 R_7 - R_2 R_4 & 0 \\ 0 & R_8 & 0 & 0 & -R_5 - R_8 \end{pmatrix}$$

and

$$V = \begin{pmatrix} V_a \\ V_b \\ V_c \\ V_d \\ V_e \end{pmatrix} \quad \text{and} \quad B = \begin{pmatrix} E_1 R_2 \\ E_1 R_3 R_5 + E_2 R_1 R_3 \\ -E_3 R_3 R_4 \\ 0 \\ -E_2 R_8 \end{pmatrix}$$

The equations would generally be solved by Gaussian elimination to obtain the node voltages and hence the branch currents.

Using the component values from Example 5.38, these equations become

$$\begin{pmatrix} 6 & -4 & 0 & -2 & 0 \\ 15 & -31 & 10 & 0 & 6 \\ 0 & 2 & -11 & 6 & 0 \\ 3 & 0 & 12 & -19 & 0 \\ 0 & 4 & 0 & 0 & -9 \end{pmatrix} \begin{pmatrix} V_a \\ V_b \\ V_c \\ V_d \\ V_e \end{pmatrix} = \begin{pmatrix} 12 \\ 57 \\ -12 \\ 0 \\ -8 \end{pmatrix}$$

Use of a computer package avoids the tedious arithmetic associated with Gaussian elimination and yields:

$$V_a = 2.969 \qquad V_b = 0.5250 \qquad V_c = 2.200 \qquad V_d = 1.858 \qquad V_e = 1.122$$

It is then straightforward to calculate the branch currents:

$$I_a = \frac{V_a - V_d}{R_2} = 278 \text{ mA} \qquad I_b = \frac{V_b - V_c}{R_3} = -558 \text{ mA} \qquad I_c = \frac{V_c - V_d}{R_4} = 342 \text{ mA}$$

$$I_d = \frac{V_c}{R_8} = 281 \text{ mA} \qquad I_e = \frac{V_c - E_3}{R_6} = -900 \text{ mA} \qquad I_f = \frac{V_d}{R_7} = 619 \text{ mA}$$

Compare these answers with those of Example 5.38.

It is possible to analyse electrical networks containing more complex elements such as capacitors, inductors, active devices, etc., using the same approach. The equations are more complicated but the technique is the same. Often it is necessary to use iterative techniques in view of the size and complexity of the problem. These are examined in the following section.

5.12 Iterative techniques for the solution of simultaneous equations

The techniques met so far for the solution of simultaneous equations are known as **direct methods**, which generally lead to the solution after a finite number of stages in the calculation process have been carried out. An alternative collection of techniques is available and these are known as **iterative methods**. They generate a sequence of approximate solutions which may converge to the required solution, and are particularly advantageous when large systems of equations are to be solved by computer. We shall study two such techniques here; Jacobi's method and Gauss–Seidel iteration.

Example 5.39

Solve the equations

$$2x + y = 4$$
$$x - 3y = -5$$

using Jacobi's iterative method.

Solution

We first rewrite the equations as:

$$2x = -y + 4$$
$$-3y = -x - 5$$

and then as

$$x = -\tfrac{1}{2}y + 2$$

$$y = \tfrac{1}{3}x + \tfrac{5}{3} \tag{5.10}$$

Jacobi's method involves 'guessing' a solution and substituting the guess in the right-hand side of the equations in (5.10). Suppose we guess $x = 0$, $y = 0$. Substitution then gives:

$$x = 2$$

$$y = \tfrac{5}{3}$$

We now use these values as estimates of the solution and resubstitute into the right-hand side of Equation (5.10). This time we find,

$$x = -\tfrac{1}{2}(\tfrac{5}{3}) + 2 = 1.1667 \quad \text{(to four decimal places)}$$

$$y = \tfrac{1}{3}(2) + \tfrac{5}{3} = 2.3333 \quad \text{(to four decimal places)}$$

The whole process is repeated in the hope that each successive application or **iteration** will give an answer closer to the required solution, that is, successive **iterates** will converge. In order to keep track of the calculations, we label the initial guess $x^{(0)}$, $y^{(0)}$, the result of the first iteration $x^{(1)}$, $y^{(1)}$ and so on. Generally, we find:

$$x^{(n+1)} = -\tfrac{1}{2}y^{(n)} + 2$$

$$y^{(n+1)} = \tfrac{1}{3}x^{(n)} + \tfrac{5}{3}$$

The results of successively applying these formulae are shown in Table 5.1. The sequence of values of $x^{(n)}$ seems to converge to 1 while that of $y^{(n)}$ seems to converge to 2. ▲

Clearly this sort of approach is simple to program and iterative techniques such as Jacobi's method are best implemented on a computer. When writing the

Table 5.1 Iterates produced by Jacobi's method.

Iteration no. (n)	$x^{(n)}$	$y^{(n)}$
0	0	0
1	2.0000	1.6667
2	1.1667	2.3333
3	0.8333	2.0556
4	0.9722	1.9444
5	1.0278	1.9907
6	1.0047	2.0093
7	0.9954	2.0016
8	0.9992	1.9985
9	1.0008	1.9997
10	1.0002	2.0003

program a test should be incorporated so that after each iteration a check for convergence is made by comparing successive iterates. In many cases, even when convergence does occur, it is slow and so other techniques are used which converge more rapidly. The Gauss–Seidel method is attractive for this reason. It uses the most recent approximation to x when calculating y leading to improved rates of convergence as the following example shows.

Example 5.40

Use the Gauss–Seidel method to solve the equations of Example 5.39.

Solution

As before we write the equations in the form:

$$x = -\tfrac{1}{2}y + 2$$
$$y = \tfrac{1}{3}x + \tfrac{5}{3}$$

With $x^{(0)} = 0$, $y^{(0)} = 0$ as our initial guess, we find:

$$x^{(1)} = -\tfrac{1}{2}(0) + 2 = 2$$

To find $y^{(1)}$ we use the most recent approximation to x available, that is, $x^{(1)}$:

$$y^{(1)} = \tfrac{1}{3}(2) + \tfrac{5}{3} = 2.3333$$

Generally, we find:

$$x^{(n+1)} = -\tfrac{1}{2}y^{(n)} + 2$$
$$y^{(n+1)} = \tfrac{1}{3}x^{(n+1)} + \tfrac{5}{3}$$

and the results of successively applying these formulae are shown in Table 5.2. As before, we see that the sequence $x^{(n)}$ seems to converge to 1 and $y^{(n)}$ seems to converge to 2, although more rapidly than before. ▲

Table 5.2 Iterates produced by the Gauss–Seidel method.

Iteration no. (n)	$x^{(n)}$	$y^{(n)}$
0	0	0
1	2.0000	2.3333
2	0.8334	1.9445
3	1.0278	2.0093
4	0.9954	1.9985
5	1.0008	2.0003
6	0.9999	2.0000

Both of these techniques generalize to larger systems of equations.

Example 5.41

Perform three iterations of Jacobi's method and three iterations of the

Gauss–Seidel method to find an approximate solution of:

$$-8x + y + z = 1$$
$$x - 5y + z = 16$$
$$x + y - 4z = 7$$

with an initial guess of $x = y = z = 0$.

Solution

We rewrite the system to make x, y and z the subject of the first, second and third equation, respectively:

$$x = \tfrac{1}{8}y + \tfrac{1}{8}z - \tfrac{1}{8}$$
$$y = \tfrac{1}{5}x + \tfrac{1}{5}z - \tfrac{16}{5} \qquad (5.11)$$
$$z = \tfrac{1}{4}x + \tfrac{1}{4}y - \tfrac{7}{4}$$

To apply Jacobi's method we substitute the initial guess $x^{(0)} = y^{(0)} = z^{(0)} = 0$ into the right-hand side of Equation (5.11) to obtain $x^{(1)}$, $y^{(1)}$ and $z^{(1)}$, and then repeat the process. In general,

$$x^{(n+1)} = \tfrac{1}{8}y^{(n)} + \tfrac{1}{8}z^{(n)} - \tfrac{1}{8}$$
$$y^{(n+1)} = \tfrac{1}{5}x^{(n)} + \tfrac{1}{5}z^{(n)} - \tfrac{16}{5}$$
$$z^{(n+1)} = \tfrac{1}{4}x^{(n)} + \tfrac{1}{4}y^{(n)} - \tfrac{7}{4}$$

We find:

$$x^{(1)} = -\tfrac{1}{8} = -0.1250$$
$$y^{(1)} = -\tfrac{16}{5} = -3.2000$$
$$z^{(1)} = -\tfrac{7}{4} = -1.7500$$

Then,

$$x^{(2)} = \tfrac{1}{8}(-3.2000) + \tfrac{1}{8}(-1.7500) - \tfrac{1}{8} = -0.7438$$
$$y^{(2)} = \tfrac{1}{5}(-0.1250) + \tfrac{1}{5}(-1.7500) - \tfrac{16}{5} = -3.5750$$
$$z^{(2)} = \tfrac{1}{4}(-0.1250) + \tfrac{1}{4}(-3.2000) - \tfrac{7}{4} = -2.5813$$

Finally,

$$x^{(3)} = \tfrac{1}{8}(-3.5750) + \tfrac{1}{8}(-2.5813) - \tfrac{1}{8} = -0.8945$$
$$y^{(3)} = \tfrac{1}{5}(-0.7438) + \tfrac{1}{5}(-2.5813) - \tfrac{16}{5} = -3.8650$$
$$z^{(3)} = \tfrac{1}{4}(-0.7438) + \tfrac{1}{4}(-3.5750) - \tfrac{7}{4} = -2.8297$$

To apply the Gauss–Seidel iteration to Equation (5.11), the most recent approximation is used at each stage leading to:

$$x^{(n+1)} = \tfrac{1}{8}y^{(n)} + \tfrac{1}{8}z^{(n)} - \tfrac{1}{8}$$

$$y^{(n+1)} = \tfrac{1}{5}x^{(n+1)} + \tfrac{1}{5}z^{(n)} - \tfrac{16}{5}$$

$$z^{(n+1)} = \tfrac{1}{4}x^{(n+1)} + \tfrac{1}{4}y^{(n+1)} - \tfrac{7}{4}$$

Starting from $x^{(0)} = y^{(0)} = z^{(0)} = 0$, we find:

$$x^{(1)} = -\tfrac{1}{8} = -0.1250$$

$$y^{(1)} = \tfrac{1}{5}(-0.1250) + \tfrac{1}{5}(0) - \tfrac{16}{5} = -3.2250$$

$$z^{(1)} = \tfrac{1}{4}(-0.1250) + \tfrac{1}{4}(-3.2250) - \tfrac{7}{4} = -2.5875$$

Then,

$$x^{(2)} = \tfrac{1}{8}(-3.2250) + \tfrac{1}{8}(-2.5875) - \tfrac{1}{8} = -0.8516$$

$$y^{(2)} = \tfrac{1}{5}(-0.8516) + \tfrac{1}{5}(-2.5875) - \tfrac{16}{5} = -3.8878$$

$$z^{(2)} = \tfrac{1}{4}(-0.8516) + \tfrac{1}{4}(-3.8878) - \tfrac{7}{4} = -2.9348$$

Finally,

$$x^{(3)} = \tfrac{1}{8}(-3.8878) + \tfrac{1}{8}(-2.9348) - \tfrac{1}{8} = -0.9778$$

$$y^{(3)} = \tfrac{1}{5}(-0.9778) + \tfrac{1}{5}(-2.9348) - \tfrac{16}{5} = -3.9825$$

$$z^{(3)} = \tfrac{1}{4}(-0.9778) + \tfrac{1}{4}(-3.9825) - \tfrac{7}{4} = -2.9901$$

For completeness, further iterations are shown in Table 5.3.

As expected the Gauss–Seidel method converges more rapidly than Jacobi's. This is generally the case because it uses the most recently calculated values at each stage.   ▲

Unfortunately, as with all iterative methods convergence is not guaranteed. However, it can be shown that if the matrix of coefficients is **diagonally dominant**, that is, each diagonal element is larger in modulus than the sum of the moduli of the other elements in its row, then the Gauss–Seidel method will converge.

Table 5.3   Comparison of the Jacobi and Gauss–Seidel methods.

Iteration no. (n)	Jacobi's method			Gauss–Seidel		
	$x^{(n)}$	$y^{(n)}$	$z^{(n)}$	$x^{(n)}$	$y^{(n)}$	$z^{(n)}$
0	0.0000	0.0000	0.0000	0.0000	0.0000	0.0000
1	−0.1250	−3.2000	−1.7500	−0.1250	−3.2250	−2.5875
2	−0.7438	−3.5750	−2.5813	−0.8516	−3.8878	−2.9348
3	−0.8945	−3.8650	−2.8297	−0.9778	−3.9825	−2.9901
4	−0.9618	−3.9448	−2.9399	−0.9966	−3.9973	−2.9985
5	−0.9856	−3.9803	−2.9767	−0.9995	−3.9996	−2.9998
6	−0.9946	−3.9924	−2.9915	−0.9999	−3.9999	−3.0000
7	−0.9980	−3.9972	−2.9968			
8	−0.9992	−3.9990	−2.9988			
9	−0.9997	−3.9996	−2.9996			

Example 5.42

Use Jacobi's method and the Gauss–Seidel method to obtain approximate solutions for the node voltages of the electrical network examined in Example 5.38.

Solution

The node voltage equations are:

$$6V_a - 4V_b - 2V_d = 12$$

$$15V_a - 31V_b + 10V_c + 6V_e = 57$$

$$2V_b - 11V_c + 6V_d = -12$$

$$3V_a + 12V_c - 19V_d = 0$$

$$4V_b - 9V_e = -8$$

These can be rearranged to give

$$V_a = \frac{2V_b + V_d + 6}{3}$$

$$V_b = \frac{15V_a + 10V_c + 6V_e - 57}{31}$$

$$V_c = \frac{2V_b + 6V_d + 12}{11}$$

$$V_d = \frac{3V_a + 12V_c}{19}$$

$$V_e = \frac{4V_b + 8}{9}$$

The results of applying Jacobi's method with an initial guess of

$$V_a^{(0)} = V_b^{(0)} = V_c^{(0)} = V_d^{(0)} = V_e^{(0)} = 0$$

are shown in Table 5.4. Convergence was achieved to within 0.001 after 44 iterations. The results of applying the Gauss–Seidel method are shown in Table 5.5. Convergence was achieved to within 0.001 after 21 iterations. Clearly the Gauss–Seidel method converges more rapidly than Jacobi's method. ▲

Table 5.4 Node voltages derived from Jacobi's method.

Iteration no. (n)	$V_a^{(n)}$	$V_b^{(n)}$	$V_c^{(n)}$	$V_d^{(n)}$	$V_e^{(n)}$
0	0.0000	0.0000	0.0000	0.0000	0.0000
1	2.0000	−1.8387	1.0909	0.0000	0.8889
⋮ 20	2.8710	0.4802	2.1379	1.8265	1.0738
⋮ 44	2.9679	0.5243	2.1990	1.8578	1.1215

Table 5.5   Node voltages derived from the Gauss–Seidel method.

Iteration no. (n)	$V_a^{(n)}$	$V_b^{(n)}$	$V_c^{(n)}$	$V_d^{(n)}$	$V_e^{(n)}$
0	0.0000	0.0000	0.0000	0.0000	0.0000
1	2.0000	−0.8710	0.9326	0.9048	0.5018
⋮					
20	2.9665	0.5226	2.1985	1.8569	1.1212
21	2.9674	0.5233	2.1989	1.8573	1.1215

EXERCISE 5.8

Perform three iterations of the methods of Jacobi and Gauss–Seidel to obtain approximate solutions of the following. In each case, use an initial guess of $x^{(0)} = y^{(0)} = z^{(0)} = 0$.

(a) $4x + y + z = -1$

$x + 6y + 2z = 0$

$x + 2y + 4z = 1$

(b) $5x + y - z = 4$

$x - 4y + z = -4$

$2x + 2y - 4z = -6$

(c) $4x + y + z = 17$

$x + 3y - z = 9$

$2x - y + 5z = 1$

Miscellaneous exercises

1. Simplify $\begin{vmatrix} \cosh\theta & \sinh\theta \\ \sinh\theta & \cosh\theta \end{vmatrix}$.

2. Given that $A = \begin{pmatrix} 0 & 2 & 3 \\ 2 & 0 & 0 \\ 1 & -1 & 0 \end{pmatrix}$ find A^{-1} and A^2. Show that $A^2 + 6A^{-1} - 7I = 0$, where I denotes the 3×3 identity matrix.

3. If $A = \begin{pmatrix} 2 & -1 & 4 \\ 1 & 0 & 0 \\ 1 & -2 & 0 \end{pmatrix}$ find

   (a) $|A|$

   (b) $\text{adj}(A)$, and

   (c) A^{-1}.

4. Find the inverse of the matrix

   $\begin{pmatrix} 1 & -2 & 0 \\ 3 & 1 & 5 \\ -1 & 2 & 3 \end{pmatrix}$

Hence solve the equations:

$x - 2y = 3$

$3x + y + 5z = 12$

$-x + 2y + 3z = 3$

5. Use Gaussian elimination to solve

$x + 2y - 3z + 2w = 2$

$2x + 5y - 8z + 6w = 5$

$3x + 4y - 5z + 2w = 4$

6. Use Jacobi's method to obtain a solution of $AX = B$ to three decimal places where

$A = \begin{pmatrix} 10 & 1 & 0 \\ 1 & 10 & 1 \\ 0 & 1 & 10 \end{pmatrix}$ and $B = \begin{pmatrix} 1 \\ 2 \\ 1 \end{pmatrix}$

7. Use a matrix method to solve

$2x + y - z = 3$

$x - y + 2z = 1$

$3x + 4y + 3z = 2$

Complex numbers 6

6.1 Introduction

Complex numbers often seem strange when first encountered but it is worth persevering with them because they provide a powerful mathematical tool for solving several engineering problems. One of the main applications is to the analysis of alternating current (a.c.) circuits. Engineers are very interested in these because the mains supply is itself a.c., and electricity generation and transportation are dominated by a.c. voltages and currents.

A great deal of signal analysis and processing uses mathematical models based on complex numbers because they allow the manipulation of sinusoidal quantities to be undertaken more easily. Furthermore, the design of filters to be used in communications equipment relies heavily on their use.

One area of particular relevance is control engineering. So much so that control engineers often prefer to think of a control system in terms of a 'complex plane' representation rather than a 'time domain' representation. We will develop these concepts in this and subsequent chapters.

6.2 Complex numbers

We have already examined quadratic equations such as

$$x^2 - x - 6 = 0 \tag{6.1}$$

and have met techniques for finding the roots of such equations. The formula for obtaining the roots of a quadratic equation $ax^2 + bx + c = 0$ is

$$x = \frac{-b \pm \sqrt{b^2 - 4ac}}{2a} \tag{6.2}$$

Applying this formula to Equation (6.1), we find:

$$x = \frac{+1 \pm \sqrt{(-1)^2 - 4(1)(-6)}}{2}$$

$$= \frac{1 \pm \sqrt{25}}{2}$$

$$= \frac{1 \pm 5}{2}$$

so that $x = 3$ and $x = -2$ are the two roots. However, if we try to apply the formula to the equation:

$$2x^2 + 2x + 5 = 0$$

we find:

$$x = \frac{-2 \pm \sqrt{-36}}{4}$$

A problem now arises in that we need to find the square root of a negative number. We know from experience that squaring both positive and negative numbers yields a positive result; thus,

$$6^2 = 36 \quad \text{and} \quad (-6)^2 = 36$$

so that there is no real number whose square is -36. In the general case, if $ax^2 + bx + c = 0$, we see by examining the square root in Equation (6.2), that this problem will always arise whenever $b^2 - 4ac < 0$. Nevertheless, it turns out to be very useful to invent a technique for dealing with such situations, leading to the theory of complex numbers.

To make progress we introduce a number, denoted j, with the property that

■ $j^2 = -1$

We have already seen that using the real number system we cannot obtain a negative number by squaring a real number so the number j is not real – we say it is **imaginary**. This imaginary number has a very useful role to play in engineering mathematics. Using it we can now formally write down an expression for the square root of any negative number. Thus,

$$\sqrt{-36} = \sqrt{36 \times (-1)}$$

$$= \sqrt{36 \times j^2}$$

$$= 6j$$

Returning to the solution of the quadratic equation $2x^2 + 2x + 5 = 0$, we find:

$$x = \frac{-2 \pm \sqrt{-36}}{4}$$

$$= \frac{-2 \pm 6j}{4}$$

$$= \frac{-1 \pm 3j}{2}$$

We have found two roots, namely, $x = -\frac{1}{2} + \frac{3}{2}j$ and $x = -\frac{1}{2} - \frac{3}{2}j$. These numbers are called **complex numbers** and we see that they are made up of two parts – a **real part** and an **imaginary part**. For the first complex number the real part is $-1/2$ and the imaginary part is $3/2$. For the second complex number the real part is $-1/2$ and the imaginary part is $-3/2$. In a more general case we usually use the letter z to denote a complex number with real part a and imaginary part b, so $z = a + bj$. We write $a = \text{Re}(z)$ and $b = \text{Im}(z)$, and denote the set of all complex numbers by $\mathbb{C}$. Note that $a, b \in \mathbb{R}$ whereas $z \in \mathbb{C}$.

■ $z = a + bj$ $z \in \mathbb{C}$

   $a = \text{Re}(z)$ $b = \text{Im}(z)$

Complex numbers which have a zero imaginary part are purely real and hence all real numbers are also complex numbers, that is, $\mathbb{R} \subset \mathbb{C}$.

6.2.1 *The complex conjugate*

If $z = a + bj$, we define its **complex conjugate** to be the number $\bar{z} = a - bj$, that is, we change the sign of the imaginary part.

Example 6.1

Write down the complex conjugates of:

(a) $-7 + j$

(b) $6 - 5j$

(c) 6

(d) j

Solutions

To find the complex conjugates of the given numbers we change the sign of the imaginary parts. A purely real number has an imaginary part 0. We find:

(a) $-7 - j$

(b) $6 + 5j$

(c) 6, there is no imaginary part to alter

(d) $-j$ ▲

We recall that the solution of the quadratic equation $2x^2 + 2x + 5 = 0$ yielded the two complex numbers $-\frac{1}{2} + \frac{3}{2}j$ and $-\frac{1}{2} - \frac{3}{2}j$, and note that these form a complex conjugate pair. This illustrates a more general result:

■ When the polynomial equation $P(x) = 0$ has real coefficients, any complex roots will always occur in **complex conjugate pairs**.

Consider the following example.

Example 6.2

Show that the equation $x^3 - 7x^2 + 19x - 13 = 0$ has a root at $x = 1$ and find the other roots.

Solution

If we let $P(x) = x^3 - 7x^2 + 19x - 13$, then $P(1) = 1 - 7 + 19 - 13 = 0$ so that $x = 1$ is a root. This means that $x - 1$ must be a factor of $P(x)$ and so we can express $P(x)$ in the form:

$$P(x) = x^3 - 7x^2 + 19x - 13 = (x - 1)(\alpha x^2 + \beta x + \gamma)$$

$$= \alpha x^3 + (\beta - \alpha)x^2 + (\gamma - \beta)x - \gamma$$

where α, β and γ are coefficients to be determined. Comparing the coefficients of x^3 we find $\alpha = 1$. Comparing the constant coefficients we find $\gamma = 13$. Finally, comparing coefficients of x we find $\beta = -6$, and hence:

$$P(x) = x^3 - 7x^2 + 19x - 13 = (x - 1)(x^2 - 6x + 13)$$

The other two roots of $P(x) = 0$ are found by solving the quadratic equation $x^2 - 6x + 13 = 0$, that is,

$$x = \frac{6 \pm \sqrt{36 - 52}}{2} = \frac{6 \pm \sqrt{-16}}{2} = 3 \pm 2j$$

and again we note that the complex roots occur as a complex conjugate pair. This illustrates the general result given in Chapter 1 that an nth degree polynomial has n roots.   ▲

Using the fact that $j^2 = -1$ we can develop other quantities.

Example 6.3

Simplify the expression j^3.

Solution

We have

$$j^3 = j^2 \times j$$
$$= (-1) \times j$$
$$= -j \quad \blacktriangle$$

6.3 Operations with complex numbers

Two complex numbers are equal *if and only if* their real parts are equal and their imaginary parts are equal.

Example 6.4

Find x and y so that $x + 6j$ and $3 - yj$ represent the same complex number.

Solution

If both quantities represent the same complex number we have:

$$x + 6j = 3 - yj$$

Since the real parts must be equal we can equate them, that is

$$x = 3$$

Similarly, we find, by equating imaginary parts:

$$6 = -y$$

so that $y = -6$. $\quad \blacktriangle$

The operations of addition, subtraction, multiplication and division can all be performed on complex numbers.

6.3.1 *Addition and subtraction*

To add two complex numbers we simply add the real parts and add the imaginary parts; to subtract a complex number from another we subtract the corresponding real parts and subtract the corresponding imaginary parts as shown in Example 6.5.

Example 6.5

If $z_1 = 3 - 4j$ and $z_2 = 4 + 2j$ find $z_1 + z_2$ and $z_1 - z_2$.

Solution

$$z_1 + z_2 = (3 - 4j) + (4 + 2j)$$
$$= (3 + 4) + (-4 + 2)j$$
$$= 7 - 2j$$
$$z_1 - z_2 = (3 - 4j) - (4 + 2j)$$
$$= (3 - 4) + (-4 - 2)j$$
$$= -1 - 6j \quad \blacktriangle$$

6.3.2 *Multiplication*

To multiply two complex numbers we use the fact that $j^2 = -1$.

Example 6.6

If $z_1 = 2 - 2j$ and $z_2 = 3 + 4j$, find $z_1 z_2$.

Solution

$$z_1 z_2 = (2 - 2j)(3 + 4j)$$

Removing brackets we find:

$$z_1 z_2 = 6 - 6j + 8j - 8j^2$$
$$= 6 - 6j + 8j + 8 \quad \text{using } j^2 = -1$$
$$= 14 + 2j \quad \blacktriangle$$

Example 6.7

If $z = 3 - 2j$ find $z\bar{z}$.

Solution

If $z = 3 - 2j$ then its conjugate is $\bar{z} = 3 + 2j$. Therefore,

$$z\bar{z} = (3 - 2j)(3 + 2j)$$
$$= 9 - 6j + 6j - 4j^2$$
$$= 9 - 4j^2$$
$$= 13$$

We see that the answer is a real number. $\quad \blacktriangle$

Whenever we multiply a complex number by its conjugate the answer is a real number. Thus if $z = a + bj$

$$z\bar{z} = (a + bj)(a - bj)$$
$$= a^2 + baj - abj - b^2j^2$$
$$= a^2 + b^2$$

■ If $z = a + bj$ then $z\bar{z} = a^2 + b^2$

6.3.3 *Division*

To divide two complex numbers it is necessary to make use of the complex conjugate. We multiply both the numerator and denominator by the conjugate of the denominator and then simplify the result.

Example 6.8

If $z_1 = 2 + 9j$ and $z_2 = 5 - 2j$ find z_1/z_2.

Solution

We seek $(2 + 9j)/(5 - 2j)$. The complex conjugate of the denominator is $5 + 2j$, so we multiply both numerator and denominator by this quantity. The effect of this is to leave the value of z_1/z_2 unaltered since we have only multiplied by 1. Therefore,

$$\frac{z_1}{z_2} = \frac{2 + 9j}{5 - 2j} = \frac{(2 + 9j)}{(5 - 2j)}\frac{(5 + 2j)}{(5 + 2j)} = \frac{10 + 45j + 4j + 18j^2}{25 + 4} = \frac{-8 + 49j}{29}$$

$$= -\frac{8}{29} + \frac{49}{29}j$$

The multiplication of two conjugates in the denominator allows a useful simplification. We see that the effect of multiplying by the conjugate of the denominator is to make the denominator of the solution purely real. ▲

EXERCISES 6.1

1. Solve the following equations:

(a) $x^2 + 1 = 0$

(b) $x^2 + 4 = 0$

(c) $3x^2 + 7 = 0$

(d) $x^2 + x + 1 = 0$

(e) $x^2/2 - x + 2 = 0$

(f) $-x^2 - 3x - 4 = 0$

(g) $2x^2 + 3x + 3 = 0$

(h) $x^2 + 3x + 4 = 0$

2. Solve the cubic equation
$3x^3 - 11x^2 + 16x - 12 = 0$, given that one of the roots is $x = 2$.

3. Express the following in the form $a + bj$:

   (a) $1/(1 + j)$

   (b) $-2/j$

   (c) $(1/j) + 1/(2 - j)$

   (d) $j/(1 + j)$

   (e) $3/(3 + 2j) + 1/(5 - j)$

4. Express the following in the form $a + bj$:

   (a) $2/(1 - j)$

   (b) $(-2 + 3j)/j$

   (c) $3j(4 - 2j)$

   (d) $(7 - 2j)(5 + 6j)$

   (e) $(5 + 3j)/(2 + 2j)$

5. Find a quadratic equation whose roots are $1 - 3j$ and $1 + 3j$.

6. If $(x + jy)^2 = 3 + 4j$, find x and y, where $x, y \in \mathbb{R}$.

7. Find the real and imaginary parts of:

   (a) $2/(4 + j) - 3/(2 - j)$

   (b) j^4

   (c) $1/j$

   (d) $1/(j^3 - 3j)$

8. Recall from Chapter 1 that the poles of a rational function $R(x) = P(x)/Q(x)$ are those values of x for which $Q(x) = 0$. Find any poles of:

   (a) $x/(x - 3)$

   (b) $3x/(x^2 + 1)$

   (c) $3/(x^2 + x + 1)$

9. Solve the equation $s^2 + 2s + 5 = 0$.

6.4 Graphical representation of complex numbers

Given a complex number $z = a + bj$ we can obtain a useful graphical interpretation of it by plotting the real part on the horizontal axis and the imaginary part on the vertical axis and obtain a unique point in the x–y plane (Figure 6.1). We call the x axis the **real axis** and the y axis the **imaginary axis,** and the whole picture an **Argand diagram.** In this context, the x–y plane is often referred to as the **complex plane.** It is often useful to exchange **Cartesian coordinates** (a, b) for **polar coordinates** r and θ as depicted in Figure 6.2.

From Figure 6.2 we note that:

$$\cos \theta = \frac{a}{r} \qquad \sin \theta = \frac{b}{r}$$

and so,

$$a = r \cos \theta \qquad b = r \sin \theta$$

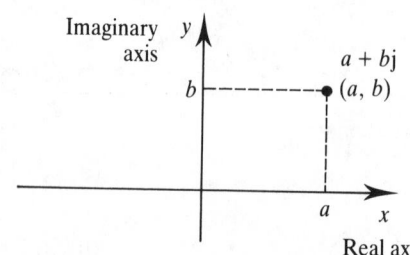

Figure 6.1 Argand diagram.

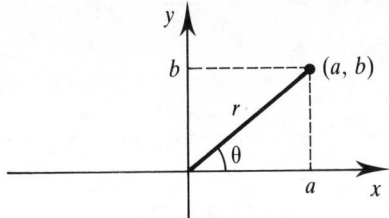

Figure 6.2 Polar and Cartesian forms of a complex number.

Furthermore,

$$\tan \theta = \frac{b}{a}$$

Using Pythagoras' theorem we obtain $r = \sqrt{a^2 + b^2}$. By finding r and θ we can express the complex number $z = a + bj$ in **polar form** as:

$$z = r \cos \theta + jr \sin \theta = r(\cos \theta + j \sin \theta)$$

which we often abbreviate to $z = r \angle \theta$. Clearly, r is the 'distance' of the point (a, b) from the origin and is called the **modulus** of the complex number z. The modulus is always a non-negative number and is denoted $|z|$. The angle is conventionally measured from the positive x axis. Angles measured in an anticlockwise sense are regarded as positive while those measured in a clockwise sense are regarded as negative. The angle θ is called the **argument** of z, denoted $\arg(z)$. Since adding or subtracting multiples of 2π from θ will result in the 'arm' in Figure 6.2 being in the same position, the argument can have many values. Usually we shall choose θ to satisfy $-\pi < \theta \leq \pi$.

■ Cartesian form: $z = a + bj$
Polar form: $z = r(\cos \theta + j \sin \theta) = r \angle \theta$

$$|z| = r = \sqrt{a^2 + b^2}$$

$$a = r \cos \theta \qquad b = r \sin \theta \qquad \tan \theta = \frac{b}{a}$$

Note that

$$r \angle (-\theta) = r(\cos(-\theta) + j \sin(-\theta))$$
$$= r(\cos \theta - j \sin \theta)$$
$$= \bar{z}$$

■ If $z = a + bj$ then $\bar{z} = a - bj$ and

$\bar{z} = r \angle (-\theta)$.

Example 6.9

Depict the complex number $z = 1 - j$ on an Argand diagram and convert it into polar form.

Solution

The real part of z is 1 and the imaginary part is -1. We therefore plot a point in the x–y plane with $x = 1$ and $y = -1$ as shown in Figure 6.3.

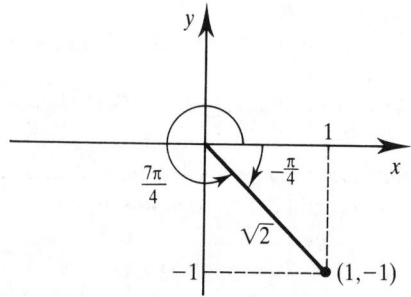

Figure 6.3 Argand diagram depicting $z = 1 - j$.

From Figure 6.3 we see that $r = \sqrt{1^2 + (-1)^2} = \sqrt{2}$ and $\theta = -45°$ or $-\pi/4$ radians. Therefore $z = 1 - j = \sqrt{2} \angle (-\pi/4)$. ▲

To express a complex number in polar form it is essential to draw an Argand diagram and not simply quote formulae, as the following example will show.

Example 6.10

Express $z = -1 - j$ in polar form.

Solution

If we use the formula $|z| = r = \sqrt{a^2 + b^2}$, we find that $r = \sqrt{2}$. Using $\tan \theta = b/a$, we find that $\tan \theta = -1/-1 = 1$ so that you may be tempted to take $\theta = \pi/4$. Figure 6.4 shows the Argand diagram and it is clear that

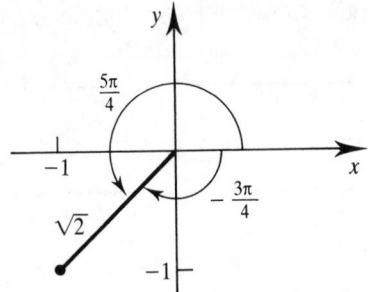

Figure 6.4 Argand diagram depicting $z = -1 - j$.

$\theta = -3\pi/4$. Therefore, $z = -1 - j = \sqrt{2} \, \angle -3\pi/4$, and we see the importance of drawing an Argand diagram. ▲

6.4.1 *Multiplication and division in polar form*

The polar form may seem more complicated than the Cartesian form but it is often more useful. For example, suppose we want to multiply the complex numbers:

$$z_1 = r_1(\cos \theta_1 + j \sin \theta_1) \qquad \text{and} \qquad z_2 = r_2(\cos \theta_2 + j \sin \theta_2)$$

We find,

$$z_1 z_2 = r_1(\cos \theta_1 + j \sin \theta_1) r_2(\cos \theta_2 + j \sin \theta_2)$$
$$= r_1 r_2 \{(\cos \theta_1 \cos \theta_2 - \sin \theta_1 \sin \theta_2) + j(\sin \theta_1 \cos \theta_2 + \sin \theta_2 \cos \theta_1)\}$$

which can be written as:

$$r_1 r_2 \{\cos(\theta_1 + \theta_2) + j \sin(\theta_1 + \theta_2)\}$$

using the trigonometrical identities of Section 1.4.6. This is a new complex number which, if we compare with the general form $r(\cos \theta + j \sin \theta)$, we see has a modulus of $r_1 r_2$ and an argument of $\theta_1 + \theta_2$. To summarize: to multiply two complex numbers we multiply their moduli and add their arguments, that is

---

■ $z_1 z_2 = r_1 r_2 \, \angle (\theta_1 + \theta_2)$

Example 6.11

If $z_1 = 3 \, \angle \pi/3$ and $z_2 = 4 \, \angle \pi/6$ find $z_1 z_2$.

Solution

Multiplying the moduli we find $r_1 r_2 = 12$, and adding the arguments we find $\theta_1 + \theta_2 = \pi/2$. Therefore $z_1 z_2 = 12 \, \angle \pi/2$. ▲

A similar development shows that to divide two complex numbers we divide their moduli and subtract their arguments, that is

■ $\dfrac{z_1}{z_2} = \dfrac{r_1}{r_2} \angle (\theta_1 - \theta_2)$

Example 6.12

If $z_1 = 3 \angle \pi/3$ and $z_2 = 4 \angle \pi/6$ find z_1/z_2.

Solution

Dividing the respective moduli, we find $r_1/r_2 = 3/4$ and subtracting the arguments, $\pi/3 - \pi/6 = \pi/6$. Hence $z_1/z_2 = 0.75 \angle \pi/6$.   ▲

EXERCISES 6.2

1. Mark on an Argand diagram points representing $z_1 = 3 - 2j$, $z_2 = -j$, $z_3 = j^2$, $z_4 = -2 - 4j$ and $z_5 = 3$. Find the modulus and argument of each complex number.

2. Express the following complex numbers in polar form:

   (a) $3 - j$

   (b) 2

   (c) $-j$

   (d) $-5 + 12j$

3. Find the modulus and argument of (a) $z_1 = -\sqrt{3} + j$ and (b) $z_2 = 4 + 4j$. Hence express $z_1 z_2$ and z_1/z_2 in polar form.

4. Express $\sqrt{2} \angle \pi/4$, $2 \angle \pi/6$ and $2 \angle -\pi/6$ in Cartesian form $a + bj$.

5. Prove the result $z_1/z_2 = r_1/r_2 \angle \theta_1 - \theta_2$.

6. Express $z = \dfrac{1}{j\omega C}$, where ω and C are real constants, in the form $a + bj$. Plot z on an Argand diagram.

7. If $z_1 = 4(\cos 40° + j \sin 40°)$ and $z_2 = 3(\cos 70° + j \sin 70°)$, express $z_1 z_2$ and z_1/z_2 in polar form.

8. Simplify:

$$\frac{(\sqrt{2} \angle (5\pi/4))^2 (2 \angle (-\pi/3))^2}{2 \angle (-\pi/6)}$$

6.5 Vectors and complex numbers

It is often convenient to represent complex numbers by vectors in the x–y plane. Figure 6.5(a) shows the complex number $z = a + jb$. Figure 6.5(b) shows the equivalent vector. Figure 6.6 shows the complex numbers $z_1 = 2 + j$ and $z_2 = 1 + 3j$.

If we now evaluate $z_3 = z_1 + z_2$ we find $z_3 = 3 + 4j$ which is also shown. If we form a parallelogram, two sides of which are the representations of z_1 and z_2, we find that z_3 is the diagonal of the parallelogram. If we regard z_1 and z_2 as vectors in the plane we see that there is a direct analogy between the triangle law of vector addition (see Section 4.3) and the addition of complex numbers.

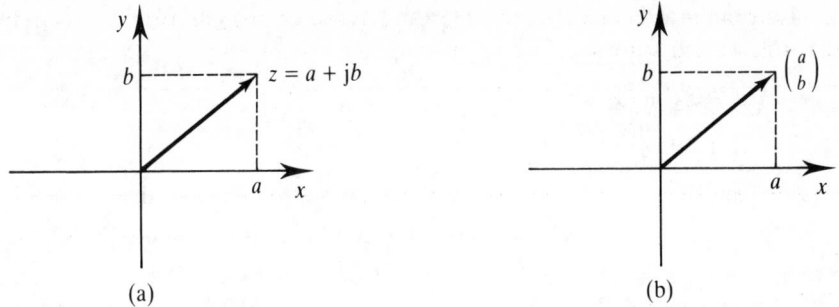

Figure 6.5 The complex number $z = a + jb$ and its equivalent vector.

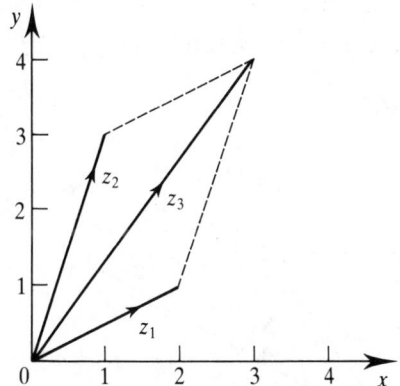

Figure 6.6 Vector addition in the complex plane.

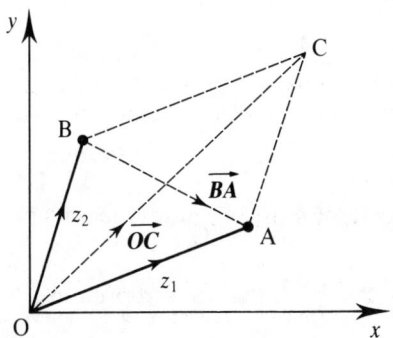

Figure 6.7 Vector addition and subtraction.

More generally, if z_1 and z_2 are any complex numbers represented on an Argand diagram by the vectors $\overrightarrow{OA}$ and $\overrightarrow{OB}$ (Figure 6.7) then upon completing the parallelogram OBCA, the sum $z_1 + z_2$ is represented by the vector $\overrightarrow{OC}$. We can also obtain a representation of the difference of two complex numbers in the following way. If we write:

$$z_3 = z_1 - z_2$$

$$= z_1 + (-z_2)$$

and note that if z_2 is represented by the vector $\overrightarrow{OB}$, then $-z_2$ is represented by the vector $-\overrightarrow{OB} = \overrightarrow{BO} = \overrightarrow{CA}$. The complex number z_1 is represented by $\overrightarrow{OA} = \overrightarrow{BC}$, so that $z_1 + (-z_2) = \overrightarrow{BC} + \overrightarrow{CA} = \overrightarrow{BA}$. Thus the difference $z_1 - z_2$ is represented by the diagonal BA. To summarize, the sum and difference of z_1 and z_2 are represented by the two diagonals of the parallelogram OBCA.

Example 6.13

Represent $z_1 = 6 + j$ and $z_2 = 3 + 4j$, and their sum and difference on an Argand diagram.

Solution

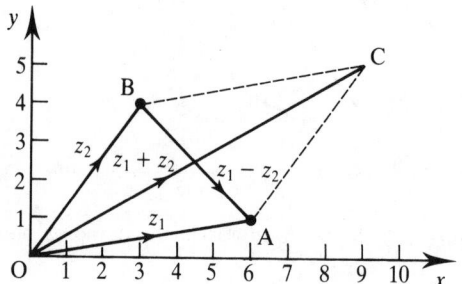

Figure 6.8   Diagram for Example 6.13.

We draw vectors $\overrightarrow{OA}$ and $\overrightarrow{OB}$ representing z_1 and z_2, respectively (Figure 6.8). Then we complete the parallelogram OACB as shown. The sum $z_1 + z_2$ is then represented by $\overrightarrow{OC}$ and the difference $z_1 - z_2$ by $\overrightarrow{BA}$. It is easy to see that the vector $\overrightarrow{BA} = \begin{pmatrix} 3 \\ -3 \end{pmatrix}$, that is, it represents the complex number $3 - 3j$, while $\overrightarrow{OC} = \begin{pmatrix} 9 \\ 5 \end{pmatrix}$, that is, it represents $9 + 5j$, so that $z_1 + z_2 = 9 + 5j$ and $z_1 - z_2 = 3 - 3j$.   ▲

6.6 The exponential form of a complex number

You will recall from Chapter 3 that many functions possess a power series expansion, that is, the function can be expressed as the sum of a sequence of terms involving integer powers of x. For example,

$$e^x = 1 + x + \frac{x^2}{2!} + \frac{x^3}{3!} + \cdots$$

and this representation is valid for any real value of x. The expression on the right-hand side is, of course, an infinite sum but its terms get smaller and smaller, and as more are included, the sum we obtain approaches e^x. Other examples of power series include:

$$\sin x = x - \frac{x^3}{3!} + \frac{x^5}{5!} - \cdots \tag{6.3}$$

and

$$\cos x = 1 - \frac{x^2}{2!} + \frac{x^4}{4!} - \cdots \tag{6.4}$$

which are also valid for any real value of x. It is useful to extend the range of applicability of these power series by allowing x to be a complex number. That is, we define the function e^z to be

$$e^z = 1 + z + \frac{z^2}{2!} + \frac{z^3}{3!} + \cdots$$

and theory beyond the scope of this book can be used to show that this representation is valid for all complex numbers z.

We have already seen that we can express a complex number in polar form:

$$z = r(\cos \theta + j \sin \theta)$$

Using Equations (6.3) and (6.4) we can write:

$$z = r\left\{ \left(1 - \frac{\theta^2}{2!} + \frac{\theta^4}{4!} - \cdots \right) + j\left(\theta - \frac{\theta^3}{3!} + \frac{\theta^5}{5!} - \cdots \right) \right\}$$

$$= r\left(1 + j\theta - \frac{\theta^2}{2!} - j\frac{\theta^3}{3!} + \frac{\theta^4}{4!} + j\frac{\theta^5}{5!} \cdots \right)$$

Furthermore, we note that $e^{j\theta}$ can be written as:

$$e^{j\theta} = 1 + j\theta + \frac{j^2\theta^2}{2!} + \frac{j^3\theta^3}{3!} + \cdots$$

$$= 1 + j\theta - \frac{\theta^2}{2!} - j\frac{\theta^3}{3!} + \cdots$$

so that,

$$z = r(\cos \theta + j \sin \theta) = re^{j\theta}$$

This is yet another form of the same complex number which we call the **exponential form**. We see that:

■ $e^{j\theta} = \cos\theta + j\sin\theta$ (6.5)

It is straightforward to show that:

■ $e^{-j\theta} = \cos\theta - j\sin\theta$

Therefore, if $\bar{z} = r(\cos\theta - j\sin\theta)$ we can equivalently write $\bar{z} = re^{-j\theta}$. The two expressions for $e^{j\theta}$ and $e^{-j\theta}$ are known as Euler's relations. From these it is easy to obtain the following useful results:

■ $\cos\theta = \dfrac{e^{j\theta} + e^{-j\theta}}{2}$ $\sin\theta = \dfrac{e^{j\theta} - e^{-j\theta}}{2j}$

Example 6.14

We saw in Chapter 1 that a waveform can be written in the form $f(t) = A\cos(\omega t + \phi)$. Consider the complex number $e^{j(\omega t + \phi)}$. We can use Euler's relations to write

$$e^{j(\omega t + \phi)} = \cos(\omega t + \phi) + j\sin(\omega t + \phi)$$

and hence,

$$f(t) = A\,\mathrm{Re}(e^{j(\omega t + \phi)}) \quad \blacktriangle$$

EXERCISES 6.3

1. Find the modulus and argument of
   (a) $3e^{j\pi/4}$, (b) $2e^{-j\pi/6}$.

2. Find the real and imaginary parts of
   (a) $5e^{j\pi/3}$, (b) $e^{j2\pi/3}$.

3. Express $z = 6(\cos 30° + j\sin 30°)$ in exponential form. Plot z on an Argand diagram and find its real and imaginary parts.

4. If σ, ω, $T \in \mathbb{R}$, find the real and imaginary parts of $e^{(\sigma + j\omega)T}$.

5. Express $z = e^{1 + j\pi/2}$ in the form $a + bj$.

6. Express $-1 - j$ in the form $re^{j\theta}$.

6.7 Phasors

Electrical engineers are often interested in analysing circuits in which there is an alternating current power supply. Almost invariably the supply waveform is sinusoidal and the resulting currents and voltages within the circuit are also sinusoidal. For example, a typical voltage is of the form

$$v(t) = V\cos(\omega t + \phi) = V\cos(2\pi f t + \phi) \tag{6.6}$$

where V is the maximum or peak value, ω is the angular frequency, f is the frequency, and ϕ is the phase relative to some reference waveform. This is known as the **time domain representation**. Each of the voltages and currents in the circuit has the same frequency as the supply but differs in magnitude and phase.

In order to analyse such circuits it is necessary to add, subtract, multiply and divide these waveforms. If the time domain representation is used then the mathematics becomes extremely tedious. An alternative approach is to introduce a waveform representation known as a **phasor**. A phasor is an entity consisting of two distinct parts: a magnitude and an angle. It is possible to represent a phasor by a complex number in polar form. The fixed magnitude of this complex number corresponds to the magnitude of the phasor and hence the amplitude of the waveform. The argument of this complex number, ϕ, corresponds to the angle of the phasor and hence the phase angle of the waveform. Figure 6.9 shows a phasor for the sinusoidal waveform of Equation (6.6).

The time dependency of the waveform is catered for by rotating the phasor anticlockwise at an angular frequency, ω. The projection of the phasor onto the real axis gives the instantaneous value of the waveform. However, the main interest of an engineer is in the phase relationships between the various sinusoids. Therefore the phasors are 'frozen' at a certain point in time. This may be chosen so that $t = 0$ or it may be chosen so that a convenient phasor, known as the **reference phasor**, aligns with the positive real axis. This approach is valid because the phase and magnitude relationships between the various phasors remain the same at all points in time once a circuit has recovered from any initial transients caused by switching.

Some textbooks refer to phasors as vectors. This can lead to confusion as it is possible to divide phasors whereas division of vectors by other vectors is not defined. In practice this is not a problem as phasors, although thought of as vectors, are manipulated as complex numbers, which can be divided. We will avoid these conceptual difficulties by introducing a different notation. We will denote a phasor

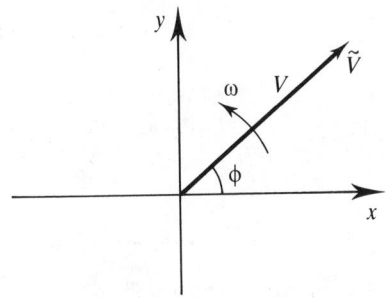

Figure 6.9 Illustration of the phasor $\tilde{V} = V \angle \phi$ where ω = angular frequency with which the phasor rotates.

by $\tilde{V}$, which corresponds to $V \angle \phi$ in complex number notation (see Figure 6.9). Thus, for example, a current $i(t) = I \cos(\omega t + \phi)$ would be written $\tilde{I}$ in phasor notation and $I \angle \phi$ in complex number notation.

Many engineers use the **root mean square** (r.m.s.) value of a sinusoid as the magnitude of a phasor. The justification for this is that it represents the value of a d.c. signal that would dissipate the same amount of power in a resistor as the sinusoid. For example, in the case of a current signal, $I_{rms}^2 R$ is the average power dissipated by the sinusoid $I \cos(\omega t + \phi)$ in a resistor R. For the case of a sinusoidal signal the r.m.s. value of the signal is $1/\sqrt{2}$ times the peak value of the signal (see Section 8.5, Example 8.33, for a proof of this). We will not adopt this approach but it is a common one.

We start by examining the phasor representation of individual circuit elements. In order to do this we need a phasor form of Ohm's law. This is:

$$\tilde{V} = \tilde{I} Z$$

(6.7)

where $\tilde{V}$ is the voltage phasor, $\tilde{I}$ is the current phasor and Z is the impedance of an element or group of elements and may be a complex quantity. Note that phasors and complex numbers are mixed together in the same equation. This is a common practice because phasors are usually manipulated as complex numbers.

Resistor

Experimentally it can be shown that if an a.c. voltage is applied to a resistor then the current is in phase with the voltage. The ratio of the magnitude of the two waveforms is equal to the resistance, R. So, given $\tilde{I} = I \angle 0$, $Z = R \angle 0$, using Equation (6.7) we have $\tilde{V} = IR \angle 0$. This is illustrated in Figure 6.10.

Inductor

For an inductor we know from experiment that the voltage leads the current by a phase of $\pi/2$, and so the phase angle of the impedance is $\pi/2$. We also know that the magnitude of the impedance is given by ωL. So, given $\tilde{I} = I \angle 0$, $Z = \omega L \angle \pi/2$, using Equation (6.7) we have $\tilde{V} = I\omega L \angle \pi/2$. An alternative way

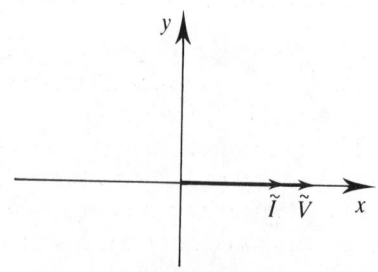

Figure 6.10 Phasor diagram for a resistor.

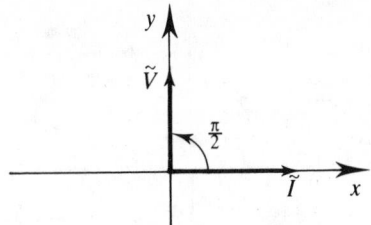

Figure 6.11 Phasor diagram for an inductor.

of representing Z for an inductor is to use the Cartesian form, that is

$$Z = \omega L e^{j\pi/2} = \omega L\left(\cos\frac{\pi}{2} + j\sin\frac{\pi}{2}\right)$$

$$= j\omega L$$

This is useful when phasors need to be added and subtracted. The phasor diagram for an inductor is illustrated in Figure 6.11.

Capacitor

For a capacitor it is known that the voltage lags the current by a phase of $\pi/2$ and the magnitude of the impedance is given by $1/(\omega C)$. So given $\tilde{I} = I\angle 0$, $Z = 1/(\omega C)\angle(-\pi/2)$, we have, using Equation (6.7), $\tilde{V} = I/(\omega C)\angle(-\pi/2)$. Alternatively,

$$Z = \frac{e^{-j\pi/2}}{\omega C} = \frac{1}{\omega C}\left(\cos\frac{\pi}{2} - j\sin\frac{\pi}{2}\right)$$

$$= -\frac{j}{\omega C}$$

Engineers often prefer to rewrite this last expression as:

$$\frac{1}{j\omega C}$$

The phasor diagram for the capacitor is illustrated in Figure 6.12.

We have shown how phasors can be multiplied by a complex number; division is very similar. Addition of phasors will now be illustrated; subtraction is similar. Consider the circuit shown in Figure 6.13 in which a resistor, capacitor and inductor are connected in series and fed by an alternating current source. As this is a series circuit, the current through each element, $\tilde{I}$, is the same by Kirchhoff's current law. By Kirchhoff's voltage law the voltage rise produced by the supply, $\tilde{V}_S$, must equal the sum of the voltage drops across the elements. Therefore:

$$\tilde{V}_S = \tilde{V}_R + \tilde{V}_C + \tilde{V}_L$$

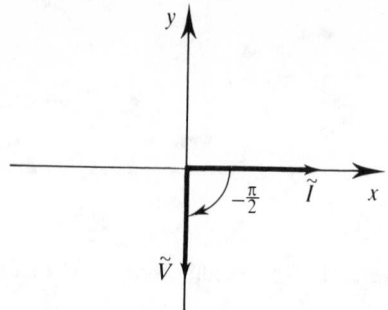

Figure 6.12 Phasor diagram for a capacitor.

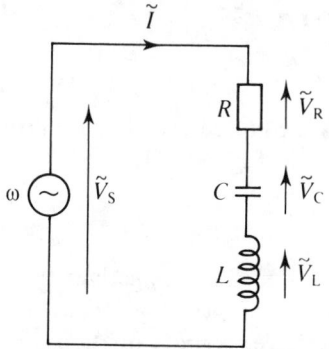

Figure 6.13 RLC circuit.

Note that this is an addition of phasors so that the voltage drops across the elements do not necessarily have the same phase. The phasor diagram for the circuit is shown in Figure 6.14, in this case with $|\tilde{V}_{L}| > |\tilde{V}_{C}|$; $\tilde{I}$ is the reference phasor.

Note the phasor addition of the element voltages gives the overall supply voltage for a particular supply current. If the magnitude of these element voltage phasors is known then it is possible to calculate the supply voltage graphically. In practice it is easier to convert the polar form of the phasors into Cartesian form and use algebra to analyse the circuit.

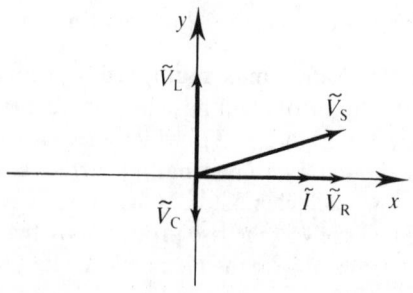

Figure 6.14 Phasor diagram for the circuit in Figure 6.13.

Now,

$$\tilde{V}_R = \tilde{I} R \angle 0 = \tilde{I} R$$

$$\tilde{V}_L = \tilde{I} \omega L \angle (\pi/2) = \tilde{I} j \omega L$$

$$\tilde{V}_C = \frac{\tilde{I}}{\omega C} \angle -\pi/2 = \frac{\tilde{I}}{j \omega C}$$

Therefore,

$$\tilde{V}_S = \tilde{V}_R + \tilde{V}_C + \tilde{V}_L$$

$$= \tilde{I} R + \tilde{I} j \omega L + \frac{\tilde{I}}{j \omega C}$$

$$= \tilde{I} \left(R + j \omega L + \frac{1}{j \omega C} \right)$$

Therefore the impedance of the circuit is $Z = R + j\omega L + 1/(j\omega C)$. We can calculate the frequency for which the impedance of the circuit has minimum magnitude:

$$Z = R + j \omega L + \frac{1}{j \omega C}$$

$$= R + j \omega L - \frac{j}{\omega C}$$

$$= R + j \left(\omega L - \frac{1}{\omega C} \right)$$

Now $|Z| = \sqrt{R^2 + (\omega L - 1/(\omega C))^2}$ and so, as ω varies $|Z|$ is a minimum when

$$\omega L - \frac{1}{\omega C} = 0$$

$$\omega^2 = \frac{1}{LC}$$

$$\omega = \sqrt{\frac{1}{LC}}$$

This minimum value is $|Z| = R$. Examining Figure 6.14 it is clear that the minimum impedance occurs when $\tilde{V}_L$ and $\tilde{V}_C$ have the same magnitude, in which case $\tilde{V}_S$ has no imaginary component. The frequency at which this occurs is known as the **resonant frequency** of the circuit.

6.8 De Moivre's theorem

A very important result in complex number theory is De Moivre's theorem which states that if $n \in \mathbb{N}$,

---

■ $(\cos \theta + j \sin \theta)^n = \cos n\theta + j \sin n\theta$ (6.8)

Example 6.15

Verify De Moivre's theorem when $n = 1$ and $n = 2$.

Solution

When $n = 1$, the theorem states:

$(\cos \theta + j \sin \theta)^1 = \cos 1\theta + j \sin 1\theta$

which is clearly true, and the theorem holds. When $n = 2$, we find:

$$(\cos \theta + j \sin \theta)^2 = (\cos \theta + j \sin \theta)(\cos \theta + j \sin \theta)$$
$$= \cos^2 \theta + j \sin \theta \cos \theta + j \cos \theta \sin \theta + j^2 \sin^2 \theta$$
$$= \cos^2 \theta - \sin^2 \theta + j(2 \sin \theta \cos \theta)$$

Recalling the trigonometrical identities $\cos 2\theta = \cos^2 \theta - \sin^2 \theta$ and $\sin 2\theta = 2 \sin \theta \cos \theta$, we can write the previous expression as:

$\cos 2\theta + j \sin 2\theta$

Therefore,

$(\cos \theta + j \sin \theta)^2 = \cos 2\theta + j \sin 2\theta$

and De Moivre's theorem has been verified when $n = 2$. ▲

The theorem also holds when n is a rational number, that is $n = p/q$ where p and q are integers. Thus we have:

■ $(\cos \theta + j \sin \theta)^{p/q} = \cos \dfrac{p}{q}\theta + j \sin \dfrac{p}{q}\theta$

In this form it can be used to obtain roots of complex numbers. For example,

$$\sqrt[3]{\cos \theta + j \sin \theta} = (\cos \theta + j \sin \theta)^{1/3} = \cos \frac{1}{3}\theta + j \sin \frac{1}{3}\theta$$

In such a case the expression obtained is only one of the possible roots. Additional roots can be found as illustrated in Example 6.16.

De Moivre's theorem is particularly important for the solution of certain types of equation.

Example 6.16

Find all complex numbers z which satisfy:

$$z^3 = 1 \tag{6.9}$$

Solution

The solution of this equation is equivalent to finding the solutions of $z = 1^{1/3}$, that is, finding the cube roots of 1. Since we are allowing z to be complex, that is $z \in \mathbb{C}$, we can write:

$$z = r(\cos \theta + j \sin \theta)$$

Then, using De Moivre's theorem:

$$z^3 = r^3(\cos \theta + j \sin \theta)^3$$
$$= r^3(\cos 3\theta + j \sin 3\theta)$$

We next convert the expression on the right-hand side of Equation (6.9) into polar form. Figure 6.15 shows the number $1 = 1 + 0j$ on an Argand diagram.

From the Argand diagram we see that its modulus is 1 and its argument is 0, or possibly $\pm 2\pi, \pm 4\pi, \ldots$, that is $2n\pi$ where $n \in \mathbb{Z}$. Consequently, we can write:

$$1 = 1(\cos 2n\pi + j \sin 2n\pi) \qquad n \in \mathbb{Z}$$

Using the polar form, Equation (6.9) becomes:

$$r^3(\cos 3\theta + j \sin 3\theta) = 1(\cos 2n\pi + j \sin 2n\pi)$$

Comparing both sides of this equation we see that

$$r^3 = 1 \qquad \text{that is} \qquad r = 1 \text{ since } r \in \mathbb{R}$$

and

$$3\theta = 2n\pi \qquad \text{that is} \qquad \theta = 2n\pi/3 \qquad n \in \mathbb{Z}$$

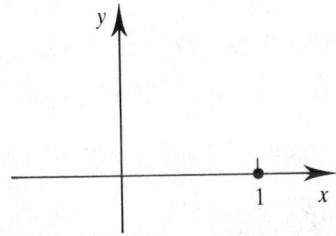

Figure 6.15 The complex number $z = 1 + 0j$.

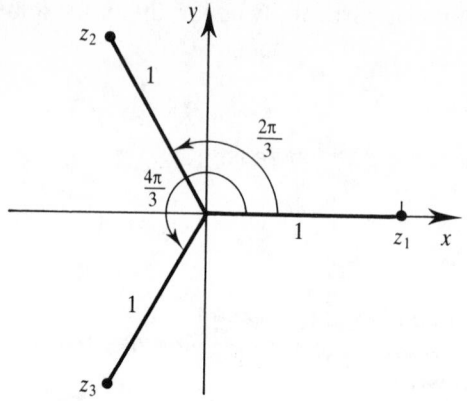

Figure 6.16   Solutions of $z^3 = 1$.

Apparently θ can take infinitely many values, but, as we shall see, the corresponding complex numbers are simply repetitions. When $n = 0$, we find $\theta = 0$, so that

$$z = z_1 = 1(\cos 0 + j \sin 0) = 1$$

is the first solution. When $n = 1$ we find $\theta = 2\pi/3$, so that:

$$z = z_2 = 1(\cos 2\pi/3 + j \sin 2\pi/3) = -\frac{1}{2} + j\frac{\sqrt{3}}{2}$$

is the second solution. When $n = 2$ we find $\theta = 4\pi/3$, so that

$$z = z_3 = 1(\cos 4\pi/3 + j \sin 4\pi/3) = -\frac{1}{2} - j\frac{\sqrt{3}}{2}$$

is the third solution. If we continue searching for solutions using larger values of n we find that we only repeat solutions already obtained. It is often useful to plot solutions on an Argand diagram and this is easily done directly from the polar form, as shown in Figure 6.16. We note that the solutions are equally spaced at angles of $2\pi/3$. ▲

Example 6.17

Find the complex numbers z which satisfy $z^2 = 4j$.

Solution

Since $z \in \mathbb{C}$ we write $z = r(\cos \theta + j \sin \theta)$. Therefore,

$$z^2 = r^2(\cos \theta + j \sin \theta)^2$$
$$= r^2(\cos 2\theta + j \sin 2\theta)$$

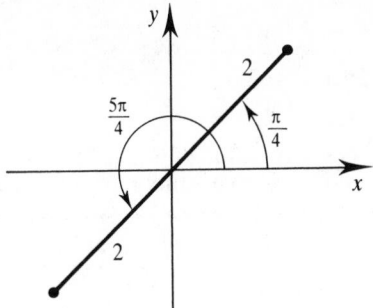

Figure 6.17   Solution of $z^2 = 4j$.

by De Moivre's theorem. Furthermore, 4j has modulus 4 and argument $\pi/2 + 2n\pi$, $n \in \mathbb{Z}$, that is

$$4j = 4\{\cos(\pi/2 + 2n\pi) + j\sin(\pi/2 + 2n\pi)\} \qquad n \in \mathbb{Z}$$

Therefore,

$$r^2(\cos 2\theta + j\sin 2\theta) = 4(\cos(\pi/2 + 2n\pi) + j\sin(\pi/2 + 2n\pi))$$

Comparing both sides of this equation, we see that:

$$r^2 = 4 \qquad \text{and so} \qquad r = 2$$

and

$$2\theta = \pi/2 + 2n\pi \qquad \text{and so} \qquad \theta = \pi/4 + n\pi$$

When $n = 0$ we find $\theta = \pi/4$, and when $n = 1$ we find $\theta = 5\pi/4$. Using larger values of n simply repeats solutions already obtained. These solutions are shown in Figure 6.17. We note that in this example the solutions are equally spaced at intervals of $2\pi/2 = \pi$. In Cartesian form,

$$z_1 = \frac{2}{\sqrt{2}} + j\frac{2}{\sqrt{2}} = \sqrt{2}(1 + j) \quad \text{and} \quad z_2 = -\frac{2}{\sqrt{2}} - j\frac{2}{\sqrt{2}} = -\sqrt{2}(1 + j) \quad \blacktriangle$$

In general, the n roots of $z^n = a + jb$ are equally spaced at angles $2\pi/n$.

Once the technique for solving equations like those in Examples 6.16 and 6.17 has been mastered, engineers find it simpler to work with the abbreviated form $r \angle \theta$. Example 6.17 reworked in this fashion becomes:

Let $z = r \angle \theta$,   then $z^2 = r^2 \angle 2\theta$

Furthermore, $4j = 4 \angle \pi/2 + 2n\pi$, and hence if $z^2 = 4j$, we have:

$$r^2 \angle 2\theta = 4 \angle \left(\frac{\pi}{2} + 2n\pi\right)$$

from which

$$r^2 = 4 \quad \text{and} \quad 2\theta = \frac{\pi}{2} + 2n\pi$$

as before. Rework Example 6.16 for yourself using this approach.

Another application of De Moivre's theorem is the derivation of trigonometric identities.

Example 6.18

Use De Moivre's theorem to show that:

$$\cos 3\theta = 4 \cos^3 \theta - 3 \cos \theta$$

and

$$\sin 3\theta = 3 \sin \theta - 4 \sin^3 \theta$$

Solution

We know that

$$(\cos \theta + j \sin \theta)^3 = \cos 3\theta + j \sin 3\theta$$

Expanding the left-hand side we find:

$$\cos^3 \theta + 3j \cos^2 \theta \sin \theta - 3 \cos \theta \sin^2 \theta - j \sin^3 \theta = \cos 3\theta + j \sin 3\theta$$

Equating the real parts gives:

$$\cos^3 \theta - 3 \cos \theta \sin^2 \theta = \cos 3\theta \tag{6.10}$$

and equating the imaginary parts gives:

$$3 \cos^2 \theta \sin \theta - \sin^3 \theta = \sin 3\theta \tag{6.11}$$

Now, writing $\sin^2 \theta = 1 - \cos^2 \theta$ in Equation (6.10) and $\cos^2 \theta = 1 - \sin^2 \theta$ in Equation (6.11), we find

$$\cos 3\theta = \cos^3 \theta - 3 \cos \theta (1 - \cos^2 \theta)$$

$$= \cos^3 \theta + 3 \cos^3 \theta - 3 \cos \theta$$

$$= 4 \cos^3 \theta - 3 \cos \theta$$

$$\sin 3\theta = 3(1 - \sin^2 \theta) \sin \theta - \sin^3 \theta$$

$$= 3 \sin \theta - 4 \sin^3 \theta$$

as required. ▲

This technique allows trigonometric functions of multiples of angles to be expressed in terms of powers. Sometimes we want to carry out the reverse process and express a power in terms of multiple angles. Consider Example 6.19.

Example 6.19

If $z = \cos \theta + j \sin \theta$ show that:

$$z + \frac{1}{z} = 2 \cos \theta \qquad z - \frac{1}{z} = 2j \sin \theta$$

and find similar expressions for $z^n + 1/z^n$ and $z^n - 1/z^n$.

Solution

Consider the complex number:

$$z = \cos \theta + j \sin \theta$$

Using De Moivre's theorem:

$$\frac{1}{z} = z^{-1} = (\cos \theta + j \sin \theta)^{-1} = \cos(-\theta) + j \sin(-\theta)$$

But $\cos(-\theta) = \cos \theta$ and $\sin(-\theta) = -\sin \theta$, so that if $z = \cos \theta + j \sin \theta$

$$\frac{1}{z} = \cos \theta - j \sin \theta$$

Consequently,

$$z + \frac{1}{z} = 2 \cos \theta \qquad \text{and} \qquad z - \frac{1}{z} = 2j \sin \theta$$

Moreover,

$$z^n = \cos n\theta + j \sin n\theta \qquad \text{and} \qquad z^{-n} = \cos n\theta - j \sin n\theta$$

so that,

$$z^n + \frac{1}{z^n} = 2 \cos n\theta \qquad \text{and} \qquad z^n - \frac{1}{z^n} = 2j \sin n\theta \quad \blacktriangle$$

$$\blacksquare \quad z^n + \frac{1}{z^n} = 2 \cos n\theta \qquad \text{and} \qquad z^n - \frac{1}{z^n} = 2j \sin n\theta$$

Example 6.20

Show that $\cos^2 \theta = \frac{1}{2}(\cos 2\theta + 1)$.

Solution

The formulae obtained in Example 6.19 allow us to obtain expressions for powers of $\cos \theta$ and $\sin \theta$. Since $2 \cos \theta = z + 1/z$, squaring both sides we have:

$$2^2 \cos^2 \theta = \left(z + \frac{1}{z} \right)^2 = z^2 + 2 + \frac{1}{z^2}$$

$$= \left(z^2 + \frac{1}{z^2} \right) + 2$$

But $z^2 + 1/z^2 = 2 \cos 2\theta$, so

$$2^2 \cos^2 \theta = 2 \cos 2\theta + 2$$

and therefore,

$$\cos^2 \theta = \tfrac{1}{2} \cos 2\theta + \tfrac{1}{2}$$

$$= \tfrac{1}{2} (\cos 2\theta + 1)$$

as required.   ▲

EXERCISES 6.4

1. Express $(\cos \theta + j \sin \theta)^9$ and $(\cos \theta + j \sin \theta)^{1/2}$ in the form $\cos n\theta + j \sin n\theta$.

2. Use De Moivre's theorem to simplify:

   (a) $(\cos 3\theta + j \sin 3\theta)(\cos 4\theta + j \sin 4\theta)$

   (b) $(\cos 8\theta + j \sin 8\theta)/(\cos 2\theta - j \sin 2\theta)$

3. Solve the equations

   (a) $z^3 + 1 = 0$

   (b) $z^4 = 1 + j$

4. Find $\sqrt[3]{2 + 2j}$ and display your solutions on an Argand diagram.

5. Prove the following trigonometric identities:

   (a) $\cos 4\theta = 8 \cos^4 \theta - 8 \cos^2 \theta + 1$

   (b) $32 \sin^6 \theta = 10 - 15 \cos 2\theta + 6 \cos 4\theta - \cos 6\theta$

6. Express $z_1 = 1 - j$ and $z_2 = (1 + j)/(\sqrt{3} - j)$ in the form $re^{j\theta}$.

7. Solve the equation $z^4 + 25 = 0$.

8. Find the fifth roots of j and depict your solutions on an Argand diagram.

9. Show that $\cos^3 \theta = \tfrac{1}{4}(\cos 3\theta + 3 \cos \theta)$.

10. Show that $\sin^4 \theta = \tfrac{1}{8}(\cos 4\theta - 4 \cos 2\theta + 3)$.

11. Express $\cos 5\theta$ in terms of powers of $\cos \theta$.

12. Express $\sin 5\theta$ in terms of powers of $\sin \theta$.

13. Given $e^{j\theta} = \cos \theta + j \sin \theta$, prove De Moivre's theorem in the form:

    $$(e^{j\theta})^n = \cos n\theta + j \sin n\theta$$

6.9 Loci and regions of the complex plane

Regions of the complex plane can often be conveniently described by means of complex numbers. For example, the points that lie on a circle of radius 2 centred at the origin (Figure 6.18) represent complex numbers all of which have a modulus of 2. The arguments are any value of θ, $-\pi < \theta \leq \pi$. We can describe all the

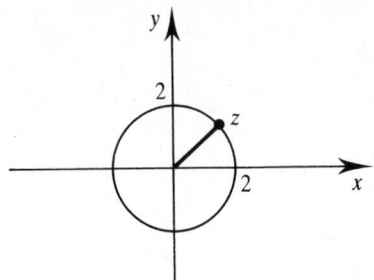

Figure 6.18 A circle, radius 2, centred at the origin.

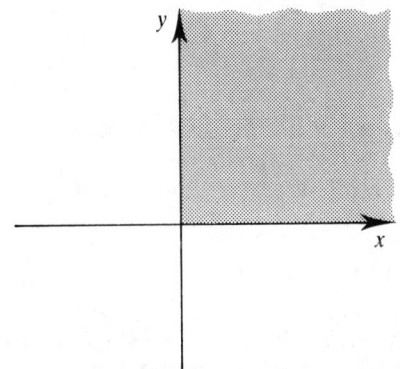

Figure 6.19 First quadrant of the x–y plane.

points on this circle by the simple expression:

$$|z| = 2$$

that is, all complex numbers with modulus 2. We say that the locus (or path) of the point z is a circle, radius 2, centred at the origin. The interior of the circle is described by $|z| < 2$ while its exterior is described by $|z| > 2$.

Similarly all points lying in the first quadrant (shaded in Figure 6.19) have arguments between 0 and $\pi/2$. This quadrant is therefore described by the expression:

$$0 < \arg(z) < \pi/2$$

Example 6.21

Sketch the locus of the point satisfying $\arg(z) = \pi/4$.

Solution

The set of points with $\arg(z) = \pi/4$ comprises complex numbers whose argument is $\pi/4$. All these complex numbers lie on the line shown in Figure 6.20. ▲

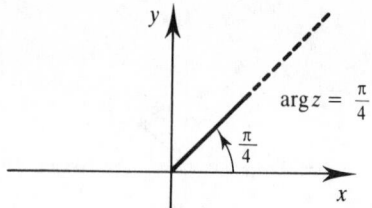

Figure 6.20   Locus of points satisfying arg$(z) = \pi/4$.

Example 6.22

Sketch the locus of the point satisfying $|z - 2| = 3$.

Solution

First mark the fixed point 2 on the Argand diagram labelling it 'A'
(Figure 6.21). Consider the complex number z represented by the point P.
From the vector triangle law of addition

$$\vec{OA} + \vec{AP} = \vec{OP}$$
$$\vec{AP} = \vec{OP} - \vec{OA}$$

Recall from Section 6.5 that the graphical representation of the sum and
difference of vectors in the plane, and the sum and difference of complex
numbers are equivalent. Since vector $\vec{OP}$ represents the complex number z,
and vector $\vec{OA}$ represents the complex number 2, $\vec{AP} = \vec{OP} - \vec{OA}$ will
represent $z - 2$. Therefore $|z - 2|$ represents the distance between A and P.
We are given that $|z - 2| = 3$, which therefore means that P can be any point
such that its distance from A is 3. This means that P can be any point on a
circle of radius 3 centred at $A(2, 0)$. The locus is shown in Figure 6.22.
$|z - 2| < 3$ represents the interior of the circle while $|z - 2| > 3$ represents
the exterior. Alternatively we can obtain the same result algebraically: given

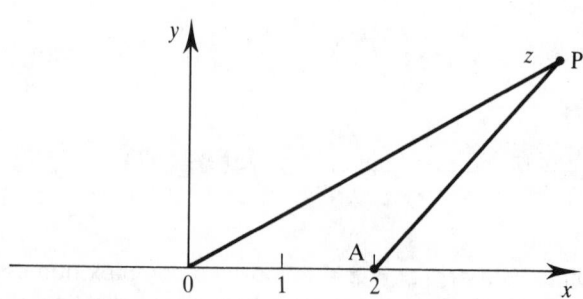

Figure 6.21   Points z and $2 + 0j$.

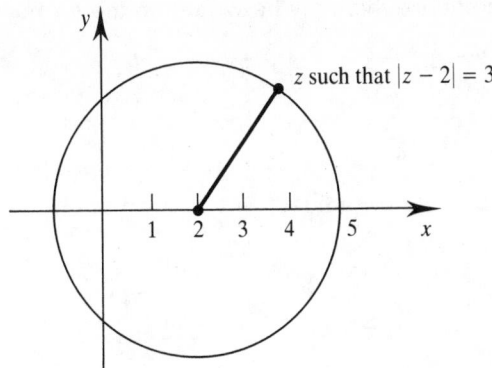

Figure 6.22 Locus of points satisfying $|z - 2| = 3$.

$|z - 2| = 3$ and also that $z = x + jy$, we can write:

$$|z - 2| = |(x - 2) + jy| = 3$$

that is

$$\sqrt{(x - 2)^2 + y^2} = 3$$

or

$$(x - 2)^2 + y^2 = 9$$

Generally, the equation $(x - a)^2 + (y - b)^2 = r^2$ represents a circle of radius r centred at (a, b), so we see that $(x - 2)^2 + y^2 = 9$ represents a circle of radius 3 centred at $(2, 0)$, as before. ▲

Example 6.23

Use the algebraic approach to find the locus of the point z which satisfies:

$$|z - 1| = \tfrac{1}{2}|z - j|$$

Solution

If $z = x + jy$, then we have:

$$|(x - 1) + jy| = \tfrac{1}{2}|x + j(y - 1)|$$

Therefore,

$$(x - 1)^2 + y^2 = \tfrac{1}{4}\{x^2 + (y - 1)^2\}$$

and so

$$4(x - 1)^2 + 4y^2 = x^2 + (y - 1)^2$$

that is

$$3x^2 - 8x + 3y^2 + 2y + 3 = 0$$

By completing the square this may be written in the form:

$$3(x - \tfrac{4}{3})^2 + 3(y + \tfrac{1}{3})^2 = \tfrac{8}{3}$$

that is

$$(x - \tfrac{4}{3})^2 + (y + \tfrac{1}{3})^2 = \tfrac{8}{9}$$

which represents a circle of radius $\sqrt{8}/3$ centred at $(4/3, -1/3)$. ▲

EXERCISES 6.5

1. Sketch the loci defined by:

   (a) $\arg(z) = 0$

   (b) $\arg(z) = \pi/2$

   (c) $\arg(z - 4) = \pi/4$

   (d) $|2z| = |z - 1|$

2. Sketch the regions defined by:

   (a) $\text{Re}(z) \geq 0$

   (b) $\text{Im}(z) < 3$

   (c) $|z| > 3$

   (d) $0 \leq \arg(z) \leq \pi/2$

   (e) $|z + 2| \leq 3$

   (f) $|z + j| > 3$

   (g) $|z - 1| < |z - 2|$

3. If $s = \sigma + j\omega$ sketch the regions defined by:

   (a) $\sigma \leq 0$

   (b) $\sigma \geq 0$

   (c) $-2 \leq \omega \leq 2$

Miscellaneous exercises

1. Show that
   $1/(\cos\theta - j\sin\theta) = \cos\theta + j\sin\theta$.

2. Simplify $(5 + 4j)/(5 - 4j)$, $1/(2 + 3j)$, $1/(2 + 3j) + 1/(2 - 3j)$, $1/(x - jy)$.

3. Find the modulus and argument of $-j$, -3, $1 + j$, $\cos\theta + j\sin\theta$.

4. Mark on an Argand diagram vectors corresponding to the following complex numbers: $-3 + 2j$, $-3 - 2j$, $\cos\pi + j\sin\pi$.

5. Express in the form $a + bj$:

   (a) $(\cos\theta + j\sin\theta)^6$

   (b) $1/(\cos\theta + j\sin\theta)^3$

   (c) $(\cos\theta + j\sin\theta)/(\cos\phi + j\sin\phi)$

6. If $z \in \mathbb{C}$, show that:

   (a) $z + \bar{z} = 2\,\text{Re}(z)$

   (b) $z - \bar{z} = 2j\,\text{Im}(z)$

   (c) $z\bar{z} = |z|^2$

7. Show that $e^{j\omega t} + e^{-j\omega t} = 2\cos\omega t$ and find an expression for $e^{j\omega t} - e^{-j\omega t}$.

8. Express $1 + e^{2j\omega t}$ in the form $a + bj$.

9. Sketch the region in the complex plane described by $|z + 2j| < 1$.

10. Express $e^{(1/2 - 6j)}$ in the form $a + bj$.

11. Solve the equation $z^4 + 1 = j\sqrt{3}$.

12. Express $s^2 + 6s + 13$ in the form $(s - a)(s - b)$ where $a, b \in \mathbb{C}$.

13. Express $2s^2 + 8s + 11$ in the form $2(s - a)(s - b)$ where $a, b \in \mathbb{C}$.

Differentiation

7

7.1 Introduction

Differentiation is a mathematical technique for analysing the way in which functions change. In particular, it determines how rapidly a function is changing at any specific point. As the function in question may represent the magnetic field of a motor, the voltage across a capacitor, the temperature of a chemical mix, etc., it is often important to know how quickly these quantities change. For example, if the voltage on an electrical supply network is falling rapidly because of a short circuit, then it is necessary to detect this in order to switch out that part of the network where the fault has occurred. However, the system should not take action for normal voltage fluctuations and so it is important to distinguish different types and rates of change. Another example would be detecting a sudden rise in the pressure of a fermentation vessel and taking appropriate action to stabilize the pressure.

Differentiation also allows us to find maximum and minimum values of a function. This can be useful in determining optimum values of particular engineering variables, for example, the minimum fuel required for a rocket to reach the moon, or the maximum efficiency of an electricity generator.

7.2 Graphical approach to differentiation

Differentiation is concerned with the rate at which a function is changing, rather than the actual change itself. We can explore the rate of change of a function by examining Figure 7.1. There are several regions to this curve corresponding to different intervals of t. In the interval $[0, 5]$ the function does not change at all. The rate of change of y is zero. From $t = 5$ to $t = 7$ the function increases slightly. Thus the rate of change of y as t increases is small. Since y is increasing, the rate of change of y is positive. From $t = 7$ to $t = 8$ there is a rapid rise in the value of

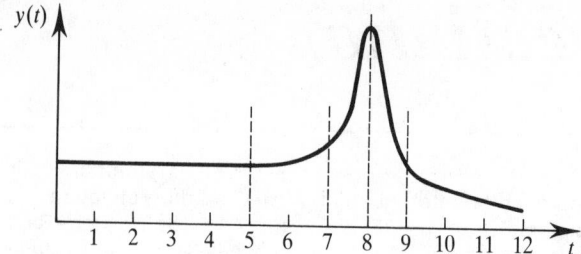

Figure 7.1   The function $y(t)$ has different rates of change over different regions of t.

the function. The rate of change of y is large and positive. From $t = 8$ to $t = 9$ the value of y decreases very rapidly. The rate of change of y is large and negative. Finally from $t = 9$ to $t = 12$ the function decreases slightly. Thus the rate of change of y is small and negative.

The aim of **differential calculus** is to specify the rate of change of a function precisely. It is not sufficient to say 'the rate of change of a function is large'. We require an exact value or expression for the rate of change. Before being able to do this we need to introduce two concepts concerning the rate of change of a function.

7.2.1 *Average rate of change of a function across an interval*

Consider Figure 7.2. When $t = t_1$, the function has a value $y(t_1)$. This is denoted by A on the curve. When $t = t_2$, the function has a value of $y(t_2)$. This point is denoted by B on the curve. The function changes by an amount $y(t_2) - y(t_1)$ over the interval $[t_1, t_2]$. Thus the average rate of change of the function over the interval is:

$$\frac{\text{change in } y}{\text{change in } t} = \frac{y(t_2) - y(t_1)}{t_2 - t_1}$$

The straight line joining A and B is known as a **chord**. Graphically, $y(t_2) - y(t_1)$

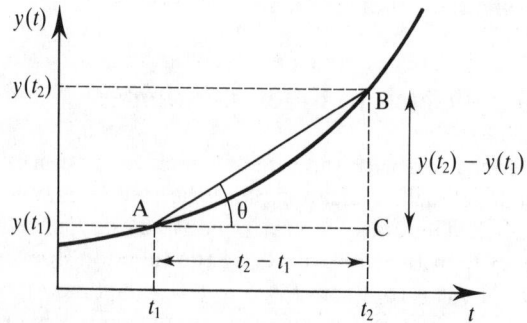

Figure 7.2   Average rate of change across an interval.

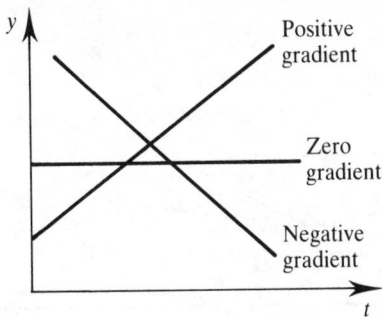

Figure 7.3   Lines can have different gradients.

is the vertical distance and $t_2 - t_1$ the horizontal distance between A and B, so that the **gradient** of the chord AB is given by

$$\frac{BC}{AC} = \frac{y(t_2) - y(t_1)}{t_2 - t_1}$$

The gradient or **slope** of a line is a measure of its steepness and lines may have positive, negative or zero gradients as shown in Figure 7.3.

Thus the gradient of the chord AB corresponds to the average rate of change of the function between A and B. To summarize,

---

■ the average rate of change of a function between two points A and B
   = the gradient of the chord AB

7.2.2 *Rate of change of a function at a point*

Consider again Figure 7.2. Suppose we require the rate of change of the function at point A. We can use the gradient of the chord AB as an approximation to this value. If B is close to A then the approximation is better than if B is not so close to A. Therefore by moving B nearer to A it is possible to improve the accuracy of this approximation (see Figure 7.4).

Suppose the chord AB is extended as a straight line on both sides of AB, and B is moved closer and closer to A until both points eventually coincide. The straight line becomes a **tangent** to the curve at A. This is the straight line that just touches the curve at A. However, the rate of change of this tangent, that is, its gradient, still corresponds to the rate of change of the function, but now it is the rate of change of the function at the point A. To summarize,

---

■ the rate of change of a function at a point A on the curve
   = the gradient of the tangent to the curve at point A

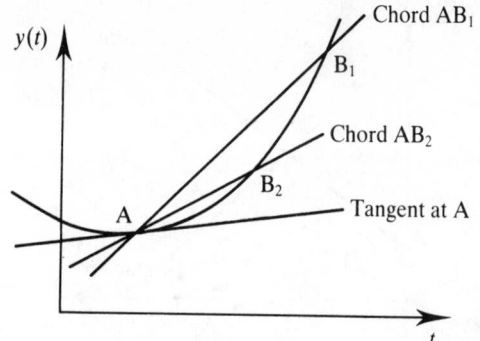

Figure 7.4 Point B is moved nearer to A to improve accuracy.

We have still to address the question of how the gradients of chords and tangents are found. This requires a knowledge of limits which is the topic of the next section.

7.3 Limits and continuity

The concept of a limit is crucial to the development of differentiation. We write $t \to c$ to denote that t approaches, or tends to, the value of c. Note carefully that this is distinct from stating $t = c$. As t tends to c we consider the value to which the function approaches and call this value the **limit** of the function as $t \to c$.

Example 7.1

If $t \to 2$, what value does:

$$f(t) = t^2 + 2t - 3$$

approach?

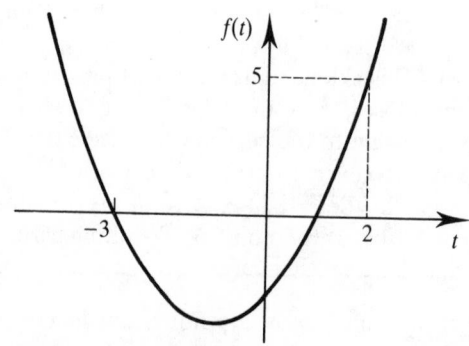

Figure 7.5 The curve $f(t) = t^2 + 2t - 3$.

Solution

Figure 7.5 shows a graph of $f(t)$. Clearly, whether $t = 2$ is approached from the left-hand side or the right-hand side the function tends to 5. That is, if $t \to 2$, then $f(t) \to 5$. We note that this is the value of $f(2)$. Informally we are saying that as t gets nearer and nearer to the value 2, so $f(t)$ gets nearer and nearer to 5. This is usually written as:

$$\lim_{t \to 2} (t^2 + 2t - 3) = 5$$

where lim is an abbreviation of 'limit'. In this example, the limit of $f(t)$ as $t \to 2$ is simply $f(2)$, but this is not true for all functions. ▲

Example 7.2

Figure 7.6 illustrates $y(x)$ defined by:

$$y(x) = \begin{cases} 1 - x & x < 0 \\ 3 & x = 0 \\ x + 1 & x > 0 \end{cases}$$

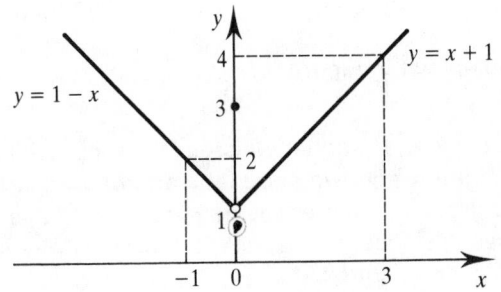

Figure 7.6   As $x \to 0$, $y \to 1$, even though $y(0) = 3$.

Evaluate:

(a) $\lim_{x \to 3} y$

(b) $\lim_{x \to -1} y$

(c) $\lim_{x \to 0} y$

Solutions

We note that this function is piecewise continuous. It has a discontinuity at $x = 0$.

(a) As $x \to 3$, then $y \to 4$. So:

$$\lim_{x \to 3} y = 4$$

(b) Similarly, as $x \to -1$, then $y \to 2$, that is

$$\lim_{x \to -1} y = 2$$

(c) As x approaches 0 what value does y approach? Note we are not evaluating $y(0)$ which actually has a value of 3. We simply ask the question 'What value is y near when x is near, but distinct from, 0?' From Figure 7.6 we see y is near to 1, that is

$$\lim_{x \to 0} y = 1 \quad \blacktriangle$$

Example 7.3

The function $y(x)$ is defined by:

$$y(x) = \begin{cases} 0 & x \leq 0 \\ x & 0 < x \leq 2 \\ x - 2 & x > 2 \end{cases}$$

(a) Sketch the function.

(b) State the limit of y as x approaches (i) 3, (ii) 2, (iii) 0.

Solutions

(a) The function is shown in Figure 7.7.

(b) (i) $\lim_{x \to 3} y = 1$.

(ii) Suppose $x < 2$ and gradually increases, approaching the value 2. Then, from the graph, we see that y approaches 2. Now, suppose $x > 2$ and gradually decreases, tending to 2. In this case y approaches 0. Hence, we cannot find the limit of y as x tends to 2. The $\lim_{x \to 2} y$ does not exist.

(iii) As x tends to 0, y tends to 0. This is true whether x approaches 0 from below, that is, from the left, or from above, that is, from the right. So,

$$\lim_{x \to 0} y = 0 \quad \blacktriangle$$

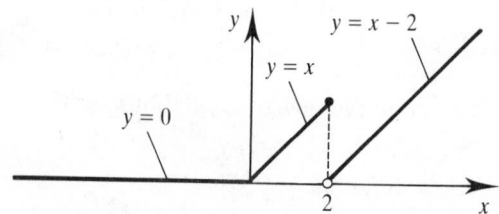

Figure 7.7 The function y has different limits as $x \to 2$ from the left and the right.

It is appropriate at this stage to introduce the concept of **left-hand** and **right-hand** limits. Referring to Example 7.3, we see that as x approaches 2 from the left, that is, from below, then y approaches 2. We say that the left-hand limit of y as x tends to 2 is 2. This is written as:

$$\lim_{x \to 2^-} y = 2$$

Similarly, the right-hand limit of y is obtained by letting x tend to 2 from above. In this case, y approaches 0. This is written as

$$\lim_{x \to 2^+} y = 0$$

Consider a point at which the left-hand and right-hand limits are equal. At such a point we say 'the limit exists at that point'.

7.3.1 *Continuous and discontinuous functions*

A function f is **continuous** at the point where $x = a$, if

$$\lim_{x \to a} f = f(a)$$

that is, the limit value matches the function value at a point of continuity. A function which is not continuous is **discontinuous**. In Example 7.3, the function is continuous at $x = 0$ because

$$\lim_{x \to 0} y = 0 = f(0)$$

but discontinuous at $x = 2$ because $\lim_{x \to 2} y$ does not exist. In Example 7.2, the function is discontinuous at $x = 0$ because $\lim_{x \to 0} y = 1$ but $y(0) = 3$. The concept of continuity corresponds to our natural understanding of a break in the graph of the function, as discussed in Chapter 1.

EXERCISES 7.1

1. The function, $f(t)$, is defined by:

$$f(t) = \begin{cases} 1 & 0 \leq t \leq 2 \\ 2 & 2 < t \leq 3 \\ 3 & t > 3 \end{cases}$$

Sketch a graph of $f(t)$ and state the following limits if they exist.

(a) $\lim_{t \to 1.5} f$

(b) $\lim_{t \to 2^+} f$

(c) $\lim_{t \to 3} f$

(d) $\lim_{t \to 0^+} f$

(e) $\lim_{t \to 3^-} f$

2. The function $g(t)$ is defined by:

$$g(t) = \begin{cases} 0 & t < 0 \\ t^2 & 0 \leq t \leq 3 \\ 2t + 3 & 3 < t \leq 4 \\ 12 & t > 4 \end{cases}$$

(a) Sketch g.

(b) State any points of discontinuity.

(c) Find, if they exist,

   (i) $\lim_{t \to 3} g$

   (ii) $\lim_{t \to 4} g$

   (iii) $\lim_{t \to 4-} g$

7.4 Rate of change at a point

We saw in Section 7.2 that the rate of change of a function at a point is the gradient of the tangent to the curve at that point. Also, we can think of a tangent at A as the limit of an extended chord AB as B → A. We now put these two ideas together to find the rate of change of a function at a point.

Example 7.4

Given $y = f(x) = 3x^2 + 2$, obtain estimates of the rate of change of y at $x = 3$ by considering the intervals:

(a) $[3, 4]$

(b) $[3, 3.1]$

(c) $[3, 3.01]$

Solutions

(a) Consider Figure 7.8.

$$y(3) = 29 \qquad y(4) = 50$$

Let A be the point on the curve $(3, 29)$. Let B be the point $(4, 50)$.

$$\text{average rate of change over the interval } [3, 4] = \frac{\text{change in } y}{\text{change in } x}$$

$$= \frac{y(4) - y(3)}{4 - 3}$$

$$= \frac{50 - 29}{4 - 3} = 21$$

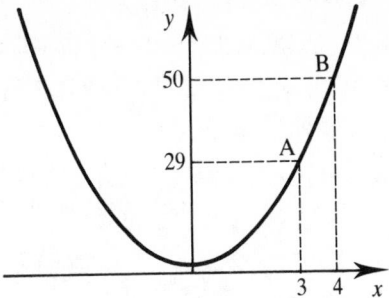

Figure 7.8 The function: $y = 3x^2 + 2$.

This is the gradient of the chord AB and is an estimate of the gradient of the tangent at A. That is, the rate of change at A is approximately 21.

(b) $y(3.1) = 30.83$ and so,

$$\text{average rate of change over the interval } [3, 3.1] = \frac{30.83 - 29}{3.1 - 3} = 18.3$$

This is a more accurate estimate of the rate of change at A.

(c) $y(3.01) = 29.1803$ and so,

$$\text{average rate of change over the interval } [3, 3.01] = \frac{29.1803 - 29}{3.01 - 3}$$

$$= 18.03$$

This is an even better estimate of the rate of change at A. Hence at A, if x increases by 1 unit then y increases by approximately 18 units. This corresponds to a steep upward slope at A. ▲

Example 7.4 illustrates the approach of estimating the rate of change at a point by using the 'shrinking interval' method. By taking smaller and smaller intervals, better and better estimates of the rate of change of the function at $x = 3$ can be obtained. However, we eventually want the interval to 'shrink' to the point $x = 3$. We introduce a small change or **increment** of x denoted by δx and consider the interval $[3, 3 + \delta x]$. By letting δx tend to zero, the interval $[3, 3 + \delta x]$ effectively shrinks to the point $x = 3$.

Example 7.5

Find the rate of change of $y = 3x^2 + 2$ at $x = 3$ by considering the interval $[3, 3 + \delta x]$ and letting δx tend to 0.

Solution

When $x = 3$, $y(3) = 29$.
When $x = 3 + \delta x$, $y(3 + \delta x) = 3(3 + \delta x)^2 + 2 = 3(\delta x)^2 + 18\delta x + 29$

$$\text{average rate of change of } y \text{ across } [3, 3 + \delta x] = \frac{\text{change in } y}{\text{change in } x}$$

$$= \frac{(3(\delta x)^2 + 18\delta x + 29) - 29}{\delta x}$$

$$= \frac{3(\delta x)^2 + 18\delta x}{\delta x}$$

$$= \frac{\delta x(3\delta x + 18)}{\delta x}$$

$$= 3\delta x + 18$$

We now let δx tend to 0, so that the interval shrinks to a point.

$$\text{Rate of change of } y \text{ when } x \text{ is } 3 = \lim_{\delta x \to 0} (3\delta x + 18) = 18$$

We have found the rate of change of y at a particular value of x, rather than across an interval. We usually write:

$$\text{rate of change of } y \text{ when } x \text{ is } 3 = \lim_{\delta x \to 0} \left(\frac{y(3 + \delta x) - y(3)}{\delta x} \right)$$

$$= \lim_{\delta x \to 0} \left(\frac{3(\delta x)^2 + 18\delta x}{\delta x} \right) = \lim_{\delta x \to 0} (3\delta x + 18)$$

$$= 18 \quad \blacktriangle$$

EXERCISES 7.2

1. Find the rate of change of $y = 3x^2 + 2$ at

   (a) $x = 4$ by considering the interval $[4, 4 + \delta x]$

   (b) $x = -2$ by considering the interval $[-2, -2 + \delta x]$

   (c) $x = 1$ by considering the interval $[1 - \delta x, 1 + \delta x]$

2. Find the rate of change of $y = 1/x$ at $x = 2$.

3. To determine the rate of change of $y = x^2 - x$ at $x = 1$ the interval $[1, 1 + \delta x]$ could be used. Equally the intervals $[1 - \delta x, 1]$ or $[1 - \delta x, 1 + \delta x]$ could be used. Show that the same answer results regardless of which interval is used.

7.5 Rate of change at a general point

Example 7.5 shows that the rate of change of a function at a particular point can be found. We will now develop a general terminology for the method. Suppose we have a function of x, $y(x)$. We wish to find the rate of change of y at a general value of x. We begin by finding the average rate of change of $y(x)$ across an interval and then allow the interval to shrink to a single point. Consider the interval $[x, x + \delta x]$. At the beginning of the interval y has a value of $y(x)$. At the end of the interval y has a value of $y(x + \delta x)$ so that the change in y is $y(x + \delta x) - y(x)$, which we denote by δy (see Figure 7.9).

$$\text{average rate of change of } y = \frac{\text{change in } y}{\text{change in } x}$$

$$= \frac{y(x + \delta x) - y(x)}{\delta x}$$

$$= \frac{\delta y}{\delta x}$$

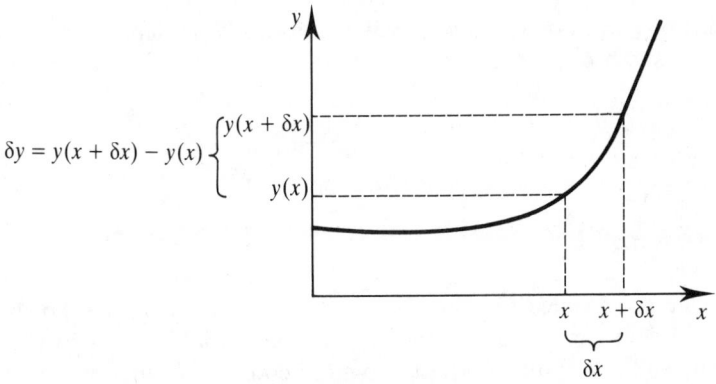

Figure 7.9 The rate of change of y at a point is found by letting $\delta x \to 0$.

Now let δx tend to 0, so that the interval shrinks to a point.

$$\text{rate of change of } y = \lim_{\delta x \to 0} \left(\frac{y(x + \delta x) - y(x)}{\delta x} \right) = \lim_{\delta x \to 0} \left(\frac{\delta y}{\delta x} \right)$$

To see how we proceed to evaluate this limit consider Example 7.6.

Example 7.6

Find the rate of change of $y(x) = 2x^2 + 3x$. Calculate the rate of change of y when $x = 2$ and when $x = -3$.

Solution

$$y(x) = 2x^2 + 3x$$

$$y(x + \delta x) = 2(x + \delta x)^2 + 3(x + \delta x)$$

$$= 2x^2 + 4x\delta x + 2(\delta x)^2 + 3x + 3\delta x$$

$$y(x + \delta x) - y(x) = 2(\delta x)^2 + 4x\delta x + 3\delta x$$

$$\text{rate of change of } y = \lim_{\delta x \to 0} \left(\frac{y(x + \delta x) - y(x)}{\delta x} \right)$$

$$= \lim_{\delta x \to 0} \left(\frac{2(\delta x)^2 + 4x\delta x + 3\delta x}{\delta x} \right)$$

$$= \lim_{\delta x \to 0} (2\delta x + 4x + 3) = 4x + 3$$

When $x = 2$, the rate of change of y is $4(2) + 3 = 11$. When $x = -3$, the rate of change of y is $4(-3) + 3 = -9$. A positive rate of change shows that the function is increasing at that particular point. A negative rate of change shows that the function is decreasing at that particular point. ▲

The rate of change of y is called the **derivative** of y. We denote $\lim_{\delta x \to 0}(\delta y/\delta x)$ by dy/dx. This is pronounced 'dee y by dee x'. Using the previous example, if:

$$y(x) = 2x^2 + 3x$$

then:

$$\frac{dy}{dx} = 4x + 3$$

dy/dx is often abbreviated to y', pronounced 'y dash' or 'y prime'. To stress that y is the dependent variable and x the independent variable we often talk of 'the rate of change of y with respect to x', or more compactly, 'the rate of change of y w.r.t. x'. The process of finding y' from y is called **differentiation**. We know that the derivative dy/dx is the gradient of the tangent to the function at a point. It is also the rate of change of the function. In many examples, the independent variable is t and we need to find the rate of change of y with respect to t, that is, find dy/dt. This is also often written as y' although $\dot{y}$, pronounced 'y dot', is also common. The reader should be aware of both notations. Finally, y' is used to denote the derivative of y whatever the independent variable may be. So dy/dz, dy/dr and dy/dw could all be represented by y'.

■   rate of change of $y(x) = \dfrac{dy}{dx} = \lim\limits_{\delta x \to 0} \left(\dfrac{y(x + \delta x) - y(x)}{\delta x} \right)$

Note that the notation dy/dx means $\lim_{\delta x \to 0} \delta y / \delta x$. If δx is very small yet finite we can state

$$\frac{dy}{dx} \approx \frac{\delta y}{\delta x}$$

This result allows an important approximation to be made. If a small change, δx, is made to the independent variable, the corresponding change in the dependent variable will be:

$$\delta y \approx \frac{dy}{dx} \delta x$$

Example 7.7

Find the gradient of the tangent to $y = x^2$ at $A(1, 1)$, $B(-1, 1)$ and $C(2, 4)$.

Solution

$$\frac{dy}{dx} = \text{gradient of a tangent to curve}$$

$$= \lim_{\delta x \to 0} \left(\frac{y(x + \delta x) - y(x)}{\delta x} \right)$$

$$= \lim_{\delta x \to 0} \left(\frac{(x + \delta x)^2 - x^2}{\delta x} \right) = \lim_{\delta x \to 0} \left(\frac{2x\delta x + (\delta x)^2}{\delta x} \right)$$

$$= \lim_{\delta x \to 0} (2x + \delta x) = 2x$$

At $(1, 1)$, $dy/dx = 2 = $ gradient of tangent at A. At $(-1, 1)$, $dy/dx = -2 = $ gradient of tangent at B. At $(2, 4)$, $dy/dx = 4 = $ gradient of tangent at C. ▲

Suppose we wish to evaluate the derivative, dy/dx, at a specific value of x, say x_0. This is denoted by:

$$\frac{dy}{dx}(x = x_0) \quad \text{or more compactly by} \quad \frac{dy}{dx}(x_0) \quad \text{or } y'(x_0)$$

An alternative notation is:

$$\left. \frac{dy}{dx} \right|_{x = x_0} \quad \text{or} \quad \left. \frac{dy}{dx} \right|_{x_0}$$

So, for Example 7.7 we could have written:

$$\frac{dy}{dx}(1) = 2 \qquad \frac{dy}{dx}(x = -1) = -2 \qquad \left. \frac{dy}{dx} \right|_{x = 2} = 4 \qquad y'(2) = 4$$

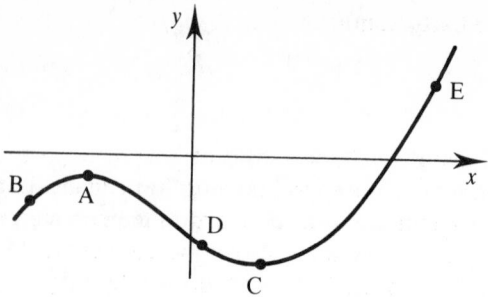

Figure 7.10 Graph for Example 7.8.

Example 7.8

Refer to Figure 7.10. By considering the gradient of the tangent at the points A, B, C, D and E state whether dy/dx is positive, negative or zero at these points.

Solution

At A and C the tangent is parallel to the x axis and so dy/dx is zero. At B and E the tangent has a positive gradient and so dy/dx is positive. At D the tangent has a negative gradient and thus dy/dx is negative. ▲

As we saw in Chapter 1, functions are used to represent physically important quantities such as voltage and current. When the current through certain devices changes, this can give rise to voltages, the magnitudes of which are proportional to the rate of change of the current. Consequently differentiation is needed to model these effects as illustrated in Examples 7.9 and 7.10.

Example 7.9 Voltage across an inductor

The voltage, v, across an inductor with inductance, L, is related to the current, i, through the inductor by:

$$v = L\frac{di}{dt}$$

This relationship is a quantification of Faraday's law which states that the voltage induced in a coil is proportional to the rate of change of magnetic flux through it. If the current in a coil is changing then this corresponds to a change in the magnetic flux through the coil. Note that if di/dt is large then v is large. ▲

Example 7.10 Current through a capacitor

The current, i, through a capacitor with capacitance, C, is related to the voltage, v, across the capacitor by

$$i = C\frac{dv}{dt}$$

It may appear confusing to talk of a current flow through a capacitor as no actual charge flows through the capacitor apart from that caused by any leakage current. Instead there is a build up of charge on the plates of the capacitor. This in turn gives rise to a voltage across the capacitor. If the current flow is large then the rate of change of this voltage will be large. The current flow through the capacitor wires is termed a **conduction current** while that between the capacitor plates is called a **displacement current.** ▲

EXERCISES 7.3

1. Calculate the gradient of the functions at the specified points.

   (a) $y = 2x^2$ at $(1, 2)$

   (b) $y = 2x - x^2$ at $(0, 0)$

   (c) $y = 1 + x + x^2$ at $(2, 7)$

   (d) $y = 2x^2 + 1$ at $(2, 9)$

2. A function, y, is such that dy/dx is constant. What can you say about the function, y?

3. For which graphs in Figure 7.11 is the derivative always (a) positive or (b) negative?

4. Find the derivative of $y(x)$ where y is

   (a) x^2

   (b) $-x^2 + 2x$

5. Differentiate $y = 2x^2 + 9$, that is, find dy/dx. What is the rate of change of y when $x = 3, -2, 1, 0$?

6. Find the rate of change of $y = 4t - t^2$. What is the value of dy/dt when $t = 2$?

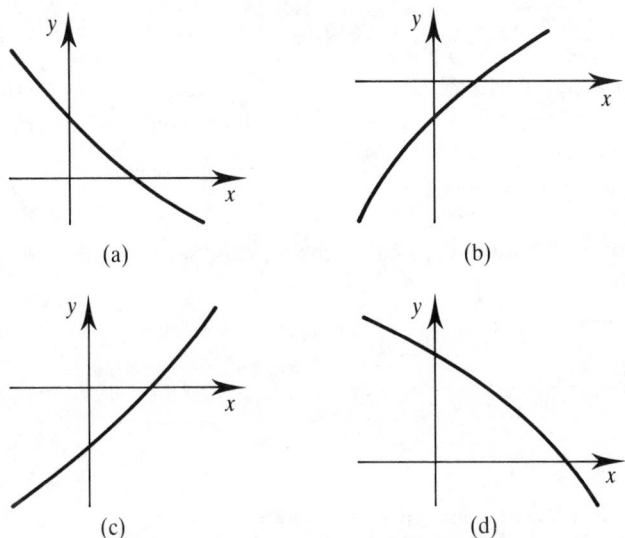

Figure 7.11 Graphs for Exercise 7.3.3.

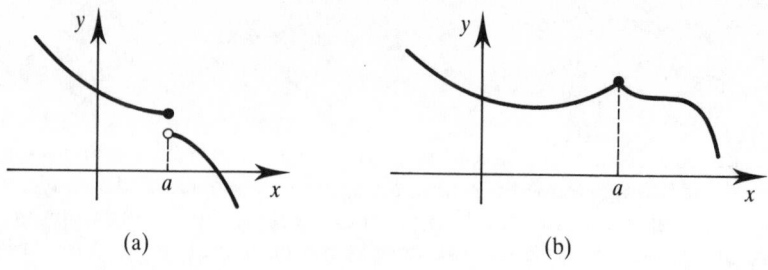

Figure 7.12   (a) The graph has a discontinuity at $x = a$; (b) The graph has a cusp at $x = a$.

7.6 Existence of derivatives

So far we have seen that the derivative, dy/dx, of a function, $y(x)$, may be viewed either algebraically or geometrically.

$$\frac{dy}{dx} = \lim_{\delta x \to 0} \left(\frac{y(x + \delta x) - y(x)}{\delta x} \right)$$

$$\frac{dy}{dx} = \text{rate of change of } y$$

$$= \text{gradient of the graph of } y$$

We now discuss briefly the existence of dy/dx. For some functions the derivative does not exist at certain points and we need to be able to recognize such points. Consider the graphs shown in Figure 7.12. Figure 7.12(a) shows a function with a discontinuity at $x = a$. The function shown in Figure 7.12(b) is continuous but has a **cusp** or **corner** at $x = a$. In both cases it is impossible to draw a tangent at $x = a$, and so dy/dx does not exist at $x = a$. It is impossible to draw a tangent to a curve at a point where the curve is not smooth. Note from Figure 7.12(b) that continuity is not sufficient to guarantee the existence of a derivative.

Example 7.11

Sketch the following functions. State the values of t for which the derivative does not exist.

(a) $y = |t|$
(b) $y = \tan t$
(c) $y = 1/t$

Solutions

(a) The graph of $y = |t|$ is shown in Figure 7.13(a). A corner exists at $t = 0$ and so the derivative does not exist here.

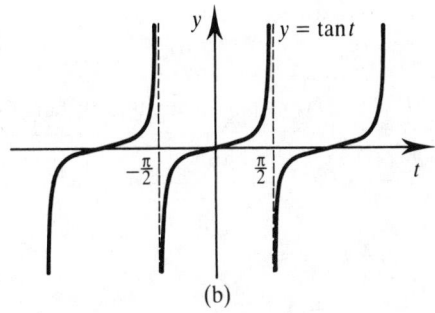

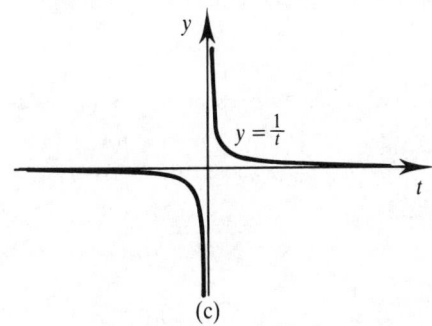

Figure 7.13 (a) There is a corner at $t = 0$; (b) tan t has discontinuities; (c) $y = 1/t$
has a discontinuity at $t = 0$.

(b) A graph of $y = \tan t$ is shown in Figure 7.13(b). There is a discontinuity
in tan t when $t = \cdots -3\pi/2, -\pi/2, \pi/2, 3\pi/2, \ldots$. No derivative exists at
these points.

(c) Figure 7.13(c) shows a graph of $y = 1/t$. The function has one
discontinuity at $t = 0$, and so the derivative does not exist here. ▲

EXERCISE 7.4

Sketch the functions and determine any
points where a derivative does not exist.

(a) $y = 1/(t - 1)$

(b) $y = |\sin t|$

(c) $y = e^t$

(d) $y = |1/t|$

(e) The unit step function $u(t) = \begin{cases} 1 & t \geq 0 \\ 0 & t < 0 \end{cases}$

(f) The ramp function $f(t) = \begin{cases} ct & t \geq 0 \\ 0 & t < 0 \end{cases}$

7.7 Common derivatives

It is time consuming to find the derivative of $y(t)$ using the 'shrinking interval' method (often referred to as **differentiation from first principles**). Consequently the derivatives of commonly used functions are listed for reference in Table 7.1. It will be helpful to memorize the most common derivatives. Note that a, b and n are constants. In the trigonometric functions, the quantity $at + b$, being an angle, must be measured in radians.

Table 7.1 Derivatives of commonly used functions.

Function, $y(t)$	Derivative, y'
constant	0
t^n	nt^{n-1}
e^{at}	ae^{at}
$\ln t$	$1/t$
$\sin(at + b)$	$a\cos(at + b)$
$\cos(at + b)$	$-a\sin(at + b)$
$\tan(at + b)$	$a\sec^2(at + b)$
$\operatorname{cosec}(at + b)$	$-a\operatorname{cosec}(at + b)\cot(at + b)$
$\sec(at + b)$	$a\sec(at + b)\tan(at + b)$
$\cot(at + b)$	$-a\operatorname{cosec}^2(at + b)$
$\sin^{-1}(at + b)$	$\dfrac{a}{\sqrt{1 - (at + b)^2}}$
$\cos^{-1}(at + b)$	$\dfrac{-a}{\sqrt{1 - (at + b)^2}}$
$\tan^{-1}(at + b)$	$\dfrac{a}{1 + (at + b)^2}$
$\sinh(at + b)$	$a\cosh(at + b)$
$\cosh(at + b)$	$a\sinh(at + b)$
$\tanh(at + b)$	$a\operatorname{sech}^2(at + b)$
$\operatorname{cosech}(at + b)$	$-a\operatorname{cosech}(at + b)\coth(at + b)$
$\operatorname{sech}(at + b)$	$-a\operatorname{sech}(at + b)\tanh(at + b)$
$\coth(at + b)$	$-a\operatorname{cosech}^2(at + b)$
$\sinh^{-1}(at + b)$	$\dfrac{a}{\sqrt{(at + b)^2 + 1}}$
$\cosh^{-1}(at + b)$	$\dfrac{a}{\sqrt{(at + b)^2 - 1}}$
$\tanh^{-1}(at + b)$	$\dfrac{a}{1 - (at + b)^2}$

Example 7.12

Use Table 7.1 to find y' when

(a) $y = e^{-7t}$
(b) $y = t^5$
(c) $y = \tan(3t - 2)$
(d) $y = \sin(\omega t + \phi)$
(e) $y = 1/\sqrt{t}$
(f) $y = 1/t^5$
(g) $y = \cosh^{-1} 5t$

Solutions

(a) From Table 7.1, we find that if:

$$y = e^{at} \qquad \text{then} \qquad y' = ae^{at}$$

In this case, $a = -7$ and so if:

$$y = e^{-7t} \qquad \text{then} \qquad y' = -7e^{-7t}$$

(b) From Table 7.1, we find that if:

$$y = t^n \qquad \text{then} \qquad y' = nt^{n-1}$$

In this case, $n = 5$ and so if:

$$y = t^5 \qquad \text{then} \qquad y' = 5t^4$$

(c) If $y = \tan(at + b)$ then $y' = a\sec^2(at + b)$. In this case, $a = 3$ and $b = -2$. Hence if:

$$y = \tan(3t - 2) \qquad \text{then} \qquad y' = 3\sec^2(3t - 2)$$

(d) If $y = \sin(at + b)$ then $y' = a\cos(at + b)$. Here $a = \omega$ and $b = \phi$, and so if:

$$y = \sin(\omega t + \phi) \qquad \text{then} \qquad y' = \omega \cos(\omega t + \phi)$$

(e) Note that $1/\sqrt{t} = t^{-1/2}$. From Table 7.1 we find that if $y = t^n$ then $y' = nt^{n-1}$. In this case, $n = -1/2$ and so if:

$$y = \frac{1}{\sqrt{t}} \qquad \text{then} \qquad y' = -\frac{1}{2}t^{-3/2}$$

(f) Note that $1/t^5 = t^{-5}$. Using Table 7.1, we find that if $y = t^{-5}$ then $y' = -5t^{-6}$.

(g) From Table 7.1, if $y = \cosh^{-1}(at + b)$ then $y' = a/\sqrt{(at + b)^2 - 1}$. In this case, $a = 5$ and $b = 0$. Hence, if

$$y = \cosh^{-1} 5t \qquad \text{then} \qquad y' = \frac{5}{\sqrt{25t^2 - 1}} \quad \blacktriangle$$

EXERCISES 7.5

1. Use Table 7.1 to find y' given:

(a) $y = t^2$

(b) $y = t^9$

(c) $y = t^{-3}$

(d) $y = t$

(e) $y = 1/t$

(f) $y = 1/t^2$

(g) $y = e^{3t}$

(h) $y = e^{-3t}$

(i) $y = 1/e^{5t}$

(j) $y = t^{1/2}$

(k) $y = \sin(2t + 3)$

(l) $y = \cos(4 - t)$

(m) $y = \tan((t/2) + 1)$

(n) $y = \operatorname{cosec}(3t + 7)$

(o) $y = \cot(1 - t)$

(p) $y = \sec(2t - \pi)$

(q) $y = \sin^{-1}(t + \pi)$

(r) $y = \pi$

(s) $y = \tan^{-1}(-2t - 1)$

(t) $y = \cos^{-1}(4t - 3)$

(u) $y = \tanh(6t)$

(v) $y = \cosh(2t + 5)$

(w) $y = \sinh((t + 3)/2)$

(x) $y = \operatorname{sech}(-t)$

(y) $y = \coth((2t/3) - 1/2)$

(z) $y = \cosh^{-1}(t + 3)$

2. Find dy/dx when

(a) $y = 1/\sqrt{x}$

(b) $y = e^{2x/3}$

(c) $y = e^{-x/2}$

(d) $y = \ln x$

(e) $y = \operatorname{cosec}((2x - 1)/3)$

(f) $y = \tan^{-1}(\pi x + 3)$

(g) $y = \tanh(2x + 1)$

(h) $y = \sinh^{-1}(-3x)$

(i) $y = \cot(\omega x + \pi)$ ω constant

(j) $y = 1/\sin(5x + 3)$

(k) $y = \cos 3x$

(l) $y = 1/\cos 3x$

(m) $y = \tan(2x + \pi)$

(n) $y = \operatorname{cosech}((x - 1)/2)$

(o) $y = \tanh^{-1}((2x + 3)/7)$

7.8 Differentiation as a linear operator

In mathematical language differentiation is a **linear operator**. This means that if we wish to differentiate the sum of two functions we can differentiate each function separately and then simply add the two derivatives, that is

derivative of $(f + g)$ = derivative of f + derivative of g

This is expressed mathematically as:

$$\blacksquare \quad \frac{d}{dx}(f + g) = \frac{df}{dx} + \frac{dg}{dx}$$

We can regard d/dx as the operation of differentiation being applied to the expression which follows it. The properties of a linear operator also make the handling of constant factors easy. To differentiate kf, where k is a constant we take k times the derivative of f, that is

derivative of $(kf) = k \times$ derivative of f

Mathematically, we would state:

■ $\dfrac{d}{dx}(kf) = k\dfrac{df}{dx}$

Table 7.1 together with these two linearity properties allows us to differentiate some quite complicated functions.

Example 7.13

Differentiate:

(a) $3t^2$ (c) 7

(b) $9t$ (d) $3t^2 + 9t + 7$

Solutions

(a) Let $y = 3t^2$, then

$$\frac{dy}{dt} = \frac{d}{dt}(3t^2)$$

$$= 3\frac{d}{dt}(t^2) \text{ using linearity}$$

$$= 3(2t) \text{ from the table}$$

$$= 6t$$

(b) Let $y = 9t$, then

$$\frac{dy}{dt} = \frac{d}{dt}(9t)$$

$$= 9\frac{d}{dt}(t) \text{ using linearity}$$

$$= 9$$

(c) Let $y = 7$, then $y' = 0$.

(d) Let $y = 3t^2 + 9t + 7$

$$\frac{dy}{dt} = \frac{d}{dt}(3t^2 + 9t + 7)$$

$$\frac{dy}{dt} = 3\frac{d}{dt}(t^2) + 9\frac{d}{dt}(t) + \frac{d}{dt}(7) \quad \text{using linearity}$$

$$= 6t + 9 \quad \blacktriangle$$

Example 7.14 Fluid flow into a tank

If fluid is being poured into a tank at a rate of q m^3 s^{-1}, then this will result in an increase in volume, V, of fluid in the tank. The rate of increase in volume, dV/dt m^3 s^{-1}, is given by:

$$\frac{dV}{dt} = q$$

This relationship follows from the principle of conservation of mass. If q is large, then dV/dt is large which corresponds to the fluid volume in the tank increasing at a fast rate. Consequently, the height of the fluid, h, also increases at a fast rate. If the cross-sectional area of the tank, A, is constant, then $V = Ah$. Therefore,

$$\frac{dV}{dt} = \frac{d}{dt}(Ah) = A\frac{dh}{dt}$$

because differentiation is a linear operator. So:

$$A\frac{dh}{dt} = q \quad \blacktriangle$$

Example 7.15

Use Table 7.1 and the linearity properties of differentiation to find y' where

(a) $y = 3e^{2x}$

(b) $y = 1/x$

(c) $y = 3\sin 4x$

(d) $y = \sin 2x - \cos 5x$

(e) $y = 3\ln x$

(f) $y = \ln 2x$

(g) $y = 3x^2 + 7x - 5$

Solutions

(a) If $y = 3e^{2x}$, then:

$$\frac{dy}{dx} = \frac{d}{dx}(3e^{2x}) = 3\frac{d}{dx}(e^{2x}) \qquad \text{using linearity}$$

$$= 3(2e^{2x}) \qquad \text{using Table 7.1}$$

$$= 6e^{2x}$$

(b) If $y = x^{-1}$, then:

$$y' = -1x^{-2} \qquad \text{from Table 7.1}$$

$$= -\frac{1}{x^2}$$

(c) If $y = 3 \sin 4x$, then:

$$\frac{dy}{dx} = \frac{d}{dx}(3 \sin 4x) = 3\frac{d}{dx}(\sin 4x) \qquad \text{using linearity}$$

$$= 3(4 \cos 4x) \qquad\qquad \text{using Table 7.1}$$

$$= 12 \cos 4x$$

(d) The linearity properties allow us to differentiate each term individually. If,

$$y = \sin 2x - \cos 5x \qquad \text{then} \qquad \frac{dy}{dx} = \frac{d}{dx}(\sin 2x) - \frac{d}{dx}(\cos 5x)$$

$$= 2 \cos 2x + 5 \sin 5x$$

(e) If $y = 3 \ln x$, then:

$$\frac{dy}{dx} = \frac{d}{dx}(3 \ln x) = 3\frac{d}{dx}(\ln x) \qquad \text{using linearity}$$

$$= \frac{3}{x} \qquad\qquad\qquad \text{using Table 7.1}$$

(f) If $y = \ln 2x$, then:

$$y = \ln 2 + \ln x \qquad \text{using laws of logarithms}$$

and so

$$\frac{dy}{dx} = 0 + \frac{1}{x} = \frac{1}{x} \qquad \text{since } \ln 2 \text{ is constant}$$

(g) Each term is differentiated:

$$y' = 6x + 7 - 0 = 6x + 7 \quad \blacktriangle$$

Example 7.16

Differentiate $y(t) = e^t$.

Solution

From Table 7.1, we find:

$$\frac{dy}{dt} = e^t = y$$

We note that the derivative of e^t is again e^t. This is the only function which reproduces itself upon differentiation. $\blacktriangle$

Example 7.17 Small signal resistance

In Chapter 1, Example 1.25, we examined the mathematical model of a semiconductor diode. At room temperature this was:

$$I = I_s(e^{40V} - 1)$$

where I is the diode current, V the applied voltage and I_s is the constant reverse saturation current. Sometimes when a diode is used in a circuit it may be **biased** to operate in a certain region of its I–V characteristic. This means that its use is restricted to a certain voltage range. The point around which it operates is known as its **operating point**. This is illustrated in Figure 7.14.

Deviations from this operating point may be small in certain cases. If so, they are known as **small signal** variations and are caused by small alternating current voltages being superimposed on the main d.c. bias voltage. In calculating how the diode will react to such small signal voltages, the slope of the diode characteristic around the operating point is more relevant than the overall ratio of current to voltage. Provided the deviations from the operating point are not large, the tangent to the I–V curve at the operating point provides an adequate model for how the diode will behave. The slope of the curve can be obtained by differentiating the diode equation. So,

$$\frac{dI}{dV} = 40I_s e^{40V}$$

It is usual to write small changes in current and voltage as δI and δV. Therefore, since

$$\frac{\delta I}{\delta V} \approx \frac{dI}{dV}$$

$$\delta I \approx 40I_s e^{40V}\, \delta V$$

This expression allows the change in diode current, δI, to be estimated given a change in diode voltage, δV, provided the operating point is known and the changes are small. ▲

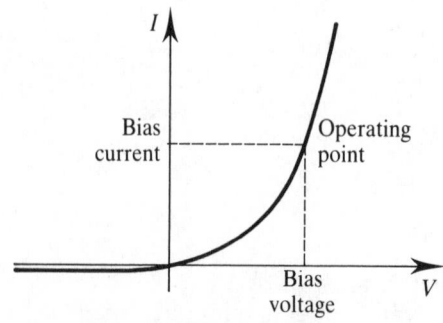

Figure 7.14 Diode characteristic showing operating point.

Example 7.18

Find the derivative of $y = e^{-t} + t^2$, when t is

(a) 1
(b) 0

Solutions

$$y = e^{-t} + t^2$$

$$y' = -e^{-t} + 2t$$

(a) When $t = 1$, $y' = -e^{-1} + 2 = 1.632$, that is $y'(1) = 1.632$.
(b) When $t = 0$, $y' = -1$, that is $y'(0) = -1$.   ▲

Example 7.19 **Obtaining a linear model for a simple fluid system**

Consider the fluid system illustrated in Figure 7.15. The pump is driven by a d.c. motor. The pump/motor can be modelled by a linear relationship in which the fluid flow rate, q_i, is proportional to the control voltage, v_{in}, that is

$$q_i = k_p v_{in}$$

where k_p = pump/motor constant, v_{in} = control voltage (V), q_i = flow rate in to the tank $(m^3 \, s^{-1})$. The valve has a non-linear characteristic given by the quadratic polynomial:

$$p = 20\,000 \, q_o^2$$

where p = pressure at the base of the tank $(N \, m^{-2})$, q_o = flow out of the tank $(m^3 \, s^{-1})$. The fluid being used is water which has a density $\rho = 998 \, kg \, m^{-3}$. Assume $g = 9.81 \, m \, s^{-2}$ and that $k_p = 0.03 \, m^3 \, s^{-1} \, V^{-1}$. Carry out the following:

(a) Calculate the flow rate out of the tank, q_o, and the control voltage, v_{in}, when the system is in equilibrium and the height of the water in the tank, h, is 0.25 m.

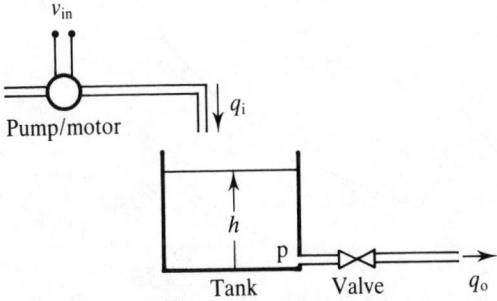

Figure 7.15 A fluid system comprising pump, tank and valve.

(b) Obtain a linear model for the system, valid for small changes about a
water height of 0.25 m. Use this model to calculate the new water height
and flow rate out of the tank when the control voltage is increased by 0.4 V.

Solutions

(a) The pressure at the bottom of the tank is given by:

$$p = \rho gh = 998 \times 9.81 \times 0.25 = 2448$$

The flow rate through the valve is given by:

$$q_o^2 = \frac{p}{20\,000}$$

$$q_o = \sqrt{\frac{p}{20\,000}} = \sqrt{\frac{2448}{20\,000}} = 0.350 \text{ m}^3 \text{ s}^{-1}$$

Now if the system is in equilibrium, then the height of the water in the
tank must have stabilized to a constant value. Therefore:

$$q_i = q_o = 0.350$$

and so,

$$v_{in} = \frac{q_i}{k_p} = \frac{0.350}{0.03} = 11.7 \text{ V}$$

(b) Before answering this part it is worth examining what is meant by a
linear model. Figure 7.16 shows the valve characteristic together with a
linear approximation around an output flow rate of $q_o = 0.350 \text{ m}^3 \text{ s}^{-1}$.
This corresponds to a water height of 0.25 m.

A linear model for the valve is one in which the relationship
between p and q_o is approximated by the straight line which forms a
tangent to the curve at the operating point. The operating point is the
point around which the model is valid. It is clear that if the straight line
approximation is used for points that are a large distance from the

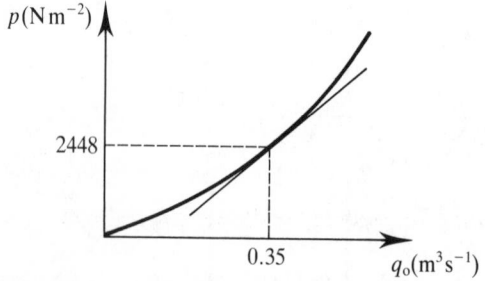

Figure 7.16 Relationship between pressure across valve (p) and flow through valve
(q_o).

operating point, then the linear model will not be very accurate. However, for small changes around the operating point the approximation is reasonably accurate. Clearly a different operating point will require a different linear approximation. In order to obtain the gradient of this line it is necessary to differentiate the function relating valve pressure to valve flow. So,

$$p = 20\,000q_o^2$$

$$\frac{dp}{dq_o} = 40\,000q_o$$

At the operating point $q_o = 0.350$, therefore,

$$\left.\frac{dp}{dq_o}\right|_{q_o = 0.350} = 40\,000 \times 0.350 = 14\,000$$

This value is the gradient of the tangent to the curve at the operating point. Small changes around an operating point are usually indicated by the notation δ. Therefore,

$$\frac{\delta p}{\delta q_o} \approx \left.\frac{dp}{dq_o}\right|_{q_o = 0.350} = 14\,000$$

$$\delta p = 14\,000\delta q_o \tag{7.1}$$

Note that equality has been assumed for the purposes of the linear model. It is easy to relate a change in pump flow to a change in control voltage because the relationship is linear and so a linear approximation is not required. So,

$$q_i = k_p v_{in}$$

Differentiating this equation w.r.t. v_{in} yields:

$$\frac{dq_i}{dv_{in}} = k_p$$

In this case

$$\frac{\delta q_i}{\delta v_{in}} = \frac{dq_i}{dv_{in}} = k_p$$

and has a constant value independent of the operating point.

$$\delta q_i = k_p \, \delta v_{in} \tag{7.2}$$

The relationship between pressure at the bottom of the tank and the water height is also linear.

$$p = \rho g h$$

$$\frac{dp}{dh} = \rho g$$

$$\frac{\delta p}{\delta h} = \frac{dp}{dh} = \rho g$$

$$\delta p = \rho g\, \delta h \qquad\qquad\qquad\qquad\qquad (7.3)$$

The change in control voltage, δv_{in}, is given as 0.4. We also know that $k_p = 0.03$. Therefore, using Equation (7.2), we have

$$\delta q_i = 0.03 \times 0.4 = 0.012$$

Now if time is allowed for the system to reach equilibrium with this increased input flow, then $\delta q_o = \delta q_i$. In other words, the output flow increases by the same amount as the input flow and the water height once again stabilizes to a fixed value. Therefore,

$$\delta q_o = \delta q_i = 0.012$$

Using Equation (7.1), we find:

$$\delta p = 14\,000\, \delta q_o = 14\,000 \times 0.012 = 168$$

Using Equation (7.3) we get

$$\delta h = \frac{\delta p}{\rho g} = \frac{168}{998 \times 9.81} = 0.0172$$

Therefore the new water height is $0.25 + 0.0172 = 0.267$ m to three significant figures. The new water flow rate is $0.35 + 0.012 = 0.362$ m^3 s^{-1}.

To recap, all the elements of the fluid system were linear apart from the valve. By obtaining a linear model for the valve, valid for values close to the operating point, it was possible to calculate the effect of changing the control voltage to the motor. It is important to stress that the linear model for the valve is only good for small changes around the operating point. In this case the increase in control voltage was approximately 3%. The model would not have been very good for predicting the effect of a 50% increase in control voltage. Linear models of non-linear systems are particularly useful when several components are non-linear, as they are much easier to analyse. We examine these concepts in more detail in Chapter 9. ▲

EXERCISES 7.6

1. Differentiate the following functions.

   (a) $y = 4x^3 - 5x^2$

   (b) $y = 3 \sin(5t) + 2e^{4t}$

   (c) $y = \sin(4t) + 3 \cos(2t) - t$

   (d) $y = \tan(3z)$

   (e) $y = 2e^{3t} + 17 - 4 \sin(2t)$

   (f) $y = 1/t^3 + (\cos 5t)/2$

   (g) $y = 2w^3/3 + e^{4w}/2$

   (h) $y = \sqrt{x} + \ln(\sqrt{x})$. *Hint:* $\sqrt{x} = x^{1/2}$, and use the laws of logarithms.

2. Evaluate the derivatives of the functions at the given value.

   (a) $y = 2t + 9 + e^{t/2}$ $t = 1$

   (b) $y = (t^2 - 4t + 6)/3$ $t = 2$

   (c) $y = \sin t + \cos t$ $t = 1$

   (d) $y = 3e^{2t} - 2 \sin(t/2)$ $t = 0$

   (e) $y = 5 \tan(2x) + 1/e^{2x}$ $x = 0.5$

   (f) $y = 3 \ln t + \sin(3t)$ $t = 0.25$

3. Find dx/dt, if:

   (a) $x = e^{\omega t}$

   · (b) $x = e^{-\omega t}$

   where ω is a constant.

4. Find the derivative of:

   (a) $y = 3 \sin^{-1}(2t) - 5 \cos^{-1}(3t)$

   (b) $y = \frac{1}{2} \tan^{-1}(t + 2) + 4 \cos^{-1}(2t - 1)$

   (c) $y = 2 \sinh(3t - 1) - 4 \cosh((t - 3)/2)$

   (d) $y = \{\operatorname{cosech}(4t) + 3 \operatorname{sech}(6t)\}/2$

   (e) $y = 2 \sinh^{-1}((t + 1)/2) - 3 \cosh^{-1}((1 - t)/2)$

   (f) $y = 3 \tanh^{-1}(2t + 3) - 2 \tanh^{-1}(3t + 2)$

5. A function, $y(t)$, is given by

   $$y(t) = \frac{t^3}{3} - \frac{5t^2}{2} + 4t + 1$$

   (a) Find dy/dt

   (b) For which values of t is the derivative zero?

6. Find the equation of the tangent to the curve

   $$y(x) = x^3 + 7x^2 - 9$$

   at the point $(2, 27)$.

7. Find values of t in the interval $[0, \pi]$ for which the tangent to $x(t) = \sin 2t$ has zero gradient.

8. Find the rate of change of

   $$z(t) = 2e^{t/2} - t^2$$

   when

   (a) $t = 0$

   (b) $t = 3$

7.9 Rules of differentiation

There are three rules which enable us to differentiate more complicated functions. They are (a) the product rule, (b) the quotient rule, (c) the chain rule. Traditionally they are written with x as the independent variable but apply in an analogous way for other independent variables.

7.9.1 *The product rule*

As the name suggests, this rule allows us to differentiate a product of functions, such as $x \sin x$, $t^2 \cos 2t$ and $e^z \ln z$.

■ If

$$y(x) = u(x)\,v(x)$$

then

$$\frac{dy}{dx} = \frac{du}{dx}v + u\frac{dv}{dx} = u'v + uv'$$

Example 7.20

Find y' given:

(a) $y = x \sin x$
(b) $y = t^2 e^t$
(c) $y = e^{-2t} \cos 3t$

Solutions

(a) $y = x \sin x = uv$. Choose $u = x$ and $v = \sin x$. Then $u' = 1$, $v' = \cos x$. Applying the product rule to y yields:

$$y' = \sin x + x \cos x$$

(b) $y = t^2 e^t = uv$. Choose $u = t^2$ and $v = e^t$. Then $u' = 2t$ and $v' = e^t$. Applying the product rule to y yields:

$$y' = 2te^t + t^2 e^t$$

(c) $y = e^{-2t} \cos 3t = uv$. Choose $u = e^{-2t}$ and $v = \cos 3t$. Then $u' = -2e^{-2t}$ and $v' = -3 \sin 3t$. Applying the product rule yields:

$$y' = -2e^{-2t} \cos 3t + e^{-2t}(-3 \sin 3t) = -e^{-2t}(2 \cos 3t + 3 \sin 3t) \quad \blacktriangle$$

7.9.2 *The quotient rule*

This rule allows us to differentiate a quotient of functions, such as $x/\sin x$, $(t^2 - t + 3)/(t + 2)$ and $e^{-3z}/(z^2 - 1)$.

■ When

$$y(x) = \frac{u(x)}{v(x)}$$

then,

$$y' = \frac{v(du/dx) - u(dv/dx)}{v^2} = \frac{vu' - uv'}{v^2}$$

Example 7.21

Find y' given:

(a) $y = \dfrac{\sin x}{x}$

(b) $y = t^2/(2t + 1)$

(c) $y = e^{2t}/(t^2 + 1)$

Solutions

(a) $y = \dfrac{\sin x}{x} = u/v$. So $u = \sin x$, $v = x$ and $u' = \cos x$, $v' = 1$. Using the

quotient rule the derivative of y is found.

$$y' = \frac{x \cos x - \sin x}{x^2}$$

(b) $y = t^2/(2t + 1) = u/v$. So $u = t^2$, $v = 2t + 1$ and $u' = 2t$, $v' = 2$. Hence,

$$y' = \frac{(2t + 1)2t - (t^2)(2)}{(2t + 1)^2} = \frac{2t(t + 1)}{(2t + 1)^2}$$

(c) $y = e^{2t}/(t^2 + 1)$. So $u = e^{2t}$, $v = t^2 + 1$ and $u' = 2e^{2t}$, $v' = 2t$. Application of the quotient rule yields:

$$y' = \frac{(t^2 + 1)2e^{2t} - e^{2t}2t}{(t^2 + 1)^2} = \frac{2e^{2t}(t^2 - t + 1)}{(t^2 + 1)^2} \quad \blacktriangle$$

7.9.3 The chain rule

This rule helps us to differentiate complicated functions, where a substitution can be used to simplify the function. Suppose $y = y(z)$ and $z = z(x)$. Then y may be considered as a function of x. For example, if $y = z^3 - z$ and $z = \sin 3x$, then $y = (\sin 3x)^3 - \sin(3x)$. Suppose we seek the derivative, dy/dx. Note that the derivative w.r.t. x is sought.

■ The chain rule states:

$$\frac{dy}{dx} = \frac{dy}{dz} \times \frac{dz}{dx}$$

Example 7.22

Given $y = z^6$ where $z = x^2 + 1$ find dy/dx.

Solution

If $y = z^6$ and $z = x^2 + 1$, then $y = (x^2 + 1)^6$. We recognize this as the composition $y(z(x))$ (see Section 1.3.5). Now $y = z^6$ and so $dy/dz = 6z^5$. Also

$z = x^2 + 1$ and so $dz/dx = 2x$. Using the chain rule:

$$\frac{dy}{dx} = \frac{dy}{dz} \times \frac{dz}{dx} = 6z^5 2x = 12x(x^2 + 1)^5 \quad \blacktriangle$$

Example 7.23

Differentiate:

(a) $y = 3e^{\sin x}$
(b) $y = (3t^2 + 2t - 9)^{10}$
(c) $y = \sqrt{1 + t^2}$
(d) $y = \ln(1 + x)$
(e) $y = \ln(1 + \cos x)$

Solutions

In these examples we must formulate the function z ourselves.

(a) Let $z(x) = \sin x$. Then $y(z) = 3e^z$ so $dy/dz = 3e^z$; $z(x) = \sin x$ and so $dz/dx = \cos x$. The chain rule is used to find dy/dx.

$$\frac{dy}{dx} = \frac{dy}{dz} \times \frac{dz}{dx} = 3e^z \cos x = 3e^{\sin x} \cos x$$

(b) Let $z(t) = 3t^2 + 2t - 9$. Then $y(z) = z^{10}$, $dy/dz = 10z^9$, $dz/dt = 6t + 2$. Using the chain rule dy/dt is found.

$$\frac{dy}{dt} = \frac{dy}{dz} \times \frac{dz}{dt} = 10z^9(6t + 2) = 20(3t + 1)(3t^2 + 2t - 9)^9$$

(c) Let $z(t) = 1 + t^2$. Then $y = \sqrt{z} = z^{1/2}$, $dy/dz = \frac{1}{2}z^{-1/2}$ and $dz/dt = 2t$. Using the chain rule, we obtain

$$\frac{dy}{dt} = \frac{dy}{dz} \times \frac{dz}{dt} = \frac{1}{2}z^{-1/2}2t = \frac{t}{\sqrt{z}} = \frac{t}{\sqrt{1 + t^2}}$$

(d) Let $z(x) = 1 + x$. Then $y(z) = \ln z$, $dy/dz = 1/z$ and $dz/dx = 1$. Using the chain rule, we find:

$$\frac{dy}{dx} = \frac{dy}{dz} \times \frac{dz}{dx} = \frac{1}{z} \times 1 = \frac{1}{z} = \frac{1}{1 + x}$$

(e) Let $z(x) = 1 + \cos x$. Then $y(z) = \ln z$, $dy/dz = 1/z$ and $dz/dx = -\sin x$. Using the chain rule, we get:

$$\frac{dy}{dx} = \frac{dy}{dz} \times \frac{dz}{dx} = \frac{-\sin x}{z} = \frac{-\sin x}{1 + \cos x} \quad \blacktriangle$$

EXERCISES 7.7

1. Use the product rule to differentiate the following functions.

   (a) $y = \sin x \cos x$

   (b) $y = \ln t \tan t$

   (c) $y = (t^3 + 1)e^{2t}$

   (d) $y = \sqrt{x}e^x$

   (e) $y = e^t \sin t \cos t$

   (f) $y = 3 \sinh 2t \cosh 3t$

   (g) $y = (1 + \sin t) \tan t$

   (h) $y = 4 \sinh(t + 1) \cosh(1 - t)$

2. Use the quotient rule to find the derivatives of the following:

   (a) $(\cos x)/\sin x$

   (b) $(\tan t)/\ln t$

   (c) $e^{2t}/(t^3 + 1)$

   (d) $(3x^2 + 2x - 9)/(x^3 + 1)$

   (e) $(x^2 + x + 1)/(1 + e^x)$

   (f) $(\sinh 2t)/\cosh 3t$

   (g) $(1 + e^t)/(1 + e^{2t})$

3. Use the chain rule to differentiate the following.

   (a) $(t^3 + 1)^{100}$

   (b) $\sin^3(3t + 2)$

   (c) $\ln(x^2 + 1)$

   (d) $(2t + 1)^{1/2}$

   (e) $3\sqrt{\cos(2x - 1)}$

   (f) $1/(t + 1)$

   (g) $(at + b)^n$, a and b constants

4. For which values of t is the derivative of $y(t) = e^{-t}t^2$, zero?

5. Find the rate of change of y at the specified values of t.

   (a) $y = \sin(1/t)$ $t = 1$

   (b) $y = (t^2 - 1)^{17}$ $t = 1$

   (c) $y = \sinh(t^2)$ $t = 2$

   (d) $y = (1 + t + t^2)/(1 - t)$ $t = 2$

   (e) $y = e^t/(t \sin t)$ $t = 1$

6. Find the equation of the tangent to:

   $$y(x) = e^{3x}(1 - x) \qquad \text{at the point } (0, 1)$$

7.10 Parametric, implicit and logarithmic differentiation

7.10.1 *Parametric differentiation*

In some circumstances both y and x depend upon a third variable, t. This third variable is often called a **parameter**. By eliminating t, y can be found as a function of x. For example, if $y = (1 + t)^2$ and $x = 2t$ then, eliminating t, we can write $y = (1 + x/2)^2$. Hence, y may be considered as a function of x, and so the derivative dy/dx can be found. However, sometimes the elimination of t is difficult or even impossible. Consider the example $y = \sin t + t$, $x = t^2 + e^t$. In this case, it is impossible to obtain y in terms of x. The derivative dy/dx can still be found using the chain rule.

■ $$\frac{dy}{dx} = \frac{dy}{dt} \times \frac{dt}{dx} = \frac{dy}{dt} \bigg/ \frac{dx}{dt}$$

Finding dy/dx by this method is known as **parametric differentiation**.

Example 7.24

Given $y = (1 + t)^2$, $x = 2t$ find dy/dx.

Solution

By eliminating t, we see

$$y = \left(1 + \frac{x}{2}\right)^2$$

and so

$$\frac{dy}{dx} = 1 + \frac{x}{2}$$

Parametric differentiation is an alternative method of finding dy/dx which does not require the elimination of t.

$$\frac{dy}{dt} = 2(1 + t) \qquad \frac{dx}{dt} = 2$$

Using the chain rule, we obtain:

$$\frac{dy}{dx} = \frac{dy}{dt} \bigg/ \frac{dx}{dt} = \frac{2(1 + t)}{2} = 1 + t = 1 + \frac{x}{2} \quad \blacktriangle$$

Example 7.25

Given $y = e^t + t$, $x = t^2 + 1$, find dy/dx using parametric differentiation.

Solution

$$\frac{dy}{dt} = e^t + 1 \qquad \frac{dx}{dt} = 2t$$

Hence,

$$\frac{dy}{dx} = \frac{dy}{dt} \bigg/ \frac{dx}{dt} = \frac{e^t + 1}{2t}$$

In this example, the derivative is expressed in terms of t. This will always be the case when t has not been eliminated between x and y. $\blacktriangle$

Example 7.26

If $x = \sin t + \cos t$ and $y = t^2 - t + 1$ find $\dfrac{dy}{dx}(t = 0)$.

Solution

$$\frac{dy}{dt} = 2t - 1 \qquad \frac{dx}{dt} = \cos t - \sin t$$

Hence,

$$\frac{dy}{dx} = \frac{2t - 1}{\cos t - \sin t}$$

When $t = 0$, $dy/dx = -1/1 = -1$. ▲

7.10.2 *Implicit differentiation*

Suppose we are told that:

$$y^3 + x^3 = 5 \sin x + 10 \cos y$$

Although y depends upon x, it is impossible to write the equation in the form $y = f(x)$. We say y is expressed **implicitly** in terms of x. The form $y = f(x)$ is an **explicit** expression for y in terms of x. However, given an implicit expression for y it is still possible to find dy/dx. Usually dy/dx will be expressed in terms of both x and y. Essentially, the chain rule is used when differentiating implicit expressions.

Example 7.27

Given

$$x^3 + y = 1 + y^3$$

find dy/dx.

Solution

Consider differentiation of the left-hand side w.r.t. x.

$$\frac{d}{dx}(x^3 + y) = \frac{d}{dx}(x^3) + \frac{dy}{dx} = 3x^2 + \frac{dy}{dx}$$

Now consider differentiation of the right-hand side w.r.t. x.

$$\frac{d}{dx}(1 + y^3) = \frac{d}{dx}(1) + \frac{d}{dx}(y^3) = \frac{d}{dx}(y^3)$$

To differentiate y^3, we proceed as follows. Let $z = y^3$, so that we seek dz/dx. Now differentiating z w.r.t. y we get:

$$\frac{dz}{dy} = 3y^2$$

We now use the chain rule:

$$\frac{dz}{dx} = \frac{dz}{dy}\frac{dy}{dx} = 3y^2\frac{dy}{dx}$$

So finally,

$$3x^2 + \frac{dy}{dx} = 3y^2\frac{dy}{dx}$$

from which

$$\frac{dy}{dx} = \frac{3x^2}{3y^2 - 1}$$

Note that dy/dx is expressed in terms of x and y. ▲

Example 7.28

Find dy/dx given

(a) $\ln y = y - x^2$

(b) $x^2y^3 - e^y = e^{2x}$

Solutions

(a) To differentiate $\ln y$ we let $z = \ln y$. Then $dz/dy = 1/y$. Using the chain rule we obtain:

$$\frac{dz}{dx} = \frac{dz}{dy}\frac{dy}{dx} = \frac{1}{y}\frac{dy}{dx}$$

Hence differentiating the given equation w.r.t. x yields:

$$\frac{1}{y}\frac{dy}{dx} = \frac{dy}{dx} - 2x$$

from which

$$\frac{dy}{dx} = \frac{2xy}{y - 1}$$

(b) Consider $\dfrac{d}{dx}(x^2y^3)$. Using the product rule we find:

$$\frac{d}{dx}(x^2y^3) = \frac{d}{dx}(x^2)y^3 + x^2\frac{d}{dx}(y^3) = 2xy^3 + x^23y^2\frac{dy}{dx}$$

Consider $\dfrac{d}{dx}(e^y)$. Let $z = e^y$ so $dz/dy = e^y$. Hence,

$$\frac{d}{dx}(e^y) = \frac{dz}{dx} = \frac{dz}{dy}\frac{dy}{dx} = e^y\frac{dy}{dx}$$

So, upon differentiating, the equation becomes:

$$2xy^3 + x^23y^2\frac{dy}{dx} - e^y\frac{dy}{dx} = 2e^{2x}$$

from which,

$$\frac{dy}{dx} = \frac{2e^{2x} - 2xy^3}{3x^2y^2 - e^y} \quad \blacktriangle$$

7.10.3 *Logarithmic differentiation*

The technique of **logarithmic differentiation** is useful when we need to differentiate a cumbersome product. The method involves taking the natural logarithm of the function to be differentiated. This is illustrated in the following examples.

Example 7.29

Given

$$y = x^3(1 + x)^9 e^{6x}$$

find dy/dx.

Solution

The product rule could be used. However, we will use logarithmic differentiation. Taking the natural logarithm of the equation and applying the laws of logarithms produces

$$\ln y = \ln(x^3(1 + x)^9 e^{6x}) = \ln x^3 + \ln(1 + x)^9 + \ln e^{6x}$$

$$\ln y = 3 \ln x + 9 \ln(1 + x) + 6x$$

This equation is now differentiated.

$$\frac{1}{y}\frac{dy}{dx} = \frac{3}{x} + \frac{9}{1 + x} + 6$$

and so,

$$\frac{dy}{dx} = y\left(\frac{3}{x} + \frac{9}{1 + x} + 6\right)$$

$$= 3x^2(1 + x)^9 e^{6x} + 9x^3(1 + x)^8 e^{6x} + 6x^3(1 + x)^9 e^{6x}$$

Note that $\ln y$ has been differentiated implicitly (see Example 7.28). $\blacktriangle$

Example 7.30

If $y = \sqrt{1 + t^2} \sin^2 t$ find y'.

Solution

$$\ln y = \tfrac{1}{2}\ln(1 + t^2) + 2 \ln \sin t$$

Differentiation yields:

$$\frac{1}{y}y' = \frac{1}{2}\frac{2t}{1+t^2} + 2\frac{\cos t}{\sin t}$$

$$y' = y\left(\frac{t}{1+t^2} + 2\cot t\right) = \sqrt{1+t^2}\sin^2 t\left(\frac{t}{1+t^2} + 2\cot t\right) \quad \blacktriangle$$

EXERCISES 7.8

1. Find dy/dx given

   (a) $2y^2 - 3x^3 = x + y$

   (b) $\sqrt{y} + \sqrt{x} = x^2 + y^3$

   (c) $\sqrt{2x+3y} = 1 + e^x$

   (d) $y = \dfrac{e^x\sqrt{1+x}}{x^2}$

   (e) $2xy^4 = x^3 + 3xy^2$

   (f) $\sin(x+y) = 1 + y$

   (g) $\ln(x^2 + y^2) = 2x - 3y$

   (h) $ye^{2y} = x^2 e^{x/2}$

2. Find dy/dx, given:

   (a) $x = t^2 \qquad y = 1 + t^3$

   (b) $x = \sin t \qquad y = e^t$

   (c) $x = (1+t)^3 \qquad y = 1 + t^3$

   (d) $x = \cos 2t \qquad y = 3t$

   (e) $x = 3/t \qquad y = e^{2t}$

   (f) $x = e^t - e^{-t} \qquad y = e^t + e^{-t}$

3. Use logarithmic differentiation to find the derivative of the following functions:

   (a) $z = t^4(1-t)^6(2+t)^4$

   (b) $y = \dfrac{(1+x^2)^3 e^{7x}}{(2+x)^6}$

   (c) $x = (1+t)^3(2+t)^4(3+t)^5$

   (d) $y = \dfrac{(\sin^4 t)(2-t^2)^4}{(1+e^t)^6}$

4. If $x = t + t^2 + t^3$ and $y = \sin 2t$, find dy/dx when $t = 1$.

5. Given $x = 1 + t^6$ and $y = 1 - t^2$, find:

   (a) the rate of change of x w.r.t. t when $t = 2$

   (b) the rate of change of y w.r.t. t when $t = 1$

   (c) the rate of change of y w.r.t. x when $t = 1$

   (d) the rate of change of x w.r.t. y when $t = 2$

6. Find the equations of the tangents to

   $$y^2 = x^2 + 6y$$

   when $x = 4$.

7.11 The Newton–Raphson technique

We often need to solve equations such as:

$$f(x) = 2x^4 - x^3 + x^2 - 10 = 0$$

$$f(t) = 2e^{-3t} - t^2 = 0$$

$$f(t) = t - \sin t = 0$$

The Newton–Raphson technique is a method of obtaining an approximate solution, or root, of such equations. It involves the use of differentiation.

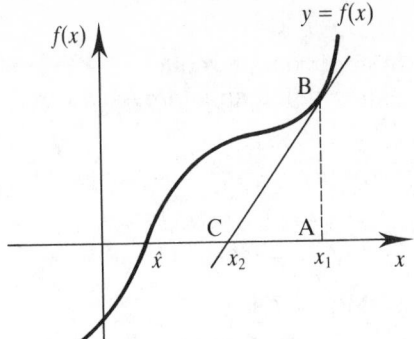

Figure 7.17 The tangent at B intersects the x axis at C.

Suppose we wish to find a root of $f(x) = 0$. Figure 7.17 illustrates the curve $y = f(x)$. Roots of the equation $f(x) = 0$ correspond to where the curve cuts the x axis. One such root is illustrated and is labelled $x = \hat{x}$. Suppose we know that $x = x_1$ is an approximate solution. Let A be the point on the x axis where $x = x_1$ and let B be the point on the curve where $x = x_1$. The tangent at B is drawn and cuts the x axis at C where $x = x_2$. Clearly $x = x_2$ is a better approximation to $\hat{x}$ than x_1. We now find x_2 in terms of the known value, x_1.

AB = distance of B above the x axis = $f(x_1)$

$CA = x_1 - x_2$

Hence,

$$\text{gradient of line CB} = \frac{AB}{CA} = \frac{f(x_1)}{x_1 - x_2}$$

But CB is a tangent to the curve at $x = x_1$ and so has gradient $f'(x_1)$. Hence,

$$f'(x_1) = \frac{f(x_1)}{x_1 - x_2}$$

$$x_1 - x_2 = \frac{f(x_1)}{f'(x_1)}$$

and therefore,

$$\blacksquare \quad x_2 = x_1 - \frac{f(x_1)}{f'(x_1)} \tag{7.4}$$

Equation (7.4) is known as the Newton–Raphson formula. Knowing an approximate root of $f(x) = 0$, that is x_1, the Newton–Raphson formula enables us to calculate an improved approximate root, x_2.

Example 7.31

Given that $x_1 = 7.5$ is an approximate root of $e^x - 6x^3 = 0$, use the Newton–Raphson technique to find an improved value.

Solution

$$x_1 = 7.5$$

$$f(x) = e^x - 6x^3 \qquad f(x_1) = -723$$

$$f'(x) = e^x - 18x^2 \qquad f'(x_1) = 796$$

Using the Newton–Raphson technique the value of x_2 is found.

$$x_2 = x_1 - \frac{f(x_1)}{f'(x_1)} = 7.5 - \frac{(-723)}{796} = 8.41$$

An improved estimate of the root of $e^x - 6x^3 = 0$ is $x = 8.41$. To two decimal places the true answer is $x = 8.05$. ▲

The Newton–Raphson technique can be used repeatedly as illustrated in Example 7.32. This generates a sequence of approximate solutions which may converge to the required root. Each application of the method is known as an iteration.

Example 7.32

A root of $3 \sin x = x$ is near to $x = 2.5$. Use two iterations of the Newton–Raphson technique to find a more accurate approximation.

Solution

The equation must first be written in the form $f(x) = 0$, that is,

$$3 \sin x - x = 0$$

Then

$$x_1 = 2.5$$

$$f(x) = 3 \sin x - x \qquad f(x_1) = -0.705$$

$$f'(x) = 3 \cos x - 1 \qquad f'(x_1) = -3.403$$

Then

$$x_2 = 2.5 - \frac{(-0.705)}{(-3.403)} = 2.293$$

The process is repeated with $x_1 = 2.293$ as the initial approximation.

$$x_1 = 2.293 \qquad f(x_1) = -0.042 \qquad f'(x_1) = -2.983$$

Then

$$x_2 = 2.293 - \frac{(-0.042)}{(-2.983)} = 2.279$$

Using two iterations of the Newton–Raphson technique, we obtain $x = 2.28$ as an improved estimate of the root. ▲

Example 7.33

An approximate root of:

$$x^3 - 2x^2 - 5 = 0$$

is $x = 3$. By using the Newton–Raphson technique repeatedly, determine the value of the root correct to two decimal places.

Solution

As a general rule, working should be carried out to one more decimal place than required. We have

$$x_1 = 3$$
$$f(x) = x^3 - 2x^2 - 5 \qquad f(x_1) = 4$$
$$f'(x) = 3x^2 - 4x \qquad f'(x_1) = 15$$

Hence

$$x_2 = 3 - \frac{4}{15} = 2.733$$

An improved estimate of the value of the root is 2.73 (2 d.p.). The method is used again, taking $x_1 = 2.73$ as the initial approximation.

$$x_1 = 2.73 \qquad f(x_1) = 0.441 \qquad f'(x_1) = 11.439$$

$$x_2 = 2.73 - \frac{0.441}{11.439} = 2.691$$

An improved estimate is $x = 2.69$ (2 d.p.). The method is used again

$$x_1 = 2.69 \qquad f(x_1) = -0.007 \qquad f'(x_1) = 10.948$$

So

$$x_2 = 2.69 - \frac{(-0.007)}{10.948} = 2.691$$

There is no change in the value of the approximate root and so to two decimal places the root of $f(x) = 0$ is $x = 2.69$. ▲

The previous examples illustrate the general Newton–Raphson formula.

■ If $x = x_n$ is an approximate root of $f(x) = 0$, then an improved estimate, x_{n+1}, is given by

$$x_{n+1} = x_n - \frac{f(x_n)}{f'(x_n)}$$

The Newton–Raphson formula is easy to program in a loop structure. Exit from the loop is usually conditional upon $|x_{n+1} - x_n|$ being smaller than some prescribed very small value. This condition shows that successive approximate roots are very close to each other. For an alternative method which avoids differentiation but usually converges less rapidly see Appendix I.

Example 7.34 Series diode–resistor circuit

Consider the circuit of Figure 7.18(a). A diode is in series with a resistor with resistance, R. The voltage across the diode is denoted by v and the current through the diode is denoted by i. The i–v relationship for the diode is non-linear and is given by:

$$i = I_s(e^{40v} - 1)$$

where I_s is the reverse saturation current of the diode. Given that the supply voltage, V_s, is 2 V, I_s is 10^{-14} A and R is 22 kΩ, calculate the steady-state values of i and v.

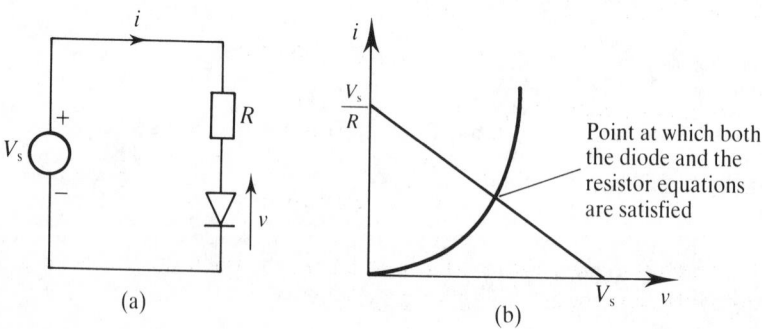

(a)

(b)

Figure 7.18 A simple non-linear circuit: (a) series diode–resistor circuit; (b) resistor load line superimposed on diode characteristic.

Solution

There are several ways to solve this problem. A difficulty exists because the diode i–v relationship is non-linear. One possibility is to draw a **load line** for the resistor superimposed on the diode i–v characteristic, as shown in Figure 7.18(b). The load line is an equation for the resistor characteristic written in terms of the voltage across the diode, v, and the current through the diode, i. It is given by:

$$V_s - v = iR$$

$$i = -\frac{1}{R}v + \frac{V_s}{R}$$

This is a straight line with slope $-1/R$ and vertical intercept V_s/R. When $v = 0$, $i = V_s/R$. This corresponds to all of the supply voltage being dropped across the resistor. When $v = V_s$, $i = 0$. This corresponds to all of the supply voltage being dropped across the diode. Therefore, these two limits correspond to the points within which the circuit must operate. The solution to the circuit can be obtained by determining the intercept of the diode characteristic and the load line. This is possible because both the resistor characteristic and diode characteristic are formulated in terms of v and i, and so any solution must have the same values of i and v for both components. If an accurate graph is used, it is possible to obtain a reasonably good solution. An alternative approach is to use the Newton–Raphson technique. Combining the two component equations gives

$$-v + V_s = RI_s(e^{40v} - 1)$$

Now $R = 2.2 \times 10^4$, $I_s = 10^{-14}$, $V_s = 2$ and so,

$$-v + 2 = 2.2 \times 10^4 \times 10^{-14}(e^{40v} - 1)$$

Now, define $f(v)$ by:

$$f(v) = 2.2 \times 10^{-10}(e^{40v} - 1) + v - 2$$

We wish to solve $f(v) = 0$. We have

$$f'(v) = 2.2 \times 10^{-10} \times 40e^{40v} + 1 = 8.8 \times 10^{-9}e^{40v} + 1$$

Choose an initial guess of $v_1 = 0.5$

$$v_2 = v_1 - \frac{f(v_1)}{f'(v_1)} = 0.5 - \frac{2.2 \times 10^{-10}(e^{20} - 1) + 0.5 - 2}{8.8 \times 10^{-9}e^{20} + 1} = 0.7644$$

With an equation of this complexity, it is better to use a computer or a programmable calculator. Doing so gives:

$$v_5 = 0.6895, \ldots, v_{10} = 0.5770, \ldots, v_{14} = 0.5650$$

which is accurate to four decimal places.

It is useful to check the solution by independently calculating the current through the diode using the two different expressions. So,

$$i = 10^{-14}(e^{40 \times 0.5650} - 1) = 6.53 \times 10^{-5} \text{ A}$$

$$i = -\frac{0.5650}{2.2 \times 10^4} + \frac{2}{2.2 \times 10^4} = 6.53 \times 10^{-5} \text{ A}$$

and therefore the solution is correct. ▲

EXERCISES 7.9

1. Use the Newton–Raphson technique to find the value of a root of the following equations correct to two decimal places. An approximate root, x_1, is given in each case.

   (a) $2 \cos x = x^2$ $x_1 = 0.8$
   (b) $3x^3 - 4x^2 + 2x - 9 = 0$ $x_1 = 2$

   (c) $e^{x/2} - 5x = 0$ $x_1 = 6$
   (d) $\ln x = 1/x$ $x_1 = 1.6$
   (e) $\sin x + 2x/\pi = 1$ $x_1 = 0.6$

2. Explain circumstances in which the Newton–Raphson technique may fail to converge to a root of $f(x) = 0$.

7.12 Higher derivatives

The derivative, y', of a function $y(x)$, is more correctly called the **first derivative** of y w.r.t. x. Since y' itself is a function of x, then it is often possible to differentiate this too. The derivative of y' is called the **second derivative** of y.

$$\text{second derivative of } y = \frac{d}{dx}\left(\frac{dy}{dx}\right)$$

which is written as d^2y/dx^2 or more compactly as y''.

Example 7.35

If $y(x) = 3x^2 + 8x + 9$, find y' and y''.

Solution

$$y' = 6x + 8$$

$$y'' = \frac{d}{dx}(6x + 8) = 6 \quad ▲$$

Example 7.36

If $y(t) = 2 \sin 3t$, find y' and y''.

Solution

$$y' = 6 \cos 3t$$

$$y'' = -18 \sin 3t$$

The first and second derivatives w.r.t. time, t, are also denoted by $\dot{y}$ and $\ddot{y}$.

Example 7.37

Find y'' given

$$1 + xy = x^2 + y^2$$

Solution

The equation is differentiated implicitly to obtain dy/dx.

$$0 + y + xy' = 2x + 2yy'$$

$$y' = \frac{2x - y}{x - 2y}$$

The quotient rule is now used with $u = 2x - y$ and $v = x - 2y$. The derivatives of u and v are:

$$\frac{du}{dx} = u' = 2 - y' \qquad \frac{dv}{dx} = v' = 1 - 2y'$$

Then,

$$y'' = \frac{(x - 2y)(2 - y') - (2x - y)(1 - 2y')}{(x - 2y)^2}$$

This is simplified to:

$$y'' = \frac{3xy' - 3y}{(x - 2y)^2}$$

Replacing y' by $(2x - y)/(x - 2y)$ and simplifying yields:

$$y'' = \frac{6(x^2 - xy + y^2)}{(x - 2y)^3}$$

Note that it is possible to simplify this further by observing that $x^2 + y^2 = 1 + xy$ as given. Therefore,

$$y'' = \frac{6(1 + xy - xy)}{(x - 2y)^3} = \frac{6}{(x - 2y)^3} \quad \blacktriangle$$

Just as the first derivative may be differentiated to obtain the second derivative, so the second derivative may be differentiated to find the third derivative and so on. A similar notation is used. The third derivative is written d^3y/dx^3 or y''' or $y^{(3)}$. The fourth derivative is written d^4y/dx^4 or y^{iv} or $y^{(4)}$. The fifth derivative is written d^5y/dx^5 or y^v or $y^{(5)}$.

Example 7.38

Find the first five derivatives of $z(t) = 2t^3 + \sin t$.

Solution

$$z' = 6t^2 + \cos t$$

$$z'' = 12t - \sin t$$

$$z''' = 12 - \cos t$$

$$z^{iv} = \sin t$$

$$z^{v} = \cos t \quad \blacktriangle$$

Example 7.39

Calculate the values of x for which $y'' = 0$, given $y = x^4 - x^3$.

Solution

$$y = x^4 - x^3$$

$$y' = 4x^3 - 3x^2$$

$$y'' = 12x^2 - 6x$$

Putting $y'' = 0$ gives

$$12x^2 - 6x = 0$$

$$6x(2x - 1) = 0$$

$$x = 0, \tfrac{1}{2} \quad \blacktriangle$$

The first and second derivatives can be used to describe the nature of increasing and decreasing functions. In Figure 7.19(a, b) the tangents to the curves have positive gradients, that is, $y' > 0$. As can be seen, as x increases the value of the function increases. Conversely, in Figure 7.19(c, d) the tangents have negative gradients ($y' < 0$) and as x increases the value of the function decreases. The sign of the first derivative tells us whether y is increasing or decreasing. However, the curves in (a) and (b) both show y increasing but, clearly, there is a difference in the way y changes.

Consider again Figure 7.19(a). The tangents at A, B and C are shown. As x increases the gradient of the tangent increases, that is, y' increases as x increases. Since y' increases as x increases then the derivative of y' is positive, that is, $y'' > 0$. (Compare with: y increases when its derivative is positive.) So for the curve shown in Figure 7.19(a), $y' > 0$ and $y'' > 0$.

For that shown in Figure 7.19(b) the situation is different. The value of y' decreases as x increases, as can be seen by considering the gradients of the tangents at A, B and C, that is, the derivative of y' must be negative. For this curve $y' > 0$ and $y'' < 0$.

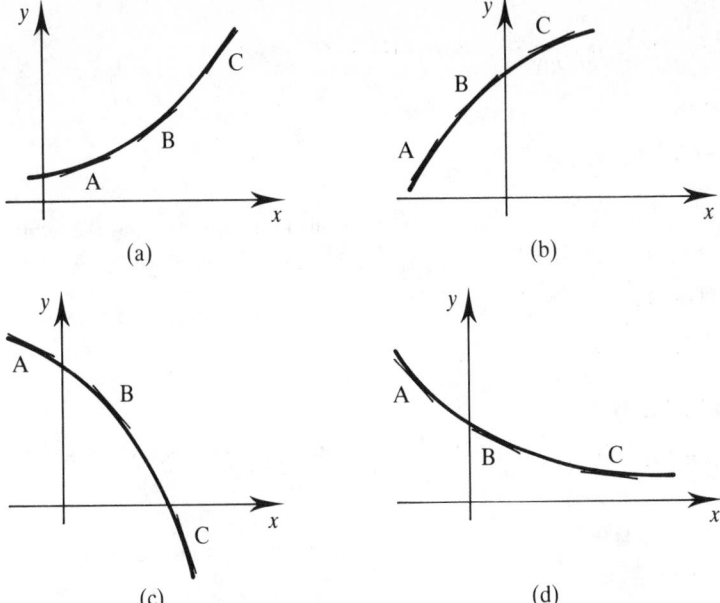

Figure 7.19 (a) *y* is concave up $(y' > 0, y'' > 0)$; (b) *y* is concave down $(y' > 0, y'' < 0)$; (c) *y* is concave down $(y' < 0, y'' < 0)$; (d) *y* is concave up $(y' < 0, y'' > 0)$.

A function is **concave down** when y' decreases and **concave up** when y' increases. Hence Figure 7.19(a) illustrates a concave up function; Figure 7.19(b) illustrates a concave down function. The sign of the second derivative can be used to distinguish between concave up and concave down functions.

Consider now the functions shown in Figure 7.19(c) and Figure 7.19(d). In both (c) and (d), *y* is decreasing and so $y' < 0$. In (c) the gradient of the tangent becomes increasingly negative, that is, it is decreasing. Hence, for the function in (c) $y'' < 0$. Conversely, for the function in (d) the gradient of the tangent is increasing as *x* increases, although it is always negative, that is, $y'' > 0$. So for the function in (c) $y' < 0$ and $y'' < 0$, that is, the function is concave down. For the function in (d) $y' < 0$ and $y'' > 0$, that is, the function is concave up. In summary, we can state:

■ When $y' > 0$, *y* is increasing. When $y' < 0$, *y* is decreasing.
When y' is increasing the function is concave up.
When y' is decreasing the function is concave down.

As will be seen in the next section, higher derivatives are used to determine the location and nature of important points called maximum points, minimum points and points of inflexion.

EXERCISES 7.10

1. Calculate dy/dt and d^2y/dt^2 given:

   (a) $y = t^2 + t$

   (b) $y = 2t^3 - t^2 + 1$

   (c) $y = \sin 2t$

   (d) $y = \sin kt$ k constant

   (e) $y = 2e^{3t} - t^2 + 1$

   (f) $y = t/(t + 1)$

   (g) $y = 4 \cos(t/2)$

   (h) $y = e^t t$

   (i) $y = \sinh 4t$

   (j) $y = \sin^2 t$

2. If

   $$y = 2x^3 + 3x^2 - 12x + 1$$

   find values of x for which $y'' = 0$.

3. If $dy/dt = 3t^2 + t$, find:

   (a) d^2y/dt^2

   (b) d^3y/dt^3

4. Find values of t at which $y'' = 0$, where:

   $$y = \frac{t^3}{3} - \frac{7t^2}{2} + 12t - 1$$

5. Determine whether the following functions are concave up or concave down.

   (a) $y = e^t$

   (b) $y = t^2$

   (c) $y = 1 + t - t^2$

6. Determine the interval on which $y = t^3$ is (a) concave up , (b) concave down.

7. Evaluate y'' at the specified value of t.

   (a) $y = 2 \cos t - t^2$ $t = 1$

   (b) $y = (\sin t + \cos t)/2$ $t = \pi/2$

   (c) $y = (1 + t)e^t$ $t = 0$

7.13 Maximum points, minimum points and points of inflexion

7.13.1 *Maximum and minimum points*

Consider Figure 7.20(a): A and B are important points on the curve. At A the function stops increasing and starts to decrease. At B it stops decreasing and starts to increase. A is a **local maximum**, B is a **local minimum**. Note that A is not the highest point on the curve, nor B the lowest point. However for that part of the

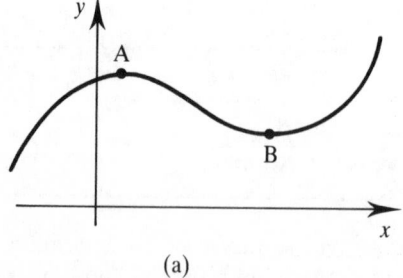

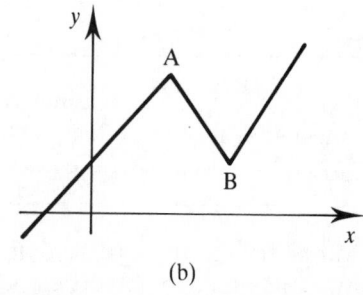

(a)   (b)

Figure 7.20   The function y has a local maximum at A and a local minimum at B.

curve near to A, A is the highest point. The word 'local' is used to stress that A is maximum in its locality. Similarly, B is the lowest point in its locality.

In Figure 7.20(a) the tangents at A and B are parallel to the x axis and so at these points dy/dx is 0. However, in Figure 7.20(b) there are corners at A and B, and so y' does not exist at these points. Hence when searching for maximum and minimum points we need only examine those points at which dy/dx is 0 or does not exist. Points at which $dy/dx = 0$ are known as turning points of the function.

■ At maximum and minimum points, either $dy/dx = 0$ or it does not exist.

To distinguish between maximum and minimum points we consider y' on either side of the point. At maximum points such as A (Figure 7.20(a)), $y' > 0$, that is, y is increasing immediately to the left of the point, and $y' < 0$, that is, y is decreasing immediately to the right of the point. At minimum points, such as B, $y' < 0$ immediately to the left of the point, and $y' > 0$ immediately to the right. Note that this first derivative test for maximum and minimum points can be used even when the derivative does not exist at the point in question.

Example 7.40

Determine the position of all maximum and minimum points of the following functions.

(a) $y = x^2$
(b) $y = -t^2 + t + 1$
(c) $y = x^3/3 + x^2/2 - 2x + 1$
(d) $y = |t|$

Solutions

(a) $y' = 2x$, $y' = 0$ at $x = 0$. Immediately to the left of $x = 0$, $y' < 0$.
Immediately to the right of the point, $y' > 0$. Hence y has a minimum at $x = 0$. A graph of $y(x)$ is given in Figure 7.21 and clearly shows a minimum at $x = 0$, $y = 0$.

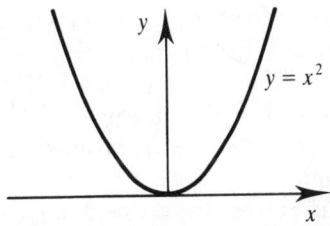

$y = x^2$

Figure 7.21   The function y has a minimum at $x = 0$.

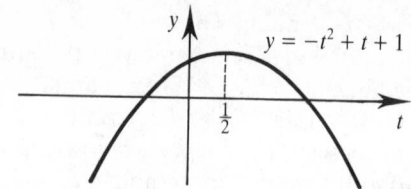

Figure 7.22   The function y has a maximum at $t = \frac{1}{2}$.

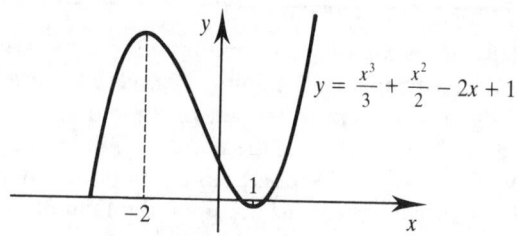

Figure 7.23   The function y has a maximum at $x = -2$ and a minimum at $x = 1$.

(b) $y' = -2t + 1$, $y' = 0$ at $t = 1/2$. Immediately to the left of $t = 1/2$, $y' > 0$. Immediately to the right, $y' < 0$, and so there is a maximum at $t = 1/2$ (see Figure 7.22).

(c) $y' = x^2 + x - 2 = (x - 1)(x + 2)$. Solving $y' = 0$ yields $x = 1, -2$.
   At $x = 1$: immediately to the left of $x = 1$, $y' < 0$. Immediately to the right, $y' > 0$ and there is a minimum at $x = 1$.
   At $x = -2$: immediately to the left of $x = -2$, $y' > 0$. Immediately to the right $y' < 0$ and so there is a maximum at $x = -2$ (see Figure 7.23).

(d) $y = |t| = \begin{cases} -t & t \le 0 \\ t & t > 0 \end{cases}$

Then $y' = -1$ for $t < 0$, and $y' = 1$ for $t > 0$. The derivative is not defined at $t = 0$. However, to the left of $t = 0$, $y' < 0$; to the right $y' > 0$ and so $t = 0$ is a minimum point. A graph of $y = |t|$ is shown in Figure 7.13(a). ▲

Rather than examine the sign of y' on both sides of the point a second derivative test may be used. On passing through a maximum point y' changes from positive to 0 to negative, as shown in Figure 7.24. Hence, y' is decreasing. If y'' is negative then this indicates y' is decreasing and the point is therefore a maximum point. Conversely, on passing through a minimum point, y' increases, going from negative to 0 to positive (see Figure 7.25). If y'' is positive then y' is increasing and this indicates a minimum point.

So, having located the points where $y' = 0$, we look at the second derivative, y''. Thus $y'' > 0$ implies a minimum point; $y'' < 0$ implies a maximum point. If

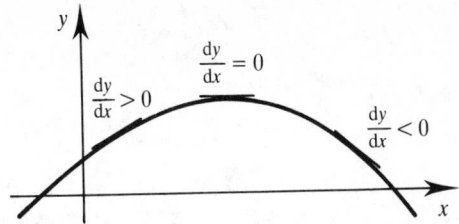

Figure 7.24 The derivative $\dfrac{dy}{dx}$ decreases on passing through a maximum point.

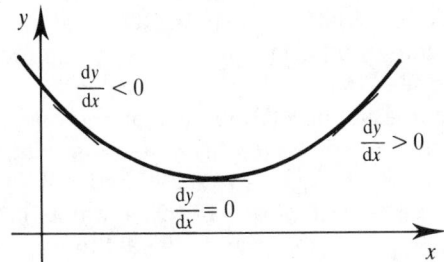

Figure 7.25 The derivative $\dfrac{dy}{dx}$ increases on passing through a minimum point.

$y'' = 0$, then we must return to the earlier more basic test of examining y' on both sides of the point. In summary:

■ $y' = 0$ and $y'' < 0$ indicates a maximum point.
     $y' = 0$ and $y'' > 0$ indicates a minimum point.
If $y' = 0$ and $y'' = 0$ then no conclusion is possible. You must examine the sign of y' on both sides of the point.

Example 7.41

Use the second derivative test to find all maximum and minimum points of the functions in Example 7.40.

Solutions

(a) $y' = 2x$, $y'' = 2$: $y' = 0$ when $x = 0$. Also, $y''(0) = 2$. Since $y''(0) > 0$ then $x = 0$ is a minimum point.

(b) $y' = -2t + 1$, $y'' = -2$: $y' = 0$ at $t = 1/2$. Also $y''(1/2) = -2$ and so $t = 1/2$ is a maximum point.

(c) $y' = x^2 + x - 2$, $y'' = 2x + 1$: $y' = 0$ at $x = 1, -2$. At $x = 1$, $y'' = 3(>0)$ and so $x = 1$ is a minimum point. At $x = -2$, $y'' = -3(<0)$ and so

$x = -2$ is a maximum point.

(d) $y' = \begin{cases} -1 & t < 0 \\ 1 & t > 0 \\ \text{undefined at } t = 0 \end{cases}$

Since $y'(0)$ is undefined, we use the first derivative test. This was employed in Example 7.40. ▲

Example 7.42 Rise time for a second order electrical system

Consider the electrical system illustrated in Figure 7.26. The input voltage, v_i, is applied to terminals a–b. The output from the system is a voltage, v_o, measured across the terminals c–d. The easiest way to determine the time response of this system to a particular input is to use the technique of Laplace transforms (see Chapter 11).

When a step input is applied to the system, the general form of the response depends on whether a quantity called the damping ratio, ζ, is such that $\zeta > 1$, $\zeta = 1$ or $\zeta < 1$. The quantity ζ itself depends upon the values of L, C and R. This is illustrated in Figure 7.27. If the damping ratio, $\zeta < 1$, then v_o overshoots its final value and the system is said to be underdamped. For this case it can be shown that:

$$v_o = U - Ue^{-\alpha t}\left(\cos(\beta t) + \frac{\alpha \sin(\beta t)}{\beta}\right) \tag{7.5}$$

where U is the height of a step input applied at $t = 0$.

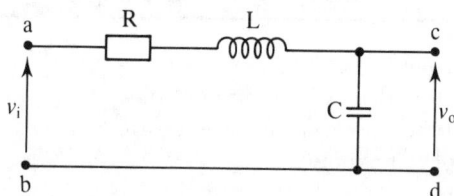

Figure 7.26   A second order electrical system.

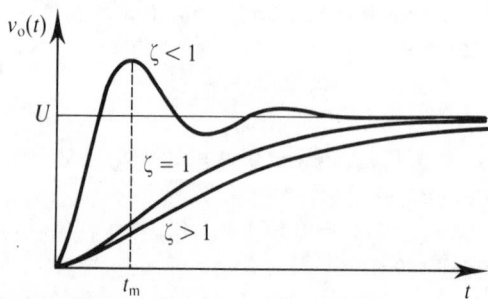

Figure 7.27   Response of a second order system to a step input.

$$\alpha = \frac{R}{2L} \tag{7.6}$$

$$\omega_r = \frac{1}{\sqrt{LC}} \qquad \text{resonant frequency} \tag{7.7}$$

$$\beta = \sqrt{\omega_r^2 - \alpha^2} \qquad \text{natural frequency} \tag{7.8}$$

Engineers are often interested in knowing how quickly a system will respond to a particular input. For many systems this is an important design criterion. One way of characterizing the speed of response of the system is the time taken for the output to reach a certain level in response to a step input. This is known as the **rise time** and is often defined as the time taken for the output to rise from 10% to 90% of its final value. However, by looking at the underdamped response illustrated in Figure 7.27 it is clear that the time, t_m, required for the output to reach its maximum value would also provide an indicator of system response time. As the derivative of a function is zero at a maximum point it is possible to calculate this time.

Differentiating Equation (7.5) and using the product rule,

$$\frac{dv_o}{dt} = \frac{d}{dt}\left(U - Ue^{-\alpha t}\left(\cos(\beta t) + \frac{\alpha \sin(\beta t)}{\beta}\right)\right) \qquad t > 0$$

$$= 0 - U\frac{d}{dt}(e^{-\alpha t}\cos(\beta t)) - U\frac{d}{dt}\left(\frac{e^{-\alpha t}\alpha \sin(\beta t)}{\beta}\right)$$

$$= -U(-\alpha e^{-\alpha t}\cos(\beta t) - e^{-\alpha t}\beta \sin(\beta t))$$

$$\quad - U\left(\frac{-\alpha e^{-\alpha t}\alpha \sin(\beta t)}{\beta} + \frac{e^{-\alpha t}\alpha\beta \cos(\beta t)}{\beta}\right)$$

$$= -Ue^{-\alpha t}\left(-\alpha \cos(\beta t) - \beta \sin(\beta t) - \frac{\alpha^2 \sin(\beta t)}{\beta} + \alpha \cos(\beta t)\right)$$

$$= Ue^{-\alpha t}\left(\beta + \frac{\alpha^2}{\beta}\right)\sin(\beta t)$$

At a turning point

$$\frac{dv_o}{dt} = 0 = Ue^{-\alpha t}\left(\frac{\beta^2 + \alpha^2}{\beta}\right)\sin(\beta t)$$

This occurs when $\sin(\beta t) = 0$, which corresponds to $t = k\pi/\beta$, $k \in \mathbb{N}$. It is now straightforward to calculate t_m, once β has been calculated, using Equations (7.6), (7.7) and (7.8) for particular values of R, L and C. You may like to show that the turning point corresponding to $k = 1$ is a maximum by calculating d^2v_o/dt^2 and carrying out the second derivative test.

It is possible to check whether or not a system is underdamped using the following formulae.

$$\zeta = \frac{R}{R_c} \qquad \text{damping ratio} \qquad (7.9)$$

$$R_c = 2\sqrt{\frac{L}{C}} \qquad \text{critical resistance} \qquad (7.10)$$

Let us look at a specific case with typical values $L = 40\,\text{mH}$, $C = 1\,\mu\text{F}$, $R = 200\,\Omega$. Using Equations (7.9) and (7.10), we find

$$R_c = 2\sqrt{\frac{L}{C}} = 2\sqrt{\frac{4 \times 10^{-2}}{1 \times 10^{-6}}} = 400$$

$$\zeta = \frac{R}{R_c} = \frac{200}{400} = 0.5$$

and therefore the system is underdamped ($\zeta < 1$).

$$\omega_r = \frac{1}{\sqrt{LC}} = 5000$$

$$\alpha = \frac{R}{2L} = \frac{200}{2 \times 4 \times 10^{-2}} = 2500$$

$$\beta = \sqrt{\omega_r^2 - \alpha^2} = \sqrt{5000^2 - 2500^2} = 4330$$

$$t_m = \frac{\pi}{4330} = 7.26 \times 10^{-4} = 726\,\mu\text{s} \qquad k = 1 \quad \blacktriangle$$

Example 7.43 Maximum power transfer

Consider the circuit of Figure 7.28 in which a non-ideal voltage source is connected to a variable load resistor with resistance R_L. The source voltage is V and its internal resistance is R_S. Calculate the value of R_L which results

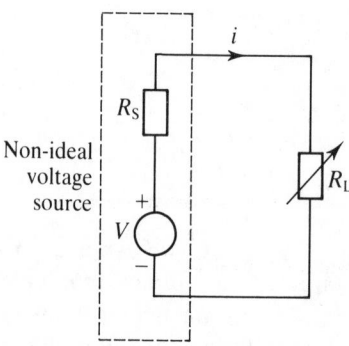

Figure 7.28 Maximum power transfer occurs when $R_L = R_S$.

in the maximum power being transferred from the voltage source to the load resistor.

Solution

Let i be the current flowing in the circuit. Using Kirchhoff's voltage law and Ohm's law gives

$$V = i(R_S + R_L)$$

Let P be the power developed in the load resistor. Then,

$$P = i^2 R_L = \frac{V^2 R_L}{(R_S + R_L)^2}$$

Clearly P depends on the value of R_L. Differentiating, we obtain

$$\frac{dP}{dR_L} = V^2 \frac{1(R_S + R_L)^2 - R_L 2(R_S + R_L)}{(R_S + R_L)^4} = V^2 \frac{R_S - R_L}{(R_S + R_L)^3}$$

Equating dP/dR_L to zero to obtain the turning point gives:

$$V^2 \frac{R_S - R_L}{(R_S + R_L)^3} = 0$$

that is,

$$R_L = R_S$$

We need to check if this is a maximum turning point so,

$$\frac{dP}{dR_L} = V^2 \frac{R_S - R_L}{(R_S + R_L)^3}$$

$$\frac{d^2 P}{dR_L^2} = V^2 \frac{-1(R_S + R_L)^3 - (R_S - R_L)3(R_S + R_L)^2}{(R_S + R_L)^6}$$

$$= V^2 \frac{-(R_S + R_L) - 3(R_S - R_L)}{(R_S + R_L)^4}$$

$$= V^2 \frac{2R_L - 4R_S}{(R_S + R_L)^4}$$

$$= 2V^2 \frac{(R_L - 2R_S)}{(R_S + R_L)^4}$$

When $R_L = R_S$, this expression is negative and so the turning point is a maximum. Therefore, maximum power transfer occurs when the load resistance equals the source resistance. ▲

Example 7.44

Find all maximum and minimum points of $y = x^3$.

Solution

$$y = x^3 \qquad y' = 3x^2 \qquad y'' = 6x$$

$y' = 0$ at $x = 0$. When $x = 0$, $y'' = 0$ and so the second derivative test does not help. We return to examine the sign of y' on both sides of $x = 0$, that is, use the first derivative test. Just to the left of $x = 0$, $y' > 0$. Just to the right of $x = 0$, $y' > 0$. On both sides of $x = 0$, $y' > 0$, and so $x = 0$ is neither a maximum nor a minimum point. This leads us to study one further kind of point: a point of inflexion. ▲

7.13.2 *Point of inflexion*

Recall that when y' is increasing (that is, $y'' > 0$) the curve is concave up, when y' is decreasing (that is, $y'' < 0$) the curve is concave down. A point at which the concavity changes from concave up to concave down, or vice versa, is called a **point of inflexion**. At such a point either $y'' = 0$ or in exceptional cases y'' does not exist. Consider again the previous example involving $y = x^3$. We have $y' = 3x^2$ so $y' = 0$ when $x = 0$. To the left of $x = 0$, y' decreases as x increases, that is, the curve is concave down to the left of $x = 0$. To the right of $x = 0$, y' is increasing as x increases and the curve is concave up. The concavity changes at $x = 0$, that is, $x = 0$ is a point of inflexion (see Figure 7.29).

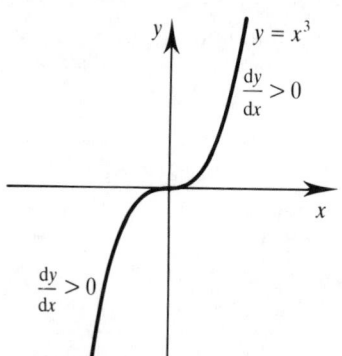

Figure 7.29 The derivative $\dfrac{dy}{dx} > 0$ for $x < 0$ and $x > 0$.

Example 7.45

Find all maximum points, minimum points and points of inflexion of $y = 1 - x^5$.

Solution

$$y' = -5x^4 \qquad y'' = -20x^3$$

Then $y' = 0$ at $x = 0$. Also $y'' = 0$ at $x = 0$ and so the second derivative test does not help. We examine the sign of y' to the left and right of $x = 0$. To the left of $x = 0$, $y' < 0$. To the right of $x = 0$, $y' < 0$. Hence $x = 0$ is neither a maximum nor a minimum point. To the left of $x = 0$, y' is increasing. Note, for example, $y'(-2) = -80$, $y'(-1) = -5$ and so y' increases as x increases. Hence the curve is concave up to the left of $x = 0$. To the right of $x = 0$, y' is decreasing as x increases, that is, the curve is concave down. The concavity changes at $x = 0$ and so this is a point of inflexion.

A common error is to state that if $y' = y'' = 0$ then there is a point of inflexion. This is not always true: consider the next example.

Example 7.46

Locate all maximum points, minimum points and points of inflexion of $y = x^4$.

Solution

$$y' = 4x^3 \qquad y'' = 12x^2$$

$y' = 0$ at $x = 0$. Also $y'' = 0$ at $x = 0$ and so the second derivative test is of no help. We return to examine y' on both sides of $x = 0$. To the left of $x = 0$, $y' < 0$; to the right $y' > 0$ and so $x = 0$ is a minimum point. Figure 7.30 illustrates this. ▲

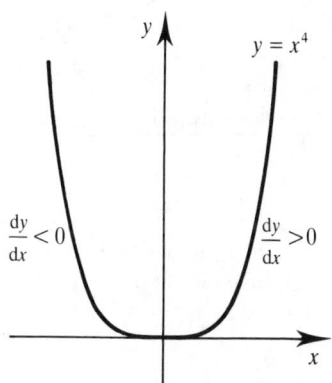

Figure 7.30 The derivative $\dfrac{dy}{dx}$ changes sign as x changes sign.

Example 7.47

Find any maximum points, minimum points and points of inflexion of
(a) $y = x^3 + 2x^2$, (b) $y = 3t^{1/5}$.

Solutions

(a) $y = x^3 + 2x^2$

$y' = 3x^2 + 4x$

Therefore $y' = 0$ when $x(3x + 4) = 0$, that is, when $x = 0$ or $x = -4/3$. Now,

$y'' = 6x + 4$

Therefore, when $x = 0$, $y'' = 4 > 0$ which corresponds to a minimum point. When $x = -4/3$, $y'' = -4 < 0$, which corresponds to a maximum point.

We seek points of inflexion by setting $y'' = 0$. This yields $x = -2/3$. Since y'' is negative when $x < -2/3$, then y' is decreasing there, that is, the function is concave down. Also, y'' is positive when $x > -2/3$ and so y' is then increasing, that is, the function is concave up. Hence there is a point of inflexion when $x = -2/3$. The graph of $y = x^3 + 2x^2$ is shown in Figure 7.31(a).

(b) $y = 3t^{1/5}$

$y' = \frac{3}{5}t^{-4/5}$

$y'' = -\frac{12}{25}t^{-9/5}$

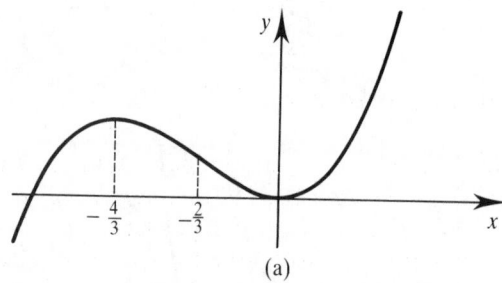

(a)

Figure 7.31   (a) There is a maximum at $x = -\frac{4}{3}$, a minimum at $x = 0$ and a point of inflexion at $x = -\frac{2}{3}$.

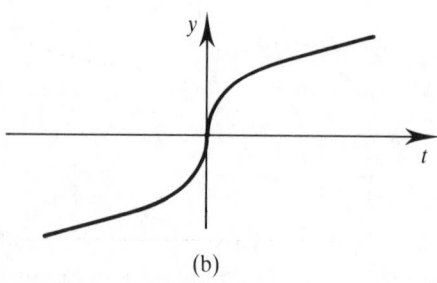

(b)

Figure 7.31   (b) The graph of $y = 3t^{1/5}$.

From Figure 7.31(b) we see y' is increasing on $(-\infty, 0)$ and decreasing on $(0, \infty)$. Hence there is a point of inflexion at $(0, 0)$. Note in this example, y'' is not defined when $t = 0$, even though $(0, 0)$ is a point of inflexion. ▲

From Examples 7.46 and 7.47 we note:

■ (1) The condition $y'' = 0$ is not sufficient to ensure a point is a point of inflexion. The concavity of the function on either side of the point where $y'' = 0$ must be considered.

(2) At a point of inflexion it is not necessary to have $y' = 0$.

(3) At a point of inflexion $y'' = 0$ or y'' does not exist.

EXERCISES 7.11

Locate the maximum points, minimum points and points of inflexion of:

(a) $y = 3t^2 + 6t - 1$

(b) $y = 4 - t - t^2$

(c) $y = x^3/3 - x^2/2 + 10$

(d) $y = x^3/3 + x^2/2 - 20x + 7$

(e) $y = t^5$

(f) $y = t^6$

(g) $y = x^4 - 2x^2$

(h) $z = t + 1/t$

(i) $y = x^5 - 5x^3/3$

(j) $y = t^{1/3}$

7.14 Differentiation of vectors

Consider Figure 7.32. If $\mathbf{r}$ represents the position vector of an object and that object moves along a curve C, then the position vector will be dependent upon the time, t. We write $\mathbf{r} = \mathbf{r}(t)$ to shown the dependence upon time. Suppose that the object is at the point P with position vector $\mathbf{r}$ at time t and at the point Q with position vector $\mathbf{r}(t + \delta t)$ at the later time $t + \delta t$ as shown in Figure 7.33. Then $\overrightarrow{PQ}$ represents the displacement vector of the object during the interval of

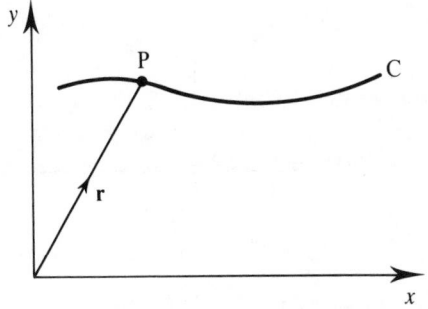

Figure 7.32 Position vector of a point P on a curve C.

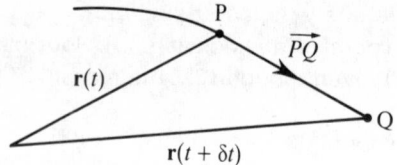

Figure 7.33 Vector $\vec{PQ}$ represents the displacement of the object during the time interval δt.

time δt. The length of the displacement vector represents the distance travelled while its direction gives the direction of motion. The average velocity during the time from t to $t + \delta t$ is the displacement vector divided by the time interval δt, that is,

$$\text{average velocity} = \frac{\vec{PQ}}{\delta t} = \frac{\mathbf{r}(t + \delta t) - \mathbf{r}(t)}{\delta t}$$

The instantaneous velocity, $\mathbf{v}$, is given by:

$$\mathbf{v} = \lim_{\delta t \to 0} \frac{\mathbf{r}(t + \delta t) - \mathbf{r}(t)}{\delta t} = \frac{d\mathbf{r}}{dt}$$

Now, since the x and y coordinates of the object depend upon the time, we can write the position vector $\mathbf{r}$ as:

$$\mathbf{r}(t) = x(t)\mathbf{i} + y(t)\mathbf{j}$$

Therefore,

$$\mathbf{r}(t + \delta t) = x(t + \delta t)\mathbf{i} + y(t + \delta t)\mathbf{j}$$

so that,

$$\mathbf{v}(t) = \lim_{\delta t \to 0} \frac{x(t + \delta t)\mathbf{i} + y(t + \delta t)\mathbf{j} - x(t)\mathbf{i} - y(t)\mathbf{j}}{\delta t}$$

$$= \lim_{\delta t \to 0} \left\{ \frac{x(t + \delta t) - x(t)}{\delta t}\mathbf{i} + \frac{y(t + \delta t) - y(t)}{\delta t}\mathbf{j} \right\}$$

$$= \frac{dx}{dt}\mathbf{i} + \frac{dy}{dt}\mathbf{j}$$

often abbreviated to $\mathbf{v} = \dot{\mathbf{r}} = \dot{x}\mathbf{i} + \dot{y}\mathbf{j}$. Recall the dot notation for derivatives w.r.t. time which is commonly used when differentiating vectors. So the velocity vector is the derivative of the position vector with respect to time. This result generalizes in an obvious way to three dimensions. If,

$$\mathbf{r}(t) = x(t)\mathbf{i} + y(t)\mathbf{j} + z(t)\mathbf{k}$$

then,

$$\dot{\mathbf{r}}(t) = \dot{x}(t)\mathbf{i} + \dot{y}(t)\mathbf{j} + \dot{z}(t)\mathbf{k}$$

The magnitude of the velocity vector gives the speed of the object. We can define the acceleration in a similar way.

$$\mathbf{a} = \frac{d\mathbf{v}}{dt} = \frac{d^2\mathbf{r}}{dt^2} = \ddot{\mathbf{r}} = \ddot{x}\mathbf{i} + \ddot{y}\mathbf{j} + \ddot{z}\mathbf{k}$$

In more general situations, we will not be dealing with position vectors but other physical quantities such as time-dependent electric or magnetic fields.

Example 7.48

If $\mathbf{a} = 3t^2\mathbf{i} + \cos 2t\mathbf{j}$, find

(a) $d\mathbf{a}/dt$

(b) $|d\mathbf{a}/dt|$

(c) $d^2\mathbf{a}/dt^2$

Solutions

(a) If $\mathbf{a} = 3t^2\mathbf{i} + \cos 2t\mathbf{j}$, then differentiation with respect to t yields:

$$\frac{d\mathbf{a}}{dt} = 6t\mathbf{i} - 2\sin 2t\mathbf{j}$$

(b) $|d\mathbf{a}/dt| = \sqrt{(6t)^2 + (-2\sin 2t)^2} = \sqrt{36t^2 + 4\sin^2 2t}$

(c) $d^2\mathbf{a}/dt^2 = 6\mathbf{i} - 4\cos 2t\mathbf{j}$ ▲

It is possible to differentiate more complicated expressions involving vectors provided certain rules are adhered to. If $\mathbf{a}$ and $\mathbf{b}$ are vectors and c is a scalar, all functions of time t, then:

■ $$\frac{d}{dt}(c\mathbf{a}) = c\frac{d\mathbf{a}}{dt} + \frac{dc}{dt}\mathbf{a}$$

$$\frac{d}{dt}(\mathbf{a} + \mathbf{b}) = \frac{d\mathbf{a}}{dt} + \frac{d\mathbf{b}}{dt}$$

$$\frac{d}{dt}(\mathbf{a} \cdot \mathbf{b}) = \mathbf{a} \cdot \frac{d\mathbf{b}}{dt} + \frac{d\mathbf{a}}{dt} \cdot \mathbf{b}$$

$$\frac{d}{dt}(\mathbf{a} \times \mathbf{b}) = \mathbf{a} \times \frac{d\mathbf{b}}{dt} + \frac{d\mathbf{a}}{dt} \times \mathbf{b}$$

Example 7.49

If $\mathbf{a} = 3t\mathbf{i} - t^2\mathbf{j}$ and $\mathbf{b} = 2t^2\mathbf{i} + 3\mathbf{j}$, verify

(a) $$\frac{d}{dt}(\mathbf{a} \cdot \mathbf{b}) = \mathbf{a} \cdot \frac{d\mathbf{b}}{dt} + \frac{d\mathbf{a}}{dt} \cdot \mathbf{b}$$

(b) $\dfrac{d}{dt}(\mathbf{a} \times \mathbf{b}) = \mathbf{a} \times \dfrac{d\mathbf{b}}{dt} + \dfrac{d\mathbf{a}}{dt} \times \mathbf{b}$

Solutions

(a) $\mathbf{a} \cdot \mathbf{b} = (3t\mathbf{i} - t^2\mathbf{j}) \cdot (2t^2\mathbf{i} + 3\mathbf{j}) = 6t^3 - 3t^2$

$\dfrac{d}{dt}(\mathbf{a} \cdot \mathbf{b}) = 18t^2 - 6t$

Also

$\dfrac{d\mathbf{a}}{dt} = 3\mathbf{i} - 2t\mathbf{j}$ $\dfrac{d\mathbf{b}}{dt} = 4t\mathbf{i}$

So,

$\mathbf{a} \cdot \dfrac{d\mathbf{b}}{dt} + \mathbf{b} \cdot \dfrac{d\mathbf{a}}{dt} = (3t\mathbf{i} - t^2\mathbf{j}) \cdot (4t\mathbf{i}) + (2t^2\mathbf{i} + 3\mathbf{j}) \cdot (3\mathbf{i} - 2t\mathbf{j})$

$= 12t^2 + 6t^2 - 6t = 18t^2 - 6t$

We have verified $\dfrac{d}{dt}(\mathbf{a} \cdot \mathbf{b}) = \mathbf{a} \cdot \dfrac{d\mathbf{b}}{dt} + \dfrac{d\mathbf{a}}{dt} \cdot \mathbf{b}$

(b) $\mathbf{a} \times \mathbf{b} = \begin{vmatrix} \mathbf{i} & \mathbf{j} & \mathbf{k} \\ 3t & -t^2 & 0 \\ 2t^2 & 3 & 0 \end{vmatrix}$

$= (9t + 2t^4)\mathbf{k}$

$\dfrac{d}{dt}(\mathbf{a} \times \mathbf{b}) = (9 + 8t^3)\mathbf{k}$

Also,

$\mathbf{a} \times \dfrac{d\mathbf{b}}{dt} = \begin{vmatrix} \mathbf{i} & \mathbf{j} & \mathbf{k} \\ 3t & -t^2 & 0 \\ 4t & 0 & 0 \end{vmatrix}$

$= 4t^3\mathbf{k}$

$\dfrac{d\mathbf{a}}{dt} \times \mathbf{b} = \begin{vmatrix} \mathbf{i} & \mathbf{j} & \mathbf{k} \\ 3 & -2t & 0 \\ 2t^2 & 3 & 0 \end{vmatrix}$

$= (9 + 4t^3)\mathbf{k}$

and so,

$\mathbf{a} \times \dfrac{d\mathbf{b}}{dt} + \dfrac{d\mathbf{a}}{dt} \times \mathbf{b} = 4t^3\mathbf{k} + (9 + 4t^3)\mathbf{k} = (9 + 8t^3)\mathbf{k} = \dfrac{d}{dt}(\mathbf{a} \times \mathbf{b})$

as required. ▲

EXERCISES 7.12

1. Given $r = \sin t\mathbf{i} + \cos t\mathbf{j}$, find:

    (a) $\dot{\mathbf{r}}$

    (b) $\ddot{\mathbf{r}}$

    (c) $|\mathbf{r}|$

    Show the position vector and velocity vector are perpendicular.

2. Show $\mathbf{r} = 3e^{-t}\mathbf{i} + (2 + t)\mathbf{j}$ satisfies

$$\ddot{\mathbf{r}} + \dot{\mathbf{r}} = \mathbf{j}$$

3. Given $\mathbf{a} = t^2\mathbf{i} - (4 - t)\mathbf{j}$, $\mathbf{b} = \mathbf{i} + t\mathbf{j}$ show

    (a) $\dfrac{d}{dt}(\mathbf{a} \times \mathbf{b}) = \left(\mathbf{a} \times \dfrac{d\mathbf{b}}{dt}\right) + \left(\dfrac{d\mathbf{a}}{dt} \times \mathbf{b}\right)$

    (b) $\dfrac{d}{dt}(\mathbf{a} \cdot \mathbf{b}) = \mathbf{a} \cdot \dfrac{d\mathbf{b}}{dt} + \mathbf{b} \cdot \dfrac{d\mathbf{a}}{dt}$

Miscellaneous exercises

1. Use the shrinking interval method to find the rate of change of $f(t) = \sin t$ at $t = 0$ by considering the interval $[0, \delta t]$.
   Hint: use the trigonometric identities in Section 1.4.6 and the small angle approximation in Section 3.5.

2. Use the shrinking interval method to find the rate of change of $f(t) = \sin t$ at a general point.

3. In Section 6.7 we showed that the impedance of an *LCR* circuit can be written as

$$Z = R + j\left(\omega L - \frac{1}{\omega C}\right)$$

    (a) Find $|Z|$.

    (b) For a given circuit, R, L and C are constants, and ω can be varied. Find $d|Z|/d\omega$.

    (c) For what value of ω will $|Z|$ have a maximum or minimum value. Does this value give a maximum or minimum value of $|Z|$?

4. If $f(i) = \ln(E - iR)$, and E and R are constants, find df/di.

5. Given $y(t) = 3 + \sin 2t$, find the average rate of change of y as t varies from 0 to 2.

6. Explain the essential difference between $\delta y/\delta x$ and dy/dx.

7. Find y' for the following functions:

    (a) $y = 2e^{-t} + 6\cos(t/2)$

    (b) $y = (-t + 2)^6$

    (c) $y = \sin 2t \cos 2t$

    (d) $y = 1/(x^2 + 1)$

8. Verify that $y(t) = A \sin \omega t + B \cos \omega t$, ω a constant, satisfies the equation:

$$y'' + \omega^2 y = 0$$

9. Use two iterations of the Newton–Raphson technique to find an improved estimate of the root of:

$$t^3 = e^t$$

    given $t = 1.8$ is an approximate root.

10. Determine the position of all maximum points, minimum points and points of inflexion of:

    (a) $y = 2t^3 - 21t^2 + 60t + 9$

    (b) $y = t(t^2 - 1)$

11. Differentiate $y = x^x$.

12. Given:

$$\mathbf{a} = (t^2 + 1)\mathbf{i} - \mathbf{j} + t\mathbf{k} \qquad \mathbf{b} = 2t\mathbf{j} - \mathbf{k}$$

    find

    (a) $\dfrac{d\mathbf{a}}{dt}$

    (b) $\dfrac{d\mathbf{b}}{dt}$

    (c) $\dfrac{d}{dt}(\mathbf{a} \cdot \mathbf{b})$

    (d) $\dfrac{d}{dt}(\mathbf{a} \times \mathbf{b})$

Integration

8

8.1 Introduction

Integration is the reverse process of differentiation; it enables us to obtain a function from a knowledge of its derivative. Integration is also used to find the area under a curve. Such an area can have various interpretations. For example, the area under a graph of power used by a motor plotted against time represents the total energy used by the motor in a particular time period. The area under a graph of current flow into a capacitor against time represents the total charge stored by the capacitor.

Circuits to carry out integration are used extensively in electronics. For example, a circuit to display the total distance travelled by a car may have a speed signal as input and may integrate this signal to give the distance travelled as output. Integrator circuits are widely used in analog computers. These computers can be used to model a physical system and observe its response to a range of inputs. The advantage of this approach is that the system parameters can be varied in order to see what effect they have on system performance. This avoids the need to build the actual system and allows design ideas to be explored relatively quickly and cheaply by an engineer.

8.2 Elementary integration

Consider the following problem: given $dy/dx = 2x$, find $y(x)$. Differentiation of the function $y(x) = x^2 + c$, where c is a constant, yields $dy/dx = 2x$ for any c. Therefore $y(x) = x^2 + c$ is a solution to the problem. As c can be any constant, there are an infinite number of different solutions. The constant c is known as a **constant of integration**. In this example, the function y has been found from a knowledge of its derivative. We say $2x$ has been **integrated**, yielding $x^2 + c$. To indicate the process of integration the symbols $\int$ and dx are used. The $\int$ sign denotes that integration is to be performed and the dx indicates that x is the independent variable. Returning to the previous problem, we write:

$$\frac{dy}{dx} = 2x$$

$$y = \int 2x \, dx = x^2 + c$$

↑   ↑   ↑
symbols for   constant of integration
integration

■ In general, if

$$\frac{dy}{dx} = f(x)$$

then

$$y = \int f(x) \, dx$$

Consider a simple example.

Example 8.1

Given dy/dx = cos $x - x$, find y.

Solution

We need to find a function which, when differentiated, yields cos $x - x$. Differentiating sin x yields cos x, while differentiating $-x^2/2$ yields $-x$. Hence,

$$y = \int (\cos x - x) \, dx = \sin x - \frac{x^2}{2} + c$$

where c is the constant of integration. Usually brackets are not used and the integral is written simply as $\int$ cos $x - x$ dx. ▲

The function to be integrated is known as the **integrand**. In Example 8.1 the integrand is cos $x - x$.

Example 8.2

Find $\dfrac{d}{dx}\left(\dfrac{x^{n+1}}{n+1} + c\right)$ and hence deduce that $\int x^n \, dx = \dfrac{x^{n+1}}{n+1} + c$.

Solution

From Table 7.1 we find

$$\frac{d}{dx}\left(\frac{x^{n+1}}{n+1} + c\right) = \frac{d}{dx}\left(\frac{x^{n+1}}{n+1}\right) + \frac{d}{dx}(c) \qquad \text{using the linearity of differentiation}$$

$$= \frac{1}{n+1}\frac{d}{dx}(x^{n+1}) + \frac{d}{dx}(c) \qquad \text{again using the linearity of differentiation}$$

$$= \frac{1}{n+1}\{(n+1)x^n\} + 0 \qquad \text{using Table 7.1}$$

$$= x^n$$

Consequently, reversing the process we find,

$$\int x^n \, dx = \frac{x^{n+1}}{n+1} + c$$

as required. Note that this result is invalid if $n = -1$ and so this result could not be applied to the integral $\int (1/x) \, dx$. ▲

Table 8.1 lists several common functions and their integrals. Although the variable x is used throughout Table 8.1, we can use this table to integrate functions of other variables, for example, t and z.

Example 8.3

Use Table 8.1 to integrate the following functions:

(a) x^4

(b) $\cos kx$, where k is a constant

(c) $\sin(3x + 2)$

(d) 5.9

(e) $\tan(6t - 4)$

(f) e^{-3z}

(g) $1/x^2$

(h) $\cos 100n\pi t$, where n is a constant

Solutions

(a) From Table 8.1, we find $\int x^n \, dx = \dfrac{x^{n+1}}{n+1} + c, n \neq -1$. To find $\int x^4 \, dx$ let $n = 4$; we obtain:

$$\int x^4 \, dx = \frac{x^5}{5} + c$$

Table 8.1 The integrals of some common functions.

$f(x)$	$\int f(x)\,dx$
k, constant	$kx + c$
x^n	$\dfrac{x^{n+1}}{n+1} + c \quad n \neq -1$
$x^{-1} = \dfrac{1}{x}$	$\ln |x| + c$
e^{ax}	$\dfrac{e^{ax}}{a} + c$
$\sin(ax + b)$	$\dfrac{-\cos(ax + b)}{a} + c$
$\cos(ax + b)$	$\dfrac{\sin(ax + b)}{a} + c$
$\tan(ax + b)$	$\dfrac{\ln|\sec(ax + b)|}{a} + c$
$\operatorname{cosec}(ax + b)$	$\dfrac{1}{a}\{\ln |\operatorname{cosec}(ax + b) - \cot(ax + b)|\} + c$
$\sec(ax + b)$	$\dfrac{1}{a}\{\ln |\sec(ax + b) + \tan(ax + b)|\} + c$
$\cot(ax + b)$	$\dfrac{1}{a}\{\ln |\sin(ax + b)|\} + c$
$\dfrac{1}{\sqrt{a^2 - x^2}}$	$\sin^{-1}\dfrac{x}{a} + c$
$\dfrac{1}{a^2 + x^2}$	$\dfrac{1}{a}\tan^{-1}\dfrac{x}{a} + c$

Note that a, b, n and c are constants.

(b) From Table 8.1, we find $\int \cos(ax + b)\,dx = \dfrac{\sin(ax + b)}{a} + c$. In this case $a = k$ and $b = 0$, and so:

$$\int \cos kx\,dx = \frac{\sin kx}{k} + c$$

(c) From Table 8.1, we find $\int \sin(ax + b)\,dx = \dfrac{-\cos(ax + b)}{a} + c$. In this case $a = 3$ and $b = 2$, and so

$$\int \sin(3x + 2)\,dx = \frac{-\cos(3x + 2)}{3} + c$$

(d) From Table 8.1, we find that if k is a constant then $\int k\,dx = kx + c$. Hence,

$$\int 5.9\,dx = 5.9x + c$$

(e) In this example, the independent variable is t but nevertheless from Table 8.1, we can deduce

$$\int \tan(at + b)\, dt = \frac{\ln|\sec(at + b)|}{a} + c.$$

Hence with $a = 6$ and $b = -4$, we obtain:

$$\int \tan(6t - 4)\, dt = \frac{\ln|\sec(6t - 4)|}{6} + c$$

(f) The independent variable is z but from Table 8.1, we can deduce $\int e^{az}\, dz = e^{az}/a + c$. Hence, taking $a = -3$ we obtain:

$$\int e^{-3z}\, dz = \frac{e^{-3z}}{-3} + c = -\frac{e^{-3z}}{3} + c$$

(g) Since $1/x^2 = x^{-2}$, we find:

$$\int \frac{1}{x^2}\, dx = \int x^{-2}\, dx = \frac{x^{-1}}{-1} + c = -\frac{1}{x} + c$$

(h) When integrating $\cos 100n\pi t$ with respect to t, note that $100n\pi$ is a constant. Hence, using part (b) we find:

$$\int \cos 100n\pi t\, dt = \frac{\sin 100n\pi t}{100n\pi} + c \quad \blacktriangle$$

8.2.1 *Integration as a linear operator*

Integration, like differentiation, is a linear operator. If f and g are two functions of x, then:

$\blacksquare$ $\displaystyle \int f + g\, dx = \int f\, dx + \int g\, dx$

This states that the integral of a sum of functions is the sum of the integrals of the individual functions. If A is a constant and f a function of x, then:

$\blacksquare$ $\displaystyle \int Af\, dx = A \int f\, dx$

Thus, constant factors can be taken through the integral sign.

If A and B are constants, and f and g are functions of x, then

■ $\displaystyle \int Af + Bg \; dx = A \int f \; dx + B \int g \; dx$

These three properties are all consequences of the fact that integration is a linear operator. Note that the first two are special cases of the third. The properties are used in Example 8.4.

Example 8.4

Use Table 8.1 and the properties of a linear operator to integrate the following expressions:

(a) $x^2 + 9$

(b) $3t^4 - \sqrt{t}$

(c) $1/x$

(d) $(t + 2)^2$

(e) $1/z + z$

(f) $4e^{2z}$

(g) $3 \sin 4t$

(h) $4 \cos(9x + 2)$

(i) $3e^{2z}$

(j) $(\sin x + \cos x)/2$

(k) $2t - e^t$

(l) $\tan\left(\dfrac{z - 1}{2}\right)$

(m) $e^t + e^{-t}$

(n) $3 \sec(4x - 1)$

(o) $2 \cot 9x$

(p) $7 \, \mathrm{cosec}(\pi/3)$

Solutions

(a) $\displaystyle \int x^2 + 9 \; dx = \int x^2 \; dx + \int 9 \; dx$ using linearity

$$= \frac{x^3}{3} + 9x + c$$ using Table 8.1

Note that only a single constant of integration is required.

(b) $\int 3t^4 - \sqrt{t}\, dt = 3\int t^4\, dt - \int t^{1/2}\, dt$ using linearity

$$= 3\left(\frac{t^5}{5}\right) - \frac{t^{3/2}}{3/2} + c \quad\quad \text{using Table 8.1}$$

$$= \frac{3t^5}{5} - \frac{2t^{3/2}}{3} + c$$

(c) $\int \dfrac{1}{x}\, dx = \ln|x| + c$ using Table 8.1

Sometimes it is convenient to use the laws of logarithms to rewrite answers involving logarithms. For example, we can write $\ln|x| + c$ as $\ln|x| + \ln|A|$ where $c = \ln|A|$. This enables us to write the integral as:

$$\int \frac{1}{x}\, dx = \ln|Ax|$$

(d) $\int (t+2)^2\, dt = \int t^2 + 4t + 4\, dt = t^3/3 + 2t^2 + 4t + c$

(e) $\int 1/z + z\, dz = \ln|z| + z^2/2 + c$

(f) $\int 4e^{2z}\, dz = 4e^{2z}/2 + c = 2e^{2z} + c$

(g) $\int 3\sin(4t)\, dt = -(3\cos 4t)/4 + c$

(h) $\int 4\cos(9x+2)\, dx = 4(\sin(9x+2))/9 + c$

(i) $\int 3e^{2z}\, dz = 3e^{2z}/2 + c$

(j) $\int (\sin x + \cos x)/2\, dx = (-\cos x + \sin x)/2 + c$

(k) $\int 2t - e^t\, dt = t^2 - e^t + c$

(l) $\int \tan\left(\dfrac{z-1}{2}\right) dz = 2\ln\left|\sec\left(\dfrac{z-1}{2}\right)\right| + c$

(m) $\int e^t + e^{-t}\, dt = e^t - e^{-t} + c$

(n) $\int 3\sec(4x-1)\, dx = \tfrac{3}{4}\ln|\sec(4x-1) + \tan(4x-1)| + c$

(o) $\int 2\cot 9x\, dx = \tfrac{2}{9}\ln|\sin 9x| + c$

(p) $\int 7\operatorname{cosec}(\pi/3)\, dx = \{7\operatorname{cosec}(\pi/3)\}x + c$ as $\operatorname{cosec}(\pi/3)$ is a constant ▲

Example 8.5   Distance travelled by a particle

The speed, v, of a particle is the rate of change of distance, s, with respect to time t, that is, $v = ds/dt$. The speed at time t is given by $3 + 2t$. Find the distance in terms of t.

Solution

We are given that:

$$v = \frac{ds}{dt} = 3 + 2t$$

and are required to find s. Therefore,

$$s = \int 3 + 2t \; dt = 3t + t^2 + c \quad \blacktriangle$$

Example 8.6   Voltage across a capacitor

The current, i, through a capacitor depends upon time, t, and is given by:

$$i = C \frac{dv}{dt}$$

where v is the voltage across the capacitor and C is the capacitance of the capacitor. Derive an expression for v.

Solution

If

$$i = C \frac{dv}{dt} \quad \text{then} \quad \frac{dv}{dt} = \frac{i}{C}$$

Therefore,

$$v = \int \frac{i}{C} \; dt = \frac{1}{C} \int i \; dt \quad \text{using linearity}$$

Note that whereas the capacitance, C, is constant, the current, i, is not and so it cannot be taken through the integral sign. In order to perform the integration we need to know i as a function of t. $\blacktriangle$

8.2.2 *Electronic integrators*

Often there is a requirement in engineering to integrate electronic signals. Various circuits are available to carry out this task. One of the simplest circuits is shown in Figure 8.1. The input voltage is v_i, the output voltage is v_o, the voltage drop across the resistor is v_R and the current flowing in the circuit is i. Applying Kirchhoff's voltage law yields:

$$v_i = v_R + v_o \tag{8.1}$$

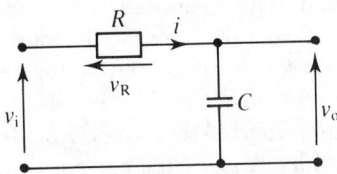

Figure 8.1   Simple integrator.

For the resistor with resistance, R,

$$v_R = iR \tag{8.2}$$

For the capacitor with capacitance, C,

$$i = C\frac{dv_o}{dt} \tag{8.3}$$

Combining Equations (8.1) to (8.3) yields

$$v_i = RC\frac{dv_o}{dt} + v_o \tag{8.4}$$

In general, v_i will be an a.c. signal consisting of a range of frequencies. The a.c. resistance of a capacitor, known as the **capacitive reactance**, or **impedance**, X_c, is given by

$$X_c = \frac{1}{2\pi f C}$$

where $f =$ frequency of the signal (Hz). It can be seen that X_c decreases with increasing frequency, f. For frequencies where X_c is small compared with R, most of the voltage drop takes place across the resistor. In other words, v_o is small compared with v_R. Examining Equation (8.1) for the case when X_c is very much less than R (written as $X_c \ll R$), and $v_o \ll v_R$, it can be seen that Equation (8.4) simplifies to:

$$v_i = RC\frac{dv_o}{dt} \tag{8.5}$$

This equation is only valid for the range of frequencies for which $X_c \ll R$. Rearranging Equation (8.5) yields:

$$\frac{dv_o}{dt} = \frac{v_i}{RC}$$

$$v_o = \frac{1}{RC}\int v_i \, dt$$

The output voltage from the circuit is an integrated version of the input voltage with a scaling factor $1/RC$.

A better electronic integrator, valid over a much greater range of frequencies, can be obtained using an operational amplifier circuit. Consider the circuit shown in Figure 8.2. One of the main characteristics of an operational amplifier is its extremely high voltage gain. This means that the input voltage must be much smaller than the output voltage if the amplifier is not to be driven into **saturation**. Saturation is a condition in which the output voltage reaches the supply voltage of the amplifier and cannot increase further. This means that point X in Figure 8.2 is virtually at a voltage of zero volts. For this reason it is known as a **virtual earth** point. For similar reasons the current drawn by the amplifier, i_a, is also very small. Using this information, it is now possible to analyse the circuit.

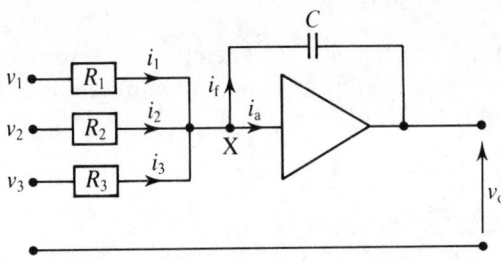

Figure 8.2   Electronic integrator.

Assuming point X is at zero volts, and using Ohm's law, gives:

$$i_1 = \frac{v_1}{R_1} \qquad i_2 = \frac{v_2}{R_2} \qquad i_3 = \frac{v_3}{R_3} \tag{8.6}$$

Assuming i_a is negligible, then:

$$i_f = i_1 + i_2 + i_3 \tag{8.7}$$

For the capacitor,

$$i_f = -C \frac{dv_o}{dt} \tag{8.8}$$

The negative sign is a result of the direction chosen for i_f. Combining Equations (8.6)–(8.8) yields:

$$\frac{v_1}{R_1} + \frac{v_2}{R_2} + \frac{v_3}{R_3} = -C \frac{dv_o}{dt}$$

$$v_o = -\int \frac{v_1}{R_1 C} + \frac{v_2}{R_2 C} + \frac{v_3}{R_3 C} \, dt$$

The circuit therefore acts as an integrator. An additional benefit is that it allows voltages to be summed before integration.

8.2.3 *Integration of trigonometric functions*

The trigonometric identities given in Section 1.4.6 together with Table 8.1 allow us to integrate a number of trigonometric functions.

Example 8.7

Evaluate

(a) $\int \cos^2 t \, dt$
(b) $\int \sin^2 t \, dt$

Solutions

Powers of trigonometric functions, for example, $\sin^2 t$, do not appear in the table of standard integrals. What we must attempt to do is rewrite the integrand to obtain a standard form.

(a) From Section 1.4.6

$$\cos^2 t = \frac{1 + \cos 2t}{2}$$

and so

$$\int \cos^2 t \, dt = \int \frac{1 + \cos 2t}{2} \, dt$$

$$= \int \frac{1}{2} \, dt + \int \frac{\cos 2t}{2} \, dt$$

$$= \frac{t}{2} + \frac{\sin 2t}{4} + c$$

(b) $\displaystyle \int \sin^2 t \, dt = \int 1 - \cos^2 t \, dt$ using the trigonometric identities

$$= \int 1 \, dt - \int \cos^2 t \, dt \qquad \text{using linearity}$$

$$= t - \left(\frac{t}{2} + \frac{\sin 2t}{4} + c \right) \qquad \text{using part (a)}$$

$$= \frac{t}{2} - \frac{\sin 2t}{4} + k \quad \blacktriangle$$

Example 8.8

Find

(a) $\int \sin 2t \cos t \, dt$

(b) $\int \sin mt \sin nt \, dt$, where m and n are constants with $m \neq n$

Solutions

(a) Using the identities in Section 1.4.6, we find

$$2 \sin A \cos B = \sin(A + B) + \sin(A - B)$$

hence sin 2t cos t can be written $\frac{1}{2}(\sin 3t + \sin t)$. Therefore,

$$\int \sin 2t \cos t \, dt = \int \frac{1}{2}(\sin 3t + \sin t) \, dt$$

$$= \frac{1}{2}\left(\frac{-\cos 3t}{3} - \cos t\right) + c$$

$$= -\frac{1}{6}\cos 3t - \frac{1}{2}\cos t + c$$

(b) Using the identity $2 \sin A \sin B = \cos(A - B) - \cos(A + B)$, we find

$$\sin mt \sin nt = \frac{1}{2}\{\cos(m - n)t - \cos(m + n)t\}$$

Therefore,

$$\int \sin mt \sin nt \, dt = \int \frac{1}{2}\{\cos(m - n)t - \cos(m + n)t\} \, dt$$

$$= \frac{1}{2}\left\{\frac{\sin(m - n)t}{m - n} - \frac{\sin(m + n)t}{m + n}\right\} + c \quad \blacktriangle$$

EXERCISES 8.1

1. Integrate the following expressions.

(a) $3 + x + 1/x$

(b) $e^{2x} - e^{-2x}$

(c) $2 \sin 3x + \cos 3x$

(d) $\sec(2t + \pi) + \cot(t/2 - \pi)$

(e) $\tan(t/2) + \operatorname{cosec}(3t - \pi)$

(f) $\sin x + x/3 + 1/e^x$

(g) $1/\cos(3x)$

(h) $(t + 1/t)^2$

(i) $1/3e^{2x}$

(j) $\tan(4t - 3) + 2 \sin(-t - 1)$

(k) $1 + 2 \cot 3x$

(l) $\sin(t/2) - 3 \cos(t/2)$

(m) $(t - 2)^2$

(n) $3e^{-t} - e^{-t/2}$

(o) $7 - 7x^6 + e^{-x}$

(p) $(k + t)^2$ k constant

(q) $k \sin t - \cos kt$ k constant

(r) $1/(25 + t^2)$

(s) $1/(\sqrt{25 - t^2})$

(t) $6/(1 + x^2) + (1 + x^2)/6$

2. The acceleration, a, of a particle is the rate of change of speed, v, with respect to time t, that is, $a = dv/dt$. The speed of the particle is the rate of change of distance, s, that is, $v = ds/dt$. If the acceleration is given by $1 + t/2$, find expressions for speed and distance.

3. By writing sinh ax and cosh ax in terms of the exponential function find:

(a) $\int \sinh ax \, dx$

(b) $\int \cosh ax \, dx$

(c) Use your results from (a) and (b) to find $\int 3 \sinh 2x + \cosh 4x \, dx$.

4. Integrate the following:

(a) $1/(x^2 + 4)$

(b) $1/(2x^2 + 4)$

(c) $3/(2x^2 + 1)$

(d) $1/\sqrt{9 - x^2}$

(e) $2/\sqrt{4 - x^2}$

(f) $-7/\sqrt{2 - 3x^2}$

(g) $1/\sqrt{1 - (x^2/2)}$

5. By writing $(3 + x)/x$ in the form $(3/x) + 1$, find $\int ((3 + x)/x)\, dx$

6. The velocity, v, of a particle is given by:

$$v = 2 + e^{-t/2}$$

(a) Given distance, s, and v are related by $ds/dt = v$ find an expression for distance.

(b) Acceleration is the rate of change of velocity with respect to t. Determine the acceleration.

7. (a) Use the product rule of differentiation to verify:

$$\frac{d}{dx}(xe^{2x}) = e^{2x} + 2xe^{2x}$$

(b) Hence show

$$\int xe^{2x}\, dx = \frac{xe^{2x}}{2} - \frac{e^{2x}}{4} + c$$

8. Integrate:

(a) $(2 - t^2)/t^3$

(b) $(4 + e^{2t})/e^{3t}$

(c) $\cos 4x/\sin 4x$

(d) $\dfrac{1}{2\sin 3x}$

(e) $(2 + x^2)/(1 + x^2)$

(f) $\sin^2 t + \cos^2 t$

8.3 Definite and indefinite integrals

All the integration solutions so far encountered have contained a constant of integration. Such integrals are known as **indefinite integrals**. Integration can be used to determine the area under curves and this gives rise to **definite integrals**.

To estimate the area under $y(x)$, it is divided into thin rectangles. The sum of the rectangular areas is an approximation to the area under the curve. Several thin rectangles will give a better approximation than a few wide ones.

Consider Figure 8.3 where the area is approximated by a large number of rectangles. Suppose each rectangle has width δx. The area of rectangle 1 is $y(x_2)\delta x$, the area of rectangle 2 is $y(x_3)\delta x$ and so on. Let $A(x_n)$ denote the total area under the curve from x_1 to x_n. Then,

$$A(x_n) \approx \text{sum of the rectangular areas} = \sum_{i=2}^{n} y(x_i)\delta x$$

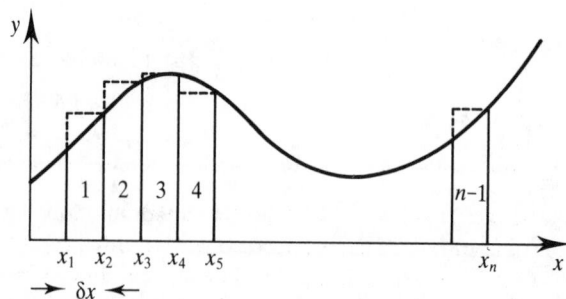

Figure 8.3 The area is approximated by $(n - 1)$ rectangles.

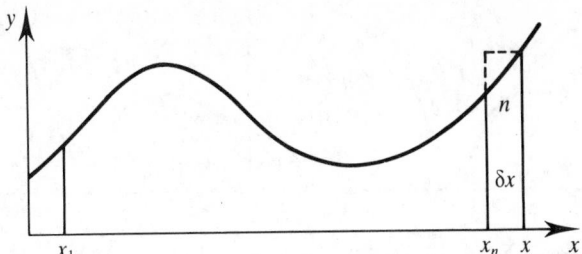

Figure 8.4 The area is extended by adding an extra rectangle.

Let the area be increased by extending the base from x_n to x. Then $A(x)$ is the total area under the curve from x_1 to x (see Figure 8.4). Then,

increase in area $= \delta A = A(x) - A(x_n) \approx y(x)\delta x$

So,

$$\frac{\delta A}{\delta x} \approx y(x)$$

In the limit as $\delta x \to 0$, we get:

$$\lim_{\delta x \to 0} \left(\frac{\delta A}{\delta x} \right) = \frac{dA}{dx} = y(x)$$

Since differentiation is the reverse of integration, we can write:

$$A = \int y(x)\, dx$$

To denote the limits of the area being considered we place values on the integral sign.

■ The area under the curve, $y(x)$, between $x = a$ and $x = b$ is denoted as:

$$\int_{x=a}^{x=b} y\, dx$$

or more compactly by

$$\int_a^b y\, dx$$

The constants a and b are known as the limits of the integral; **lower** and **upper**, respectively. Since an area has a specific value, such an integral is called a definite integral. The area under the curve up to the vertical line $x = b$ is $A(b)$ (see Figure 8.5). Similarly $A(a)$ is the area up to the vertical line $x = a$. So the area between $x = a$ and $x = b$ is $A(b) - A(a)$, as shown in Figure 8.6.

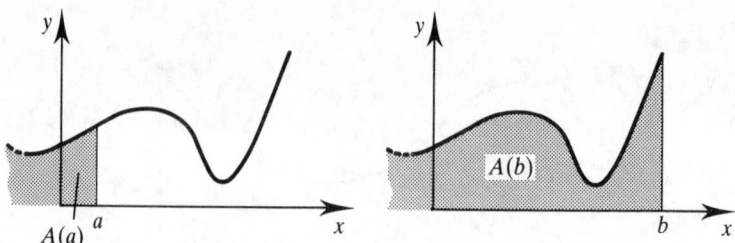

Figure 8.5 The area depends on the limits *a* and *b*.

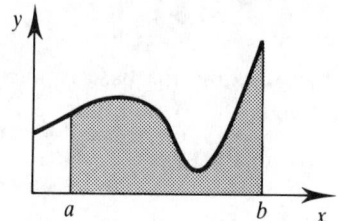

Figure 8.6 The area between $x = a$ and $x = b$ is $A(b) - A(a)$.

■ The area between $x = a$ and $x = b$ is given by:

$$\text{Area} = \int_a^b y \, dx = A(b) - A(a)$$

The integral is evaluated at the upper limit, *b* and at the lower limit, *a*, and the difference between these gives the required area.

The expression $A(b) - A(a)$ is often written as $[A(x)]_a^b$. Note that since

$$\int_a^b y \, dx = A(b) - A(a)$$

then

$$\int_b^a y \, dx = A(a) - A(b) = -\{A(b) - A(a)\}$$

that is,

■ $$\int_a^b y \, dx = -\int_b^a y \, dx$$

Interchanging the limits changes the sign of the integral.

The evaluation of definite integrals is illustrated in the following examples.

Example 8.9

Evaluate

(a) $\displaystyle\int_1^2 x^2 + 1\ dx$ (c) $\displaystyle\int_0^\pi \sin x\ dx$

(b) $\displaystyle\int_2^1 x^2 + 1\ dx$

Solutions

(a) Let I stand for $\displaystyle\int_1^2 x^2 + 1\ dx$

$$I = \int_1^2 x^2 + 1\ dx = \left[\frac{x^3}{3} + x\right]_1^2$$

The integral is now evaluated at the upper and lower limits. The difference gives the value required.

$$I = \left(\frac{2^3}{3} + 2\right) - \left(\frac{1^3}{3} + 1\right) = \frac{8}{3} + 2 - \frac{4}{3} = \frac{10}{3}$$

(b) Because interchanging the limits of integration changes the sign of the integral, we find:

$$\int_2^1 x^2 + 1\ dx = -\int_1^2 x^2 + 1\ dx = -\frac{10}{3}$$

(c) $\displaystyle\int_0^\pi \sin x\ dx = [-\cos x]_0^\pi = (-\cos \pi) - (-\cos 0) = 1 + (+1) = 2$

Figure 8.7 illustrates this area. ▲

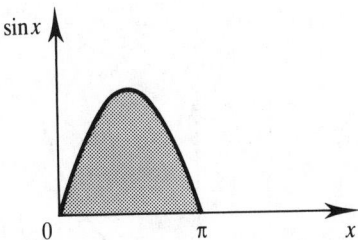

Figure 8.7 The area is given by a definite integral.

Note:

(1) The integrated function is evaluated at the upper and lower limits, and the difference found.

(2) No constant of integration is needed.

(3) Any angles are measured in radians.

Example 8.10

Find the area under $z(t) = e^{2t}$ from $t = 1$ to $t = 3$.

Solution

$$\text{area} = \int_1^3 z\, dt = \int_1^3 e^{2t}\, dt = \left[\frac{e^{2t}}{2}\right]_1^3$$

$$= \left[\frac{e^6}{2}\right] - \left[\frac{e^2}{2}\right] = 198 \quad \blacktriangle$$

If the evaluation of an area by integration yields a negative quantity this means that some or all of the corresponding area is below the horizontal axis. This is illustrated in Example 8.11.

Example 8.11

Find the area bounded by $y = x^3$ and the x axis from $x = -3$ to $x = -2$.

Solution

Figure 8.8 illustrates the required area.

$$\int_{-3}^{-2} x^3\, dx = \left[\frac{x^4}{4}\right]_{-3}^{-2}$$

$$= \left\{\frac{(-2)^4}{4}\right\} - \left\{\frac{(-3)^4}{4}\right\} = -\frac{65}{4}$$

The area is 65/4 square units; the negative sign indicates that it is below the x axis. $\blacktriangle$

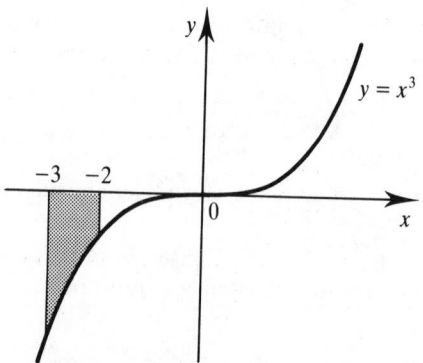

Figure 8.8 Areas below the x axis are classed as negative.

Example 8.12

Evaluate $\int_{-\pi}^{\pi} \sin x \, dx$ and comment on your findings.

Solution

$$\int_{-\pi}^{\pi} \sin x \, dx = [-\cos x]_{-\pi}^{\pi} = -\cos(\pi) + \cos(-\pi) = 0$$

The positive and negative contributions have cancelled each other out, that is, the area above the x axis is equal in size to the area below (see Figure 8.9). ▲

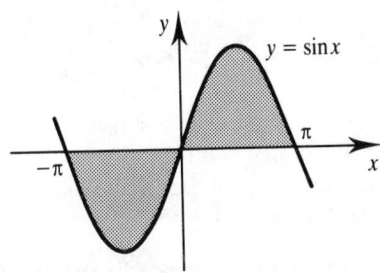

Figure 8.9 The positive and negative areas cancel each other out.

If an area contains parts both above and below the horizontal axis then calculating an integral will give the net area. If the total area is required, then the relevant limits must first be found. A sketch of the function often clarifies the situation.

Example 8.13

Find the area contained by $y = \sin x$ from $x = 0$ to $x = 3\pi/2$.

Solution

Figure 8.10 illustrates the required area. From this we see that there are parts both above and below the x axis and the crossover point occurs when $x = \pi$.

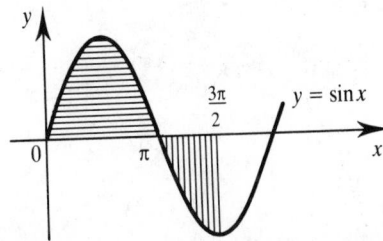

Figure 8.10 The positive and negative areas are calculated separately.

$$\int_0^\pi \sin x \, dx = [-\cos x]_0^\pi$$

$$= -\cos \pi + \cos 0 = 2$$

$$\int_\pi^{3\pi/2} \sin x \, dx = [-\cos x]_\pi^{3\pi/2}$$

$$= -\cos\left(\frac{3\pi}{2}\right) + \cos \pi = -1$$

The total area is three square units. Note, however, that the single integral over 0 to $3\pi/2$ evaluates to 1, that is, it gives the net value of 2 and -1.

$$\int_0^{3\pi/2} \sin x \, dx = [-\cos x]_0^{3\pi/2} = -\cos\left(\frac{3\pi}{2}\right) + \cos 0 = 1 \quad \blacktriangle$$

The need to evaluate the area under a curve is a common requirement in engineering. Often the rate of change of an engineering variable with time is known and it is required to calculate the value of the engineering variable. This corresponds to calculating the area under a curve.

Example 8.14 Energy used by an electric motor

Consider a small d.c. electric motor being used to drive an electric screwdriver. The amount of power that is supplied to the motor by the batteries depends on the load on the screwdriver. Therefore the power supplied to the screwdriver is a function that varies with time. Figure 8.11 shows a typical curve of power versus time. Now,

$$P = \frac{dE}{dt}$$

where P = power (W), E = energy (J). Therefore, to calculate the energy used by the motor between times t_1 and t_2, we can write:

$$E = \int_{t_1}^{t_2} P \, dt$$

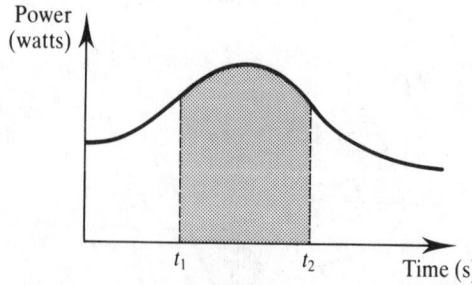

Figure 8.11 Shaded area represents the energy used to drive the motor during the time interval, $t_1 \leq t \leq t_2$.

This is equivalent to evaluating the area under the curve, $P(t)$, between t_1 and t_2, which is shown as the shaded region in Figure 8.11. ▲

Example 8.15 Capacitance of a concentric cable

A concentric cable has an inner conductor with a diameter of 15 mm and an outer conductor with an internal diameter of 25 mm, as shown in Figure 8.12. The insulator separating the two conductors has a relative permittivity of 2.5. Let us calculate the capacitance of the cable per metre length.

Before solving this problem it is instructive to derive the formula for the capacitance of a concentric cable. Imagine that the inner conductor has a charge of $+Q$ per metre length and that the outer conductor has a charge of $-Q$ per metre length. Further assume the cable is long and a central section is being analysed in order that end effects can be ignored.

Consider an imaginary cylindrical surface, radius r and length l, within the insulator (or dielectric). Gauss' theorem states that the electric flux out of any closed surface is equal to the charge enclosed by the surface. In this case, because of symmetry, the electric flux points radially outwards and so no flux is directed through the ends of the imaginary cylinder, that is, end effects can be neglected. The curved surface area of the cylinder is $2\pi rl$. Therefore, using Gauss' theorem:

$$D \times 2\pi rl = Ql$$

where D = electric flux density.

When an insulator or dielectric is present then $D = \varepsilon_r \varepsilon_0 E$, E = electric field strength, ε_r = relative permittivity, ε_0 = permittivity of free space = $8.854 \times 10^{-12}\,\mathrm{F\,m^{-1}}$. Therefore,

$$\varepsilon_r \varepsilon_0 E2\pi rl = Ql$$

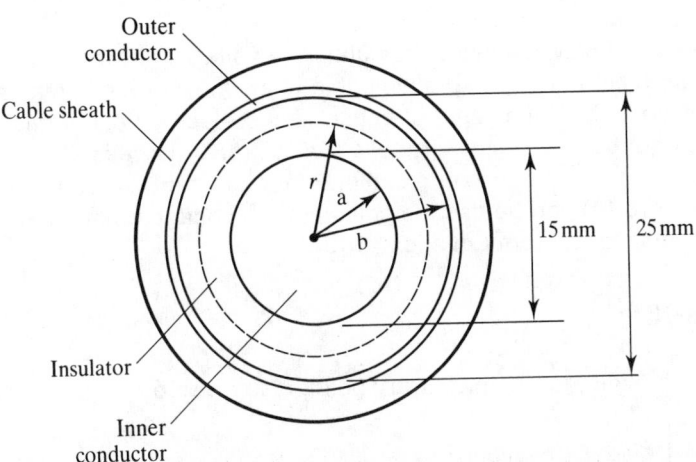

Figure 8.12 Cross-section of a concentric cable.

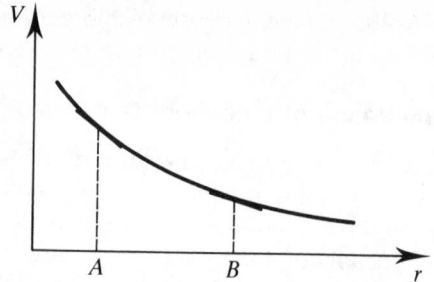

Figure 8.13   The gradient of the curve, d*V*/d*r*, is proportional to the magnitude of the electric field.

that is,

$$E = \frac{Q}{2\pi\varepsilon_r\varepsilon_0 r} \tag{8.9}$$

This equation gives a value for the electric field within the dielectric. In order to calculate the capacitance of the cable it is necessary to calculate the voltage difference between the two conductors. Let V_a represent the voltage of the inner conductor and V_b the voltage of the outer conductor.

The electric field is a measure of the rate of change of the voltage with position. In other words, if the voltage is changing rapidly with position then this corresponds to a large magnitude of the electric field. This is illustrated in Figure 8.13. The magnitude of the electric field at point A is larger than at point B. As a positive electric field, E, corresponds to a decrease in voltage, V, with position the relationship between E and V, in general, is

$$E = -\frac{dV}{dr} \tag{8.10}$$

This expression can be used to calculate the voltage difference arising as a result of an electric field. In practice, this is a simplified equation and is only valid provided r is in the same direction as the electric field. If this is not the case, then a modified vector form of Equation (8.10) is needed.

In the case of the concentric cable, E is in the same direction as r and so Equation (8.10) can be used to calculate the voltage difference between the two conductors. From Equation (8.10)

$$\frac{dV}{dr} = -E$$

Therefore the voltage at an arbitrary point, r, is given by

$$V = -\int E \, dr$$

Consequently, the voltage difference between points $r = b$ and $r = a$ is given by

$$V_b - V_a = -\int_a^b E \, dr$$

$$= -\frac{Q}{2\pi\varepsilon_r\varepsilon_0} \int_a^b \frac{1}{r} \, dr \qquad \text{using Equation (8.9)}$$

$$= -\frac{Q}{2\pi\varepsilon_r\varepsilon_0} [\ln r]_a^b$$

$$= -\frac{Q}{2\pi\varepsilon_r\varepsilon_0} \ln\left(\frac{b}{a}\right)$$

This gives the voltage of the outer conductor relative to the inner one. Thus the voltage of the inner conductor relative to the outer one is

$$V_a - V_b = \frac{Q}{2\pi\varepsilon_r\varepsilon_0} \ln\left(\frac{b}{a}\right)$$

More generally capacitance is defined as $C = Q/V$, where V is the voltage difference. Therefore

$$C = \frac{Q}{V_a - V_b} = \frac{2\pi\varepsilon_r\varepsilon_0}{\ln\left(\dfrac{b}{a}\right)}$$

Note that this is the capacitance per unit length of cable. Using $\varepsilon_r = 2.5$, $a = 7.5 \times 10^{-3}$ m, $b = 12.5 \times 10^{-3}$ m, we get

$$C = \frac{2\pi \times 2.5 \times 8.854 \times 10^{-12}}{\ln\left(\dfrac{12.5 \times 10^{-3}}{7.5 \times 10^{-3}}\right)} = 2.72 \times 10^{-10} = 272 \text{ pF m}^{-1} \quad \blacktriangle$$

8.3.1 *Use of a dummy variable*

Consider the following integrals, I_1 and I_2:

$$I_1 = \int_0^1 t^2 \, dt \qquad I_2 = \int_0^1 x^2 \, dx$$

Then,

$$I_1 = \left[\frac{t^3}{3}\right]_0^1 = \left(\frac{1}{3}\right) - (0) = \frac{1}{3}$$

$$I_2 = \left[\frac{x^3}{3}\right]_0^1 = \left(\frac{1}{3}\right) - (0) = \frac{1}{3}$$

So clearly $I_1 = I_2$. The value of I_1 does not depend upon t, and the value of I_2 does not depend upon x. In general,

$$I = \int_a^b f(t)\, dt = \int_a^b f(x)\, dx$$

Because the value of I is the same, regardless of what the integrating variable may be, we say x and t are **dummy variables**. Indeed we could write:

$$I = \int_a^b f(z)\, dz = \int_a^b f(r)\, dr = \int_a^b f(y)\, dy$$

Then z, r and y are dummy variables.

EXERCISES 8.2

1. Calculate the total area between $f(t) = \cos 2t$ and the t axis as t varies from:

   (a) 0 to $\pi/4$

   (b) $\pi/4$ to $\pi/2$

   (c) 0 to $\pi/2$

2. Evaluate the following integrals:

   (a) $\displaystyle\int_0^1 t^2 + 0.5t - 6 \; dt$

   (b) $\displaystyle\int_2^3 \frac{3}{2x} + \frac{2x}{3} \; dx$

   (c) $\displaystyle\int_1^2 e^{-2x} - 3e^{-x} \; dx$

   (d) $\displaystyle\int_0^2 3\sin(4t - \pi) + 5\cos(3t + \pi/2) \; dt$

   (e) $\displaystyle\int_0^1 2 \tan t \; dt$

   (f) $\displaystyle\int_0^{1.5} \frac{2}{1 + x^2} - \frac{1}{\sqrt{4 - x^2}} \; dx$

3. Calculate the area enclosed by the curves $y = x^2$ and $y = x$.

4. Calculate the area enclosed by the graphs of the functions $y = t^2 + 5$ and $y = 6$.

5. Evaluate the following definite integrals:

   (a) $\int_1^{1.5} 1/t + 1/e^t + 1/\sin t \; dt$

   (b) $\int_1^4 5/(8 + 3x^2) \; dx$

   (c) $\int_{-1}^1 \sinh x \; dx$

   (d) $\int_{-1}^1 \cosh x \; dx$

6. Calculate the area between $y = 2 \tan t$ and the t axis for $-1 \le t \le 1.4$.

7. Find the area between $y = \sin t$, $y = \cos t$ and the y axis, for $t \ge 0$.

8. The velocity, v, of a particle is given by:

   $$v = (1 + t)^2$$

   Find the distance travelled by the particle from $t = 1$ to $t = 4$, that is, evaluate $\int_1^4 v \; dt$.

9. Evaluate the area under the function $x = 1/t$ for $1 \le t \le 10$.

10. Evaluate:

    (a) $\int_3^2 x^{1.4} \; dx$

    (b) $\int_0^1 (e^t)^2 \; dt$

    (c) $\int_0^\pi \sin x \cos x \; dx$

    (d) $\int_1^2 \sinh^2 x \; dx$

8.4 Techniques of integration

The previous section allowed us to integrate functions which matched the list of standard integrals given in Table 8.1. Clearly, it is impossible to list all possible functions in the table and so some general techniques are required. Integration techniques may be classified as **analytical**, that is, exact, or **numerical**, that is, approximate. We will now study three analytical techniques

(1) integration by parts;
(2) integration by substitution;
(3) integration using partial fractions.

8.4.1 *Integration by parts*

This technique is used to integrate a product, and is derived from the product rule for differentiation. Let u and v be functions of x. Then the product rule of differentiation states:

$$\frac{d}{dx}(uv) = \frac{du}{dx}v + u\frac{dv}{dx}$$

that is,

$$u\frac{dv}{dx} = \frac{d}{dx}(uv) - v\frac{du}{dx}$$

Integrating this equation yields:

$$\int u\frac{dv}{dx}\,dx = \int\frac{d}{dx}(uv)\,dx - \int v\frac{du}{dx}\,dx$$

Recognizing that integration and differentiation are inverse processes allows

$$\int\frac{d(uv)}{dx}\,dx$$

to be simplified to uv. Hence,

■ $$\int u\left(\frac{dv}{dx}\right)dx = uv - \int v\left(\frac{du}{dx}\right)dx$$

This is the formula for integration by parts.

Example 8.16

Find $\int x \sin x \, dx$.

Solution

We recognize the integrand as a product of the functions x and $\sin x$. Let $u = x$, $dv/dx = \sin x$. Then $du/dx = 1$, $v = -\cos x$. Using the integration by parts formula we get:

$$\int x \sin x \, dx = x(-\cos x) - \int (-\cos x)1 \, dx$$

$$= -x \cos x + \sin x + c \quad \blacktriangle$$

When dealing with definite integrals the corresponding formula for integration by parts is:

■ $\displaystyle\int_a^b u\left(\frac{dv}{dx}\right) dx = [uv]_a^b - \int_a^b v\left(\frac{du}{dx}\right) dx$

Example 8.17

Evaluate $\displaystyle\int_0^2 x^2 e^x \, dx$.

Solution

Let $u = x^2$ and $dv/dx = e^x$. Then $du/dx = 2x$ and $v = e^x$. Hence,

$$\int_0^2 x^2 e^x \, dx = [x^2 e^x]_0^2 - \int_0^2 2x e^x \, dx$$

$$= 4e^2 - 2\int_0^2 x e^x \, dx$$

Now $\int_0^2 x e^x \, dx$ may be evaluated using integration by parts. Let $u = x$ and $dv/dx = e^x$, so that $du/dx = 1$, $v = e^x$. Then,

$$\int_0^2 x e^x \, dx = [x e^x]_0^2 - \int_0^2 e^x \, dx$$

$$= 2e^2 - [e^x]_0^2 = 2e^2 - e^2 + 1 = e^2 + 1$$

So,

$$\int_0^2 x^2 e^x \, dx = 4e^2 - 2[e^2 + 1] = 2e^2 - 2 = 12.78 \quad ▲$$

Example 8.18

Find $I = \displaystyle\int e^t \sin t \, dt$.

Solution

We let $u = e^t$, $dv/dt = \sin t$, $du/dt = e^t$, $v = -\cos t$

then

$$I = \int e^t \sin t \, dt = -e^t \cos t - \int (-\cos t) e^t \, dt + c$$

$$= -e^t \cos t + \int e^t \cos t \, dt + c$$

Applying integration by parts again yields:

$$I = -e^t \cos t + (e^t \sin t - \int e^t \sin t \, dt) + c$$

$$= e^t(\sin t - \cos t) - \int e^t \sin t \, dt + c$$

$$= e^t(\sin t - \cos t) - I + c$$

Finally, solving for I gives

$$I = \frac{e^t(\sin t - \cos t) + c}{2} \quad \blacktriangle$$

8.4.2 *Integration by substitution*

This technique is the integral equivalent of the chain rule. It is best illustrated by examples.

Example 8.19

Find $\int (3x + 1)^{2.7} \, dx$.

Solution

Let $z = 3x + 1$, so that $dz/dx = 3$, that is, $dx = dz/3$. Writing the integral in terms of z, it becomes:

$$\int z^{2.7} \frac{1}{3} \, dz = \frac{1}{3} \int z^{2.7} \, dz = \frac{1}{3}\left(\frac{z^{3.7}}{3.7}\right) + c = \frac{1}{3} \frac{(3x + 1)^{3.7}}{3.7} + c \quad \blacktriangle$$

Example 8.20

Evaluate $\displaystyle\int_2^3 t \sin(t^2) \, dt$.

Solution

Let $v = t^2$ so $dv/dt = 2t$, that is,

$$dt = \frac{1}{2t} \, dv.$$

When changing the integral from one in terms of t to one in terms of v, the limits must also be changed. When $t = 2$, $v = 4$; when $t = 3$, $v = 9$. Hence, the integral becomes:

$$\int_4^9 \frac{\sin v}{2} \, dv = \frac{1}{2}[-\cos v]_4^9 = \frac{1}{2}[-\cos 9 + \cos 4] = 0.129 \quad \blacktriangle$$

Sometimes the substitution can involve a trigonometric function.

Example 8.21

Evaluate $I = \displaystyle\int_1^2 \sin t \cos^2 t \, dt$.

Solution

Put $z = \cos t$ so that $dz/dt = -\sin t$, that is, $\sin t \, dt = -dz$. When $t = 1$, $z = \cos 1$, when $t = 2$, $z = \cos 2$. Hence:

$$I = -\int_{\cos 1}^{\cos 2} z^2 \, dz = -\left[\frac{z^3}{3}\right]_{\cos 1}^{\cos 2}$$

$$= \frac{\cos^3 1 - \cos^3 2}{3} = 0.0766 \quad \blacktriangle$$

Example 8.22

Find $I = \int e^{\tan x}/\cos^2 x \, dx$.

Solution

Put $z = \tan x$. Then $dz/dx = \sec^2 x$, $dz = dx/\cos^2 x$. Hence,

$$I = \int e^z \, dz = e^z + c = e^{\tan x} + c \quad \blacktriangle$$

Integration by substitution allows functions of the form $(df/dx)/f$ to be integrated.

Example 8.23

Find $I = \int (3x^2 + 1)/(x^3 + x + 2) \, dx$.

Solution

Put $z = x^3 + x + 2$, then $dz/dx = 3x^2 + 1$, that is, $dz = (3x^2 + 1) \, dx$. Hence,

$$I = \int \frac{dz}{z} = \ln |z| + c = \ln |x^3 + x + 2| + c \quad \blacktriangle$$

Example 8.24

Find $I = \displaystyle\int \frac{(df/dx)}{f} \, dx$.

Solution

Put $z = f$. Then $dz/dx = df/dx$, that is,

$$dz = \frac{df}{dx} dx.$$

Hence,

$$I = \int \frac{dz}{z} = \ln|z| + c = \ln|f| + c$$

and so,

$$\int \frac{(df/dx)}{f} dx = \ln|f| + c \quad \blacktriangle$$

The result of Example 8.24 is particularly important.

■ $\displaystyle \int \frac{(df/dx)}{f} dx = \ln|f| + c$

Example 8.25

Evaluate $I = \displaystyle \int_2^4 \frac{3t^2 + 2t}{t^3 + t^2 + 1} dt$.

Solution

The numerator is the derivative of the denominator and so:

$$I = [\ln|t^3 + t^2 + 1|]_2^4 = \ln 81 - \ln 13 = 1.83 \quad \blacktriangle$$

Example 8.26

Find:

(a) $I = \int 4/(5x - 7) \, dx$
(b) $I = \int t/(t^2 + 1) \, dt$
(c) $I = \int e^{t/2}/(e^{t/2} + 1) \, dt$

Solutions

The integrands are rewritten so that the numerator is the derivative of the denominator.

(a) $\displaystyle I = \int \frac{4}{5x - 7} \, dx = \frac{4}{5} \int \frac{5}{5x - 7} \, dx = \frac{4}{5} \ln|5x - 7| + c$

(b) $\displaystyle I = \int \frac{t}{t^2 + 1} \, dt = \frac{1}{2} \int \frac{2t}{t^2 + 1} \, dt$

$$= \frac{1}{2} \ln|t^2 + 1| + c$$

(c) $\displaystyle I = \int \frac{e^{t/2}}{e^{t/2} + 1} \, dt = 2 \int \frac{\frac{1}{2}e^{t/2}}{e^{t/2} + 1} \, dt = 2 \ln|e^{t/2} + 1| + c \quad \blacktriangle$

8.4.3 *Integration using partial fractions*

The technique of expressing a rational function as the sum of its partial fractions has been covered in Section 1.4.3. Expressions which at first sight cannot be integrated may in fact be integrable when expressed as their partial fractions.

Example 8.27

Find:

(a) $I = \int 1/(x^3 + x) \, dx$

(b) $I = \int (13x - 4)/(6x^2 - x - 2) \, dx$

Solutions

(a) First express the integrand in partial fractions:

$$\frac{1}{x^3 + x} = \frac{1}{x(x^2 + 1)} = \frac{A}{x} + \frac{Bx + C}{x^2 + 1}$$

Then,

$$1 = A(x^2 + 1) + x(Bx + C)$$

Equating the constant terms: $1 = A$ so that $A = 1$.
Equating the coefficients of x: $0 = C$ so that $C = 0$.
Equating the coefficients of x^2: $0 = A + B$ and hence $B = -1$.
Then,

$$I = \int \frac{1}{x^3 + x} \, dx = \int \frac{1}{x} - \frac{x}{x^2 + 1} \, dx$$

$$= \ln |x| - \frac{1}{2} \ln |x^2 + 1| + c = \ln \left| \frac{x}{\sqrt{x^2 + 1}} \right| + c$$

(b)

$$I = \int \frac{13x - 4}{6x^2 - x - 2} \, dx = \int \frac{13x - 4}{(2x + 1)(3x - 2)} \, dx$$

$$= \int \frac{3}{2x + 1} + \frac{2}{3x - 2} \, dx \qquad \text{using partial fractions}$$

$$= \frac{3}{2} \int \frac{2}{2x + 1} \, dx + \frac{2}{3} \int \frac{3}{3x - 2} \, dx$$

$$= \frac{3}{2} \ln |2x + 1| + \frac{2}{3} \ln |3x - 2| + c \quad \blacktriangle$$

Example 8.28

Evaluate $I = \displaystyle\int_0^1 \frac{4t^3 - 2t^2 + 3t - 1}{2t^2 + 1} \, dt$.

Solution

Using partial fractions we may write:

$$\frac{4t^3 - 2t^2 + 3t - 1}{2t^2 + 1} = 2t - 1 + \frac{t}{2t^2 + 1}$$

Hence,

$$I = \int_0^1 2t - 1 + \frac{t}{2t^2 + 1} \, dt = \left[t^2 - t + \frac{1}{4} \ln |2t^2 + 1| \right]_0^1$$

$$= \left[1 - 1 + \frac{1}{4} \ln 3 \right] - \left[0 - 0 + \frac{1}{4} \ln 1 \right] = 0.275 \quad \blacktriangle$$

EXERCISES 8.3

1. Use the given substitution to find the following integrals:

   (a) $\int_0^1 (9t + 2)^{10} \, dt$ $z = 9t + 2$
   (b) $\int_3^5 (-t + 1)^6 \, dt$ $z = -t + 1$
   (c) $\int_6^3 (4x - 1)^{27} \, dx$ $z = 4x - 1$
   (d) $\int \sqrt{3t + 1} \, dt$ $z = 3t + 1$
   (e) $\int (9y - 2)^{17} \, dy$ $z = 9y - 2$
   (f) $\int_0^2 3/(2z + 5)^6 \, dz$ $y = 2z + 5$
   (g) $\int t^2 \sin(t^3) \, dt$ $z = t^3$
   (h) $\int x^2 e^{x^3 + 1} \, dx$ $z = x^3 + 1$
   (i) $\int_0^{0.5} \sin(2t) e^{\cos(2t)} \, dt$ $z = \cos(2t)$
   (j) $\int_0^\pi \sin t \cos^2 t \, dt$ $z = \cos t$
   (k) $\int \cos t \sqrt{\sin t} \, dt$ $z = \sin t$

2. Use integration by parts to find:

   (a) $\int x \ln x \, dx$
   (b) $\int t \cos t \, dt$
   (c) $\int_0^{\pi/2} e^{2x} \cos x \, dx$
   (d) $\int \ln x \, dx$ *Hint:* $u = \ln x$, $dv/dx = 1$
   (e) $\int t^2 \sin(2t) \, dt$

3. Use partial fractions to find:

   (a) $\int_2^3 (3x + 2)/(x^2 - 1) \, dx$

   (b) $\int (t + 3)/(t^2 + 2t + 1) \, dt$
   (c) $\int (2t^2 + 3t + 1)/(t^3 + t) \, dt$
   (d) $\int (6t + 3)/(2t^2 - 5t + 2) \, dt$

4. Evaluate:

   (a) $\int t^2/(t^3 + 1) \, dt$
   (b) $\int_0^{\pi/3} \sin t \cos t \, dt$
   (c) $\int_1^3 4/e^{2t} \, dt$
   (d) $\displaystyle \int \frac{3x - 7}{(x - 2)(x - 3)(x - 4)} \, dx$
   (e) $\int e^x/(e^x + 1) \, dx$

5. Evaluate:

   (a) $\displaystyle \int_0^1 \frac{3}{(e^t)^2} + \sin t \cos^2 t \, dt$
   (b) $\int_0^1 4t^2 e^{t^3} + t(1 + t^2)^{12} \, dt$
   (c) $\int_0^\pi \sin^2 \omega t + \cos^2 \omega t + \omega \, dt$ ω constant
   (d) $\displaystyle \int_1^2 \frac{1 + t + t^2}{t(1 + t^2)} \, dt$
   (e) $\int_0^1 (t + e^t) \sin t \, dt$
   (f) $\int_1^3 (1 + 4x)/(2x + 4x^2) \, dx$

6. Calculate the area under
   $y(x) = (1 + 2e^{2x})/(x + e^{2x})$ from $x = 1$ to $x = 3$.

8.5 Average value and root mean square value of a function

8.5.1 *Average value of a function*

Suppose $f(t)$ is a function defined on $a \leq t \leq b$. The area, A, under f is given by:

$$A = \int_a^b f \, dt$$

A rectangle with base spanning the interval $[a, b]$ and height h has an area of $h(b - a)$. Suppose the height, h, is chosen so that the area under f and the area of the rectangle are equal. This means:

$$h(b - a) = \int_a^b f \, dt$$

$$h = \frac{\int_a^b f \, dt}{b - a}.$$

Then h is called the **average value** of the function across the interval $[a, b]$ and is illustrated in Figure 8.14.

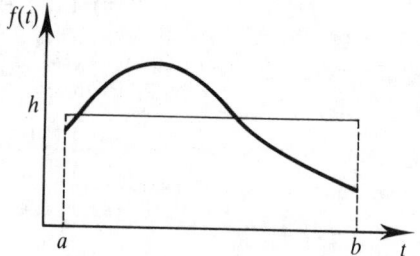

Figure 8.14   The area under the curve from $t = a$ to $t = b$ and the area of the rectangle are equal.

■ average value $= \dfrac{\int_a^b f \, dt}{b - a}$

Example 8.29

Find the average value of $f(t) = t^2$ across

(a) [1, 3]

(b) [2, 5]

Solutions

(a)
$$\text{average value} = \frac{\int_1^3 t^2\, dt}{3-1} = \frac{1}{2}\left[\frac{t^3}{3}\right]_1^3 = \frac{13}{3}$$

(b)
$$\text{average value} = \frac{\int_2^5 t^2\, dt}{5-2} = \frac{1}{3}\left[\frac{t^3}{3}\right]_2^5 = 13 \quad \blacktriangle$$

If the interval of integration is changed then, as Example 8.29 illustrates, the average value can change.

Example 8.30 A thyristor firing circuit

Figure 8.15 shows a simple circuit to control the voltage across a load resistor, R_L. This circuit has many uses, one of which is to adjust the level of lighting in a room. The circuit has an alternating current power supply with peak voltage, V_s. The main control element is the thyristor. This device is similar in many ways to a diode. It has a very high resistance when it is reverse biased and a low resistance when it is forward biased. However, unlike a diode, this low resistance depends on the thyristor being 'switched on' by the application of a gate current. The point at which the thyristor is switched on can be varied by varying the resistor, R_G. Figure 8.16 shows a typical waveform of the voltage, v_L, across the load resistor.
 The point at which the thyristor is turned on in each cycle is characterized by the quantity, αT, where $0 \leq \alpha \leq 0.25$ and T is the period of the waveform. This restriction on α reflects the fact that if the thyristor has not turned on when the supply voltage has peaked in the forward direction then it will never turn on.
 Calculate the average value of the waveform over a period and comment on the result.

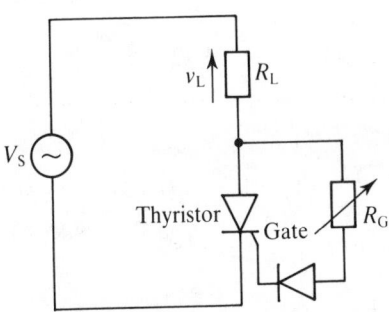

Figure 8.15 A thyristor firing circuit.

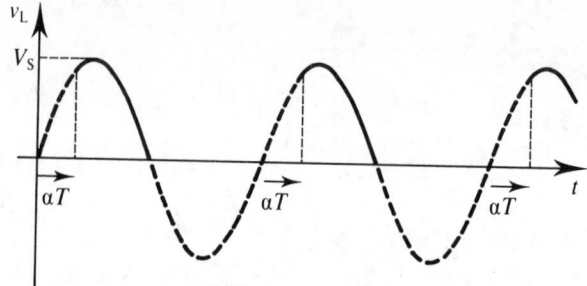

Figure 8.16 Load voltage waveform.

Solution

The average value of load voltage is:

$$\frac{1}{T}\int_0^T v_L \, dt = \frac{1}{T}\int_{\alpha T}^{T/2} V_s \sin\left(\frac{2\pi t}{T}\right) dt$$

$$= \frac{V_s}{T}\frac{T}{2\pi}\left[-\cos\left(\frac{2\pi t}{T}\right)\right]_{\alpha T}^{T/2} = \frac{V_s}{2\pi}(1 + \cos 2\pi\alpha)$$

If $\alpha = 0$, then the average value is V_s/π; the maximum value for this circuit. If $\alpha = 0.25$, then the average value is $V_s/2\pi$ which illustrates that delaying the turning on of the thyristor reduces the average value of the load voltage. ▲

8.5.2 *Root mean square value of a function*

If $f(t)$ is defined on $[a, b]$, the **root mean square** (r.m.s.) value is:

■ r.m.s. $= \sqrt{\dfrac{\int_a^b (f(t))^2 \, dt}{b - a}}$

Example 8.31

Find the r.m.s. value of $f(t) = t^2$ across $[1, 3]$.

Solution

$$\text{r.m.s.} = \sqrt{\frac{\int_1^3 (t^2)^2 \, dt}{3 - 1}} = \sqrt{\frac{\int_1^3 t^4 \, dt}{2}} = \sqrt{\frac{[t^5/5]_1^3}{2}} = \sqrt{\frac{242}{10}} = 4.92 \quad ▲$$

Example 8.32

Calculate the r.m.s. of $f(t) = A \sin t$ across $[0, 2\pi]$.

Solution

$$\text{r.m.s.} = \sqrt{\frac{\int_0^{2\pi} A^2 \sin^2 t \, dt}{2\pi}}$$

$$= \sqrt{\frac{A^2 \int_0^{2\pi} (1 - \cos 2t)/2 \, dt}{2\pi}}$$

$$= \sqrt{\frac{A^2}{4\pi} \left[t - \frac{\sin 2t}{2} \right]_0^{2\pi}}$$

$$= \sqrt{\frac{A^2 2\pi}{4\pi}} = \frac{A}{\sqrt{2}} = 0.707 \, A$$

Thus the r.m.s. value is 0.707 × the amplitude. ▲

Example 8.33

Calculate the r.m.s. value of $f(t) = A \sin(\omega t + \phi)$ across $[0, 2\pi/\omega]$.

Solution

$$\text{r.m.s.} = \sqrt{\frac{\int_0^{2\pi/\omega} A^2 \sin^2(\omega t + \phi) \, dt}{2\pi/\omega}}$$

$$= \sqrt{\frac{A^2 \omega}{4\pi} \int_0^{2\pi/\omega} 1 - \cos 2(\omega t + \phi) \, dt}$$

$$= \sqrt{\frac{A^2 \omega}{4\pi} \left[t - \frac{\sin 2(\omega t + \phi)}{2\omega} \right]_0^{2\pi/\omega}}$$

$$= \sqrt{\frac{A^2 \omega}{4\pi} \left(\frac{2\pi}{\omega} - \frac{\sin 2(2\pi + \phi)}{2\omega} + \frac{\sin 2\phi}{2\omega} \right)}$$

Now $\sin 2(2\pi + \phi) = \sin(4\pi + 2\phi)$ and since $\sin(t + \phi)$ has period 2π we see $\sin(4\pi + 2\phi) = \sin 2\phi$. Hence,

$$\text{r.m.s.} = \sqrt{\frac{A^2 \omega}{4\pi} \frac{2\pi}{\omega}} = \sqrt{\frac{A^2}{2}} = \frac{A}{\sqrt{2}} = 0.707A \quad ▲$$

Note that $\sin(\omega t + \phi)$ has period $2\pi/\omega$. The result of Example 8.33 illustrates a general result:

■ The r.m.s. value of any sinusoidal waveform taken across an interval of length one period, is

0.707 × amplitude of the waveform

EXERCISES 8.4

Calculate the average and r.m.s. values of
the following functions:

(a) $f(t) = 1 + t$ across [0, 2]

(b) $f(t) = \sin 2t$ across [0, π]

(c) $f(t) = \sqrt{t + 1}$ across [0, 3]

(d) $f(t) = \sin \omega t$ across [0, π]

(e) $f(t) = e^t$ across [−1, 1]

(f) $f(t) = \cos(t/2)$ across [0, 1]

(g) $f(t) = 2 + 3t$ across [0, 2]

(h) $f(t) = 1/t$ across [1, 3]

(i) $f(t) = 1 + e^t$ across [−1, 1]

(j) $f(t) = \sin \omega t + \cos \omega t$ across [0, 1]

8.6 Orthogonal functions

Two functions $f(x)$ and $g(x)$ are orthogonal over the interval $[a, b]$ if:

$$\int_a^b f(x)g(x)\,dx = 0$$

Example 8.34

Show that $f(x) = x$ and $g(x) = x - 1$ are orthogonal on [0, 3/2].

Solution

$$\int_0^{3/2} x(x - 1)\,dx = \left[\frac{x^3}{3} - \frac{x^2}{2}\right]_0^{3/2} = \frac{9}{8} - \frac{9}{8} = 0$$

Hence f and g are orthogonal over the interval [0, 3/2]. ▲

Clearly functions may be orthogonal over one interval but not orthogonal over
others. For example,

$$\int_0^1 x(x - 1)\,dx \neq 0$$

and so x and $x - 1$ are not orthogonal over [0, 1].

Example 8.35

Show $f(t) = 1$, $g(t) = \sin t$ and $h(t) = \cos t$ are mutually orthogonal over $[-\pi, \pi]$.

Solution

We are required to show that any pair of functions is orthogonal over $[-\pi, \pi]$.

$$\int_{-\pi}^{\pi} 1 \sin t\,dt = [-\cos t]_{-\pi}^{\pi} = -\cos \pi + \cos(-\pi)$$

$$= -(-1) + (-1) = 0$$

$$\int_{-\pi}^{\pi} 1 \cos t\,dt = [\sin t]_{-\pi}^{\pi} = \sin \pi - \sin(-\pi) = 0$$

Using the trigonometric identity $\sin 2A = 2 \sin A \cos A$, we can write:

$$\int_{-\pi}^{\pi} \sin t \cos t \, dt = \int_{-\pi}^{\pi} \frac{1}{2} \sin(2t) \, dt = -\left[\frac{\cos(2t)}{4}\right]_{-\pi}^{\pi}$$

$$= -\frac{\cos(2\pi) - \cos(-2\pi)}{4} = 0$$

Hence the functions 1, $\sin t$, $\cos t$ form an orthogonal set over $[-\pi, \pi]$. ▲

The set of Example 8.35 may be extended to:

$$\{1, \sin t, \cos t, \sin(2t), \cos(2t), \sin(3t), \cos(3t), \ldots, \sin(nt), \cos(nt)\} \qquad n \in \mathbb{N}.$$

Example 8.36

Verify that $\{1, \sin t, \cos t, \sin(2t), \cos(2t), \ldots\}$ forms an orthogonal set over $[-\pi, \pi]$.

Solution

Suppose $n, m \in \mathbb{N}$. We must show that all combinations of 1, $\sin nt$, $\sin mt$, $\cos nt$ and $\cos mt$ are orthogonal.

$$\int_{-\pi}^{\pi} 1 \sin(nt) \, dt = \left[\frac{-\cos(nt)}{n}\right]_{-\pi}^{\pi} = \frac{-\cos(n\pi) + \cos(-n\pi)}{n} = 0$$

In a similar manner, it is easy to show:

$$\int_{-\pi}^{\pi} 1 \cos(nt) \, dt = 0$$

Also, using the trigonometric identities in Section 1.4.6:

$$\int_{-\pi}^{\pi} \cos(nt) \sin(mt) \, dt = \frac{1}{2} \int_{-\pi}^{\pi} \sin(n + m)t - \sin(n - m)t \, dt$$

We have seen that $\int_{-\pi}^{\pi} \sin nt \, dt = 0$ for any $n \in \mathbb{N}$. Noting that $(n + m) \in \mathbb{N}$ and $(n - m) \in \mathbb{N}$, we see that:

$$\int_{-\pi}^{\pi} \sin(n + m)t - \sin(n - m)t \, dt = 0$$

It is left as an exercise for the reader to show that:

$$\int_{-\pi}^{\pi} \sin nt \sin mt \, dt = 0 \qquad n \neq m$$

$$\int_{-\pi}^{\pi} \cos nt \cos mt \, dt = 0 \qquad n \neq m$$

The functions thus form an orthogonal set across $[-\pi, \pi]$. ▲

The result can be extended to the set:

$$\left\{ 1, \sin\left(\frac{2\pi t}{T}\right), \cos\left(\frac{2\pi t}{T}\right), \sin\left(\frac{4\pi t}{T}\right), \cos\left(\frac{4\pi t}{T}\right), \ldots \right\}$$

which is orthogonal over $[-T/2, T/2]$. This set is important in Fourier analysis (see Chapter 13).

To complete the groundwork for Fourier analysis we include the following example.

Example 8.37

Find

(a) $\int_{-\pi}^{\pi} \sin^2(nt)\, dt \qquad n \in \mathbb{Z} \qquad n \neq 0$

(b) $\int_{-\pi}^{\pi} \cos^2(nt)\, dt \qquad n \in \mathbb{Z} \qquad n \neq 0$

Solutions

(a) We use the trigonometric identity

$$\sin^2 nt = \frac{1 - \cos 2nt}{2}$$

to get

$$I = \int_{-\pi}^{\pi} \frac{1 - \cos 2nt}{2}\, dt = \int_{-\pi}^{\pi} \frac{1}{2}\, dt = \pi$$

using the orthogonal properties of $\cos(nt)$

(b)

$$\int_{-\pi}^{\pi} \cos^2(nt)\, dt = \int_{-\pi}^{\pi} 1 - \sin^2(nt)\, dt$$

$$= 2\pi - \pi = \pi \quad \blacktriangle$$

It is possible to extend this to show:

$$\int_{-T/2}^{T/2} \sin^2\left(\frac{2n\pi t}{T}\right) dt = \int_{-T/2}^{T/2} \cos^2\left(\frac{2n\pi t}{T}\right) dt = \frac{T}{2} \qquad n \in \mathbb{Z} \qquad n \neq 0$$

8.7 Improper integrals

There are two cases when evaluation of an integral needs special care:

(1) one, or both, of the limits of an integral are infinite;

(2) the integrand becomes infinite at one, or more, points of the interval of integration.

If either (1) or (2) is true the integral is called an **improper integral**. Evaluation of improper integrals involves the use of limits.

Example 8.38

Evaluate $\int_2^\infty 1/t^2 \, dt$.

Solution

$$\int_2^\infty \frac{1}{t^2} \, dt = \left[-\frac{1}{t} \right]_2^\infty$$

To evaluate $-1/t$ at the upper limit we consider $\lim_{t \to \infty} (-1/t)$. Clearly the limit is 0. Hence,

$$\int_2^\infty \frac{1}{t^2} \, dt = 0 - \left(-\frac{1}{2} \right) = \frac{1}{2} \quad \blacktriangle$$

Example 8.39

Evaluate $\int_{-\infty}^1 e^{2x} \, dx$.

Solution

$$\int_{-\infty}^1 e^{2x} \, dx = \left[\frac{e^{2x}}{2} \right]_{-\infty}^1$$

We need to evaluate $\lim_{x \to -\infty} (e^{2x}/2)$. This limit is 0. So,

$$\int_{-\infty}^1 e^{2x} \, dx = \left[\frac{e^{2x}}{2} \right]_{-\infty}^1 = \frac{e^2}{2} - 0 = 3.69 \quad \blacktriangle$$

Example 8.40   Capacitors in series

Derive an expression for the equivalent capacitance of two capacitors connected together in series (see Figure 8.17).

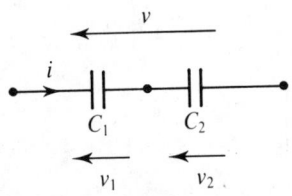

Figure 8.17   Two capacitors connected in series.

Solution

In Example 8.6 we obtained an expression for the voltage across a capacitor. This was,

$$v = \frac{1}{C} \int i \, dt$$

This can be written as a definite integral to give the voltage expression across the capacitor at a general point in time, t. The expression is:

$$v = \frac{1}{C} \int_{-\infty}^{t} i \, dt$$

Now consider the situation depicted in Figure 8.17. Writing an equation for each of the capacitors gives:

$$v_1 = \frac{1}{C_1} \int_{-\infty}^{t} i \, dt \qquad v_2 = \frac{1}{C_2} \int_{-\infty}^{t} i \, dt$$

By Kirchhoff's voltage law, $v = v_1 + v_2$ and so:

$$v = \frac{1}{C_1} \int_{-\infty}^{t} i \, dt + \frac{1}{C_2} \int_{-\infty}^{t} i \, dt = \left(\frac{1}{C_1} + \frac{1}{C_2} \right) \int_{-\infty}^{t} i \, dt$$

Therefore, the two capacitors can be replaced by an equivalent capacitance, C, given by:

$$\frac{1}{C} = \frac{1}{C_1} + \frac{1}{C_2} = \frac{C_1 + C_2}{C_1 C_2}$$

$$C = \frac{C_1 C_2}{C_1 + C_2} \quad \blacktriangle$$

Example 8.41

Evaluate $\int_3^{\infty} 2/(2t + 1) - 1/t \, dt$.

Solution

$$\int_3^{\infty} \frac{2}{2t + 1} - \frac{1}{t} \, dt = [\ln |2t + 1| - \ln |t|]_3^{\infty}$$

$$= \left[\ln \left| \frac{2t + 1}{t} \right| \right]_3^{\infty} = \left[\ln \left| 2 + \frac{1}{t} \right| \right]_3^{\infty}$$

$$= \lim_{t \to \infty} \left[\ln \left(2 + \frac{1}{t} \right) \right] - \ln \frac{7}{3}$$

$$= \ln 2 - \ln \frac{7}{3} = \ln \frac{6}{7} = -0.1542 \quad \blacktriangle$$

Example 8.42

Evaluate $\int_1^\infty \sin t \, dt$.

Solution

$$\int_1^\infty \sin t \, dt = [-\cos t]_1^\infty$$

Now $\lim_{t \to \infty}(-\cos t)$ does not exist, that is, the function, $\cos t$ does not approach a limit as $t \to \infty$, and so the integral cannot be evaluated. We say the integral **diverges.** ▲

Example 8.43

Evaluate $\int_0^1 1/\sqrt{x} \, dx$.

Solution

The integrand, $1/\sqrt{x}$, becomes infinite when $x = 0$, which is in the interval of integration. The point $x = 0$ is 'removed' from the interval. We consider $\int_b^1 1/\sqrt{x} \, dx$ where b is slightly greater than 0, and then let $b \to 0^+$. Now,

$$\int_b^1 \frac{1}{\sqrt{x}} \, dx = [2\sqrt{x}]_b^1 = 2 - 2\sqrt{b}$$

Then,

$$\int_0^1 \frac{1}{\sqrt{x}} \, dx = \lim_{b \to 0^+} \int_b^1 \frac{1}{\sqrt{x}} \, dx = \lim_{b \to 0^+} (2 - 2\sqrt{b}) = 2$$

The improper integral exists and has value 2. ▲

Example 8.44

Determine whether the integral $\int_0^2 1/x \, dx$ exists or not.

Solution

As in Example 8.43 the integrand is not defined at $x = 0$, so we consider $\int_b^2 1/x \, dx$ for $b > 0$ and then let $b \to 0^+$.

$$\int_b^2 \frac{1}{x} \, dx = [\ln |x|]_b^2 = \ln 2 - \ln b$$

So,

$$\lim_{b \to 0} \left(\int_b^2 \frac{1}{x} \, dx \right) = \lim_{b \to 0} (\ln 2 - \ln b)$$

Since $\lim_{b \to 0} \ln b$ does not exist the integral diverges. ▲

Example 8.45

Evaluate $\int_{-1}^{2} 1/x\, dx$ if possible.

Solution

We 'remove' the point $x = 0$ where the integrand becomes infinite and consider two integrals: $\int_{-1}^{b} 1/x\, dx$ where b is slightly smaller than 0, and $\int_{c}^{2} 1/x\, dx$ where c is slightly larger than 0. If these integrals exist as $b \to 0^-$ and $c \to 0^+$ then $\int_{-1}^{2} 1/x\, dx$ converges. If either of the integrals fails to converge then $\int_{-1}^{2} 1/x\, dx$ diverges. Now,

$$\int_{-1}^{b} \frac{1}{x}\, dx = \ln|b| - \ln|-1| = \ln|b|$$

$$\lim_{b \to 0^-} \int_{-1}^{b} \frac{1}{x}\, dx = \lim_{b \to 0^-} (\ln|b|)$$

This limit fails to exist and so $\int_{-1}^{2} 1/x\, dx$ diverges.   ▲

Example 8.46   Energy stored in a capacitor

Consider the circuit in Figure 8.18.

A capacitor has a voltage, V, across it as a result of stored charge. We wish to calculate the amount of energy stored in the capacitor. The switch is closed at $t = 0$ and a current, i, flows in the circuit. We have already seen in Chapter 1 that for such a case the time varying voltage across the capacitor decays exponentially and is given by,

$$v = Ve^{-t/RC}$$

So, using Ohm's law

$$i = \frac{v}{R} = \frac{Ve^{-t/RC}}{R}$$

Now the effect of closing the switch is to allow the energy stored in the capacitor to be dissipated in the resistor. Therefore, if the total energy dissipated in the resistor is calculated then this will allow the energy stored in the capacitor to be obtained. However, the energy dissipation rate, that is,

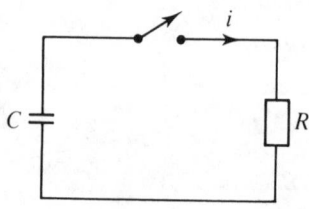

Figure 8.18 The capacitor is discharged by closing the switch.

power dissipated, is not a constant for the resistor but depends on the current flowing through it. The total energy dissipated, E, is given by:

$$E = \int_0^\infty P(t)\, dt$$

where $P(t)$ is the power dissipated in the resistor at time t. This equation has been discussed in Example 8.14. Now,

$$P = i^2R = \frac{RV^2 e^{-2t/RC}}{R^2} = \frac{V^2 e^{-2t/RC}}{R}$$

$$E = \int_0^\infty \frac{V^2 e^{-2t/RC}}{R}\, dt = \frac{V^2}{R} \int_0^\infty e^{-2t/RC}\, dt$$

$$= \frac{V^2 RC}{-2R} [e^{-2t/RC}]_0^\infty$$

Now

$$\lim_{t \to \infty} e^{-2t/RC} = 0$$

and so the energy stored in the capacitor is given by:

$$E = \frac{CV^2}{2} \quad \blacktriangle$$

Example 8.47

Find

$$I = \int_0^\infty e^{-st} \sin t\, dt \qquad \text{for } s > 0$$

Solution

Using integration by parts, we have:

$$I = \int_0^\infty e^{-st} \sin t\, dt = [-e^{-st} \cos t]_0^\infty - s \int_0^\infty e^{-st} \cos t\, dt$$

Since $s > 0$ then $\lim_{t \to \infty} e^{-st} \cos t = 0$. Note that if $s < 0$ then $\lim_{t \to \infty} e^{-st} \cos t$ does not exist and the integral would diverge.

$$I = 1 - s \int_0^\infty e^{-st} \cos t\, dt$$

Integrating by parts again we find:

$$I = 1 - s\left(\left[e^{-st} \sin t \right]_0^\infty + s \int_0^\infty e^{-st} \sin t\, dt \right)$$

$$= 1 - s^2 \int_0^\infty e^{-st} \sin t\, dt$$

$$= 1 - s^2 I$$

Hence,

$$I(1 + s^2) = 1$$

$$I = \frac{1}{1 + s^2} \quad \blacktriangle$$

Example 8.48   Integral properties of the delta function

The delta function, $\delta(t - d)$, was introduced in Chapter 1. The function is defined to be a rectangle whose area is 1, in the limit as the base length tends to 0 and as the height tends to infinity. Sometimes we need to integrate the delta function. In particular, we consider the improper integral

$$I_1 = \int_{-\infty}^{\infty} \delta(t - d) \, dt$$

The integral gives the area under the function and this is defined to be 1. Hence,

$$\int_{-\infty}^{\infty} \delta(t - d) \, dt = 1$$

In Chapter 13 we need to consider the improper integral

$$I_2 = \int_{-\infty}^{\infty} f(t) \delta(t - d) \, dt$$

where $f(t)$ is some known function of time. The delta function, $\delta(t - d)$ is zero everywhere except at $t = d$. When $t = d$, then $f(t)$ has a value $f(d)$. Hence,

$$I_2 = \int_{-\infty}^{\infty} f(t) \delta(t - d) \, dt = \int_{-\infty}^{\infty} f(d) \delta(t - d) \, dt = f(d) \int_{-\infty}^{\infty} \delta(t - d) \, dt$$

since $f(d)$ is a constant. But $\int_{-\infty}^{\infty} \delta(t - d) \, dt = 1$ and hence,

$$I_2 = f(d) \quad \blacktriangle$$

$\blacksquare$ $\displaystyle\int_{-\infty}^{\infty} \delta(t - d) \, dt = 1$

$\displaystyle\int_{-\infty}^{\infty} f(t) \delta(t - d) \, dt = f(d)$

EXERCISES 8.5

1. Evaluate, if possible:

(a) $\int_0^{\infty} e^{-t} \, dt$

(b) $\int_0^{\infty} e^{-kt} \, dt$ k is a constant

(c) $\int_1^{\infty} 1/x \, dx$

(d) $\int_1^{\infty} 1/x^2 \, dx$

(e) $\int_1^3 1/(x - 2) \, dx$

2. Evaluate:

(a) $\int_{-\infty}^{\infty} e^t \delta(t) \, dt$

(b) $\int_{-\infty}^{\infty} e^t \delta(t-4) \, dt$

(c) $\int_{-\infty}^{\infty} e^t \delta(t+3) \, dt$

(d) $4 \int_{-\infty}^{\infty} t^2 \delta(t-3) \, dt$

(e) $\int_{-\infty}^{\infty} \dfrac{(1+t)\delta(t-1)}{2} \, dt$

(f) $\int_{-\infty}^{\infty} e^{-kt} \delta(t) \, dt$

(g) $\int_{-\infty}^{\infty} e^{-kt} \delta(t-a) \, dt$

(h) $\int_{-\infty}^{\infty} e^{-k(t-a)} \delta(t-a) \, dt$

8.8 Integration of piecewise continuous functions

Integration of piecewise continuous functions is illustrated in Example 8.49. If a discontinuity occurs within the limits of integration then the interval is divided into sub-intervals so that the integrand is continuous on each sub-interval.

Example 8.49

Given

$$f(t) = \begin{cases} 2 & 0 \leq t < 1 \\ t^2 & 1 < t \leq 3 \end{cases}$$

evaluate $\int_0^3 f(t) \, dt$.

Solution

The function, $f(t)$, is piecewise continuous with a discontinuity at $t = 1$. The function is illustrated in Figure 8.19. The discontinuity occurs within the limits of integration. We split the interval of integration at the discontinuity thus:

$$\int_0^3 f(t) \, dt = \int_0^1 f(t) \, dt + \int_1^3 f(t) \, dt$$

On the intervals $(0, 1)$ and $(1, 3)$, $f(t)$ is continuous, so

$$\int_0^3 f \, dt = \int_0^1 2 \, dt + \int_1^3 t^2 \, dt$$

$$= \left[2t \right]_0^1 + \left[\frac{t^3}{3} \right]_1^3 = 2 + \left[9 - \frac{1}{3} \right] = \frac{32}{3} \quad \blacktriangle$$

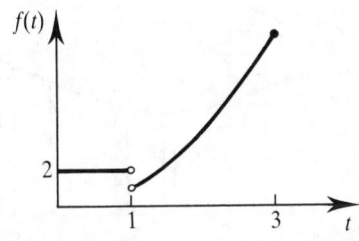

Figure 8.19 The function $f(t) = \begin{cases} 2 & 0 \leq t < 1 \\ t^2 & 1 < t \leq 3. \end{cases}$

EXERCISES 8.6

1. Given

$$f(t) = \begin{cases} 3 & -1 \leq t < 1 \\ 2t & 1 \leq t \leq 2 \\ t^2 & 2 < t \leq 3 \end{cases}$$

evaluate:

(a) $\int_{-1}^{1} f \, dt$

(b) $\int_{-0.5}^{1.5} f \, dt$

(c) $\int_{0}^{2.5} f \, dt$

(d) $\int_{-1}^{3} f \, dt$

2. Given

$$g(t) = \begin{cases} 3t & 0 \leq t < 3 \\ 15 - 2t & 3 \leq t < 4 \\ 6 & 4 \leq t \leq 6 \end{cases}$$

evaluate:

(a) $\int_{0}^{2} g(t) \, dt$

(b) $\int_{2}^{4} g(t) \, dt$

(c) $\int_{3}^{5} g(t) \, dt$

(d) $\int_{0}^{6} g(t) \, dt$

(e) $\int_{3.5}^{4.5} g(t) \, dt$

3. Given $u(t)$ is the unit step function, evaluate:

(a) $\int_{0}^{4} u(t) \, dt$

(b) $\int_{-3}^{2} u(t) \, dt$

(c) $\int_{-2}^{4} 2u(t + 1) \, dt$

(d) $\int_{-1}^{2} t \, u(t) \, dt$

(e) $\int_{0}^{4} e^{kt} \, u(t - 3) \, dt$　　k constant

8.9 Numerical techniques

Many functions, for example, $\sin x^2$ and e^x/x cannot be integrated analytically. Integration of such functions must be performed numerically. This section outlines two simple numerical techniques – the trapezium rule and Simpson's rule. More sophisticated ones exist and there are many excellent software packages available which implement these methods.

8.9.1 *Trapezium rule*

We wish to find the area under $y(x)$, from $x = a$ to $x = b$, that is, we wish to evaluate $\int_{a}^{b} y \, dx$. The required area is divided into n strips, each of width h. Each strip is then approximated by a trapezium. A typical trapezium is shown in Figure 8.20. The area of the trapezium is $\frac{1}{2}h[y_i + y_{i+1}]$. Summing the areas of all the trapezia will yield an approximation to the total area.

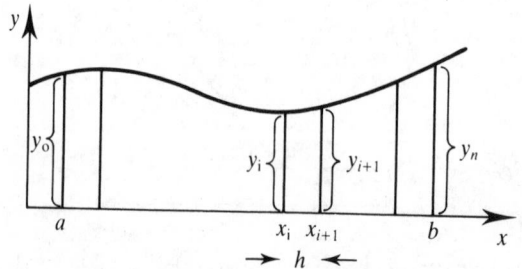

Figure 8.20　Each strip is approximated by a trapezium.

■ area of trapezia $= \dfrac{h}{2}(y_0 + y_1) + \dfrac{h}{2}(y_1 + y_2) + \cdots + \dfrac{h}{2}(y_{n-1} + y_n)$

$\phantom{■ \text{area of trapezia}} = \dfrac{h}{2}(y_0 + 2y_1 + 2y_2 + 2y_3 + \cdots + 2y_{n-1} + y_n)$

If the number of strips is increased, that is, h is decreased, then the accuracy of the approximation is increased.

Example 8.50

Use the trapezium rule to estimate $\displaystyle\int_{0.5}^{1.3} e^{x^2}\, dx$. Use a strip width of 0.1.

Solution

Using Table 8.2, we find:

sum of areas of trapezia $= \dfrac{0.1}{2}\{1.2840 + 2(1.4333) + 2(1.6323) + \cdots$

$$+ 2(4.2207) + 5.4195\}$$

$$= 2.085$$

Hence $\displaystyle\int_{0.5}^{1.3} e^{(x^2)}\, dx \approx 2.085.$ ▲

Table 8.2   Evaluating $\int_{0.5}^{1.3} e^{x^2}\, dx$ using the trapezium rule.

x	$y = e^{x^2}$
$x_0 = 0.5$	$1.2840 = y_0$
$x_1 = 0.6$	$1.4333 = y_1$
$x_2 = 0.7$	$1.6323 = y_2$
$x_3 = 0.8$	$1.8965 = y_3$
$x_4 = 0.9$	$2.2479 = y_4$
$x_5 = 1.0$	$2.7183 = y_5$
$x_6 = 1.1$	$3.3535 = y_6$
$x_7 = 1.2$	$4.2207 = y_7$
$x_8 = 1.3$	$5.4195 = y_8$

8.9.2 *Simpson's rule*

In the trapezium rule the curve $y(x)$ is approximated by a series of straight line segments. In Simpson's rule the curve is approximated by a series of quadratic curves as shown in Figure 8.21. The area is divided into an even number of strips of equal width. Consider the first pair of strips. A quadratic curve is fitted through

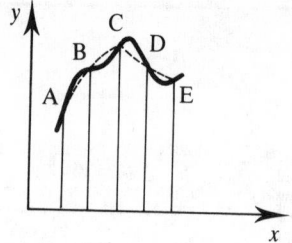

Figure 8.21 In Simpson's rule an even number of strips is used. The curve is approximated by quadratic curves.

the points A, B and C. Another quadratic curve is fitted through the points C, D and E. After some analysis an expression for approximating the area is found.

■ Simpson's rule states:

$$\text{area} \approx \frac{h}{3}(y_0 + 4y_1 + 2y_2 + 4y_3 + 2y_4 + \cdots + 2y_{n-2} + 4y_{n-1} + y_n)$$

$$= \frac{h}{3}\{y_0 + 4(y_1 + y_3 + \cdots) + 2(y_2 + y_4 + \cdots) + y_n\}$$

where n is an even number

Example 8.51

Estimate $\displaystyle\int_{0.5}^{1.3} e^{(x^2)}\, dx$ using Simpson's rule with eight strips.

Solution

The table of values is the same as in the previous example:

$$\text{approximate area} = \frac{0.1}{3}\{1.2840 + 4(1.4333 + 1.8965 + 2.7183 + 4.2207)$$

$$+ 2(1.6323 + 2.2479 + 3.3535) + 5.4195\}$$

$$= 2.075 \quad \blacktriangle$$

Example 8.52

Estimate $\int_1^2 \sqrt{1 + x^3}\, dx$ using Simpson's rule with ten strips.

Solution

With ten strips $h = 0.1$. Using Table 8.3, we find:

Table 8.3 Using Simpson's rule to estimate $\int_1^2 \sqrt{1 + x^3}\, dx$.

x	$y = \sqrt{1 + x^3}$
$x_0 = 1.0$	$1.4142 = y_0$
$x_1 = 1.1$	$1.5268 = y_1$
$x_2 = 1.2$	$1.6517 = y_2$
$x_3 = 1.3$	$1.7880 = y_3$
$x_4 = 1.4$	$1.9349 = y_4$
$x_5 = 1.5$	$2.0917 = y_5$
$x_6 = 1.6$	$2.2574 = y_6$
$x_7 = 1.7$	$2.4317 = y_7$
$x_8 = 1.8$	$2.6138 = y_8$
$x_9 = 1.9$	$2.8034 = y_9$
$x_{10} = 2.0$	$3.0000 = y_{10}$

$$\text{area} \approx \frac{0.1}{3}\{1.4142 + 4(1.5268 + 1.7880 + 2.0917 + 2.4317 + 2.8034)$$

$$+ 2(1.6517 + 1.9349 + 2.2574 + 2.6138) + 3.000\}$$

$$= 2.130 \quad \blacktriangle$$

In some cases the numerical values are not derived from a function but from actual measurements. Numerical methods can still be applied in an identical manner.

Example 8.53

Measurements of a variable, f, were made at one second intervals and are given in Table 8.4. Estimate $\int_0^6 f\, dt$ using:

(a) the trapezium rule

(b) Simpson's rule.

Table 8.4 Measurements used to estimate the integral.

t	Measurement (f)
0	4
1	4.7
2	4.9
3	5.3
4	6.0
5	5.3
6	5.9

Solutions

(a) The sum of the areas of the trapezia is:

$$\tfrac{1}{2}[4 + 2(4.7 + 4.9 + 5.3 + 6.0 + 5.3) + 5.9] = 31.2$$

(b) The area has been divided into six strips and so Simpson's rule can be applied.

$$\text{Approximate value of integral} = \tfrac{1}{3}[4 + 4(4.7 + 5.3 + 5.3)$$
$$+ 2(4.9 + 6.0) + 5.9] = 31.0. \quad \blacktriangle$$

EXERCISES 8.7

1. Estimate the following definite integrals using the trapezium rule.

   (a) $\int_0^1 \sin(t^2)\, dt$ use $h = 0.2$

   (b) $\int_1^{1.2} e^x/x\, dx$ use five strips

2. Estimate the values of the following integrals using Simpson's rule.

   (a) $\int_2^3 \ln(x^3 + 1)\, dx$ use ten strips

   (b) $\int_1^{2.6} \sqrt{t}e^t\, dt$ use eight strips

3. Evaluate, using the trapezium rule and Simpson's rule:

   (a) $\int_0^1 (x^2 + 1)^{3/2}\, dx$ use four strips

   (b) $\displaystyle\int_1^{1.6} \frac{\sin 2t}{t}\, dt$ use six strips

8.10 Integration of vectors

If a vector depends upon time t, it is often necessary to integrate it with respect to time. Recall that $\mathbf{i}$, $\mathbf{j}$ and $\mathbf{k}$ are constant vectors and are treated thus in any integration. Hence the integral,

$$\mathbf{I} = \int (f(t)\mathbf{i} + g(t)\mathbf{j} + h(t)\mathbf{k})\, dt$$

is simply evaluated as three scalar integrals, and so:

$$\mathbf{I} = \left(\int f(t)\, dt \right)\mathbf{i} + \left(\int g(t)\, dt \right)\mathbf{j} + \left(\int h(t)\, dt \right)\mathbf{k}$$

Example 8.54

If $\mathbf{r} = 3t\mathbf{i} + t^2\mathbf{j} + (1 + 2t)\mathbf{k}$, evaluate $\int_0^1 \mathbf{r}\, dt$.

Solution

$$\int_0^1 \mathbf{r}\, dt = \left(\int_0^1 3t\, dt \right)\mathbf{i} + \left(\int_0^1 t^2\, dt \right)\mathbf{j} + \left(\int_0^1 1 + 2t\, dt \right)\mathbf{k}$$

$$= \left[\frac{3t^2}{2} \right]_0^1 \mathbf{i} + \left[\frac{t^3}{3} \right]_0^1 \mathbf{j} + \left[t + t^2 \right]_0^1 \mathbf{k} = \frac{3}{2}\mathbf{i} + \frac{1}{3}\mathbf{j} + 2\mathbf{k} \quad \blacktriangle$$

EXERCISES 8.8

1. Given $\mathbf{r} = 3 \sin t\mathbf{i} - \cos t\mathbf{j} + (2 - t)\mathbf{k}$, evaluate $\int_0^\pi \mathbf{r}\, dt$.

2. Given $\mathbf{v} = \mathbf{i} - 3\mathbf{j} + \mathbf{k}$, evaluate:

   (a) $\int_0^1 \mathbf{v}\, dt$

   (b) $\int_0^2 \mathbf{v}\, dt$

3. The vector, $\mathbf{a}$, is defined by:

   $$\mathbf{a} = t^2\mathbf{i} + e^{-t}\mathbf{j} + t\mathbf{k}$$

Evaluate:

   (a) $\int_0^1 \mathbf{a}\, dt$

   (b) $\int_2^3 \mathbf{a}\, dt$

   (c) $\int_1^4 \mathbf{a}\, dt$

4. Let $\mathbf{a}$ and $\mathbf{b}$ be two three-dimensional vectors. Is the following true?

   $$\int_{t_1}^{t_2} \mathbf{a}\, dt \times \int_{t_1}^{t_2} \mathbf{b}\, dt = \int_{t_1}^{t_2} \mathbf{a} \times \mathbf{b}\, dt$$

   Recall that $\times$ denotes the vector product.

Miscellaneous exercises

1. Find $\int L/(E - iR)\, di$ where L, E and R are constants.

2. Find $\int te^{Rt/L}\, dt$ where R and L are constants.

3. If we denote by I_n the integral $\int_0^\infty x^n e^{-x}\, dx$, for $n \in \mathbb{N}^+$, so that, for example, I_3 means $\int_0^\infty x^3 e^{-x}\, dx$, show, using integration by parts, that $I_n = nI_{n-1}$. Hence, evaluate I_4. The equation, $I_n = nI_{n-1}$ is known as a **reduction formula**.

4. Find the average and root mean square values of $A \cos t + B \sin t$ across

   (a) $[0, 2\pi]$

   (b) $[0, \pi]$

5. Evaluate the following integrals:

   (a) $\int (-2t + 0.1)^4\, dt$

   (b) $\int (1 + x) \sin x\, dx$

   (c) $\int_1^2 x \sin(1 + x)\, dx$

   (d) $\int_3^6 t/\sqrt{t^2 + 1}\, dt$

   (e) $\int 1/(t^3 + 2t^2 + t)\, dt$

   (f) $\int_1^5 1/(1 + e^t)\, dt$

6. If $f(t) = \sqrt{t^2 + 1}$ find $\int_1^2 f(t)\, dt$ using

   (a) the trapezium rule with $h = 0.25$

   (b) Simpson's rule using eight strips.

7. The function $g(t)$ is piecewise continuous and defined by:

   $$g(t) = \begin{cases} 2t & 0 \leq t \leq 1 \\ 3 & 1 < t \leq 2 \end{cases}$$

   Evaluate:

   (a) $\int_0^1 g(t)\, dt$

   (b) $\int_0^{1.5} g(t)\, dt$

   (c) $\int_{0.5}^{1.7} g(t)\, dt$

   (d) $\int_1^2 g(t)\, dt$

   (e) $\int_{1.3}^{1.5} g(t)\, dt$

8. Evaluate if possible:

   (a) $\int_0^3 1/(x^2 - 4)\, dx$

   (b) $\int_0^{\pi/2} 1/(1 + t^2)\, dt$

   (c) $\int_{-\infty}^1 e^{-z}\, dz$

9. If $\mathbf{a} = t\mathbf{i} - 2t\mathbf{j} + 3t\mathbf{k}$, find $\int_0^1 \mathbf{a}\, dt$.

10. Find the total area between $f(t) = t^2 - 4$ and the t axis on the following intervals:

    (a) $[-4, -3]$

    (b) $[-3, -1]$

    (c) $[0, 3]$

    (d) $[-3, 3]$

Taylor polynomials, Taylor series and Maclaurin series

9

9.1 Introduction
9.2 Linearization using first order Taylor polynomials
9.3 Second order Taylor polynomials

9.4 Taylor polynomials of the nth order
9.5 Taylor and Maclaurin series
Miscellaneous exercises

9.1 Introduction

Often the value of a function and the values of its derivatives are known at a particular point and from this information it is desired to obtain values of the function around that point. The Taylor polynomials and Taylor series allow engineers to make such estimates. One application of this is in obtaining linearized models of non-linear systems. The great advantage of a linear model is that it is much easier to analyse than a non-linear one. It is possible to make use of the principle of superposition: this allows the effects of multiple inputs to a system to be considered separately, and the resultant output to be obtained by summing the individual outputs.

Often a system may contain only a few components that are non-linear. By linearizing these it is then possible to produce a linear model for the system. We saw an example of this when we analysed a fluid system in Section 7.8. Although electrical systems are often linear, mechanical, thermal and fluid systems, or systems containing a mixture of these, are likely to contain some non-linear components. Unfortunately it may not be possible to obtain a sufficiently accurate linear model for every non-linear system as we shall see in this chapter.

9.2 Linearization using first order Taylor polynomials

Suppose we know that y is a function of x and we know the values of y and y' when $x = a$, that is, $y(a)$ and $y'(a)$ are known. We can use $y(a)$ and $y'(a)$ to determine a linear polynomial which approximates to $y(x)$. Let this polynomial be

$$p_1(x) = c_0 + c_1 x$$

We choose the constants c_0 and c_1 so that,

$$p_1(a) = y(a)$$

$$p_1'(a) = y'(a)$$

that is, the values of p_1 and its first derivative evaluated at $x = a$ match the values of y and its first derivative evaluated at $x = a$. Then,

$$p_1(a) = y(a) = c_0 + c_1 a$$

$$p_1'(a) = y'(a) = c_1$$

Solving for c_0 and c_1 yields:

$$c_0 = y(a) - ay'(a) \qquad c_1 = y'(a)$$

Thus,

$$p_1(x) = y(a) - ay'(a) + y'(a)x$$

■ $\quad p_1(x) = y(a) + y'(a)(x - a)$

$p_1(x)$ is the **first order Taylor polynomial** generated by y at $x = a$.

Note that $p_1(x)$ and its first derivative evaluated at $x = a$ agree with $y(x)$ and its first derivative evaluated at $x = a$.

First order Taylor polynomials can also be viewed from a graphical perspective. Figure 9.1 shows the function, $y(x)$, and a tangent at Q where $x = a$. Let the equation of the tangent at $x = a$ be

$$p(x) = mx + c$$

The gradient of the tangent is, by definition, the derivative of y at $x = a$, that is, $y'(a)$. So:

$$p(x) = y'(a)x + c$$

The tangent passes through the point $(a, y(a))$, and so:

$$y(a) = y'(a)a + c$$

that is, $c = y(a) - y'(a)a$. The equation of the tangent is thus:

$$p(x) = y'(a)x + y(a) - y'(a)a$$

$$p(x) = y(a) + y'(a)(x - a)$$

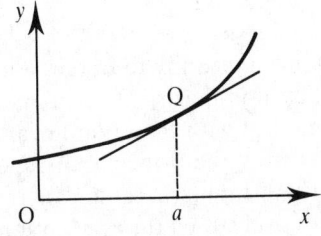

Figure 9.1 Graphical representation of a first order Taylor polynomial.

This is the first order Taylor polynomial. We see that the first order Taylor polynomial is simply the equation of the tangent to $y(x)$ where $x = a$.

Clearly, for values of x near to $x = a$ the value of $p_1(x)$ will be near to $y(x)$; $p_1(x)$ is a linear approximation to $y(x)$. In the neighbourhood of $x = a$, $p_1(x)$ closely approximates $y(x)$, but being linear is a much easier function to deal with.

Example 9.1

A function, y, and its first derivative are evaluated at $x = 2$.

$$y(2) = 1 \qquad y'(2) = 3$$

(a) State the first order Taylor polynomial generated by y at $x = 2$.
(b) Estimate $y(2.5)$.

Solutions

(a) $p_1(x) = y(2) + y'(2)(x - 2) = 1 + 3(x - 2) = -5 + 3x$

(b) We use the first order Taylor polynomial to estimate $y(2.5)$

$$p_1(2.5) = -5 + 3(2.5) = 2.5$$

Hence, $y(2.5) \approx 2.5$. ▲

Example 9.2

Find a linear approximation to $y(t) = t^2$ near $t = 3$.

Solution

We require the equation of the tangent to $y = t^2$ at $t = 3$, that is, the first order Taylor polynomial about $t = 3$.

$$p_1(t) = y(a) + y'(a)(t - a) = y(3) + y'(3)(t - 3)$$

$$= 9 + 6(t - 3)$$

$$= 6t - 9$$

At $t = 3$, $p_1(t)$ and $y(t)$ have an identical value. Near to $t = 3$, $p_1(t)$ and $y(t)$ have similar values, for example, $p_1(2.8) = 7.8$, $y(2.8) = 7.84$. ▲

9.2.1 *Linearization*

It is a frequent requirement in engineering to obtain a linear mathematical model of a system which is basically non-linear. Mathematically and computationally linear models are far easier to deal with than non-linear models. The main reason for this is that linear models obey the principle of superposition. It follows that if the application, separately, of inputs $u_1(t)$ and $u_2(t)$ to the system produces outputs $y_1(t)$ and $y_2(t)$, respectively, then the application of an input $u_1(t) + u_2(t)$ will produce an output $y_1(t) + y_2(t)$. This is only true for linear systems.

The value of this principle is that the effect of several inputs to a system can be calculated merely by adding together the effects of the individual inputs. This allows the effect of simple individual inputs to the system to be analysed and then combined to evaluate the effect of more complicated combinations of inputs. A few examples will help clarify these points.

Example 9.3 A d.c. electrical network

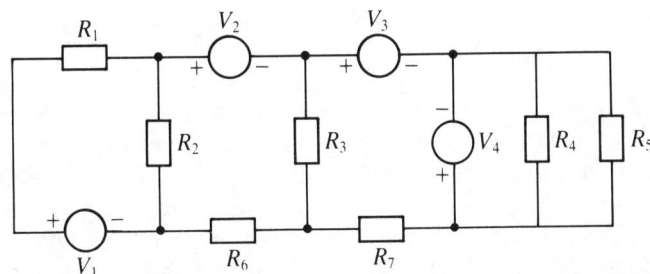

Figure 9.2   A d.c. electrical network.

Consider the d.c. network of Figure 9.2. This network is a linear system. This is because the voltage/current characteristic of a resistor is linear provided a certain voltage is not exceeded. Recall Ohm's law which is given by:

$V = IR$

where V is the voltage across the resistor, I is the current through the resistor and R is the resistance. This makes the analysis of the network relatively easy. The voltage sources, V_1, V_2, V_3, V_4, can be thought of as the inputs to the system. It is possible to analyse the effect of each of these sources separately, for example, the voltage drop across the resistor, R_5, resulting from the voltage source V_1, and then combine these effects to obtain the total effect on the system. The voltage drop across R_5 when all sources are considered would be the sum of each of the voltage drops due to the individual sources V_1, V_2, V_3 and V_4.   ▲

Example 9.4 A gravity feed water supply

Consider the water supply network of Figure 9.3. The network consists of three source reservoirs and a series of connecting pipes. Water is taken from the network at two points, S_1 and S_2. In a practical network, reservoirs are usually several kilometres away from the points at which water is taken from the network and so the effect of pressure drops along the pipes is significant. For this reason many networks require pumps to boost the pressure.

The main problem with analysing this network is that it is non-linear. This is because the relationship between pressure drop along a pipe and water flow through a pipe is not linear: a doubling of pressure does not lead to a doubling of flow. For this reason, it is not possible to use the principle

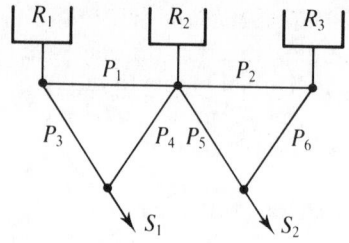

Figure 9.3 A water supply network.

of superposition when analysing the network. For example, if the effect of the pressure at S_1 for a given flow rate was calculated separately for each of the inputs to the system – the reservoirs R_1, R_2, R_3 – then the effect of all the reservoirs could not be obtained by adding the individual effects. It is not possible to obtain a linear model for this system except under very restrictive conditions and so the analysis of water networks is very complicated. ▲

Having demonstrated the value of linear models it is worth analysing how and when a non-linear system can be linearized. The first thing to note is that many systems may contain a mixture of linear and non-linear components and so it is only necessary to linearize certain parts of the system. A system of this type has been studied in Chapter 7, Example 7.19. Therefore linearization involves deciding which components of a system are non-linear, deciding whether it would be valid to linearize the components and, if so, then obtaining linearized models of the components.

Consider again Figure 9.1. Imagine it illustrates a component characteristic. The actual component characteristic is unimportant for the purposes of this discussion. For instance it could be the pressure/flow relationship of a valve or the voltage/current relationship of an electronic device. The main factor in deciding whether a valid linear model can be obtained is the range of values over which the component is required to operate. If an operating point Q were chosen and deviations from this operating point were small then it is clear from Figure 9.1 that a linear model – corresponding to the tangent to the curve at point Q – would be an appropriate model.

Obtaining a linear model is relatively straightforward. It consists of calculating the first order Taylor polynomial centred around the operating point Q. This is given by:

$$p_1(x) = y(a) + y'(a)(x - a)$$

Then $p_1(x)$ is used as the linearized model of the component with characteristic $y(x)$. It is valid provided that it is only used for values of x such that $|x - a|$ is sufficiently small. As stated, $p_1(x)$ is also the equation of the tangent to the curve at point Q. The range of values for which the model is valid depends on the curvature of the characteristic and the accuracy required.

Example 9.5 Power dissipation in a resistor

The power dissipated in a resistor varies with the current. Derive a linear model for this power variation valid for an operating point of 0.5 A. The resistor has a resistance of 10 Ω.

Solution

For a resistor

$$P = I^2 R$$

where
P = power dissipated (W)
I = current (A)
R = resistance (Ω).

The first order polynomial, valid around an operating point, $I = 0.5$, is

$$p_1(I) = P(0.5) + P'(0.5)(I - 0.5)$$

Now $P(0.5) = (0.5)^2 10 = 2.5$, $P'(0.5) = 2(0.5)(10) = 10$, and so

$$p_1(I) = 2.5 + 10(I - 0.5) = 10I - 2.5$$

It is interesting to compare this linear approximation with the true curve for values of I around the operating point. Table 9.1 shows some typical values. Notice that the linear approximation is quite good when close to the operating point but deteriorates further away. ▲

Table 9.1 A comparison of linear approximations with true values.

$I(A)$	True value of $P(W)$	Approximate value of $P(W)$
0.5	2.5	2.5
0.499	2.49001	2.49
0.501	2.51001	2.51
0.49	2.401	2.4
0.51	2.601	2.6
0.4	1.6	1.5
0.6	3.6	3.5
1.0	10	7.5

9.3 Second order Taylor polynomials

Suppose that in addition to $y(a)$ and $y'(a)$, we also have a value of $y''(a)$. With this information a second order Taylor polynomial can be found, which provides a quadratic approximation to $y(x)$. Let

$$p_2(x) = c_0 + c_1 x + c_2 x^2$$

We require

$$p_2(a) = y(a)$$
$$p_2'(a) = y'(a)$$
$$p_2''(a) = y''(a)$$

that is, the polynomial and its first two derivatives evaluated at $x = a$ match the function and its first two derivatives evaluated at $x = a$. Hence

$$p_2(a) = c_0 + c_1 a + c_2 a^2 = y(a) \tag{9.1}$$
$$p_2'(a) = c_1 + 2c_2 a = y'(a) \tag{9.2}$$
$$p_2''(a) = 2c_2 = y''(a) \tag{9.3}$$

Solving for c_0, c_1 and c_2 yields:

$$c_2 = \frac{y''(a)}{2} \quad \text{from Equation (9.3)}$$

$$c_1 = y'(a) - ay''(a) \quad \text{from Equation (9.2)}$$

$$c_0 = y(a) - ay'(a) + \frac{a^2}{2}y''(a) \quad \text{from Equation (9.1)}$$

Hence,

$$p_2(x) = y(a) - ay'(a) + \frac{a^2}{2}y''(a)$$

$$+ \{y'(a) - ay''(a)\}x + \frac{y''(a)}{2}x^2$$

Finally,

■ $$p_2(x) = y(a) + y'(a)(x - a) + y''(a)\frac{(x - a)^2}{2}$$

$p_2(x)$ is the **second order Taylor polynomial** generated by y about $x = a$

Example 9.6

Given $y(1) = 0$, $y'(1) = 1$, $y''(1) = -2$, estimate:

(a) $y(1.5)$

(b) $y(2)$

(c) $y(0.5)$

using the second order Taylor polynomial.

Solutions

The second order Taylor polynomial is $p_2(x)$.

$$p_2(x) = y(1) + y'(1)(x - 1) + y''(1)\frac{(x - 1)^2}{2}$$

$$= x - 1 - 2\frac{(x - 1)^2}{2} = x - 1 - (x - 1)^2 = -x^2 + 3x - 2$$

We use $p_2(x)$ as an approximation to $y(x)$.

(a) The value of $y(1.5)$ is approximated by $p_2(1.5)$.

$$y(1.5) \approx p_2(1.5) = 0.25$$

(b) The value of $y(2)$ is approximated by $p_2(2)$.

$$y(2) \approx p_2(2) = 0$$

(c) The value of $y(0.5)$ is approximated by $p_2(0.5)$.

$$y(0.5) \approx p_2(0.5) = -0.75 \quad \blacktriangle$$

Example 9.7 Quadratic approximation to a diode characteristic

In Chapter 7, Example 7.17 we derived a linear approximation to a diode characteristic suitable for small signal variations around an operating point. Sometimes it is not possible to use a linear approximation because the variations are too large to maintain sufficient accuracy. Even so, an approximate model may be desirable. In general, a higher order Taylor polynomial will give a more accurate model than that of a lower order polynomial. We will consider a quadratic model for a diode characteristic. Recall that for a diode at room temperature,

$$I = I(V) = I_s(e^{40V} - 1)$$

Given an operating point, V_a, the second order Taylor polynomial is:

$$p_2(V) = I(V_a) + I'(V_a)(V - V_a) + I''(V_a)\frac{(V - V_a)^2}{2}$$

Now

$$I'(V) = 40I_s e^{40V} \qquad I''(V) = 1600I_s e^{40V}$$

so,

$$p_2(V) = I_s(e^{40V_a} - 1) + 40I_s e^{40V_a}(V - V_a) + 1600I_s e^{40V_a}\frac{(V - V_a)^2}{2}$$

The coefficients only need to be calculated once. After that the calculation of a current value only involves evaluating a quadratic. $\quad \blacktriangle$

9.4 Taylor polynomials of the *n*th order

If we know y and its first n derivatives evaluated at $x = a$, that is, $y(a)$, $y'(a)$, $y''(a), \ldots, y^{(n)}(a)$, then the **$n$th order Taylor polynomial**, $p_n(x)$, may be written as:

$$■ \quad p_n(x) = y(a) + y'(a)(x - a) + y''(a)\frac{(x - a)^2}{2!} + y^{(3)}(a)\frac{(x - a)^3}{3!}$$

$$+ \cdots + y^{(n)}(a)\frac{(x - a)^n}{n!}$$

This provides an approximation to $y(x)$. The polynomial and its first n derivatives evaluated at $x = a$ match the values of $y(x)$ and its first n derivatives evaluated at $x = a$, that is,

$$p_n(a) = y(a)$$

$$p_n'(a) = y'(a)$$

$$p_n''(a) = y''(a)$$

$$\vdots$$

$$p_n^{(n)}(a) = y^{(n)}(a)$$

Example 9.8

Given $y(0) = 1$, $y'(0) = 1$, $y''(0) = 1$, $y^{(3)}(0) = -1$, $y^{(4)}(0) = 2$, obtain a fourth order Taylor polynomial generated by y about $x = 0$. Estimate $y(0.2)$.

Solution

In this example $a = 0$ and hence:

$$p_4(x) = y(0) + y'(0)x + y''(0)\frac{x^2}{2!} + y^{(3)}(0)\frac{x^3}{3!} + y^{(4)}(0)\frac{x^4}{4!}$$

$$= 1 + x + \frac{x^2}{2} - \frac{x^3}{6} + \frac{x^4}{12}$$

The Taylor polynomial can be used to estimate $y(0.2)$.

$$p_4(0.2) = 1 + 0.2 + \frac{(0.2)^2}{2} - \frac{(0.2)^3}{6} + \frac{(0.2)^4}{12} = 1.2188$$

$$y(0.2) \approx 1.2188 \quad ▲$$

Example 9.9

Given that y satisfies the equation:

$$y'' - (y')^2 + 2y = x^2 \qquad (9.4)$$

and also the conditions

$$y(0) = 1 \qquad y'(0) = 2$$

use a third order Taylor polynomial to estimate $y(0.5)$.

Solution

To write down the third order Taylor polynomial about $x = 0$ we require $y(0)$, $y'(0)$, $y''(0)$ and $y^{(3)}(0)$. From Equation (9.4),

$$y''(x) = x^2 + \{y'(x)\}^2 - 2y(x) \qquad (9.5)$$

So,

$$y''(0) = 0 + \{y'(0)\}^2 - 2y(0) = 2$$

To find $y^{(3)}(0)$, Equation (9.5) is differentiated w.r.t. x.

$$y^{(3)}(x) = 2x + 2y'(x)y''(x) - 2y'(x)$$
$$y^{(3)}(0) = 2y'(0)y''(0) - 2y'(0) = 4$$

The Taylor polynomial may now be written as:

$$p_3(x) = y(0) + y'(0)x + y''(0)\frac{x^2}{2} + y^{(3)}(0)\frac{x^3}{6}$$

$$= 1 + 2x + x^2 + \frac{2x^3}{3}$$

We use $p_3(x)$ as an approximation to $y(x)$.

$$p_3(0.5) = 1 + 1 + 0.25 + 0.0833 = 2.333$$

that is,

$$y(0.5) \approx 2.333 \quad \blacktriangle$$

9.5 Taylor and Maclaurin series

We have seen that y and its first n derivatives evaluated at $x = a$ match $p_n(x)$ and its first n derivatives evaluated at $x = a$. The Taylor polynomials have been used to estimate the values of y at various x values. It is reasonable to ask:

'How accurately do Taylor polynomials generated by y at $x = a$ approximate to y at values of x other than a?'

'If more and more terms are used in the Taylor polynomial will this produce a better and better approximation to y?'

To answer these questions we introduce the **Taylor series**. As more and more terms are included in the Taylor polynomial, we obtain an infinite series, known as a Taylor series. We denote this infinite series by $p(x)$.

■ Taylor series

$$p(x) = y(a) + y'(a)(x - a) + y''(a)\frac{(x - a)^2}{2!} + y^{(3)}(a)\frac{(x - a)^3}{3!}$$
$$+ \cdots + y^{(n)}(a)\frac{(x - a)^n}{n!} + \cdots$$

For some Taylor series, the value of the series equals the value of the generating function for every value of x. For example, the Taylor series for e^x, $\sin x$ and $\cos x$ equal the values of e^x, $\sin x$ and $\cos x$ for every value of x. However, some functions have a Taylor series which equals the function only for a limited range of x values. Example 9.11 gives a case of a function which equals its Taylor series only when $-1 < x < 1$.

A special, commonly used, case of a Taylor series occurs when $a = 0$. This is known as the **Maclaurin series**.

■ Maclaurin series

$$p(x) = y(0) + y'(0)x + y''(0)\frac{x^2}{2!} + y^{(3)}(0)\frac{x^3}{3!} + \cdots + y^{(n)}(0)\frac{x^n}{n!} + \cdots$$

Example 9.10

Determine the Maclaurin series for $y = e^x$.

Solution

In this example $y(x) = e^x$ clearly $y'(x) = e^x$ too. Similarly,

$$y''(x) = y^{(3)}(x) = \cdots = y^{(n)}(x) = e^x$$

for all values of n. Evaluating at $x = 0$, yields:

$$y(0) = y'(0) = y''(0) = \cdots = y^{(n)}(0) = 1$$

and so,

$$p(x) = 1 + x + \frac{x^2}{2!} + \frac{x^3}{3!} + \cdots + \frac{x^n}{n!} + \cdots$$

Hence,

$$e^x = 1 + x + \frac{x^2}{2!} + \frac{x^3}{3!} + \cdots$$

for all values of x, that is,

$$e^x = \sum_{n=0}^{\infty} \frac{x^n}{n!} \quad \blacktriangle$$

Example 9.11

Determine the Maclaurin series for $y(x) = 1/(1 + x)$.

Solution

The value of y and its derivatives at $x = 0$ are found.

$$y(x) = \frac{1}{1 + x} \qquad y(0) = 1$$

$$y'(x) = \frac{-1}{(1 + x)^2} \qquad y'(0) = -1$$

$$y''(x) = \frac{2!}{(1 + x)^3} \qquad y''(0) = 2!$$

$$y'''(x) = \frac{-3!}{(1 + x)^4} \qquad y'''(0) = -3!$$

$$\vdots \qquad \vdots$$

$$y^{(n)}(x) = \frac{(-1)^n n!}{(1 + x)^{n+1}} \qquad y^{(n)}(0) = (-1)^n n!$$

Hence using the formula for the Maclaurin series we find:

$$\frac{1}{1 + x} = 1 - 1(x) + 2!\frac{x^2}{2!} - 3!\frac{x^3}{3!} + \cdots (-1)^n n!\frac{x^n}{n!} + \cdots$$

$$= 1 - x + x^2 - x^3 + x^4 + \cdots + (-1)^n x^n + \cdots$$

It can be shown, using the ratio test, that this series converges for $|x| < 1$. Hence,

$$\frac{1}{1 + x} = 1 - x + x^2 - x^3 + \cdots = \sum_{0}^{\infty} (-1)^n x^n \qquad \text{for } |x| < 1$$

For values of x outside $(-1, 1)$ the values of $1/(1 + x)$ and $\Sigma(-1)^n x^n$ are simply not equal; try evaluating the left-hand side and right-hand side with, say, $x = -2$. $\quad \blacktriangle$

Example 9.12

Find the Taylor series for $y(x) = e^{-x}$ about $x = 1$.

Solution

$$y = e^{-x} \qquad y(1) = e^{-1}$$
$$y' = -e^{-x} \qquad y'(1) = -e^{-1}$$
$$y'' = e^{-x} \qquad y''(1) = e^{-1}$$

and so on. Hence,

$$e^{-x} = e^{-1} - (e^{-1})(x-1) + e^{-1}\frac{(x-1)^2}{2!} - \left\{e^{-1}\frac{(x-1)^3}{3!} + \cdots\right\}$$

$$= e^{-1}\left\{1 - (x-1) + \frac{(x-1)^2}{2!} - \frac{(x-1)^3}{3!} + \cdots\right\}$$

$$e^{-x} = e^{-1}\sum_0^\infty (-1)^n\frac{(x-1)^n}{n!} \qquad \blacktriangle$$

Example 9.13

Obtain the Maclaurin series for $\sin t$.

Solution

$$y(t) = \sin t \qquad y(0) = 0$$
$$y'(t) = \cos t \qquad y'(0) = 1$$
$$y''(t) = -\sin t \qquad y''(0) = 0$$
$$y^{(3)}(t) = -\cos t \qquad y^{(3)}(0) = -1$$
$$y^{(4)}(t) = \sin t \qquad y^{(4)}(0) = 0$$

$$\sin t = y(0) + y'(0)t + \frac{y''(0)t^2}{2!} + \frac{y^{(3)}(0)t^3}{3!} + \cdots$$

$$= t - \frac{t^3}{3!} + \frac{t^5}{5!} - \frac{t^7}{7!} + \cdots$$

$$= \sum_{i=0}^\infty \frac{(-1)^i t^{2i+1}}{(2i+1)!} \qquad \blacktriangle$$

An alternative form of Taylor series is often used in numerical analysis. We know that the Taylor series for $y(x)$ generated about $x = a$ is given by

$$y(x) = y(a) + y'(a)(x-a) + y''(a)\frac{(x-a)^2}{2!} + y^{(3)}(a)\frac{(x-a)^3}{3!} + \cdots$$

Replacing a by x_0 we obtain

$$y(x) = y(x_0) + y'(x_0)(x-x_0) + y''(x_0)\frac{(x-x_0)^2}{2!} + y^{(3)}(x_0)\frac{(x-x_0)^3}{3!} + \cdots$$

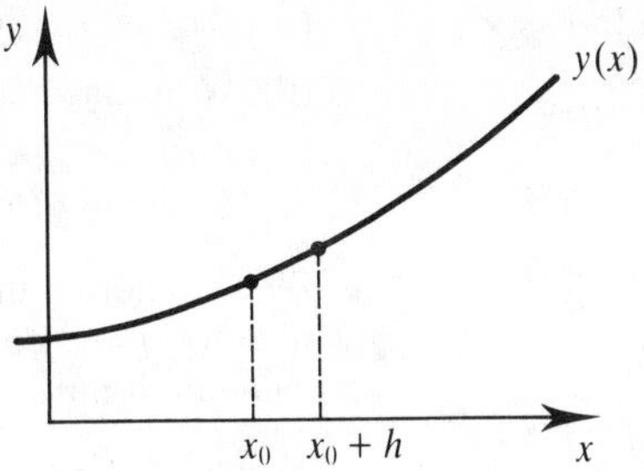

Figure 9.4 The value $y(x_0 + h)$ can be estimated using values of y and its derivatives at $x = x_0$.

If we now let $x - x_0 = h$, we see that

$$y(x_0 + h) = y(x_0) + y'(x_0)h + y''(x_0)\frac{h^2}{2!} + y^{(3)}(x_0)\frac{h^3}{3!} + \cdots$$

To interpret this form of Taylor series we refer to Figure 9.4.

If y and its derivatives are known when $x = x_0$, then the Taylor series can be used to find y at a nearby point, where $x = x_0 + h$. This form of Taylor series is used in numerical methods of solving differential equations.

EXERCISES 9.1

1. A function, $f(t)$, and its first and second derivatives are evaluated at $t = 3$. Find the second order Taylor polynomial generated by f about $t = 3$.

$$f(3) = 2 \qquad f'(3) = -1 \qquad f''(3) = 1$$

2. Given that y satisfies the equation:

$$\frac{dy}{dx} + \frac{y^2}{2} = xy \qquad y(1) = 2$$

   (a) Calculate $y'(1)$, $y''(1)$ and $y^{(3)}(1)$.

   (b) State the third order Taylor polynomial generated by $y(x)$ about $x = 1$.

   (c) Estimate $y(1.25)$.

3. Find the Maclaurin series for

   (a) $\cos t$

   (b) $\ln(1 + t)$

4. (a) Given that $y(x) = x^2$,

   (i) Calculate the Taylor series of $y(x)$ about $x = a$

   (ii) Calculate the Maclaurin series of $y(x)$

   (b) Given that $f(x) = \sin x + x^2$, find the Maclaurin series for $f(x)$.

Miscellaneous exercises

1. A function, $s(t)$, and its first derivative are evaluated at $t = 3$; $s(3) = 4$, $s'(3) = -1$.

   (a) Find the first order Taylor polynomial generated by s at $t = 3$.

   (b) Estimate $s(2.9)$ and $s(3.2)$.

   (c) If additionally we know $s''(3) = 1$, use a second order Taylor polynomial to estimate $s(2.9)$ and $s(3.2)$.

2. Find a quadratic approximation to $y = x^3$ near $x = 2$.

3. Find linear approximations to

   (a) $z(t) = e^t$ near $t = 1$

   (b) $w(t) = \sin 3t$ near $t = 1$

   (c) $v(t) = e^t + \sin 3t$ near $t = 1$

4. Find the Maclaurin expansion of $y(x) = \ln(2 + x)$ up to and including the term in x^4.

5. The function, $y(x)$, satisfies the equation

   $$y'' + xy' - 3y = x^2 + 1$$

   $$y(0) = 1 \qquad y'(0) = 2$$

   (a) Evaluate $y''(0)$

   (b) Differentiate the equation

   (c) Evaluate $y^{(3)}(0)$

   (d) Write down a cubic approximation for $y(x)$

   (e) Estimate $y(0.5)$.

6. By considering the Maclaurin expansions of $\sin(kx)$ and $\cos(kx)$, k constant, evaluate if possible:

   (a) $\lim\limits_{x \to 0} \dfrac{\sin(kx)}{x}$

   (b) $\lim\limits_{x \to 0} \dfrac{\cos(kx) - 1}{x}$

   (c) $\lim\limits_{x \to 0} \dfrac{\sin(kx)}{1 - \cos(kx)}$

7. Obtain a linear approximation to
   $$y(t) = at^4 + bt^3 + ct^2 + dt + e \text{ around } t = 1.$$

Ordinary differential equations

10

10.1 Introduction

The solution of problems concerning the motion of objects, the flow of charged particles, heat transport, etc., often involves discussion of relations of the form:

$$\frac{dx}{dt} = f(x, t) \quad \text{or} \quad \frac{dq}{dt} = g(q, t)$$

In the first equation, x might represent distance, then dx/dt is the rate of change of distance with respect to time, t, that is, speed. In the second equation, q might be charge and dq/dt the rate of flow of charge, that is, current. These are examples of **differential equations**, so called because they are equations involving the derivatives of various quantities. Such equations arise out of situations in which change is occurring. To solve such a differential equation means to find the function $x(t)$ or $q(t)$ when we are given the differential equation. Solutions to these equations may be **analytical** in that we can write down an answer in terms of common elementary functions such as e^t, $\sin t$ and so on. Alternatively, the problem may be so difficult that only **numerical methods** are available, which produce approximate solutions.

In engineering, differential equations are most commonly used to model **dynamic systems**. These are systems which change with time. Examples include an electronic circuit with time-dependent currents and voltages, a chemical production line in which pressures, tank levels, flow rates, etc., vary with time, and a semiconductor device in which hole and electron densities change with time.

10.2 Basic definitions

In order to solve a differential equation it is important to identify certain features. Recall from Chapter 1 that in a function such as $y = x^2 + 3x$ we say x is the independent variable and y is the dependent variable since the value of y depends upon the choice we have made for x. In differential equations such as:

$$\frac{dx}{dt} = f(x, t)$$

t is the independent variable and x the dependent variable. Similarly if $dy/dx = f(y, x)$, x is the independent variable and y the dependent variable. We see that the variable being differentiated is the dependent variable. Before classifying differential equations let us derive a simple one.

Example 10.1 An RC charging circuit

Consider the RC circuit of Figure 10.1. Suppose we wish to derive a differential equation which models the circuit so that we can determine the voltage across the capacitor at any time, t. Clearly there are two different cases corresponding to the switch being open and the switch being closed. We will concentrate on the latter and for convenience assume that the switch is closed at $t = 0$. This is equivalent to applying a step voltage at $t = 0$ to the RC network. From Kirchhoff's voltage law we have:

$$v_R + v_C = V_S \qquad \text{that is} \qquad v_R = V_S - v_C$$

where V_S is the voltage of the supply, v_C is the voltage across the capacitor, and v_R is the voltage across the resistor. Using Ohm's law for the resistor then gives:

$$i = \frac{V_S - v_C}{R}$$

where i is the current flowing in the circuit after the switch is closed. For the capacitor

$$i = C\frac{dv_C}{dt}$$

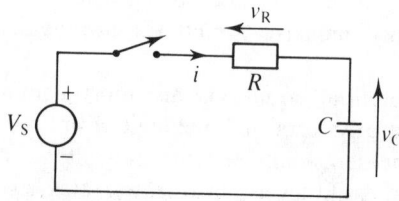

Figure 10.1   An RC charging circuit.

Combining these equations gives:

$$C\frac{dv_C}{dt} = \frac{V_S - v_C}{R}$$

and hence

$$RC\frac{dv_C}{dt} + v_C = V_S$$

This is the differential equation which models the variation in voltage across the capacitor with time. Here v_C is the dependent variable and t is the independent variable, and when we are required to solve this differential equation we must attempt to find v_C as a function of t. ▲

Differential equations which have features in common are often grouped together and given certain classifications and it is usually the case that appropriate methods of solution depend upon the classifications. Some important terminology is now given.

10.2.1 *Order*

The **order** of a differential equation is the order of its highest derivative.

Example 10.2

State the order of:

(a) $\dfrac{d^2y}{dx^2} + \dfrac{dy}{dx} = x$

(b) $\dfrac{dx}{dt} = (xt)^5$

Solutions

(a) The highest derivative is d^2y/dx^2 – a second derivative. The order is therefore two.

(b) The only derivative appearing is dx/dt – a first derivative. The order is therefore one. ▲

10.2.2 *Degree*

The **degree** of a differential equation is the degree to which the highest derivative is raised.

Example 10.3

Find the degree of the following equations.

(a) $\left(\dfrac{dy}{dx}\right)^3 + y = \sin x$

(b) $\dfrac{dx}{dt} + x^2 = 1$

(c) $\left(\dfrac{d^2y}{dx^2}\right)^2 + \left(\dfrac{dy}{dx}\right)^3 = 3$

Solutions

(a) The highest derivative is dy/dx. This term is raised to the power 3. Therefore the degree of the equation is three.

(b) The highest derivative is dx/dt. This term is raised to the power 1 and so the degree of the equation is one.

(c) The highest derivative is d^2y/dx^2. This term is raised to the power 2 and so the degree of the equation is two. ▲

10.2.3 *Linear*

A differential equation is linear if the dependent variable and all its derivatives occur only to the first power, that is, occur linearly. If an equation is not linear then it is said to be non-linear. A product of terms involving the dependent variable, for example, $y \, dy/dx$, is non-linear.

Example 10.4

Decide whether or not the following equations are linear.

(a) $\sin x \dfrac{dy}{dx} + y = x$

(b) $\dfrac{dx}{dt} + x = t^3$

(c) $\dfrac{d^2y}{dx^2} + y^2 = 0$

(d) $\dfrac{dy}{dx} + \sin y = 0$

Solutions

In (a), (c) and (d) the dependent variable is y, and the independent variable is x. In (b) the dependent variable is x and the independent variable is t.

(a) This equation is linear.

(b) This equation is linear. It does not matter that the term in t, the independent variable, is raised to the power 3.

(c) This equation is non-linear, the non-linearity arising through the term y^2.

(d) This equation is non-linear, the non-linearity arising through the term $\sin y$. ▲

10.2.4 *The solution of a differential equation*

The **solution** of a differential equation is a relationship between the dependent and independent variables such that the differential equation is satisfied for all values of the independent variable over a specified domain.

Example 10.5

Verify that $y = e^x$ is a solution of the differential equation

$$\frac{dy}{dx} = y$$

Solution

If $y = e^x$ then $dy/dx = e^x$. For all values of x, we see that $dy/dx = y$ and so $y = e^x$ is a solution. Note also that this equation is first order, first degree and linear. ▲

There are frequently many different functions which satisfy a differential equation, that is, there are many solutions. The **general solution** embraces all of these and all possible solutions can be obtained from it.

Example 10.6

Verify that $y = Ce^x$ is a solution of $dy/dx = y$, where C is any constant.

Solution

If $y = Ce^x$, then $dy/dx = Ce^x$. Therefore, for all values of x, $dy/dx = y$ and the equation is satisfied for any constant C; C is called an **arbitrary constant** and by varying it, all possible solutions can be obtained. For example, by choosing C to be 1, we obtain the solution of the previous example. In fact, $y = Ce^x$ is the general solution of $dy/dx = y$. ▲

More generally, to determine C we require more information given in the form of a **condition**. For example, if we are told that at $x = 0$, $y = 4$ then from $y = Ce^x$ we have:

$$4 = Ce^0 = C$$

so that $C = 4$. Therefore $y = 4e^x$ is the solution of the differential equation which additionally satisfies the condition $y(0) = 4$. This is called a **particular solution**. In general, application of conditions to the general solution yields the particular solution. Later we shall see that some general solutions contain more than one arbitrary constant. To obtain a particular solution, the number of given independent conditions must be the same as the number of constants.

EXERCISES 10.1

1. Verify that $y = 3 \sin 2x$ is a solution of $d^2y/dx^2 + 4y = 0$.

2. Verify that $3e^x$, Axe^x, $Axe^x + Be^x$, where A, B are constants, all satisfy the differential equation:

$$\frac{d^2y}{dx^2} - 2\frac{dy}{dx} + y = 0$$

3. Verify that $x = t^2 + A \ln t + B$ is a solution of:

$$t\frac{d^2x}{dt^2} + \frac{dx}{dt} = 4t$$

4. Verify that $y = A \cos x + B \sin x$ satisfies the differential equation:

$$\frac{d^2y}{dx^2} + y = 0$$

Verify also that $y = A \cos x$ and $y = B \sin x$ each individually satisfy the equation.

5. If $y = Ae^{2x}$ is the general solution of $dy/dx = 2y$, find the particular solution satisfying $y(0) = 3$. What is the particular solution satisfying $dy/dx = 2$ when $x = 0$?

6. Identify the dependent and independent variables of the following differential equations. Give the order and degree of the equations and state which are linear.

(a) $\dfrac{dy}{dx} + 9y = 0$

(b) $\left(\dfrac{dy}{dx}\right)\left(\dfrac{d^2y}{dx^2}\right) + 3\dfrac{dy}{dx} = 0$

(c) $\dfrac{d^3x}{dt^3} + 5\dfrac{dx}{dt} = \sin x$

7. Show that $x(t) = 7 \cos 3t - 2 \sin 2t$ is a solution of:

$$\frac{d^2x}{dt^2} + 2x = -49 \cos 3t + 4 \sin 2t$$

10.3 Initial and boundary conditions

We have seen that conditions are required to obtain values for the arbitrary constants appearing in the general solution of a differential equation. These conditions can be given in different ways.

10.3.1 *One-point conditions*

By way of example, consider the second order linear differential equation:

$$\frac{d^2y}{dx^2} + y = 0$$

the general solution of which is $y = A \cos x + B \sin x$. Because there are two arbitrary constants, two conditions are necessary to obtain a particular solution. We may be given two conditions at a single point. For example, when $x = 0$, $y = 0$ and $dy/dx = 1$. Application of the first condition to the general solution gives:

$$0 = A \cos 0 + B \sin 0 = A$$

Therefore $A = 0$. This condition fixes A and so the solution becomes simply $y = B \sin x$. To apply the second condition, $dy/dx = 1$ when $x = 0$, we must differentiate y.

$$\frac{dy}{dx} = B \cos x$$

Then, applying the second condition, we get:

$$1 = B \cos 0$$

$$= B$$

so that $B = 1$. Finally, the particular solution is $y = \sin x$.

If we seek a solution over the interval $0 \leq x \leq 1$ say, then the two conditions given at the point $x = 0$ are termed **initial conditions** for obvious reasons.

10.3.2 *Two-point conditions*

Suppose we seek the particular solution of:

$$\frac{d^2 y}{dx^2} + y = 0$$

satisfying $y = 0$ when $x = 0$ and $y = 0$ when $x = \pi$. Applying the first condition to the general solution $y = A \cos x + B \sin x$ gives:

$$0 = A \cos 0 + B \sin 0 = A$$

Therefore $A = 0$, and the solution becomes $y = B \sin x$. Applying the second condition $y = 0$ when $x = \pi$, we find:

$$0 = B \sin \pi$$

However, $\sin \pi = 0$ which means that the second condition is satisfied whatever the value of B, and the solution satisfying the two given conditions is:

$$y = B \sin x$$

for any B. We note that this solution is not unique.

If the solution is sought over the interval $0 \leq x \leq \pi$, say, and the conditions are given at both ends of the interval they are termed **boundary conditions**. We use the term **initial value problem** for the problem of solving a differential equation subject to initial conditions, and the term **boundary value problem** when the conditions imposed are boundary conditions.

Example 10.7

Given that the general solution of:

$$\frac{d^2y}{dx^2} + y = 0$$

is

$$y = A \cos x + B \sin x$$

find the particular solution which satisfies the boundary conditions $y = 0$ when $x = 0$ and $y = 3$ when $x = \pi/2$.

Solution

Applying the first condition gives:

$$0 = A \cos 0 + B \sin 0 = A$$

and hence $A = 0$. Applying the second condition gives:

$$3 = B \sin \frac{\pi}{2} = B$$

and hence $B = 3$. Therefore the required particular solution is

$$y = 3 \sin x \quad \blacktriangle$$

EXERCISES 10.2

1. The general solution of the equation $d^2x/dt^2 - 3\,dx/dt + 2x = 0$ is given by

   $$x = Ae^t + Be^{2t}$$

   Find the particular solution which satisfies $x = 3$ and $dx/dt = 5$ when $t = 0$.

2. The general solution of $d^2y/dx^2 - 2dy/dx + y = 0$ is $y = Axe^x + Be^x$. Find the particular

   solution satisfying $y(0) = 0$, $\dfrac{dy}{dx}(0) = 1$.

3. The general solution of $d^2x/dt^2 = -\omega^2 x$ is $x = Ae^{j\omega t} + Be^{-j\omega t}$, where $j^2 = -1$. Verify that this is indeed a solution. What is the particular solution satisfying $x(0) = 0$, $\dfrac{dx}{dt}(0) = 1$? Express the general solution and the particular solution in terms of trigonometric functions.

10.4 First order equations – separation of variables

The simplest first order equations to deal with are those of the form:

$$\frac{dy}{dx} = f(x)$$

where the right-hand side is a function of the independent variable only. No special treatment is necessary and direct integration yields y as a function of x, that is,

$$y = \int f(x)\,dx$$

Example 10.8

Find the general solution of $dy/dx = 3\cos 2x$.

Solution

Given that $dy/dx = 3\cos 2x$, then $y = \int 3\cos 2x\,dx = \frac{3}{2}\sin 2x + C$. This is the required general solution. ▲

When the function f on the right-hand side of the equation depends upon both independent and dependent variables this approach is not possible. However, first order equations which can be written in the form:

$$\frac{dy}{dx} = f(x)g(y) \tag{10.1}$$

form an important class known as **separable equations**. To obtain a solution we first divide both sides of Equation (10.1) by $g(y)$ to give:

$$\frac{1}{g(y)}\frac{dy}{dx} = f(x)$$

Integrating both sides with respect to x yields

$$\int \frac{1}{g(y)}\frac{dy}{dx}\,dx = \int \frac{1}{g(y)}\,dy = \int f(x)\,dx$$

The equation is then said to be separated. If the last two integrals can be found, we obtain a relationship between y and x, although it is not always possible to write y explicitly in terms of x as the following examples will show.

Example 10.9

Solve $dy/dx = e^{-x}/y$.

Solution

Here $f(x) = e^{-x}$ and $g(y) = 1/y$. Multiplication through by y yields:

$$y\frac{dy}{dx} = e^{-x}$$

Integration of both sides with respect to x gives:

$$\int y\,dy = \int e^{-x}\,dx$$

so that

$$\frac{y^2}{2} = -e^{-x} + C$$

Note that the constants arising from the two integrals have been combined to give a single constant, C. Finally we can rearrange this expression to give y in terms of x:

$$y^2 = -2e^{-x} + 2C$$

that is,

$$y = \pm\sqrt{D - 2e^{-x}}$$

where $D = 2C$. ▲

It is important to stress that the constant of integration must be inserted at the stage at which the integration is actually carried out, and not simply added to the answer at the end.

Example 10.10

Solve $dy/dx = 3x^2 e^{-y}$ subject to $y(0) = 1$.

Solution

Here $g(y) = e^{-y}$ and $f(x) = 3x^2$. Separating the variables and integrating we find:

$$\int e^y \, dy = \int 3x^2 \, dx$$

so that $e^y = x^3 + C$. Imposing the initial condition $y(0) = 1$ we find:

$$e^1 = (0)^3 + C$$

so that $C = e$. Therefore,

$$e^y = x^3 + e$$

Note that since the exponential function is always positive, the solution will be valid only for $x^3 + e > 0$. Taking natural logarithms gives the particular solution explicitly:

$$y = \ln(x^3 + e)$$ ▲

Example 10.11

Solve

$$\frac{dx}{dt} = \frac{t^2 + 1}{x^2 + 1}$$

Solution

Separating the variables and integrating we find:

$$\int (x^2 + 1)\,dx = \int (t^2 + 1)\,dt$$

Therefore,

$$\frac{x^3}{3} + x = \frac{t^3}{3} + t + C$$

which is the general solution. Here we note that x has not been obtained explicitly in terms of t, although we have found a relationship between x and t which satisfies the differential equation. To obtain the value of x at any given t it would be necessary to solve the cubic equation. ▲

Sometimes, equations which are not immediately separable can be reduced to separable form by an appropriate substitution as the following example shows.

Example 10.12

By means of the substitution $z = y/x$, solve the equation

$$\frac{dy}{dx} = \frac{y^2}{x^2} + \frac{y}{x} + 1 \tag{10.2}$$

Solution

If $z = y/x$ then $y = zx$. Because the solution, y, is a function of x the variable z depends upon x also. The product rule gives $dy/dx = z + x\,dz/dx$, so that Equation (10.2) becomes:

$$z + x\frac{dz}{dx} = z^2 + z + 1$$

that is,

$$x\frac{dz}{dx} = z^2 + 1$$

This new equation has independent variable x and dependent variable z, and is separable. We find:

$$\int \frac{dz}{z^2 + 1} = \int \frac{dx}{x}$$

so that $\tan^{-1} z = \ln|x| + C$. Writing $C = \ln|D|$ we have

$$\tan^{-1} z = \ln|x| + \ln|D| = \ln|Dx|$$

so that $z = \tan(\ln|Dx|)$. Returning to the original variables we see that the general solution is

$$y = zx = x \tan(\ln|Dx|) \quad \blacktriangle$$

Example 10.13 *RL* circuit with step input

Write down the differential equation governing the current, i, flowing in the RL circuit shown in Figure 10.2 when a step voltage of magnitude E is applied to the circuit at $t = 0$. Solve this differential equation to obtain $i(t)$. Assume that when $t = 0$, $i = 0$.

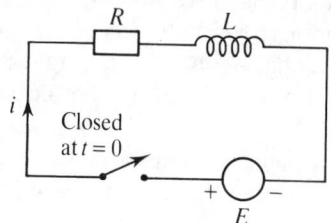

Figure 10.2   A step voltage is applied to the circuit at $t = 0$.

Solution

Applying Kirchhoff's voltage law and Ohm's law to the circuit we find:

$$iR + L\frac{di}{dt} = E \qquad t \geq 0$$

that is,

$$L\frac{di}{dt} = E - iR$$

so that

$$\int \frac{L}{E - iR}\,di = \int dt$$

Note in particular that in this equation L, E and R are constants and so the variables i and t have been separated. If the applied voltage, E, varied with time this would not have been the case since the left-hand side would contain terms dependent upon t. Integrating, we find

$$\int \frac{L}{E - iR}\,di = \frac{-L}{R}\int \frac{-R}{E - iR}\,di = -\frac{L}{R}\ln(E - iR) = t + C$$

To find the constant of integration, C, a condition is required. The physical condition $i = 0$ at $t = 0$ provides this. Applying $i = 0$ when $t = 0$ we find:

$$C = -\frac{L}{R}\ln E$$

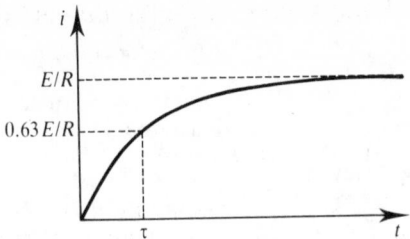

Figure 10.3 Response of the circuit of Figure 10.2 to a step input.

so that

$$\frac{L}{R}\ln\frac{E}{E-iR}=t$$

Then

$$\ln\frac{E}{E-iR}=\frac{Rt}{L}$$

hence

$$\frac{E}{E-iR}=e^{Rt/L}$$

Rearrangement for i then gives

$$i=\frac{E}{R}(1-e^{-Rt/L})$$

The graph of this current against time is shown in Figure 10.3. We note that as $t\to\infty$, $i\to E/R$. The rate at which the current increases towards its final value depends upon the values of the components R and L. It is common to define a **time constant**, τ, for the circuit. In this case $\tau=L/R$ and the equation for the current can be written as

$$i=\frac{E}{R}(1-e^{-t/\tau})$$

The smaller the value of τ, the more rapidly the current reaches its final value. It is possible to estimate a value of τ from a laboratory test curve by noting that after one time constant, that is, $t=\tau$, i has reached $1-e^{-1}\approx0.63$ of its final value. ▲

EXERCISES 10.3

1. Find the general solution of $dx/dt = \ln t$.
   Find the particular solution satisfying
   $x(1) = 1$.

2. Find the general solutions of the following
   equations:

   (a) $dy/dx = kx,$ k constant
   (b) $dy/dx = -ky,$ k constant
   (c) $dy/dx = y^2$
   (d) $y\, dy/dx = \sin x$
   (e) $y\, dy/dx = x + 2$

   (f) $x^2\, dy/dx = 2y^2 + yx$
   (g) $dx/dt = t^4/x^5$

3. Find the general solutions of the following
   equations:

   (a) $dx/dt = xt$
   (b) $dy/dx = x/y$
   (c) $t\, dx/dt = \tan x$
   (d) $dx/dt = (x^2 - 1)/t$

4. Find the general solution of the equation
   $dx/dt = t(x - 2)$. Find the particular
   solution which satisfies $x(0) = 5$.

10.5 First order equations – use of an integrating factor

The integrating factor method can be applied to equations of the form:

$$\frac{dy}{dx} + P(x)y = Q(x) \tag{10.3}$$

where P and Q are two functions of x, that is, first order linear equations. The idea behind the method is to multiply through by some, as yet unknown, function of x, $\mu(x)$ to obtain:

$$\mu(x)\frac{dy}{dx} + \mu(x)P(x)y = \mu(x)Q(x)$$

in such a way that the left-hand side can be equivalently written as:

$$\frac{d}{dx}(\mu(x)y)$$

thus yielding

$$\frac{d}{dx}(\mu(x)y) = \mu(x)Q(x) \tag{10.4}$$

In this form the equation is said to be **exact** and $\mu(x)$ is called an **integrating factor**. Now, because the derivative of $\mu(x)y$ is $\mu(x)Q(x)$ it follows that:

$$\mu(x)y = \int \mu(x)Q(x)\, dx$$

Provided this integral can be evaluated we can then determine y. As stated earlier we require

$$\frac{d}{dx}(\mu y) = \mu\frac{dy}{dx} + \mu Py$$

Using the product rule,

$$\frac{d}{dx}(\mu y) = \mu\frac{dy}{dx} + \frac{d\mu}{dx}y$$

and so we must have:

$$\mu\frac{dy}{dx} + \frac{d\mu}{dx}y = \mu\frac{dy}{dx} + \mu Py$$

Equating coefficients of y we find

$$\frac{d\mu}{dx} = \mu P$$

This is separable and yields:

$$\int\frac{d\mu}{\mu} = \int P\,dx$$

Then,

$$\ln\mu - \ln K = \int P\,dx \qquad -\ln K \text{ is the constant of integration}$$

Therefore,

$$\ln\frac{\mu}{K} = \int P\,dx$$

In practice we choose $K = 1$ because if all terms of the equation are multiplied by $\mu(x)$, a constant multiplicative factor is irrelevant.

Finally, we have the following result:

■ The integrating factor for

$$\frac{dy}{dx} + P(x)y = Q(x)$$

is given by

$$\mu(x) = Ke^{\int P(x)\,dx}$$

Example 10.14

Solve the differential equation $dy/dx + y/x = 1$ using the integrating factor method.

Solution

Referring to the standard first order linear equation:

$$\frac{dy}{dx} + P(x)y = Q(x)$$

we see that $P(x) = 1/x$ and $Q(x) = 1$. Using the previous formula for $\mu(x)$, we find:

$$\mu(x) = e^{\int (1/x)\,dx}$$

$$= e^{\ln x}$$

$$= x$$

Then from Equation (10.4) we have:

$$\frac{d}{dx}(xy) = xQ(x) = x \qquad \text{since } Q(x) = 1$$

Therefore,

$$xy = \int x\,dx = \frac{x^2}{2} + C$$

and finally $y = x/2 + C/x$ is the required general solution. ▲

Example 10.15 RL circuit with ramp input

The differential equation governing current flow, $i(t)$, in a series RL circuit when a voltage $u(t)t$ is applied, is given by:

$$iR + L\frac{di}{dt} = t \qquad t \geq 0 \qquad i(0) = 0$$

Show that this equation can be written in the form of Equation (10.3). Hence use the integrating factor method to find $i(t)$.

Solution

The equation can be written as:

$$\frac{di}{dt} + \frac{R}{L}i = \frac{1}{L}t \qquad t \geq 0$$

which is a first order linear equation. Comparing with Equation (10.3) we take $P(t) = R/L$ and $Q(t) = t/L$. Then the integrating factor, μ, is given by:

$$\mu = e^{\int P(t)\,dt}$$

$$= e^{\int (R/L)\,dt}$$

$$= e^{Rt/L}$$

From Equation (10.4) we have

$$\frac{d}{dt}(e^{Rt/L}i) = e^{Rt/L}\frac{t}{L}$$

so that,

$$ie^{Rt/L} = \frac{1}{L}\int te^{Rt/L}\,dt$$

$$= \frac{1}{L}\left(\frac{te^{Rt/L}}{R/L} - \int\frac{e^{Rt/L}}{R/L}\,dt\right)$$

$$= \frac{te^{Rt/L}}{R} - \frac{L}{R^2}e^{Rt/L} + K$$

where K is the constant of integration. The general solution is:

$$i = \frac{t}{R} - \frac{L}{R^2} + Ke^{-Rt/L}$$

When $t = 0$, $i = 0$. This gives the initial condition required to find K. Applying $i(0) = 0$ gives $K = L/R^2$ so that the particular solution is:

$$i = \frac{t}{R} + \frac{L}{R^2}(e^{-Rt/L} - 1) \quad \blacktriangle$$

EXERCISES 10.4

1. Find the general solution of $dx/dt = 2x + 4t$. What is the particular solution which satisfies $x(1) = 2$?

2. Find the general solution of $dy/dx + y = 2x + 5$.

3. Solve $dx/dt = t - tx$, $x(0) = 0$.

4. Use an integrating factor to obtain the general solution of $iR + L\,di/dt = \sin \omega t$ where R, L and ω are constants.

5. Solve $x\,dy/dx + y = x^4$.

6. Use an integrating factor to find the general solution of $t\,dx/dt + x = 3t$.

7. Find the general solution of $dx/dt + 2xt = t$. Find the particular solution satisfying the condition $x(0) = -1$.

8. Find the general solution of $t\dot{x} + 3x = e^t/t^2$.

10.6 Second order linear equations

The general form of a second order linear ordinary differential equation is:

$$p(x)\frac{d^2y}{dx^2} + q(x)\frac{dy}{dx} + r(x)y = f(x) \tag{10.5}$$

where $p(x)$, $q(x)$, $r(x)$ and $f(x)$ are functions of x only.

An important relative of this equation is

$$p(x)\frac{d^2y}{dx^2} + q(x)\frac{dy}{dx} + r(x)y = 0 \tag{10.6}$$

which is obtained from Equation (10.5) by ignoring the term which is independent of y. Equation (10.6) is said to be a **homogeneous** equation – all its terms contain y or its derivatives. Equation (10.5) is said to be **inhomogeneous.**

To denote a general linear differential equation we sometimes use the symbol:

$$L\{y\}$$

to stand for all the terms in the equation involving y or its derivatives. Hence if $L\{y\}$ stands for:

$$p(x)\frac{d^2y}{dx^2} + q(x)\frac{dy}{dx} + r(x)y$$

we can write the general second order linear inhomogeneous equation as:

$$L\{y\} = f(x)$$

and the corresponding homogeneous equation as

$$L\{y\} = 0$$

For example, if

$$\frac{d^2y}{dx^2} - 4y = x^3$$

we could write this as:

$$L\{y\} = x^3$$

where

$$L\{y\} = \frac{d^2y}{dx^2} - 4y$$

When $L\{y\} = f(x)$ is a linear differential equation, L is a **linear differential operator** on y. Throughout the book, considerable use is made of the concept of a linear operator. This is because a wide variety of apparently dissimilar techniques rely on a common set of properties. You will recall from Chapters 7 and 8 that the processes of differentiation and integration are linear. Any linear operator, L, carries out an operation on functions f_1 and f_2 in the following ways:

■ (1) $L\{f_1 + f_2\} = L\{f_1\} + L\{f_2\}$

(2) $L\{af_1\} = aL\{f_1\}$, where a is constant

(3) $L\{af_1 + bf_2\} = aL\{f_1\} + bL\{f_2\}$, where a, b are constants

---

Note that (1) and (2) are special cases of (3). These properties are illustrated in the following example.

Example 10.16

If,

$$L\{y\} = \frac{d^2y}{dx^2} + 3x\frac{dy}{dx} - 2y$$

show that:

(a) $L\{y_1 + y_2\} = L\{y_1\} + L\{y_2\}$

(b) $L\{ay\} = aL\{y\}$

Solutions

Note that $L\{y\} = d^2y/dx^2 + 3x\,dy/dx - 2y = f(x)$ is a linear differential equation and so L is a linear differential operator.

(a) Applying the given linear operator to the sum $y_1 + y_2$, we have:

$$L\{y_1 + y_2\} = \frac{d^2}{dx^2}(y_1 + y_2) + 3x\frac{d}{dx}(y_1 + y_2) - 2(y_1 + y_2)$$

$$= \frac{d^2y_1}{dx^2} + \frac{d^2y_2}{dx^2} + 3x\frac{dy_1}{dx} + 3x\frac{dy_2}{dx} - 2y_1 - 2y_2$$

because differentiation is a linear operator. Rearranging, we find:

$$L\{y_1 + y_2\} = \frac{d^2y_1}{dx^2} + 3x\frac{dy_1}{dx} - 2y_1 + \frac{d^2y_2}{dx^2} + 3x\frac{dy_2}{dx} - 2y_2$$

$$= L\{y_1\} + L\{y_2\}$$

as required.

(b) $$L\{ay\} = \frac{d^2(ay)}{dx^2} + 3x\frac{d(ay)}{dx} - 2(ay)$$

$$= a\frac{d^2y}{dx^2} + 3ax\frac{dy}{dx} - 2ay$$

$$= a\left(\frac{d^2y}{dx^2} + 3x\frac{dy}{dx} - 2y\right)$$

$$= aL\{y\}$$

as required. ▲

It is important to note that the properties stated only apply to linear operators.

The following properties of linear equations, are necessary for finding solutions of second order linear equations. They follow from the properties of a linear operator.

10.6.1 *Property 1*

If $y_1(x)$ and $y_2(x)$ are any two linearly independent solutions of $L\{y\} = 0$, then the general solution, $y_H(x)$, is

■ $y_H(x) = Ay_1(x) + By_2(x)$

where A, B are constants.

We see that the second order linear ordinary differential equation has two arbitrary constants in its general solution. The functions $y_1(x)$ and $y_2(x)$ are **linearly independent** if one is not simply a multiple of the other.

10.6.2 *Property 2*

Let $y_H(x)$ be the general solution of the homogeneous equation $L\{y\} = 0$. Let $y_P(x)$ be any solution of $L\{y\} = f(x)$. The general solution of $L\{y\} = f(x)$ is then given by:

■ $y(x) = y_H(x) + y_P(x)$

In other words, to find the general solution of an inhomogeneous equation we must first find the general solution of the corresponding homogeneous problem, and then add to it any solution of the inhomogeneous equation.

The function $y_H(x)$ is known as the **complementary function** and $y_P(x)$ is called the **particular integral**. Clearly the complementary function of a homogeneous problem is the same as its general solution; we shall often write $y(x)$ for both. If conditions are given they are applied to the general solution of the inhomogeneous equation to determine any unknown constants. This yields the particular solution satisfying the given conditions.

Example 10.17

Verify that $y_1(x) = x$ and $y_2(x) = 1$ both satisfy $d^2y/dx^2 = 0$. Write down the general solution of this equation and verify that this indeed satisfies the equation.

Solution

If $y_1(x) = x$ then $dy_1/dx = 1$ and $d^2y_1/dx^2 = 0$, so that y_1 satisfies $d^2y/dx^2 = 0$. If $y_2(x) = 1$, then $dy_2/dx = 0$ and $d^2y_2/dx^2 = 0$, so that y_2 satisfies $d^2y/dx^2 = 0$. From property 1, the general solution of $d^2y/dx^2 = 0$ is:

$$y_H(x) = Ax + B(1)$$

$$= Ax + B$$

To verify that this satisfies the equation proceed as follows:

$$\frac{dy_H}{dx} = A$$

$$\frac{d^2y_H}{dx^2} = 0$$

and so $y_H(x)$ satisfies $d^2y/dx^2 = 0$. ▲

Example 10.18

Given:

$$L\{y\} = \frac{d^2y}{dx^2} + y = x \tag{10.7}$$

(a) Show that $y_H = A \cos x + B \sin x$ is a solution of the corresponding homogeneous equation $L\{y\} = 0$

(b) Verify that $y_P = x$ is a particular integral

(c) Verify that $y_H + y_P$ does indeed satisfy the inhomogeneous equation $L\{y\} = x$

Solutions

(a) If $y_H = A \cos x + B \sin x$, then

$$y'_H = -A \sin x + B \cos x \qquad y''_H = -A \cos x - B \sin x$$

We see immediately that $y_H + y''_H = 0$ so that y_H is a solution of the homogeneous equation.

(b) If $y_P = x$ then $y'_P = 1$ and $y''_P = 0$. Substitution into the inhomogeneous equation shows that y_P satisfies Equation (10.7), that is, $y_P = x$ is a particular integral.

(c) Writing:

$$y = A \cos x + B \sin x + x$$

we have,

$$y' = -A \sin x + B \cos x + 1 \qquad y'' = -A \cos x - B \sin x$$

Substitution into the left-hand side of Equation (10.7) gives:

$$(-A \cos x - B \sin x) + (A \cos x + B \sin x + x)$$

which equals x, and so the complementary function plus the particular integral is indeed a solution of the inhomogeneous equation. ▲

10.6.3 *Constant coefficient equations*

We now proceed to study in detail those second order linear equations which have constant coefficients. The general form of such an equation is:

$$L\{y\} = a \frac{d^2 y}{dx^2} + b \frac{dy}{dx} + cy = f(x) \tag{10.8}$$

where a, b, c are constants. The homogeneous form of Equation (10.8) is

$$L\{y\} = a \frac{d^2 y}{dx^2} + b \frac{dy}{dx} + cy = 0 \tag{10.9}$$

Equations of this form arise in the analysis of circuits. Consider the following example.

Example 10.19 The LCR circuit

Write down the differential equation governing the current flowing in the series LCR circuit shown in Figure 10.4.

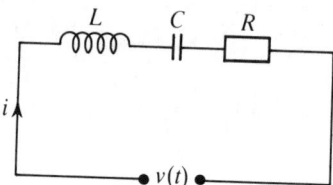

Figure 10.4   An LCR circuit.

Solution

Using Kirchhoff's voltage law and the individual laws for each component we find

$$L \frac{di}{dt} + iR + \frac{1}{C} \int i \, dt = v(t)$$

If this equation is now differentiated with respect to t we find:

$$L \frac{d^2 i}{dt^2} + R \frac{di}{dt} + \frac{1}{C} i = \frac{dv(t)}{dt}$$

This is an inhomogeneous second order differential equation, with the inhomogeneity arising from the term dv/dt. When the circuit components L, R and C are constants we have what is termed a **linear time-invariant system**, and the differential equation then has constant coefficients. ▲

■ A linear time-invariant system has components whose properties do not vary with time and as such can be modelled by a linear constant coefficient differential equation.

10.6.4 *Finding the complementary function*

As stated in Property 2, finding the general solution of $L\{y\} = ay'' + by' + cy = f$ is a two-stage process. The first task is to determine the complementary function. This is the general solution of the corresponding homogeneous equation, that is, $L\{y\} = ay'' + by' + cy = 0$. We now focus attention on the solution of such equations.

Example 10.20

Verify that $y_1 = e^{4x}$ and $y_2 = e^{2x}$ both satisfy the constant coefficient homogeneous equation:

$$\frac{d^2y}{dx^2} - 6\frac{dy}{dx} + 8y = 0 \qquad (10.10)$$

Write down the general solution of this equation.

Solution

If $y_1 = e^{4x}$, differentiation yields:

$$\frac{dy_1}{dx} = 4e^{4x}$$

and similarly,

$$\frac{d^2y_1}{dx^2} = 16e^{4x}$$

Substitution into Equation (10.10) gives:

$$16e^{4x} - 6(4e^{4x}) + 8e^{4x} = 0$$

so that $y_1 = e^{4x}$ is indeed a solution. Similarly if $y_2 = e^{2x}$, then dy_2/d$x = 2e^{2x}$ and d$^2y_2$/d$x^2 = 4e^{2x}$. Substitution into Equation (10.10) gives:

$$4e^{2x} - 6(2e^{2x}) + 8e^{2x} = 0$$

so that $y_2 = e^{2x}$ is also a solution of Equation (10.10). Now e^{2x} and e^{4x} are linearly independent functions. So, from Property 1 we have:

$$y_H(x) = Ae^{4x} + Be^{2x}$$

as the general solution of Equation (10.10). ▲

Example 10.21

Find values of k so that $y = e^{kx}$ is a solution of:

$$\frac{d^2y}{dx^2} - \frac{dy}{dx} - 6y = 0$$

Hence state the general solution.

Solution

As suggested we try a solution of the form $y = e^{kx}$. Differentiating we find $dy/dx = ke^{kx}$ and $d^2y/dx^2 = k^2e^{kx}$. Substitution into the given equation yields:

$$k^2e^{kx} - ke^{kx} - 6e^{kx} = 0$$

that is,

$$(k^2 - k - 6)e^{kx} = 0$$

The only way this equation can be satisfied for all values of x is if

$$k^2 - k - 6 = 0 \tag{10.11}$$

that is,

$$(k - 3)(k + 2) = 0$$

so that $k = 3$ or $k = -2$. That is to say, if $y = e^{kx}$ is to be a solution of the differential equation k must be either 3 or -2. We therefore have found two solutions

$$y_1(x) = e^{3x} \quad \text{and} \quad y_2(x) = e^{-2x}$$

These two functions are linearly independent and we can therefore apply Property 1 to give the general solution:

$$y_H(x) = Ae^{3x} + Be^{-2x}$$

Equation (10.11) for determining k is called the **auxiliary equation**. ▲

Example 10.22

Find the auxiliary equation of the differential equation:

$$a\frac{d^2y}{dx^2} + b\frac{dy}{dx} + cy = 0$$

Solution

We try a solution of the form $y = e^{kx}$ so that $dy/dx = ke^{kx}$ and $d^2y/dx^2 = k^2e^{kx}$. Substitution into the given differential equation yields:

$$ak^2e^{kx} + bke^{kx} + ce^{kx} = 0$$

that is,

$$(ak^2 + bk + c)e^{kx} = 0$$

Since this equation is to be satisfied for all values of x, then

$$ak^2 + bk + c = 0$$

is the required auxiliary equation.   ▲

■ The auxiliary equation of $a\,d^2y/dx^2 + b\,dy/dx + cy = 0$ is

$$ak^2 + bk + c = 0$$

Solution of this quadratic equation gives the values of k which we seek. Clearly the nature of the roots will depend upon the values of a, b and c. If $b^2 > 4ac$ the roots will be real and distinct. The two values of k thus obtained, k_1 and k_2, will allow us to write down two independent solutions.

$$y_1(x) = e^{k_1x} \qquad y_2(x) = e^{k_2x}$$

and so the general solution of the differential equation will be:

$$y(x) = Ae^{k_1x} + Be^{k_2x}$$

■ If the auxiliary equation has real, distinct roots k_1 and k_2, the complementary function will be:

$$y(x) = Ae^{k_1x} + Be^{k_2x}$$

On the other hand, if $b^2 = 4ac$ the two roots of the auxiliary equation will be equal and this method will therefore only yield one independent solution. In this case, special treatment is required. If $b^2 < 4ac$ the two roots of the auxiliary equation will be complex, that is, k_1 and k_2 will be complex numbers. The procedure for dealing with such cases will become apparent in the following examples.

Example 10.23

Find the general solution of:

$$\frac{d^2y}{dx^2} + 3\frac{dy}{dx} - 10y = 0$$

Find the particular solution which satisfies the conditions $y(0) = 1$ and $y'(0) = 1$.

Solution

By letting $y = e^{kx}$, so that $dy/dx = ke^{kx}$ and $d^2y/dx^2 = k^2e^{kx}$, the auxiliary equation is found to be:

$$k^2 + 3k - 10 = 0$$

Therefore,

$$(k - 2)(k + 5) = 0$$

so that $k = 2$ and $k = -5$. Thus there exist two solutions $y_1 = e^{2x}$ and $y_2 = e^{-5x}$. From Property 1 we can write the general solution as:

$$y = Ae^{2x} + Be^{-5x}$$

To find the particular solution we must now impose the given conditions:

$$y(0) = 1 \quad \text{gives} \quad 1 = A + B \quad = 2 + 5$$

$$y'(0) = 1 \quad \text{gives} \quad 1 = 2A - 5B$$

from which $A = 6/7$ and $B = 1/7$. Finally, the required particular solution is $y = \frac{6}{7}e^{2x} + \frac{1}{7}e^{-5x}$. ▲

Example 10.24

Find the general solution of:

$$\frac{d^2y}{dx^2} + 4y = 0$$

Solution

As before, let $y = e^{kx}$ so that $dy/dx = ke^{kx}$ and $d^2y/dx^2 = k^2e^{kx}$. The auxiliary equation is easily found to be:

$$k^2 + 4 = 0$$

that is,

$$k^2 = -4$$

so that,

$$k = \pm 2j$$

that is, we have complex roots. The two independent solutions of the equation are thus

$$y_1(x) = e^{2jx} \quad \text{and} \quad y_2(x) = e^{-2jx}$$

so that the general solution can be written in the form

$$y(x) = Ae^{2jx} + Be^{-2jx}$$

However, in cases such as this, it is usual to rewrite the solution in the following way. Recall from Chapter 6 that Euler's relations give:

$$e^{2jx} = \cos 2x + j \sin 2x$$

and

$$e^{-2jx} = \cos 2x - j \sin 2x$$

so that

$$y(x) = A(\cos 2x + j \sin 2x) + B(\cos 2x - j \sin 2x)$$

If we now relabel the constants such that:

$$A + B = C \quad \text{and} \quad Aj - Bj = D$$

we can write the general solution in the form:

$$y(x) = C \cos 2x + D \sin 2x \quad \blacktriangle$$

Example 10.25

Given $ay'' + by' + cy = 0$, write down the auxiliary equation. If the roots of the auxiliary equation are complex and are denoted by:

$$k_1 = \alpha + \beta j \qquad k_2 = \alpha - \beta j$$

show that the general solution is:

$$y(x) = e^{\alpha x}(A \cos \beta x + B \sin \beta x)$$

Solution

Substitution of $y = e^{kx}$ into the differential equation yields:

$$(ak^2 + bk + c)e^{kx} = 0$$

and so,

$$ak^2 + bk + c = 0$$

This is the auxiliary equation. If $k_1 = \alpha + \beta j$, $k_2 = \alpha - \beta j$ then the general solution is

$$y = Ce^{(\alpha + \beta j)x} + De^{(\alpha - \beta j)x}$$

where C and D are arbitrary constants. Using the laws of indices this is rewritten as:

$$y = Ce^{\alpha x}e^{\beta j x} + De^{\alpha x}e^{-\beta j x} = e^{\alpha x}(Ce^{\beta j x} + De^{-\beta j x})$$

Then, using Euler's relations, we obtain:

$$y = e^{\alpha x}(C \cos \beta x + Cj \sin \beta x + D \cos \beta x - Dj \sin \beta x)$$
$$= e^{\alpha x}\{(C + D) \cos \beta x + (Cj - Dj) \sin \beta x\}$$

Writing $A = C + D$ and $B = Cj - Dj$, we find:

$$y = e^{\alpha x}(A \cos \beta x + B \sin \beta x)$$

This is the required solution. ▲

■ If the auxiliary equation has complex roots, $\alpha + \beta j$ and $\alpha - \beta j$, then the complementary function is:

$$y = e^{\alpha x}(A \cos \beta x + B \sin \beta x)$$

Note that Example 10.24 is a special case of Example 10.25 with $\alpha = 0$ and $\beta = 2$.

Example 10.26

Find the general solution of $y'' + 2y' + 4y = 0$.

Solution

The auxiliary equation is $k^2 + 2k + 4 = 0$. This equation has complex roots given by:

$$k = \frac{-2 \pm \sqrt{4 - 16}}{2}$$

$$= \frac{-2 \pm \sqrt{12}\,j}{2}$$

$$= -1 \pm \sqrt{3}\,j$$

Using the result of Example 10.25 with $\alpha = -1$ and $\beta = \sqrt{3}$ we find the general solution is

$$y = e^{-x}(A \cos \sqrt{3}x + B \sin \sqrt{3}x) \quad ▲$$

Example 10.27

The auxiliary equation of $ay'' + by' + cy = 0$ is $ak^2 + bk + c = 0$. Suppose this equation has equal roots $k = k_1$. Verify that $y = xe^{k_1 x}$ is a solution of the differential equation.

Solution

We have:

$$y = xe^{k_1 x} \qquad y' = e^{k_1 x}(1 + k_1 x) \qquad y'' = e^{k_1 x}(k_1^2 x + 2k_1)$$

Substitution into the left-hand side of the differential equation yields:

$$e^{k_1 x}\{a(k_1^2 x + 2k_1) + b(1 + k_1 x) + cx\} = e^{k_1 x}\{(ak_1^2 + bk_1 + c)x + 2ak_1 + b\}$$

But $ak_1^2 + bk_1 + c = 0$ since k_1 satisfies the auxiliary equation. Also,

$$k_1 = \frac{-b \pm \sqrt{b^2 - 4ac}}{2a}$$

but since the roots are equal, then $b^2 - 4ac = 0$ hence $k_1 = -b/2a$. So $2ak_1 + b = 0$. We conclude that $y = xe^{k_1 x}$ is a solution of $ay'' + by' + cy = 0$ when the roots of the auxiliary equation are equal. ▲

■ If the auxiliary equation has two equal roots, k_1, the complementary function is:

$$y = Ae^{k_1 x} + Bxe^{k_1 x}$$

Example 10.28

Obtain the general solution of the equation:

$$\frac{d^2 y}{dx^2} + 8\frac{dy}{dx} + 16y = 0$$

Solution

As before, a trial solution of the form $y = e^{kx}$ yields an auxiliary equation.

$$k^2 + 8k + 16 = 0$$

This equation factorizes so that:

$$(k + 4)(k + 4) = 0$$

and we obtain equal roots, that is, $k = -4$ (twice). If we proceed as before, writing $y_1(x) = e^{-4x}$, $y_2(x) = e^{-4x}$, it is clear that the two solutions are not independent. To apply Property 1 we need to find a second independent solution. Using the result of Example 10.27 we conclude that, because the roots of the auxiliary equation are equal, the second independent solution is $y_2 = xe^{-4x}$. The general solution is then:

$$y(x) = Ae^{-4x} + Bxe^{-4x} \quad ▲$$

EXERCISES 10.5

1. Obtain the general solutions, that is, the complementary functions, of the following homogeneous equations.

   (a) $d^2y/dx^2 - 3\,dy/dx + 2y = 0$

   (b) $d^2y/dx^2 - 2\,dy/dx + y = 0$

   (c) $d^2y/dx^2 + 9y = 0$

   (d) $d^2y/dx^2 - 2\,dy/dx = 0$

2. Find the auxiliary equation for the differential equation

$$L\frac{d^2i}{dt^2} + R\frac{di}{dt} + \frac{1}{C}i = 0$$

Hence write down the complementary function.

3. Find the complementary function of the equation

$$\frac{d^2y}{dx^2} + \frac{dy}{dx} + y = 0$$

10.6.5 *Finding a particular integral*

We stated in Property 2 that the general solution of an inhomogeneous equation is the sum of the complementary function and a particular integral. We have seen how to find the complementary function in the case of a constant coefficient equation. We shall now deal with the problem of finding a particular integral. Recall that the particular integral is any solution of the inhomogeneous equation. There are a number of advanced techniques available for finding such solutions but these are beyond the scope of this book. We shall adopt a simpler strategy. Since any solution will do we shall try to find such a solution by a combination of educated guesswork and trial and error.

Example 10.29

Find the general solution of the equation

$$\frac{d^2y}{dx^2} - \frac{dy}{dx} - 6y = e^{2x} \tag{10.12}$$

Solution

The complementary function for this equation has already been shown in Example 10.21 to be:

$$y_H = Ae^{3x} + Be^{-2x}$$

We shall attempt to find a solution of the inhomogeneous problem by trying a function of the same form as that on the right-hand side. In particular, let us try $y_P(x) = \alpha e^{2x}$, where α is a constant that we shall now determine. If $y_P(x) = \alpha e^{2x}$ then $dy_P/dx = 2\alpha e^{2x}$ and $d^2y_P/dx^2 = 4\alpha e^{2x}$. Substitution in Equation (10.12) gives:

$$4\alpha e^{2x} - 2\alpha e^{2x} - 6\alpha e^{2x} = e^{2x}$$

that is,

$$-4\alpha e^{2x} = e^{2x}$$

so that y_P will be a solution if α is chosen so that $-4\alpha = 1$, that is, $\alpha = -1/4$. Therefore the particular integral is $y_P(x) = -e^{2x}/4$. From Property 2 the general solution of the inhomogeneous equation is found by summing this particular integral and the complementary function:

$$y(x) = Ae^{3x} + Be^{-2x} - \tfrac{1}{4}e^{2x} \quad \blacktriangle$$

Example 10.30

Obtain a particular integral of the equation:

$$\frac{d^2y}{dx^2} - 6\frac{dy}{dx} + 8y = x$$

Solution

In the last example, we found that a fruitful approach was to assume a solution in the same form as that on the right-hand side. Suppose we assume a solution $y_P(x) = \alpha x$ and proceed to determine α. This approach will actually fail, but let us see why. If $y_P(x) = \alpha x$ then $dy_P/dx = \alpha$ and $d^2y_P/dx^2 = 0$. Substitution into the differential equation yields:

$$0 - 6\alpha + 8\alpha x = x$$

and α ought now to be chosen so that this expression is true for all x. If we equate the coefficients of x we find $8\alpha = 1$ so that $\alpha = 1/8$, but with this value of α the constant terms are inconsistent. Clearly a particular integral of the form αx is not possible. The problem arises because differentiation of the term αx produces constant terms which are unbalanced on the right-hand side. So, we try a solution of the form:

$$y_P(x) = \alpha x + \beta$$

with α, β constants. Proceeding as before $dy_P/dx = \alpha$, $d^2y_P/dx^2 = 0$. Substitution in the differential equation now gives:

$$0 - 6\alpha + 8(\alpha x + \beta) = x$$

Equating coefficients of x we find:

$$8\alpha = 1 \tag{10.13}$$

and equating constant terms we find:

$$-6\alpha + 8\beta = 0 \tag{10.14}$$

From Equation (10.13), $\alpha = 1/8$ and then from Equation (10.14)

$$-6(1/8) + 8\beta = 0$$

so that,

$$8\beta = 3/4$$

that is,

$$\beta = 3/32$$

The required particular integral is $y_P(x) = x/8 + 3/32$. $\quad \blacktriangle$

Experience leads to the trial solutions suggested in Table 10.1.

Table 10.1   Trial solutions to find the particular integral.

$f(x)$	Trial solution
constant	constant
polynomial in x of degree r	polynomial in x of degree r
$\cos kx$	$a \cos kx + b \sin kx$
$\sin kx$	$a \cos kx + b \sin kx$
ae^{kx}	αe^{kx}

Example 10.31

Find a particular integral for the equation:

$$\frac{d^2y}{dx^2} - 6\frac{dy}{dx} + 8y = 3 \cos x$$

Solution

We shall try a solution of the form:

$$y_P(x) = \alpha \cos x + \beta \sin x$$

Differentiating, we find:

$$\frac{dy_P}{dx} = -\alpha \sin x + \beta \cos x$$

$$\frac{d^2y_P}{dx^2} = -\alpha \cos x - \beta \sin x$$

Substitution into the differential equation gives:

$$(-\alpha \cos x - \beta \sin x) - 6(-\alpha \sin x + \beta \cos x) + 8(\alpha \cos x + \beta \sin x)$$

$$= 3 \cos x$$

Equating coefficients of $\cos x$ we find:

$$-\alpha - 6\beta + 8\alpha = 3 \tag{10.15}$$

while those of $\sin x$ give:

$$-\beta + 6\alpha + 8\beta = 0 \tag{10.16}$$

Solving Equations (10.15) and (10.16) simultaneously we find $\alpha = 21/85$ and $\beta = -18/85$, so that the particular integral is:

$$y_P(x) = \tfrac{21}{85} \cos x - \tfrac{18}{85} \sin x \quad \blacktriangle$$

Example 10.32 An LC circuit with sinusoidal input

The differential equation governing the flow of current in a series LC circuit when subject to an applied voltage $v(t) = V_0 \sin \omega t$ is

$$L\frac{d^2 i}{dt^2} + \frac{1}{C}i = \omega V_0 \cos \omega t$$

Derive this equation and then obtain its general solution.

Solution

Kirchhoff's voltage law and the component laws give:

$$L\frac{di}{dt} + \frac{1}{C}\int i \, dt = V_0 \sin \omega t$$

To avoid processes of differentiation and integration in the same equation let us differentiate this equation with respect to t. This yields

$$L\frac{d^2 i}{dt^2} + \frac{1}{C}i = \omega V_0 \cos \omega t$$

as required.

The homogeneous equation is $L \, d^2 i/dt^2 + i/C = 0$. Letting $i = e^{kt}$ we find the auxiliary equation is $Lk^2 + 1/C = 0$ so that $k = \pm j/\sqrt{LC}$. Therefore, using the result of Example 10.25, with $\alpha = 0$ and $\beta = 1/\sqrt{LC}$, the complementary function is:

$$i = A \cos \frac{t}{\sqrt{LC}} + B \sin \frac{t}{\sqrt{LC}}$$

To find a particular integral, try $i = E \cos \omega t + F \sin \omega t$, where E and F are constants. We find:

$$\frac{di}{dt} = -\omega E \sin \omega t + \omega F \cos \omega t$$

$$\frac{d^2 i}{dt^2} = -\omega^2 E \cos \omega t - \omega^2 F \sin \omega t$$

Substitution into the inhomogeneous equation yields:

$$L(-\omega^2 E \cos \omega t - \omega^2 F \sin \omega t) + \frac{1}{C}(E \cos \omega t + F \sin \omega t) = V_0 \omega \cos \omega t$$

Equating coefficients of $\sin \omega t$ gives:

$$-\omega^2 LF + \frac{F}{C} = 0$$

Equating coefficients of cos ωt gives:

$$-\omega^2 LE + \frac{E}{C} = V_0\omega$$

so that $F = 0$ and $E = CV_0\omega/(1 - \omega^2 LC)$. Hence the particular integral is $i = \{CV_0\omega/(1 - \omega^2 LC)\}\cos\omega t$. Finally, the general solution is:

$$i = A\cos\frac{t}{\sqrt{LC}} + B\sin\frac{t}{\sqrt{LC}} + \frac{CV_0\omega\cos\omega t}{1 - \omega^2 LC} \quad \blacktriangle$$

Example 10.33 Transmission lines

A transmission line is an arrangement of electrical conductors for transporting electromagnetic waves. Although this definition could be applied to most electrical cables, it is usually restricted to cables used to transport high frequency electromagnetic waves. There are several different types of transmission lines. The most familiar one is the coaxial cable which is used to carry the signal from a television aerial to a television set (see Figure 10.5). When a high frequency wave is being carried by a cable several effects become important which can usually be neglected when dealing with a low frequency wave. These are,

(1) the capacitance, C, between the two conductors,

(2) the series inductance, L, of the two conductors,

(3) the leakage current through the insulation layer that separates the two conductors.

The electrical parameters of a coaxial cable are evenly distributed along its length. This is true for transmission lines in general and so it is usual to specify per unit length values for the parameters. When constructing a mathematical model of a transmission line it is easier to think in terms of lumped components spanning a distance δz and then allow δz to tend to zero (see Figure 10.6). The leakage between the two conductors is conventionally modelled by a conductance, G, as this simplifies the mathematics and avoids confusion with the line resistance, R. Note that C, G, L and R are per unit length values for the transmission line.

For most transmission lines of interest the signal that is being carried varies sinusoidally with time. Therefore the voltage and current depend on both position along the line, z, and time, t. However, it is common to separate

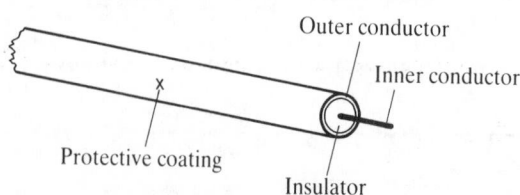

Outer conductor

Inner conductor

Protective coating

Insulator

Figure 10.5   A coaxial cable.

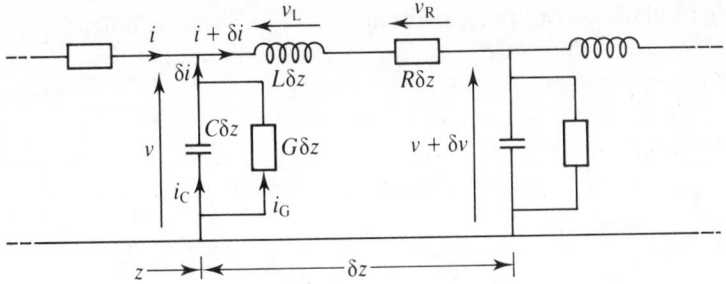

Figure 10.6   A section of a transmission line.

the time dependence from the voltage and current expressions in order to simplify the analysis. It must be remembered that any voltage and current variation with position has superimposed upon it a sinusoidal variation with time. Therefore, ignoring the time-dependent element we write the voltage as v and the current as i, knowing that they are functions of z.

Consider the circuit of Figure 10.6 which represents a section of the transmission line of length δz. Applying Kirchhoff's voltage law to the circuit yields:

$$v + \delta v - v + v_L + v_R = 0$$

$$\delta v = -v_L - v_R$$

where v_L is the voltage across the inductor and v_R is the voltage across the resistor. Using the individual component laws for the inductor and resistor gives:

$$\delta v = -ij\omega L\, \delta z - iR\, \delta z$$

$$= -i(R + j\omega L)\, \delta z$$

Note that δi has been ignored because it is small compared to i. Now consider the parallel combination of the capacitor and resistor (with units of conductance). Applying Kirchhoff's current law to this combination yields:

$$\delta i = i_C + i_G$$

where i_C is the current through the capacitor and i_G is the current through the resistor. Using the individual component laws for the capacitor and resistor gives:

$$\delta i = -vj\omega C\, \delta z - vG\, \delta z$$

$$= -v(G + j\omega C)\, \delta z$$

Dividing these two circuit equations by δz yields:

$$\frac{\delta v}{\delta z} = -i(R + j\omega L)$$

$$\frac{\delta i}{\delta z} = -v(G + j\omega C)$$

In order to model a continuous transmission line with evenly distributed parameters, δz is allowed to tend to zero. In the limit the two circuit equations become:

$$\frac{dv}{dz} = -i(R + j\omega L) \qquad (10.17)$$

$$\frac{di}{dz} = -v(G + j\omega C) \qquad (10.18)$$

Differentiating Equation (10.17) yields:

$$\frac{d^2v}{dz^2} = -(R + j\omega L)\frac{di}{dz}$$

Substituting for di/dz from Equation (10.18) yields:

$$\frac{d^2v}{dz^2} = (R + j\omega L)(G + j\omega C)v$$

This is usually written as:

$$\frac{d^2v}{dz^2} = \gamma^2 v \qquad \text{where} \qquad \gamma^2 = (R + j\omega L)(G + j\omega C) \qquad (10.19)$$

This is the differential equation that describes the variation of the voltage, v, with position, z, along the transmission line. The general solution of this equation is easily shown to be:

$$v = v_1 e^{-\gamma z} + v_2 e^{\gamma z} \qquad (10.20)$$

where v_1 and v_2 are constants that depend on the initial conditions for the transmission line. It is useful to write $\gamma = \alpha + j\beta$ thus separating the real and imaginary parts of γ. Equation (10.20) can then be written as:

$$v = v_1 e^{-\alpha z} e^{-j\beta z} + v_2 e^{\alpha z} e^{j\beta z} \qquad (10.21)$$

The quantity $v_1 e^{-\alpha z} e^{-j\beta z}$ represents the forward wave on the transmission line. It consists of a decaying exponential multiplied by a sinusoidal term. The decaying exponential represents a gradual attenuation of the wave caused by losses as it travels along the transmission line. The quantity $v_2 e^{\alpha z} e^{j\beta z}$ represents the backward wave produced by reflection. Reflection occurs if the transmission line is not matched with its load. As the wave is travelling in the opposite direction to the forward wave, $e^{\alpha z}$ still represents an attenuation but in this case an attenuation as z decreases.

A **loss-less** line is one in which the attenuation is negligible. This case corresponds to $\alpha = 0$, and so $\gamma = j\beta$. If $\gamma = j\beta$ then $\gamma^2 = -\beta^2$ so that, from Equation (10.19), $(R + j\omega L)(G + j\omega C)$ must be real and negative. We see that this is the case when $R = 0$ and $G = 0$. This agrees with what would be expected in practice as it is the resistive and conductive terms that lead to energy dissipation. ▲

10.6.6 *Inhomogeneous term appears in the complementary function*

In some examples, terms which form part of the complementary function also appear in the inhomogeneous term. This gives rise to an additional complication. Consider Example 10.34.

Example 10.34

Consider the equation $y'' - y' - 6y = e^{3x}$. It is straightforward to show that the complementary function is:

$$y = Ae^{3x} + Be^{-2x}$$

Find a particular integral and deduce the general solution.

Solution

Suppose we try to find a particular integral by using a trial solution of the form $y_P = \alpha e^{3x}$. Substitution into the left-hand side of the inhomogeneous equation yields:

$$9\alpha e^{3x} - 3\alpha e^{3x} - 6\alpha e^{3x} = 0$$

so that αe^{3x} is clearly not a solution of the inhomogeneous equation. The reason is that e^{3x} is part of the complementary function and so causes the left-hand side to vanish. To obtain a particular integral in such a case, we carry out the procedure required when the auxiliary equation has equal roots. That is, we try $y_P = \alpha x\, e^{3x}$. We find:

$$y' = \alpha e^{3x}(3x + 1) \qquad y'' = \alpha e^{3x}(9x + 6)$$

Substitution into the inhomogeneous equation yields:

$$\alpha e^{3x}(9x + 6) - \alpha e^{3x}(3x + 1) - 6\alpha x e^{3x} = e^{3x}$$

Most terms cancel, leaving

$$5\alpha e^{3x} = e^{3x}$$

so that $\alpha = 1/5$. Finally, the required particular integral is $y_P = xe^{3x}/5$. The general solution is then $y = Ae^{3x} + Be^{-2x} + xe^{3x}/5$. ▲

EXERCISES 10.6

1. Find the general solution of
$$d^2x/dt^2 - 2\,dx/dt - 3x = 6.$$

2. Find a particular integral for the equation
$$\frac{d^2x}{dt^2} - 3\frac{dx}{dt} + 2x = 5e^{3t}$$

3. Find a particular integral for the equation
$$\frac{d^2x}{dt^2} - x = 4e^{-t}$$

4. Obtain the general solution of
$$y'' - y' - 2y = 6.$$

5. Obtain the general solution of the equation

$$\frac{d^2y}{dx^2} + 3\frac{dy}{dx} + 2y = 10 \cos 2x$$

Find the particular solution satisfying

$$y(0) = 1, \frac{dy}{dx}(0) = 0.$$

6. Find a particular integral for the equation

$$\frac{d^2y}{dx^2} + \frac{dy}{dx} + y = 1 + x$$

7. Find the general solution of

(a) $d^2x/dt^2 - 6\,dx/dt + 5x = 3$

(b) $d^2x/dt^2 - 2\,dx/dt + x = e^t$

8. For the circuit shown in Figure 10.7 show that:

$$RCL\frac{d^2i_2}{dt^2} + L\frac{di_2}{dt} + Ri_2 = E(t)$$

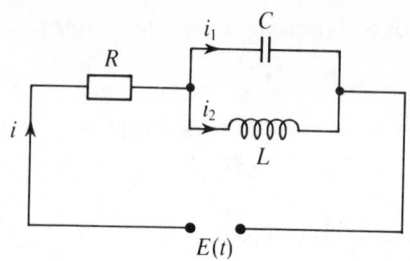

Figure 10.7 Circuit for Exercise 10.6.8.

If $L = 1$ mH, $R = 10\,\Omega$, $C = 1\,\mu$F and $E(t) = 2 \sin 100\pi t$, find the complementary function.

9. Find the general solution of

$$\frac{d^2i}{dt^2} + 8\frac{di}{dt} + 25i = 48 \cos 3t - 16 \sin 3t$$

10.7 Analog simulation

It is possible to solve differential equations using electronic circuits based on operational amplifiers. The advantage of this approach is the ease with which the coefficients of the differential equation can be adjusted and the effect on the solution observed. The technique is known as **analog simulation**. This is a common approach in engineering design because an engineer is often required to analyse many different mathematical models. Special purpose computers, known as analog computers, are available with the electronic circuits already incorporated, thus making it easier to simulate a particular differential equation.

Three basic types of circuit are required to enable ordinary differential equations with constant coefficients to be simulated. The first type is an integrator which has already been discussed in Section 8.2. The usual symbol for such a circuit is shown in Figure 10.8.

If an analog computer is used it is common for there to be a gain of only one or ten on the input voltage. A gain above ten is better achieved by linking together

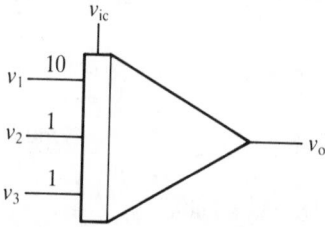

Figure 10.8 An integrator.

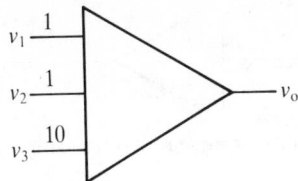

Figure 10.9 A summer.

two circuits in series. In addition there is usually a facility to allow initial conditions to be set. The equation for the circuit of Figure 10.8 is:

$$v_o = -\int_0^t (10v_1 + v_2 + v_3)\, dt - v_{ic} \tag{10.22}$$

where v_{ic} indicates the initial output voltage. Note that the gain is usually written on the input line.

The second type of circuit is the summer. It has the symbol shown in Figure 10.9. The equation for the circuit of Figure 10.9 is:

$$v_o = -(v_1 + v_2 + 10v_3) \tag{10.23}$$

Finally, a circuit is required to allow gains to be varied. This is simply a potentiometer and is illustrated together with its symbol in Figure 10.10. The equation for the circuit of Figure 10.10 is simply:

$$v_o = kv_i \qquad 0 \le k \le 1 \tag{10.24}$$

The potentiometer only allows the gain to be varied between 0 and 1. If a variable gain greater than 1 is required, the potentiometer must be placed in series with a circuit that increases gain, for example, a one-input summer with a gain of 10.

These circuits can be combined to obtain the solutions of differential equations. This is best illustrated by means of an example. Consider the general form of a second order differential equation with constant coefficients.

$$a\frac{d^2y}{dt^2} + b\frac{dy}{dt} + cy = f(t) \tag{10.25}$$

The general approach to obtaining the circuit for a particular differential equation is to assume a certain point in the circuit corresponds to a particular term and then arrange to connect that point to the correctly synthesized value. It is more

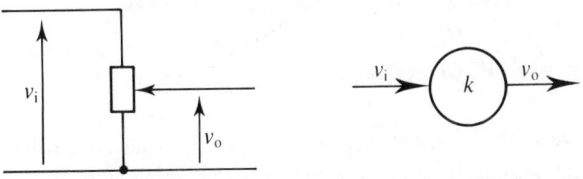

Figure 10.10 A potentiometer.

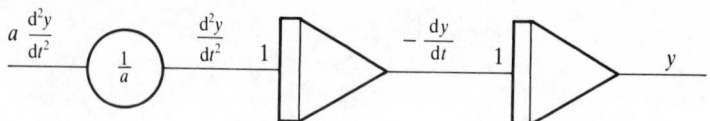

Figure 10.11   First stage in synthesizing Equation 10.25.

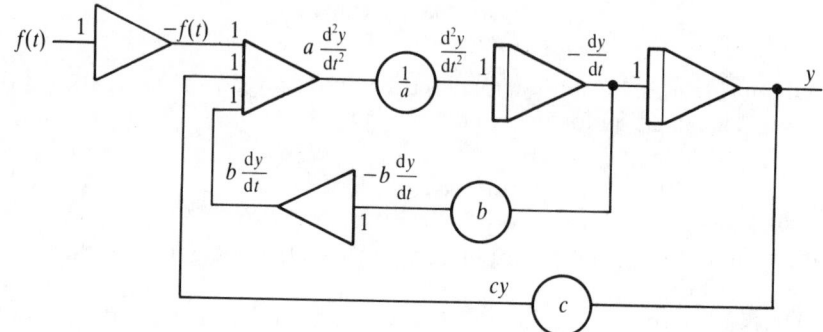

Figure 10.12   The complete circuit to synthesize Equation 10.25.

straightforward if the assumed point corresponds to the highest derivative. Rearranging Equation (10.25) we find:

$$a\frac{d^2y}{dt^2} = f(t) - b\frac{dy}{dt} - cy \tag{10.26}$$

Assuming $a\, d^2y/dt^2$ is already available, a potentiometer can be used to obtain d^2y/dt^2. Integrators can then be used to obtain the variables dy/dt and y. This is shown in Figure 10.11. The next stage is to obtain the expression corresponding to the right hand side of Equation (10.26). This requires a summer to add together the individual terms and potentiometers to allow individual coefficients to be obtained. An invertor is also required to obtain the correct sign for the variable dy/dt. This simply consists of a summer with one input and a gain of one. The final circuit is shown in Figure 10.12. The three potentiometers allow different values of a, b and c to be obtained. If gains greater than one were required then extra summers would be needed, or alternatively the input gains of the existing summers could be adjusted. The input to the circuit, $f(t)$, can be simulated by applying a signal at the appropriate point in the circuit. The output of the circuit, $y(t)$, corresponds to the solution of the differential equation. It may be necessary to include initial conditions corresponding to $y(0)$ and $\dfrac{dy}{dt}(0)$.

10.8 Higher order equations

In this section we shall consider second and higher order equations and show how they can be represented as a set of simultaneous first order equations. The main reason for doing this is that when a computer solution is required it is useful to express an equation in this form. Details of the analytical solution of such systems are not considered here although one technique is discussed in Chapter 11.

It is possible to express a second order differential equation as two first order equations. Thus if we have:

$$\frac{d^2y}{dx^2} = f\left(\frac{dy}{dx}, y, x\right) \tag{10.27}$$

we can introduce the new dependent variables y_1 and y_2 such that $y_1 = y$ and $y_2 = dy/dx$. Equation (10.27) then becomes:

$$\frac{dy_1}{dx} = y_2 \qquad \frac{dy_2}{dx} = f(y_2, y_1, x) \tag{10.28}$$

These first order simultaneous differential equations are often referred to as **coupled** equations.

Example 10.35

Express the equation

$$\frac{d^2y}{dx^2} - 7\frac{dy}{dx} + 3y = 0$$

as a set of first order equations.

Solution

Letting $y_1 = y$, and $y_2 = dy/dx$ we find $dy_2/dx = d^2y/dx^2$. Therefore the differential equation becomes

$$\frac{dy_1}{dx} = y_2 \qquad \frac{dy_2}{dx} - 7y_2 + 3y_1 = 0$$

We note that these equations can also be written as:

$$\frac{dy_1}{dx} = y_2$$

$$\frac{dy_2}{dx} = -3y_1 + 7y_2$$

or, in matrix form,

$$\begin{pmatrix} y_1' \\ y_2' \end{pmatrix} = \begin{pmatrix} 0 & 1 \\ -3 & 7 \end{pmatrix} \begin{pmatrix} y_1 \\ y_2 \end{pmatrix}$$

where $'$ denotes d/dx. ▲

Higher order differential equations can be reduced to a set of first order equations in a similar way:

Example 10.36

Express the equation

$$\frac{d^3x}{dt^3} - 7\frac{d^2x}{dt^2} + 3\frac{dx}{dt} + 2x = 0$$

as a set of first order equations.

Solution

Letting $x_1 = x$, $x_2 = dx/dt$ and $x_3 = d^2x/dt^2$ we find:

$$\frac{dx_1}{dt} = x_2$$

$$\frac{dx_2}{dt} = x_3$$

$$\frac{dx_3}{dt} - 7x_3 + 3x_2 + 2x_1 = 0$$

is the set of first order equations representing the given differential equation. ▲

Example 10.37

(a) Express the coupled first order equations:

$$\frac{dy_1}{dx} = y_1 + y_2$$

$$\frac{dy_2}{dx} = 4y_1 - 2y_2$$

as a second order ordinary differential equation, and obtain its general solution.

(b) Express the given equations in the form:

$$\begin{pmatrix} y_1' \\ y_2' \end{pmatrix} = A \begin{pmatrix} y_1 \\ y_2 \end{pmatrix}$$

where A is a 2×2 matrix.

Solutions

(a) Differentiating the second of the given equations we have:

$$\frac{d^2y_2}{dx^2} = 4\frac{dy_1}{dx} - 2\frac{dy_2}{dx}$$

and then, using the first we find

$$\frac{d^2y_2}{dx^2} = 4(y_1 + y_2) - 2\frac{dy_2}{dx}$$

But from the second given equation $4y_1 = dy_2/dx + 2y_2$, and therefore:

$$\frac{d^2y_2}{dx^2} = \frac{dy_2}{dx} + 2y_2 + 4y_2 - 2\frac{dy_2}{dx}$$

that is,

$$\frac{d^2y_2}{dx^2} + \frac{dy_2}{dx} - 6y_2 = 0$$

Writing y for y_2 we find:

$$\frac{d^2y}{dx^2} + \frac{dy}{dx} - 6y = 0$$

To solve this we let $y = e^{kx}$ to obtain:

$$k^2 + k - 6 = 0$$

$$(k - 2)(k + 3) = 0$$

Therefore,

$$k = 2, -3$$

The general solution is then $y = y_2 = Ae^{2x} + Be^{-3x}$. It is straightforward to show that y_1 satisfies the same second order differential equation and hence has the same general solution, but with different arbitrary constants.

(b) The first order equations can be written as:

$$\begin{pmatrix} y_1' \\ y_2' \end{pmatrix} = \begin{pmatrix} 1 & 1 \\ 4 & -2 \end{pmatrix}\begin{pmatrix} y_1 \\ y_2 \end{pmatrix}$$

Therefore $A = \begin{pmatrix} 1 & 1 \\ 4 & -2 \end{pmatrix}$. ▲

EXERCISE 10.7

Express

$$\frac{dy_1}{dt} = 2y_1 + 6y_2 \quad \text{and}$$

$$\frac{dy_2}{dt} = -2y_1 - 5y_2$$

as a single second order equation. Solve this equation and hence find y_1 and y_2. Express the equations in the form $y' = Ay$.

10.9 State–space models

There are several ways to model linear time-invariant systems mathematically. One way, which we have already examined, is to use linear differential equations with constant coefficients. A second method is to use transfer functions which will be discussed in Chapter 11. A third type of model is the state–space model. The state–space technique is particularly useful for modelling complex engineering systems in which there are several inputs and outputs. It also has a convenient form for solution by means of a digital computer.

The basis of the state–space technique is the representation of a system by means of a set of first order coupled differential equations. The number of first order differential equations required to model a system defines the **order** of the system. For example, if three differential equations are required then the system is a third order system. Associated with the first order differential equations are a set of **state variables**; the same number as there are differential equations.

The concept of a state variable lies at the heart of the state–space technique. A system is defined by means of its state variables. Provided the initial values of these state variables are known, it is possible to predict the behaviour of the system with time by means of the first order differential equations. One complication is that the choice of state variables to characterize a system is not unique. Many different choices of a set of state variables for a particular system are often possible. However, a system of order n only requires n state variables to specify it. Introducing more state variables than this only introduces redundancy.

The choice of state variables for a system is, to some extent, dependent on experience but there are certain guidelines that can be followed to obtain a valid choice of variables. One thing that is particularly important is that the state variables are independent of each other. For example, for the electrical system of Figure 10.13, the choice of v_L, and $3v_L$ would lead to the two state variables being dependent on each other. A valid choice would be v_L and i which are not directly dependent on each other. This is a second order system and so only requires two state variables to model its behaviour.

Many engineering systems may have a high order and so require several differential equations to model their behaviour. For this reason, a standard way of laying out these equations has evolved to reduce the chance of making errors. There is also an added advantage in that the standard layout makes it easier to present the equations to a digital computer for solution. Before introducing the standard form, an example will be presented to illustrate the state variable method.

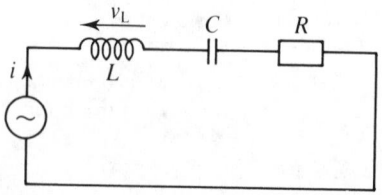

Figure 10.13 The circuit could be modelled using v_L and i as the state variables.

Example 10.38

Consider the mechanical system illustrated in Figure 10.14. A mass rests on a frictionless surface and is connected to a fixed wall by means of an ideal spring and an ideal damper. A force $f(t)$, is applied to the mass and the position of the mass is w.

By considering the forces acting on the mass, M, it is possible to devise a differential equation that models the behaviour of the system. The force produced by the spring is Kw (K is the spring stiffness) and opposes the forward motion. The force produced by the damper is $B \, dw/dt$ (B is the damping coefficient) and this also opposes the forward motion. Since the mass is constant, Newton's second law of motion states that the net force on the mass is equal to the product of the mass and its acceleration. Thus we find:

$$f(t) - B\frac{dw}{dt} - Kw = M\frac{d^2w}{dt^2} \qquad (10.29)$$

Note that this is a second order differential equation and so the system is a second order system.

In order to obtain a state–space model the state variables have to be chosen. There are several possible choices. A logical one would be the position, w, of the mass. A second state variable is required as the system is second order. In this case the velocity of the mass will be chosen. The velocity of the mass is not directly dependent on its position and so the two variables are independent. Another possible choice would have been $dw/dt - w$. This may seem a clumsy choice but for certain problems such choices may lead to simplifications in the state variable equations.

It is customary to use the symbols $x_1, x_2, x_3, \ldots$ to represent variables for reasons that will become clear shortly. So,

$$x_1 = w \qquad (10.30a)$$

$$x_2 = \frac{dw}{dt} \qquad (10.30b)$$

Because of the particular choice of state variables, it is easy to obtain the first of the first order differential equations – thus illustrating the need for experience when choosing state variables.

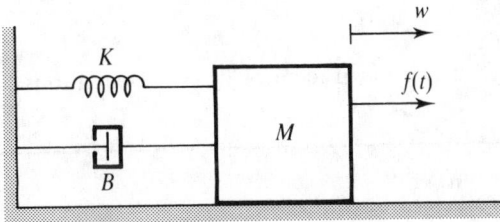

Figure 10.14 A second order mechanical system.

Differentiating Equation (10.30a) gives:

$$\frac{dx_1}{dt} = \frac{dw}{dt} = x_2$$

This is the first state equation although it is usually written as,

$$\dot{x}_1 = x_2$$

where $\dot{x}_1$ denotes dx_1/dt. The second of the first order equations is obtained by rearranging Equation (10.29)

$$\frac{d^2w}{dt^2} = -\frac{K}{M}w - \frac{B}{M}\frac{dw}{dt} + \frac{1}{M}f(t) \tag{10.31}$$

However, differentiating Equation (10.30b) we get:

$$\frac{d^2w}{dt^2} = \dot{x}_2$$

Then, using Equation (10.31) we obtain:

$$\dot{x}_2 = -\frac{K}{M}x_1 - \frac{B}{M}x_2 + \frac{1}{M}f(t)$$

Finally, it is usual to arrange the state equations in a particular way,

$$\dot{x}_1 = \qquad\qquad + x_2$$

$$\dot{x}_2 = -\frac{K}{M}x_1 - \frac{B}{M}x_2 + \frac{1}{M}f(t)$$

$$w = x_1$$

Note that it is conventional to relate the output of the system to the state variables. Assume that for this system the required output variable is the position of the mass, that is, w.

It is straightforward to rewrite these equations in matrix form:

$$\begin{pmatrix} \dot{x}_1 \\ \dot{x}_2 \end{pmatrix} = \begin{pmatrix} 0 & 1 \\ -K/M & -B/M \end{pmatrix}\begin{pmatrix} x_1 \\ x_2 \end{pmatrix} + \begin{pmatrix} 0 \\ 1/M \end{pmatrix}f(t)$$

$$w = (1 \quad 0)\begin{pmatrix} x_1 \\ x_2 \end{pmatrix} \quad \blacktriangle$$

More generally, the standard form of the state equations for a linear system is given by:

■ $\dot{\mathbf{x}}(t) = A\mathbf{x}(t) + B\mathbf{u}(t)$

$\mathbf{y}(t) = C\mathbf{x}(t) + D\mathbf{u}(t)$

For a system with n state variables, r inputs and p outputs:

$\mathbf{x}(t)$ is an n component column vector representing the states of the nth order system. It is usually called the **state vector**.

$\mathbf{u}(t)$ is an r component column vector composed of the input functions to the system. It is usually called the **input vector**.

$\mathbf{y}(t)$ is a p component column vector composed of the defined outputs of the system. It is referred to as the **output vector**.

$A = (n \times n)$ matrix, known as the **state matrix**.

$B = (n \times r)$ matrix, known as the **input matrix**.

$C = (p \times n)$ matrix, known as the **output matrix**.

$D = (p \times r)$ matrix, which represents the direct coupling between the input and the output.

A, B, C and D have constant elements if the system is time invariant. When presented in this form the equations often appear to be extremely complicated. In fact, the problem is only one of notation. The general nature of the notation allows any linear system to be specified but in many cases the matrices are zero or have simple coefficient values. For example, the matrix D is often zero for a system as it is unusual to have direct coupling between the input and the output of a system. In this format it would be straightforward to present the equations to a digital computer for solution.

Example 10.39 An armature controlled d.c. motor

Derive a state–space model for an armature controlled d.c. motor connected to a mechanical load with combined moment of inertia J, and viscous friction coefficient B. The arrangement is shown in Figure 10.15.

v_a = applied armature voltage

i_a = armature current

R_a = armature resistance

L_a = armature inductance

e_b = back e.m.f. of the motor

ω_m = angular speed of the motor

T = torque generated by the motor.

Solution

Let us assume that the system input is the armature voltage, v_a, and the system output is the angular speed of the motor, ω_m. This is a second order system and so two state variables are required. We will choose the armature current and the angular speed of the motor. So,

$$x_1 = i_a \qquad x_2 = \omega_m$$

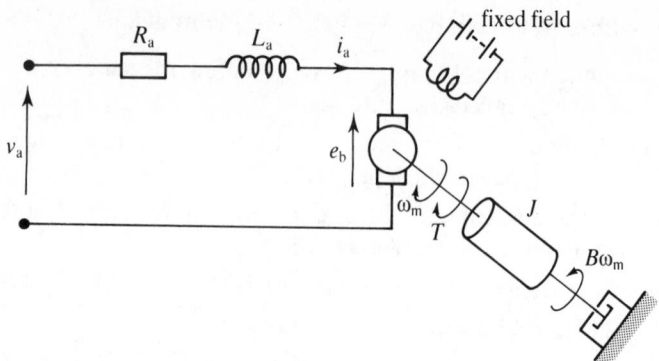

Figure 10.15   An armature controlled d.c. motor.

The next stage is to obtain a mathematical model for the system. Using Kirchhoff's voltage law and the component laws for the resistor and inductor we obtain, for the armature circuit,

$$v_a = i_a R_a + L_a \frac{di_a}{dt} + e_b$$

Now for a d.c. motor the back e.m.f. is proportional to the speed of the motor and is given by $e_b = K_e \omega_m$, K_e constant. So,

$$v_a = i_a R_a + L_a \frac{di_a}{dt} + K_e \omega_m$$

$$\frac{di_a}{dt} = -\frac{R_a}{L_a} i_a - \frac{K_e}{L_a} \omega_m + \frac{1}{L_a} v_a \tag{10.32}$$

Let us now turn to the mechanical part of the system. If G is the net torque about the axis of rotation then the rotational form of Newton's second law of motion states $G = J\, d\omega/dt$, where J is the moment of inertia, and $d\omega/dt$ is the angular acceleration. In this example, the torques are that generated by the motor, T, and a frictional torque $B\omega_m$ which opposes the motion, so that:

$$T - B\omega_m = J \frac{d\omega_m}{dt}$$

For a d.c. motor, the torque developed by the motor is proportional to the armature current and is given by $T = K_T i_a$, where K_T is a constant. So,

$$K_T i_a - B\omega_m = J \frac{d\omega_m}{dt}$$

$$\frac{d\omega_m}{dt} = \frac{K_T}{J} i_a - \frac{B}{J} \omega_m \tag{10.33}$$

Equations (10.32) and (10.33) are the state equations for the system. They can be arranged in matrix form to give:

$$\begin{pmatrix} \dot{i}_a \\ \dot{\omega}_m \end{pmatrix} = \begin{pmatrix} -R_a/L_a & -K_e/L_a \\ K_T/J & -B/J \end{pmatrix} \begin{pmatrix} i_a \\ \omega_m \end{pmatrix} + \begin{pmatrix} 1/L_a \\ 0 \end{pmatrix} v_a$$

Alternatively the notation $\mathbf{x} = \begin{pmatrix} x_1 \\ x_2 \end{pmatrix} = \begin{pmatrix} i_a \\ \omega_m \end{pmatrix}$ and $u = v_a$, can be used.

However, when there is no confusion it is better to retain the original symbols because it makes it easier to see at a glance what the state variables are.

Finally, an output equation is needed. In this case it is trivial as the output variable is the same as one of the state variables. So,

$$\omega_m = (0 \quad 1) \begin{pmatrix} i_a \\ \omega_m \end{pmatrix} \quad \blacktriangle$$

Example 10.40 Coupled tanks

Derive a state–space model for the coupled tank system shown in Figure 10.16. The tanks have cross-sectional areas A_1 and A_2, valve resistances R_1 and R_2, fluid heights h_1 and h_2, input flows q_1 and q_2. Additionally, there is a flow, q_o, out of tank 2. Assume that the valves can be modelled as linear elements and let the density of the fluid in the tanks be ρ. Let q_i be the intermediate flow between the two tanks.

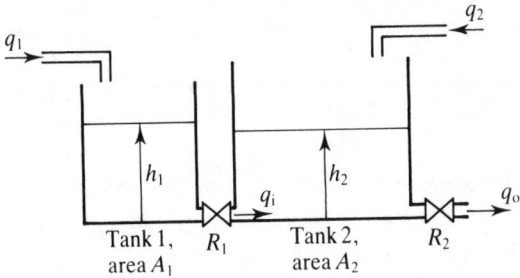

Figure 10.16 Coupled tanks.

Solution

A convenient choice of state variables is the height of the fluid in each of the tanks, although a perfectly acceptable choice would be the volume of fluid in each of the tanks. For tank 1, conservation of mass gives:

$$q_1 - q_i = A_1 \frac{dh_1}{dt}$$

For the resistance element, R_1, the pressure difference across the valve is equal to the product of the flow through the valve and the valve resistance.

This can be thought of as a fluid equivalent of Ohm's law. Note that atmospheric pressure has been ignored as it is the same on both sides of the valve. So,

$$\rho g h_1 - \rho g h_2 = q_i R_1$$

$$q_i = \frac{\rho g}{R_1}(h_1 - h_2) \tag{10.34}$$

Combining these two equations gives

$$\frac{dh_1}{dt} = -\frac{\rho g}{R_1 A_1} h_1 + \frac{\rho g}{R_1 A_1} h_2 + \frac{1}{A_1} q_1 \tag{10.35}$$

For tank 2,

$$q_i + q_2 - q_o = A_2 \frac{dh_2}{dt}$$

$$\rho g h_2 = R_2 q_o \tag{10.36}$$

Combining these equations and using Equation (10.34) to eliminate q_i and q_o gives:

$$\frac{dh_2}{dt} = \frac{\rho g}{R_1 A_2} h_1 - \left(\frac{\rho g}{R_1 A_2} + \frac{\rho g}{R_2 A_2}\right) h_2 + \frac{1}{A_2} q_2 \tag{10.37}$$

Equations (10.35) and (10.37) are the state–space equations for the system and can be written in matrix form as:

$$\begin{pmatrix} \dot{h}_1 \\ \dot{h}_2 \end{pmatrix} = \begin{pmatrix} \dfrac{-\rho g}{R_1 A_1} & \dfrac{\rho g}{R_1 A_1} \\ \dfrac{\rho g}{R_1 A_2} & -\dfrac{\rho g}{R_1 A_2} - \dfrac{\rho g}{R_2 A_2} \end{pmatrix} \begin{pmatrix} h_1 \\ h_2 \end{pmatrix} + \begin{pmatrix} \dfrac{1}{A_1} & 0 \\ 0 & \dfrac{1}{A_2} \end{pmatrix} \begin{pmatrix} q_1 \\ q_2 \end{pmatrix}$$

Note that in this case the input vector is two dimensional as there are two inputs to the system. The output equation is given by:

$$q_o = (0 \quad \rho g / R_2) \begin{pmatrix} h_1 \\ h_2 \end{pmatrix}$$

and is obtained directly from Equation (10.36). ▲

10.10 Numerical methods

All the techniques we have so far met for solving differential equations are known as analytical methods, and these methods give rise to a solution in terms of elementary functions such as $\sin x$, e^x, x^3, etc. In practice, relatively few differential equations can be solved in terms of elementary functions and it is therefore necessary to resort to numerical techniques which result in approximate solutions being obtained at a sequence of values of the independent variable x. We shall consider the first order equation:

$$\frac{dy}{dx} = f(x, y)$$

subject to the initial condition $y(x_0) = y_0$. Usually the solution is obtained at equally spaced values of x, and we call this spacing the **step size**, denoted by h. Unlike analytical methods which give the solution over a continuous range of values of x, numerical methods only give an approximate solution at discrete points although we do have control over the step size h.

It is important to emphasize that numerical methods give approximate solutions. We shall write y_n for this approximate solution at $x = x_n$, whereas we write $y(x_n)$ for the true solution there. In general, these values will not be the same although we hope their difference will be small for obvious reasons.

The simplest numerical method for the solution of:

$$\frac{dy}{dx} = f(x, y) \qquad y(x_0) = y_0$$

is Euler's method which we shall now study.

10.11 Euler's method

You will recall, from Chapter 7, that given a function $y(x)$, the quantity dy/dx represents the gradient of that function. So if $dy/dx = f(x, y)$ and we seek $y(x)$, we see that the differential equation tells us the gradient of the required function. Given the initial condition $y = y_0$ when $x = x_0$ we can picture this single point as shown in Figure 10.17. Moreover we know the gradient of the solution here. Because $dy/dx = f(x, y)$ we see that

$$\left.\frac{dy}{dx}\right|_{x = x_0} = f(x_0, y_0)$$

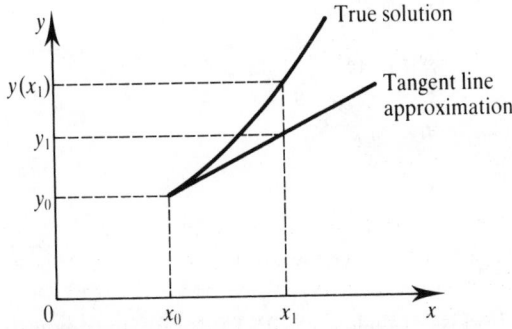

Figure 10.17 Approximation used in Euler's method.

which equals the gradient of the solution at $x = x_0$. Thus the exact solution passes through (x_0, y_0) and has gradient $f(x_0, y_0)$ there. We can draw a straight line through this point with the required gradient to approximate the solution as shown in Figure 10.17. This straight line approximates the true solution, but only near (x_0, y_0) because in general, the gradient is not constant but changes. So, in practice we only extend it a short distance, h, along the x axis to where $x = x_1$. The y coordinate at this point is then taken as y_1. We now develop an expression for y_1. The straight line has gradient $f(x_0, y_0)$ and passes through (x_0, y_0). It can be shown that its equation is therefore:

$$y = y_0 + (x - x_0)f(x_0, y_0)$$

When $x = x_1$ the y coordinate is then given by:

$$y_1 = y_0 + (x_1 - x_0)f(x_0, y_0)$$

and since $x_1 - x_0 = h$ we find:

$$y_1 = y_0 + hf(x_0, y_0)$$

This equation can be used to find y_1. We then regard (x_1, y_1) as known. From this known point the whole process is then repeated using the formula:

■ $y_{i+1} = y_i + hf(x_i, y_i)$

---

and we can therefore generate a whole sequence of approximate values of y. Naturally, the accuracy of the solution will depend upon the step size, h. In fact, for Euler's method, the error incurred is roughly proportional to h, so that by halving the step size we roughly halve the error.

An alternative way of deriving Euler's method is to use a Taylor series expansion. Recall from Chapter 9 that:

$$y(x_0 + h) = y(x_0) + hy'(x_0) + \frac{h^2}{2!}y''(x_0) + \cdots$$

If we truncate after the second term we find:

$$y(x_0 + h) \approx y(x_0) + hy'(x_0)$$

that is,

$$y_1 = y_0 + hy'(x_0) = y_0 + hf(x_0, y_0)$$

so that Euler's method is equivalent to the Taylor series truncated after the second term.

Example 10.41

Use Euler's method with $h = 0.25$ to obtain a numerical solution of:

$$\frac{dy}{dx} = -xy^2$$

subject to $y(0) = 2$ giving approximate values of y for $0 \leq x \leq 1$. Work throughout to three decimal places and determine the exact solution for comparison.

Solution

We need to calculate y_1, y_2, y_3 and y_4. The corresponding x values are $x_1 = 0.25$, $x_2 = 0.5$, $x_3 = 0.75$ and $x_4 = 1.0$. Euler's method becomes:

$$y_{i+1} = y_i + 0.25(-x_i y_i^2) \qquad \text{with} \qquad x_0 = 0 \qquad y_0 = 2$$

We find:

$$y_1 = 2 - 0.25(0)(2^2) = 2.000$$

$$y_2 = 2 - 0.25(0.25)(2^2) = 1.750$$

$$y_3 = 1.750 - 0.25(0.5)(1.750^2) = 1.367$$

$$y_4 = 1.367 - 0.25(0.75)(1.367^2) = 1.017$$

The exact solution can be found by separating the variables:

$$\int \frac{dy}{y^2} = -\int x \, dx$$

so that

$$-\frac{1}{y} = -\frac{x^2}{2} + C$$

Imposing $y(0) = 2$ gives $C = -1/2$ so that:

$$-\frac{1}{y} = -\frac{x^2}{2} - \frac{1}{2}$$

Finally,

$$y = \frac{2}{x^2 + 1}$$

Table 10.2 summarizes the numerical and exact solutions. In this example only one correct significant figure is obtained. In practice, for most equations a very small step size is necessary which means that computation is extremely time consuming. An improvement to Euler's method, which usually yields more accurate solutions, is given in Section 10.12. ▲

Table 10.2 Comparison of numerical solution by Euler's method with exact solution.

i	x_i	y_i numerical	$y(x_i)$ exact
0	0.000	2.000	2.000
1	0.250	2.000	1.882
2	0.500	1.750	1.600
3	0.750	1.367	1.280
4	1.000	1.017	1.000

Example 10.42

Obtain a solution for values of x between 1 and 2, of:

$$\frac{dy}{dx} = \frac{y}{x}$$

subject to $y = 1$ when $x = 1$ using Euler's method. Use a step size of $h = 0.2$, working throughout to two decimal places of accuracy. Compare your answer with the analytical solution. Comment upon the approximate and exact solutions.

Solution

Here we have $f(x, y) = y/x$, $y_0 = 1$, $x_0 = 1$ and $h = 0.20$. With a step size of 0.2, $x_1 = 1.2$, $x_2 = 1.4, \ldots, x_5 = 2.0$. We need to calculate $y_1, y_2, \ldots, y_5$. Euler's method, $y_{i+1} = y_i + hf(x_i, y_i)$, reduces to:

$$y_{i+1} = y_i + 0.2\left(\frac{y_i}{x_i}\right)$$

Therefore,

$$y_1 = y_0 + 0.2\left(\frac{y_0}{x_0}\right) = 1 + 0.2\left(\frac{1}{1}\right) = 1.20$$

Similarly,

$$y_2 = y_1 + 0.2\left(\frac{y_1}{x_1}\right) = 1.20 + 0.2\left(\frac{1.20}{1.20}\right) = 1.40$$

Continuing in a similar fashion we obtain the results shown in Table 10.3. To obtain the analytical solution we separate the variables to give:

$$\int \frac{dy}{y} = \int \frac{dx}{x}$$

Table 10.3 The
solution to Example
10.42 by Euler's
method.

i	x_i	y_i
0	1.00	1.00
1	1.20	1.20
2	1.40	1.40
3	1.60	1.60
4	1.80	1.80
5	2.00	2.00

so that,

$$\ln y = \ln x + \ln D = \ln Dx$$

Therefore,

$$y = Dx$$

When $x = 1$, $y = 1$ so that $D = 1$, and the analytical solution is therefore $y = x$. We see that in this example the numerical solution by Euler's method produces the exact solution. This will always be the case when the exact solution is a linear function and exact arithmetic is employed. ▲

EXERCISES 10.8

1. Using Euler's method estimate $y(3)$ given:

$$y' = \frac{x + y}{x} \qquad y(2) = 1$$

Use (a) $h = 0.5$, (b) $h = 0.25$. Solve this equation analytically and compare your numerical solutions with the true solution.

2. Find $y(0.5)$ if $y' = x + y$, $y(0) = 0$. Use

$h = 0.25$ and $h = 0.1$. Find the true solution for comparison.

3. Use Euler's method to find $v(0.01)$ given:

$$10^{-2}\frac{dv}{dt} + v = \sin 100\pi t \qquad v(0) = 0$$

Take $h = 0.005$ and $h = 0.002$. Find the analytical solution for comparison.

10.12 Improved Euler method

You will recall that Euler's method is obtained by first finding the slope of the solution at (x_0, y_0) and imposing a straight-line approximation (often called a tangent-line approximation) there as indicated in Figure 10.17. If we knew the gradient of the solution at $x = x_1$ in addition to the gradient at $x = x_0$, a better approximation to the gradient over the whole interval might be the mean of the two. Unfortunately, the gradient at $x = x_1$ which is:

$$\frac{dy}{dx}\bigg|_{x=x_1} = f(x_1, y_1)$$

cannot be obtained until we know y_1. What we can do, however, is use the value of y_1 obtained by Euler's method in estimating the gradient at $x = x_1$. This gives rise to the improved Euler method.

$$y_1 = y_0 + h \times (\text{average of gradients at } x_0 \text{ and } x_1)$$

$$= y_0 + h \times \left\{ \frac{f(x_0, y_0) + f(x_1, y_1)}{2} \right\}$$

$$= y_0 + \frac{h}{2}\{f(x_0, y_0) + f(x_1, y_0 + hf(x_0, y_0))\}$$

Then, knowing y_1 the whole process is started again to find y_2, etc. Generally,

■ $y_{i+1} = y_i + \dfrac{h}{2}\{f(x_i, y_i) + f(x_{i+1}, y_i + hf(x_i, y_i))\}$

It can be shown that, like Euler's method, the improved Euler method is equivalent to truncating the Taylor series expansion, in this case after the third term.

Example 10.43

Use the improved Euler method to solve the differential equation, $y' = -xy^2$, $y(0) = 2$, in Example 10.41. As before take $h = 0.25$, but work throughout to four decimal places.

Solution

Here $f(x, y) = -xy^2$, $y_0 = 2$, $x_0 = 0$. We find $f(x_0, y_0) = f(0, 2) = -0(2^2) = 0$. The improved Euler method:

$$y_1 = y_0 + \frac{h}{2}\{f(x_0, y_0) + f(x_1, y_0 + hf(x_0, y_0))\}$$

yields:

$$y_1 = 2 + \frac{0.25}{2}\{0 + f(0.25, 2)\}$$

Now $f(0.25, 2) = -0.25(2^2) = -1$, so that:

$$y_1 = 2 + 0.125(-1) = 1.875$$

We shall set out the calculations required in Table 10.4. You should work through the next few stages yourself. Comparing the values obtained in this

Table 10.4 Applying the improved Euler method to Example 10.43.

i	x_i	y_i	$f(x_i, y_i)$	$y_i + hf(x_i, y_i)$	$f(x_{i+1}, y_i + hf(x_i, y_i))$	y_{i+1}
0	0	2	0	2	−1	1.8750
1	0.25	1.8750	−0.8789	1.6553	−1.3700	1.5939
2	0.5	1.5939	−1.2703	1.2763	−1.2217	1.2824
3	0.75	1.2824	−1.2334	0.9741	−0.9489	1.0096

Table 10.5 Comparison of the improved Euler method with the exact solution.

i	x_i	y_i	$y(x_i)$
0	0.000	2.0000	2.0000
1	0.2500	1.8750	1.8824
2	0.5000	1.5939	1.6000
3	0.7500	1.2824	1.2800
4	1.0000	1.0096	1.0000

way with the exact solution, we see that, over the interval of interest, our solution is usually correct to two decimal places (Table 10.5). ▲

Example 10.44

Apply both Euler's method and the improved Euler method to the solution of:

$$\frac{dy}{dx} = 2x \qquad y = 1 \text{ when } x = 0$$

for $0 \leq x \leq 0.5$ using $h = 0.1$. Compare your answers with the analytical solution. Work throughout to three decimal places.

Solution

We have $dy/dx = 2x$, $x_0 = 0$, $y_0 = 1$. Therefore, using Euler's method we find:

$$y_{i+1} = y_i + hf(x_i, y_i)$$

$$= y_i + 0.1(2x_i)$$

$$= y_i + 0.2x_i$$

Therefore,

$$y_1 = y_0 + 0.2x_0 = 1 + (0.2)(0) = 1$$

$$y_2 = y_1 + 0.2x_1 = 1 + (0.2)(0.1) = 1.02$$

Table 10.6 The solution of Example 10.44 using Euler's method, the improved Euler method and the exact solution.

x_i	y_i (Euler)	y_i (improved Euler)	$y(x_i)$ (exact)
0	1.000	1.000	1.000
0.1	1.000	1.010	1.010
0.2	1.020	1.040	1.040
0.3	1.060	1.090	1.090
0.4	1.120	1.160	1.160
0.5	1.200	1.250	1.250

The complete solution appears in Table 10.6. Check the values given in the table for yourself. Using the improved Euler method we have:

$$y_{i+1} = y_i + \frac{h}{2}\{f(x_i, y_i) + f(x_{i+1}, y_i + hf(x_i, y_i))\}$$

$$= y_i + 0.05(2x_i + 2x_{i+1})$$

$$= y_i + 0.1(x_i + x_{i+1})$$

Therefore,

$$y_1 = y_0 + 0.1(x_0 + x_1) = 1 + 0.1(0 + 0.1) = 1.01$$

$$y_2 = y_1 + 0.1(x_1 + x_2) = 1.01 + 0.1(0.1 + 0.2) = 1.04$$

and so on. The complete solution appears in Table 10.6. Check this for yourself. The analytical solution is:

$$y = \int 2x \, dx = x^2 + c$$

Applying the condition $y(0) = 1$ gives $c = 1$, and so $y = x^2 + 1$. Values of this function are also shown in the table and we note the marked improvement given by the improved Euler method. ▲

EXERCISE 10.9

Apply the improved Euler method to Exercises 10.8.

10.13 Runge–Kutta method of order 4

The 'Runge–Kutta' methods are a large family of methods used for solving differential equations. The Euler and improved Euler methods are special cases of this family. The order of the method refers to the highest power of h included in the Taylor series expansion. We shall now present the fourth order Runge–Kutta method. The derivation of this is beyond the scope of this book.

To solve the equation $dy/dx = f(x, y)$ subject to $y = y_0$ when $x = x_0$ we generate the sequence of values, y_i, from the formula:

$$y_{i+1} = y_i + \frac{h}{6}(k_1 + 2k_2 + 2k_3 + k_4) \qquad (10.38)$$

where

$$k_1 = f(x_i, y_i)$$

$$k_2 = f\left(x_i + \frac{h}{2}, y_i + \frac{h}{2}k_1\right)$$

$$k_3 = f\left(x_i + \frac{h}{2}, y_i + \frac{h}{2}k_2\right)$$

$$k_4 = f(x_i + h, y_i + hk_3)$$

Example 10.45

Use the Runge–Kutta method to solve $dy/dx = -xy^2$, for $0 \le x \le 1$, subject to $y(0) = 2$. Use $h = 0.25$ and work to four decimal places.

Solution

We shall show the calculations required to find y_1. You should follow this through and then verify the results in Table 10.7. The exact solution is shown for comparison.

$$f(x, y) = -xy^2 \qquad h = 0.25 \qquad x_0 = 0 \qquad y_0 = 2$$

Taking $i = 0$ in Equation (10.38), we have:

$$k_1 = f(x_0, y_0) = -0(2)^2 = 0$$

$$k_2 = f(0.125, 2) = -0.125(2)^2 = -0.5$$

$$k_3 = f\left(0.125, 2 + \frac{0.25}{2}(-0.5)\right) = f(0.125, 1.9375)$$

$$= -0.125(1.9375)^2 = -0.4692$$

$$k_4 = f(0.25, 2 + 0.25(-0.4692)) = f(0.25, 1.8827)$$

$$= -0.25(1.8827)^2 = -0.8861$$

Therefore,

$$y_1 = 2 + \frac{0.25}{6}(0 + 2(-0.5) + 2(-0.4692) + (-0.8861)) = 1.8823 \quad \blacktriangle$$

Table 10.7 Comparison of the Runge–Kutta method with exact solutions.

i	x_i	y_i	$y(x_i)$
0	0.0	2.0	2.0
1	0.25	1.8823	1.8824
2	0.5	1.5999	1.6000
3	0.75	1.2799	1.2800
4	1.0	1.0000	1.0000

Example 10.46

Use the fourth order Runge–Kutta method to obtain a solution of:

$$\frac{dy}{dx} = x^2 + x - y$$

subject to $y = 0$ when $x = 0$, for $0 \leq x \leq 0.6$ with $h = 0.2$. Work throughout to four decimal places.

Solution

We have

$$y_{i+1} = y_i + \frac{h}{6}(k_1 + 2k_2 + 2k_3 + k_4)$$

where $k_1 = f(x_i, y_i)$, $k_2 = f\left(x_i + \frac{h}{2}, y_i + \frac{h}{2}k_1\right)$, $k_3 = f\left(x_i + \frac{h}{2}, y_i + \frac{h}{2}k_2\right)$, $k_4 = f(x_i + h, y_i + hk_3)$. In this example, $f(x, y) = x^2 + x - y$ and $x_0 = 0$, $y_0 = 0$. The first stage in the solution is given by:

$k_1 = f(x_0, y_0) = 0$

$k_2 = f(0.1, 0) = 0.11$

$k_3 = f(0.1, 0 + (0.1)(0.11)) = f(0.1, 0.011) = 0.099$

$k_4 = f(0.2, 0 + (0.2)(0.099)) = f(0.2, 0.0198) = 0.2202$

Therefore,

$$y_1 = 0 + \frac{0.2}{6}(0 + 2(0.11) + 2(0.099) + 0.2202) = 0.0213$$

which, in fact, is correct to 4 decimal places. Check the next stage for yourself. The complete solution is shown in Table 10.8. ▲

Table 10.8
The solution to
Example 10.46.

x_i	y_i
0.0	0.0
0.2	0.0213
0.4	0.0897
0.6	0.2112

10.13.1 *Higher order equations*

The techniques discussed for the solution of single first order equations generalize readily to higher order equations such as those described in Section 10.8 and Section 10.9. It is obvious that a computer solution is essential and there are a wide variety of computer packages available to solve such equations.

EXERCISES 10.10

1. Find $y(0.4)$ if $y' = (x + y)^2$ and $y(0) = 1$ using the Runge–Kutta method of order 4. Take (a) $h = 0.2$ and (b) $h = 0.1$.

2. Repeat Exercises 10.8 using the fourth order Runge–Kutta method.

Miscellaneous exercises

1. Find the general solutions of the following equations:

   (a) $dx/dt = 2x$

   (b) $(1 + t)\, dx/dt = 3$

   (c) $dy/dx = y^2 \cos x$

2. Solve $dy/dx = 2$, subject to $y(0) = 3$.

3. Find the general solution of $t\dot{x} + x = 2t$.

4. The technique used for solving linear second order equations can be applied to first order equations. If $L\{y\} = f(x)$ is a first order linear differential equation its general solution can be written as $y(x) = y_H(x) + y_P(x)$ where $L\{y_H\} = 0$ and $L\{y_P\} = f(x)$. Apply this technique to solve:

   $$L\{y\} = \frac{dy}{dx} + 4y = x$$

5. Solve $dx/dt + 2x = e^{2t} \cos t$

   (a) by using an integrating factor

   (b) by finding its complementary function and a particular integral.

6. Find the general solution of $y'' + 16y = x^2$.

7. Find the particular solution of $y'' + 3y' - 4y = e^x$, $y(0) = 2$, $y'(0) = 0$.

8. Use Euler's method with $h = 0.1$ to estimate $x(0.4)$ given $dx/dt = x^2 - 2xt$, $x(0) = 2$.

9. The generalization of Euler's method to the two coupled equations:

   $$\frac{dy_1}{dx} = f(x, y_1, y_2) \qquad \frac{dy_2}{dx} = g(x, y_1, y_2)$$

is given by

$$y_{1_{(i+1)}} = y_{1_{(i)}} + hf(x_i, y_{1_{(i)}}, y_{2_{(i)}})$$

$$y_{2_{(i+1)}} = y_{2_{(i)}} + hg(x_i, y_{1_{(i)}}, y_{2_{(i)}})$$

Given the coupled equations:

$$\frac{dy_1}{dx} = xy_1 + y_2 \qquad \frac{dy_2}{dx} = xy_2 + y_1$$

estimate $y_1(0.3)$ and $y_2(0.3)$, if $y_1(0) = 1$ and $y_2(0) = -1$. Take $h = 0.1$.

10. A particle moves in a straight line such that its displacement from the origin O is x, where x satisfies the differential equation:

$$\frac{d^2x}{dt^2} + 16x = 0$$

(a) Find the general solution of this equation.

(b) If $x(\pi/4) = -12$, and $\dot{x}(\pi/4) = 20$, find the displacement of the particle when $t = \pi/2$.

11. Use an integrating factor to solve the differential equation:

$$\frac{dx}{dt} + x \cot t = \cos 3t$$

The Laplace transform 11

11.1 Introduction

The Laplace transform is used to solve linear constant coefficient differential equations. This is achieved by transforming them to algebraic equations. The algebraic equations are solved, then an inverse Laplace transform is used to obtain a solution in terms of the original variables. This technique can be applied to both single and simultaneous differential equations and so is extremely useful given that differential equation models are common as we saw in Chapter 10.

The Laplace transform is also used to produce transfer functions for the elements of an engineering system. These are represented in diagrammatic form as blocks. The various blocks of the system, corresponding to the system elements, are connected together and the result is a block diagram for the whole system. By breaking a system down in this way it is much easier to visualize how the various parts of the system interact and so a transfer function model is complementary to a time-domain model and is a valuable way of viewing an engineering system. Transfer functions are useful in many areas of engineering, but are particularly important in the design of control systems.

11.2 Definition of the Laplace transform

Let $f(t)$ be a function of time t. In many real problems only values of $t \geq 0$ are of interest. Hence $f(t)$ is given for $t \geq 0$, and for all $t < 0$, $f(t)$ is taken to be 0.

■ The Laplace transform of $f(t)$ is $F(s)$, defined by:

$$F(s) = \int_0^\infty e^{-st} f(t) \, dt$$

427

The variable s may be real or complex. As the integral is improper restrictions may need to be placed on s to ensure that the integral does not diverge. Let $\text{Re}(s) > k$, for some constant k. For the integral to exist it requires that:

$$\lim_{t \to \infty} e^{-kt}|f(t)| = 0$$

For example, if $f(t) = e^{4t}$, then:

$$e^{-kt}|f(t)| = e^{(-k+4)t}$$

and

$$\lim_{t \to \infty} e^{-kt}|f(t)| = 0$$

only if $k > 4$. Therefore the Laplace transform of e^{4t} exists only if $\text{Re}(s) > 4$.

The Laplace transform changes, or transforms the function $f(t)$ into a different function $F(s)$. Note also that whereas $f(t)$ is a function of t, $F(s)$ is a function of s. To denote the Laplace transform of $f(t)$ we write $\mathcal{L}\{f(t)\}$.

Example 11.1

Find the Laplace transforms of

(a) 1

(b) sin t

Solutions

(a) $\mathcal{L}\{1\} = \displaystyle\int_0^\infty e^{-st}1 \, dt = \left[\dfrac{e^{-st}}{-s}\right]_0^\infty = \dfrac{1}{s} = F(s)$

This transform exists provided $\text{Re}(s) > 0$.

(b) $\mathcal{L}\{\sin t\} = \displaystyle\int_0^\infty e^{-st} \sin t \, dt$

This integration was performed in Example 8.47. Hence,

$$\mathcal{L}\{\sin t\} = \frac{1}{s^2 + 1} = F(s) \qquad \text{Re}(s) > 0 \quad \blacktriangle$$

Throughout the chapter it is assumed that s has a value such that all integrals exist.

11.3 Laplace transforms of some common functions

Determining the Laplace transform of a given function, $f(t)$, is essentially an exercise in integration. In order to save effort a look-up table is often used. Table 11.1 lists some common functions and their corresponding Laplace transforms.

Table 11.1 The Laplace transforms of some common functions.

Function, $f(t)$	Laplace transform, $F(s)$
1	$1/s$
t	$1/s^2$
t^2	$2/s^3$
t^n	$n!/s^{n+1}$
e^{-at}	$1/(s+a)$
$t^n e^{-at}$	$n!/(s+a)^{n+1}$
$\sin bt$	$b/(s^2 + b^2)$
$\cos bt$	$s/(s^2 + b^2)$
$e^{-at} \sin bt$	$b/[(s+a)^2 + b^2]$
$e^{-at} \cos bt$	$(s+a)/[(s+a)^2 + b^2]$
$\sinh bt$	$b/(s^2 - b^2)$
$\cosh bt$	$s/(s^2 - b^2)$
$e^{-at} \sinh bt$	$b/[(s+a)^2 - b^2]$
$e^{-at} \cosh bt$	$(s+a)/[(s+a)^2 - b^2]$
$t \sin bt$	$2bs/(s^2 + b^2)^2$
$t \cos bt$	$(s^2 - b^2)/(s^2 + b^2)^2$
$u(t)$ unit step	$1/s$
$u(t-d)$	e^{-sd}/s
$\delta(t)$	1
$\delta(t-d)$	e^{-sd}

Example 11.2

Use Table 11.1 to determine the Laplace transform of each of the following functions.

(a) t^3

(b) t^7

(c) $\sin 4t$

(d) e^{-2t}

(e) $\cos(t/2)$

(f) $\sinh 3t$

(g) $\cosh 5t$

(h) $t \sin 4t$

(i) $e^{-t} \sin 2t$

(j) $e^{3t} \cos t$

Solutions

From Table 11.1 we find the following results:

	$f(t)$	$F(s)$
(a)	t^3	$6/s^4$
(b)	t^7	$7!/s^8$
(c)	$\sin 4t$	$4/(s^2 + 16)$
(d)	e^{-2t}	$1/(s + 2)$
(e)	$\cos(t/2)$	$s/(s^2 + 0.25)$
(f)	$\sinh 3t$	$3/(s^2 - 9)$
(g)	$\cosh 5t$	$s/(s^2 - 25)$
(h)	$t \sin 4t$	$8s/(s^2 + 16)^2$
(i)	$e^{-t} \sin 2t$	$\dfrac{2}{(s + 1)^2 + 4}$
(j)	$e^{3t} \cos t$	$\dfrac{s - 3}{(s - 3)^2 + 1}$

▲

EXERCISES 11.1

1. Determine the Laplace transforms of the following functions.

   (a) $\sin 6t$

   (b) $\cos 4t$

   (c) $\sin(2t/3)$

   (d) $\cos(4t/3)$

   (e) t^4

   (f) $t^2 t^3$

   (g) e^{-3t}

   (h) e^{3t}

   (i) $1/e^{4t}$

   (j) $t \cos 3t$

   (k) $t \sin t$

   (l) $e^{-t} \sin 3t$

   (m) $\dfrac{\cos 7t}{e^{5t}}$

2. Show from the definition of the Laplace transform that
   $$\mathscr{L}\{u(t - d)\} = e^{-sd}/s, \ d > 0.$$

11.4 Properties of the Laplace transform

There are some useful properties of the Laplace transform that can be exploited. They allow us to find the Laplace transforms of more difficult functions. The properties we shall examine are:

(1) Linearity
(2) Shift theorems
(3) Initial value theorem
(4) Final value theorem

11.4.1 *Linearity*

Let f and g be two functions of t and let k be a constant which may be negative. Then

---

■ $\mathcal{L}\{f + g\} = \mathcal{L}\{f\} + \mathcal{L}\{g\}$

$\mathcal{L}\{kf\} = k\mathcal{L}\{f\}$

The first property states that to find the Laplace transform of a sum of functions, we simply sum the Laplace transforms of the individual functions. The second property says that if we multiply a function by a constant k, then the corresponding transform is also multiplied by k. Both of these properties follow directly from the definition of the Laplace transform and linearity properties of integrals, and mean that the Laplace transform is a linear operator. Using the linearity properties and Table 11.1, we can find the Laplace transforms of more complicated functions.

Example 11.3

Find the Laplace transforms of the following functions.

(a) $3 + 2t$
(b) $5t^2 - 2e^t$

Solutions

(a) $\mathcal{L}\{3 + 2t\} = \mathcal{L}\{3\} + \mathcal{L}\{2t\}$

$= 3\mathcal{L}\{1\} + 2\mathcal{L}\{t\}$

$= \dfrac{3}{s} + \dfrac{2}{s^2}$

(b) $\mathcal{L}\{5t^2 - 2e^t\} = \mathcal{L}\{5t^2\} + \mathcal{L}\{-2e^t\}$

$= 5\mathcal{L}\{t^2\} - 2\mathcal{L}\{e^t\}$

$= \dfrac{10}{s^3} - \dfrac{2}{s - 1}$ ▲

With a little practice, some of the intermediate steps may be excluded.

Example 11.4

Find the Laplace transforms of the following:

(a) $5 \cos 3t + 2 \sin 5t - 6t^3$

(b) $-e^{-t} + \frac{1}{2}(\sin t + \cos t)$

Solutions

(a) $\mathcal{L}\{5 \cos 3t + 2 \sin 5t - 6t^3\} = \dfrac{5s}{s^2 + 9} + \dfrac{10}{s^2 + 25} - \dfrac{36}{s^4}$

(b) $\mathcal{L}\left\{-e^{-t} + \dfrac{\sin t + \cos t}{2}\right\} = \dfrac{-1}{s + 1} + \dfrac{1}{2(s^2 + 1)} + \dfrac{s}{2(s^2 + 1)}$

$\qquad\qquad\qquad = \dfrac{-1}{s + 1} + \dfrac{1 + s}{2(s^2 + 1)} \quad \blacktriangle$

11.4.2 *First shift theorem*

■ If $\mathcal{L}\{f(t)\} = F(s)$ then

$\qquad \mathcal{L}\{e^{-at}f(t)\} = F(s + a) \qquad a$ a constant

We obtain $F(s + a)$ by replacing every 's' in $F(s)$ by 's + a'. The variable s has been shifted by an amount a.

Example 11.5

The Laplace transform of a function, $f(t)$, is given by:

$$F(s) = \dfrac{2s + 1}{s(s + 1)}$$

State the Laplace transform of:

(a) $e^{-2t}f(t)$

(b) $e^{3t}f(t)$

Solutions

(a) Use the first shift theorem with $a = 2$.

$\qquad \mathcal{L}\{e^{-2t}f(t)\} = F(s + 2)$

$\qquad\qquad\qquad = \dfrac{2(s + 2) + 1}{(s + 2)(s + 2 + 1)}$

$\qquad\qquad\qquad = \dfrac{2s + 5}{(s + 2)(s + 3)}$

(b) Use the first shift theorem with $a = -3$.

$$\mathscr{L}\{e^{3t}f(t)\} = F(s - 3)$$

$$= \frac{2(s - 3) + 1}{(s - 3)(s - 3 + 1)}$$

$$= \frac{2s - 5}{(s - 3)(s - 2)} \quad \blacktriangle$$

11.4.3 *Second shift theorem*

■ If $\mathscr{L}\{f(t)\} = F(s)$ then

$$\mathscr{L}\{u(t - d)f(t - d)\} = e^{-sd}F(s) \qquad d > 0$$

The function, $u(t - d)f(t - d)$, is obtained by moving $u(t)f(t)$ to the right by an amount d. This is illustrated in Figure 11.1. Note that because $f(t)$ is defined to be 0 for $t < 0$, then $f(t - d) = 0$ for $t < d$. It may appear that $u(t - d)$ is redundant. However, it is necessary for inversion of the Laplace transform, which will be covered in Section 11.6.

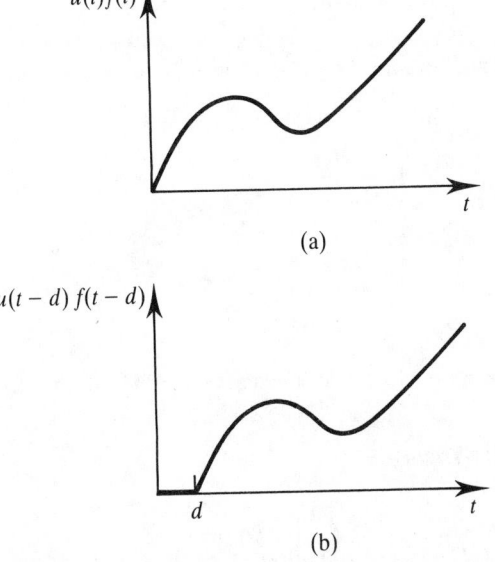

(a)

(b)

Figure 11.1   Shifting the function $u(t)f(t)$ to the right by an amount d yields the function $u(t - d)f(t - d)$.

Example 11.6

Given $\mathcal{L}\{f(t)\} = 2s/(s+9)$, find $\mathcal{L}\{u(t-2)f(t-2)\}$

Solution

Use the second shift theorem with $d = 2$.

$$\mathcal{L}\{u(t-2)f(t-2)\} = \frac{2se^{-2s}}{s+9} \quad \blacktriangle$$

Example 11.7

The Laplace transform of a function is e^{-3s}/s^2. Find the function.

Solution

The exponential term in the transform suggests that the second shift theorem is used. Let

$$e^{-sd}F(s) = \frac{e^{-3s}}{s^2}$$

so that $d = 3$ and $F(s) = 1/s^2$. If we let

$$\mathcal{L}\{f(t)\} = F(s) = \frac{1}{s^2}$$

then $f(t) = t$ and so $f(t-3) = t - 3$. Now, using the second shift theorem

$$\mathcal{L}\{u(t-3)f(t-3)\} = \mathcal{L}\{u(t-3)(t-3)\} = \frac{e^{-3s}}{s^2}$$

Hence the required function is $u(t-3)(t-3)$ as shown in Figure 11.2. $\quad \blacktriangle$

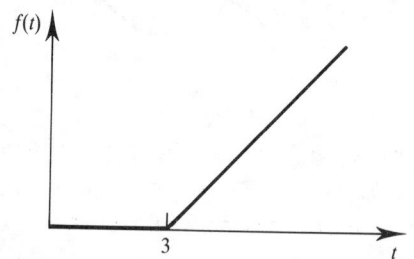

Figure 11.2   The function $f(t) = u(t-3)(t-3)$.

11.4.4 *Initial value theorem*

This theorem applies only to real values of s. First we need to define exponential order. A function, $f(t)$, is of **exponential order**, v, if there are positive constants M and T such that

$$|f(t)| \leq Me^{vt} \qquad \text{for all} \qquad t \geq T$$

Roughly speaking, $f(t)$ is of exponential order v if it grows no faster than a function of the form $M e^{vt}$. The initial value theorem can now be stated.

■ If $f(t)$ is continuous on $[0, \infty)$, $f'(t)$ is piecewise continuous on $[0, \infty)$ and both are of exponential order v then the initial value theorem states:

$$\lim_{s \to \infty} sF(s) = f(0)$$

Example 11.8

Verify the initial value theorem for:

(a) $\sin t$

(b) $\cos t$

Solutions

(a) Let $f(t) = \sin t$. As $f(t)$ and $f'(t)$ are continuous functions they satisfy the continuity requirements. Now $|f(t)| = |\sin t|$ which is always less than or equal to 1. So, $|f(t)| \leq e^{vt}$ for any non-negative value of v. For example, choosing $v = 1$ we see that $\sin t$ is of exponential order 1. Similarly $f'(t)$ is of exponential order 1 also. We have,

$$F(s) = \frac{1}{s^2 + 1}$$

and so,

$$\lim_{s \to \infty} sF(s) = \lim_{s \to \infty} \frac{s}{s^2 + 1} = 0$$

Furthermore

$$f(0) = \sin 0 = 0$$

Hence, $\lim_{s \to \infty} sF(s) = f(0)$ and the theorem is verified.

(b) Let $f(t) = \cos t$. The functions $f(t)$ and $f'(t)$ satisfy the continuity requirements and both are of exponential order 1.

$$F(s) = \frac{s}{s^2 + 1}$$

$$\lim_{s \to \infty} sF(s) = \lim_{s \to \infty} \frac{s^2}{s^2 + 1} = 1$$

Furthermore

$$f(0) = \cos 0 = 1$$

Hence $\lim_{s \to \infty} sF(s) = f(0)$ and the theorem is verified. ▲

11.4.5 *Final value theorem*

This theorem applies only to real values of s and for functions, $f(t)$, which possess a limit as $t \to \infty$.

■ The final value theorem states:

$$\lim_{s \to 0} sF(s) = \lim_{t \to \infty} f(t)$$

Some care is needed when applying the theorem. The Laplace transform of some functions exists only for $\text{Re}(s) > 0$ and for these functions taking the limit as $s \to 0$ is not sensible.

Example 11.9

Verify the final value theorem for $f(t) = e^{-2t}$.

Solution

$\int_0^\infty e^{-st} e^{-2t} \, dt$ exists provided $\text{Re}(s) > -2$ and so,

$$F(s) = \frac{1}{s + 2} \qquad \text{Re}(s) > -2$$

Since we only require $\text{Re}(s) > -2$, it is permissible to let $s \to 0$.

$$\lim_{s \to 0} sF(s) = \lim_{s \to 0} \frac{s}{s + 2} = 0$$

Furthermore

$$\lim_{t \to \infty} e^{-2t} = 0$$

So $\lim_{s \to 0} sF(s) = \lim_{t \to \infty} f(t)$ and the theorem is verified. ▲

EXERCISES 11.2

1. Find the Laplace transforms of the following functions.

   (a) $3t^2 - 4$

   (b) $2 \sin 4t + 11 - t$

   (c) $2 - t^2 + 2t^4$

   (d) $3e^{2t} + 4 \sin t$

   (e) $\frac{1}{3} \sin 3t - 4 \cos\left(\dfrac{t}{2}\right)$

   (f) $3t^4 e^{5t} + t$

   (g) $\sinh 2t + 3 \cosh 2t$

   (h) $e^{-t} \sin 3t + 4e^{-t} \cos 3t$

2. The Laplace transform of $f(t)$ is given as

$$F(s) = \frac{3s^2 - 1}{s^2 + s + 1}$$

   Find the Laplace transform of

   (a) $e^{-t} f(t)$

(b) $e^{3t}f(t)$

(c) $e^{-t/2}f(t)$

(c) $\mathcal{L}\left\{u(t-4)\dfrac{f(t-4)}{2}\right\}$

3. Given

4. Verify the initial value theorem for:

(a) $f(t) = t^2$

$$\mathcal{L}\{f(t)\} = \frac{4s}{s^2 + 1}$$

(b) $f(t) = e^{3t}$

5. Verify the final value theorem for:

find

(a) $f(t) = e^{-t}\sin t$

(a) $\mathcal{L}\{u(t-1)f(t-1)\}$

(b) $f(t) = e^{-t} + 1$

(b) $\mathcal{L}\{3u(t-2)f(t-2)\}$

11.5 Laplace transform of derivatives and integrals

In later sections we shall use the Laplace transform to solve differential equations. In order to do this we need to be able to find the Laplace transform of derivatives of functions. Let $f(t)$ be a function of t, and f' and f'' the first and second derivatives of f. The Laplace transform of $f(t)$ is $F(s)$. Then

■ $\mathcal{L}\{f'\} = sF(s) - f(0)$

    $\mathcal{L}\{f''\} = s^2F(s) - sf(0) - f'(0)$

where $f(0)$ and $f'(0)$ are the initial values of f and f'. The general case for the Laplace transform of an nth derivative is:

    $\mathcal{L}\{f^{(n)}\} = s^nF(s) - s^{n-1}f(0) - s^{n-2}f'(0) - \cdots - f^{(n-1)}(0)$

Another useful result is:

■ $\mathcal{L}\left\{\displaystyle\int_0^t f(t)\,dt\right\} = \dfrac{1}{s}F(s)$

Example 11.10

The Laplace transform of $x(t)$ is $X(s)$. Given $x(0) = 2$ and $x'(0) = -1$, write expressions for the Laplace transforms of

(a) $2x'' - 3x' + x$

(b) $-x'' + 2x' + x$

Solutions

    $\mathcal{L}\{x'\} = sX(s) - x(0) = sX(s) - 2$

    $\mathcal{L}\{x''\} = s^2X(s) - sx(0) - x'(0) = s^2X(s) - 2s + 1$

(a) $\mathscr{L}\{2x'' - 3x' + x\} = 2(s^2X(s) - 2s + 1) - 3(sX(s) - 2) + X(s)$

$$= (2s^2 - 3s + 1)X(s) - 4s + 8$$

(b) $\mathscr{L}\{-x'' + 2x' + x\} = -(s^2X(s) - 2s + 1) + 2(sX(s) - 2) + X(s)$

$$= (-s^2 + 2s + 1)X(s) + 2s - 5 \quad \blacktriangle$$

Example 11.11 Voltage across a capacitor

The voltage, $v(t)$, across a capacitor of capacitance C is given by:

$$v(t) = \frac{1}{C}\int_0^t i(t) \, dt$$

Taking Laplace transforms yields

$$V(s) = \frac{1}{Cs}I(s)$$

where $V(s) = \mathscr{L}\{v(t)\}$ and $I(s) = \mathscr{L}\{i(t)\}$. $\quad \blacktriangle$

EXERCISES 11.3

1. The Laplace transform of $y(t)$ is $Y(s)$, $y(0) = 3$, $y'(0) = 1$. Find the Laplace transforms of the following expressions.

   (a) y'

   (b) y''

   (c) $3y'' - y'$

   (d) $y'' + 2y' + 3y$

   (e) $3y'' - y' + 2y$

   (f) $-4y'' + 5y' - 3y$

2. Given the Laplace transform of $f(t)$ is $F(s)$, $f(0) = 2$, $f'(0) = 3$ and $f''(0) = -1$,

   find the Laplace transforms of:

   (a) $3f' - 2f$

   (b) $3f'' - f' + f$

   (c) f'''

   (d) $2f''' - f'' + 4f' - 2f$

3. (a) If $F(s) = \mathscr{L}\{f(t)\} = \int_0^\infty e^{-st}f(t) \, dt$, show using integration by parts that

   (i) $\mathscr{L}\{f'(t)\} = sF(s) - f(0)$

   (ii) $\mathscr{L}\{f''(t)\} = s^2F(s) - sf(0) - f'(0)$

   (b) If $F(s) = \mathscr{L}\{f(t)\}$ prove that
   $\mathscr{L}\{e^{-at}f(t)\} = F(s + a)$

   Deduce $\mathscr{L}\{te^{-t}\}$, given $\mathscr{L}\{t\} = 1/s^2$.

11.6 Inverse Laplace transforms

As mentioned in Section 11.5, the Laplace transform can be used to solve differential equations. However, before such an application can be put into practice, we must study the inverse Laplace transform. So far in this chapter we have been given functions of t and found their Laplace transforms. We now consider the problem of finding a function $f(t)$, having been given the Laplace transform, $F(s)$. Clearly

Table 11.1 and the properties of Laplace transforms will help us to do this. If $\mathcal{L}\{f(t)\} = F(s)$ we write:

$$f(t) = \mathcal{L}^{-1}\{F(s)\}$$

and call $\mathcal{L}^{-1}$ the inverse Laplace transform. Like $\mathcal{L}$, $\mathcal{L}^{-1}$ can be shown to be a linear operator.

Example 11.12

Find the inverse Laplace transforms of the following:

(a) $2/s^3$

(b) $16/s^3$

(c) $s/(s^2 + 1)$

(d) $1/(s^2 + 1)$

(e) $(s + 1)/(s^2 + 1)$

Solutions

(a) We need to find a function of t which has a Laplace transform of $2/s^3$.
Using Table 11.1 we see $\mathcal{L}^{-1}\{2/s^3\} = t^2$.

(b) $\mathcal{L}^{-1}\left\{\dfrac{16}{s^3}\right\} = 8\mathcal{L}^{-1}\left\{\dfrac{2}{s^3}\right\} = 8t^2$

(c) $\mathcal{L}^{-1}\left\{\dfrac{s}{s^2 + 1}\right\} = \cos t$

(d) $\mathcal{L}^{-1}\left\{\dfrac{1}{s^2 + 1}\right\} = \sin t$

(e) $\mathcal{L}^{-1}\left\{\dfrac{s + 1}{s^2 + 1}\right\} = \mathcal{L}^{-1}\left\{\dfrac{s}{s^2 + 1}\right\} + \mathcal{L}^{-1}\left\{\dfrac{1}{s^2 + 1}\right\} = \cos t + \sin t$

In parts (a), (c) and (d) we obtained the inverse Laplace transform by referring directly to the table. In (b) and (e) we used the linearity properties of the inverse transform, and then referred to Table 11.1. ▲

Example 11.13

Find the inverse Laplace transforms of the following functions.

(a) $\dfrac{10}{(s + 2)^4}$

(b) $\dfrac{(s + 1)}{(s + 1)^2 + 4}$

(c) $\dfrac{15}{(s - 1)^2 - 9}$

Solutions

(a) $\mathcal{L}^{-1}\left\{\dfrac{10}{(s+2)^4}\right\} = \dfrac{10}{6}\mathcal{L}^{-1}\left\{\dfrac{6}{(s+2)^4}\right\} = \dfrac{5t^3 e^{-2t}}{3}$

(b) $\mathcal{L}^{-1}\left\{\dfrac{s+1}{(s+1)^2+4}\right\} = \mathcal{L}^{-1}\left\{\dfrac{s+1}{(s+1)^2+2^2}\right\} = e^{-t}\cos 2t$

(c) $\mathcal{L}^{-1}\left\{\dfrac{15}{(s-1)^2-9}\right\} = 5\mathcal{L}^{-1}\left\{\dfrac{3}{(s-1)^2-3^2}\right\} = 5e^t\sinh 3t$

The function is written to match exactly the standard forms given in Table 11.1, with possibly a constant factor being present. Often the denominator needs to be written in standard form as illustrated in the next example. ▲

Example 11.14

Find the inverse Laplace transforms of the following functions.

(a) $\dfrac{s+3}{s^2+6s+13}$

(b) $\dfrac{2s+3}{s^2+6s+13}$

(c) $\dfrac{s-1}{2s^2+8s+11}$

Solutions

(a) By completing the square we can write

$$s^2+6s+13 = (s+3)^2+4 = (s+3)^2+2^2$$

Hence we may write:

$$\mathcal{L}^{-1}\left\{\frac{s+3}{s^2+6s+13}\right\} = \mathcal{L}^{-1}\left\{\frac{s+3}{(s+3)^2+2^2}\right\} = e^{-3t}\cos 2t$$

(b) $\dfrac{2s+3}{s^2+6s+13} = \dfrac{2s+3}{(s+3)^2+2^2} = \dfrac{2s+6}{(s+3)^2+2^2} - \dfrac{3}{(s+3)^2+2^2}$

$$= 2\left(\frac{s+3}{(s+3)^2+2^2}\right) - \frac{3}{2}\left(\frac{2}{(s+3)^2+2^2}\right)$$

The expressions in brackets are standard forms so their inverse Laplace transforms can be found from Table 11.1.

$$\mathcal{L}^{-1}\left\{\frac{2s+3}{s^2+6s+13}\right\} = 2e^{-3t}\cos 2t - \frac{3e^{-3t}\sin 2t}{2}$$

(c) We write the expression using standard forms.

$$\frac{s-1}{2s^2 + 8s + 11} = \frac{1}{2}\frac{s-1}{s^2 + 4s + 5.5} = \frac{1}{2}\frac{s-1}{(s+2)^2 + 1.5}$$

$$= \frac{1}{2}\left(\frac{s+2}{(s+2)^2 + 1.5} - \frac{3}{(s+2)^2 + 1.5}\right)$$

$$= \frac{1}{2}\left(\frac{s+2}{(s+2)^2 + 1.5} - \frac{3}{\sqrt{1.5}}\frac{\sqrt{1.5}}{(s+2)^2 + 1.5}\right)$$

Having written the expression in terms of standard forms, the inverse Laplace transform can now be found.

$$\mathcal{L}^{-1}\left\{\frac{s-1}{2s^2 + 8s + 11}\right\} = \frac{1}{2}\left(e^{-2t}\cos\sqrt{1.5}t - \frac{3}{\sqrt{1.5}}e^{-2t}\sin\sqrt{1.5}t\right)$$

In every case the given function of s is written as a linear combination of standard forms contained in Table 11.1. ▲

EXERCISES 11.4

By using the standard forms in Table 11.1 find the inverse Laplace transforms of the following functions.

(a) $\dfrac{3}{2s}$

(b) $\dfrac{4}{s} - \dfrac{1}{s^3}$

(c) $\dfrac{30}{s^2}$

(d) $\dfrac{1}{3(s+2)}$

(e) $\dfrac{3s-7}{s^2+9}$

(f) $\dfrac{s-6}{s-4}$

(g) $\dfrac{s+4}{(s+4)^2+1}$

(h) $\dfrac{5}{(s+4)^2+1}$

(i) $\dfrac{6s+17}{(s+4)^2+1}$

(j) $\dfrac{s}{s^2+2s+7}$

(k) $0.5/(s+0.5)^2$

11.7 Using partial fractions to find the inverse Laplace transform

The inverse Laplace transform of a fraction is often best found by expressing it as its partial fractions, and finding the inverse transform of these. (See Section 1.4.3 for a treatment of partial fractions.)

Example 11.15

Find the inverse Laplace transform of $\dfrac{3s^2 + 6s + 2}{s^3 + 3s^2 + 2s}$.

Solution

Using Example 1.19 we have:

$$\frac{3s^2 + 6s + 2}{s^3 + 3s^2 + 2s} = \frac{1}{s} + \frac{1}{s+1} + \frac{1}{s+2}$$

The inverse Laplace transform of the partial fractions is easily found.

$$\mathcal{L}^{-1}\left\{\frac{3s^2 + 6s + 2}{s^3 + 3s^2 + 2s}\right\} = 1 + e^{-t} + e^{-2t} \quad \blacktriangle$$

Example 11.16

Find the inverse Laplace transform of $\dfrac{3s^2 + 11s + 14}{s^3 + 2s^2 - 11s - 52}$

Solution

From Example 1.20 we have:

$$\frac{3s^2 + 11s + 14}{s^3 + 2s^2 - 11s - 52} = \frac{2}{s-4} + \frac{s+3}{s^2 + 6s + 13}$$

We find the inverse Laplace transforms of the partial fractions.

$$\mathcal{L}^{-1}\left\{\frac{2}{s-4}\right\} = 2e^{4t} \qquad \mathcal{L}^{-1}\left\{\frac{s+3}{s^2 + 6s + 13}\right\} = e^{-3t}\cos 2t$$

So

$$\mathcal{L}^{-1}\left\{\frac{3s^2 + 11s + 14}{s^3 + 2s^2 - 11s - 52}\right\} = 2e^{4t} + e^{-3t}\cos 2t \quad \blacktriangle$$

EXERCISES 11.5

Express the following fractions as partial fractions and hence find their inverse Laplace transforms.

(a) $\dfrac{3s + 3}{(s-1)(s+2)}$

(b) $\dfrac{5s}{(s+1)(2s-3)}$

(c) $\dfrac{2s + 5}{s+2}$

(d) $\dfrac{s^2 + 4s + 4}{s^3 + 2s^2 + 5s}$

(e) $\dfrac{1 - s}{(s+1)(s^2 + 2s + 2)}$

(f) $\dfrac{s + 4}{s^2 + 4s + 4}$

(g) $\dfrac{2(s^3 - 3s^2 + s - 1)}{(s^2 + 4s + 1)(s^2 + 1)}$

(h) $\dfrac{3s^2 - s + 8}{(s^2 - 2s + 3)(s+2)}$

11.8 Finding the inverse Laplace transform using complex numbers

In Sections 11.6 and 11.7 we found the inverse Laplace transform using standard forms and partial fractions. We now look at a method of finding inverse Laplace transforms using complex numbers. Essentially the method is one using partial fractions, but where all the factors in the denominator are linear, that is, there are no quadratic factors. We illustrate the method using the examples from Section 11.6.

Example 11.17

Find the inverse Laplace transforms of the following functions.

(a) $\dfrac{s + 3}{s^2 + 6s + 13}$

(b) $\dfrac{2s + 3}{s^2 + 6s + 13}$

(c) $\dfrac{s - 1}{2s^2 + 8s + 11}$

Solutions

(a) We first factorize the denominator. To do this we solve $s^2 + 6s + 13 = 0$ using the formula:

$$s = \frac{-6 \pm \sqrt{36 - 4(13)}}{2} = \frac{-6 \pm \sqrt{-16}}{2} = \frac{-6 \pm 4j}{2} = -3 \pm 2j$$

It then follows that the denominator can be factorized as $(s - a)(s - b)$ where $a = -3 + 2j$ and $b = -3 - 2j$. Then, using the partial fractions method:

$$\frac{s + 3}{s^2 + 6s + 13} = \frac{s + 3}{(s - a)(s - b)} = \frac{A}{s - a} + \frac{B}{s - b}$$

The unknown constants A and B can now be found.

$$s + 3 = A(s - b) + B(s - a)$$

Put $s = a = -3 + 2j$

$$2j = A(-3 + 2j - b) = A(4j)$$

$$A = \frac{1}{2}$$

Equate the coefficients of s

$$1 = A + B$$

$$B = \frac{1}{2}$$

So,

$$\frac{s + 3}{s^2 + 6s + 13} = \frac{1}{2}\left(\frac{1}{s - a} + \frac{1}{s - b}\right)$$

The inverse Laplace transform can now be found.

$$\mathcal{L}^{-1}\left\{\frac{s+3}{s^2+6s+13}\right\} = \frac{1}{2}\mathcal{L}^{-1}\left\{\frac{1}{s-a}+\frac{1}{s-b}\right\}$$

$$= \frac{1}{2}(e^{at}+e^{bt}) = \frac{1}{2}(e^{(-3+2j)t}+e^{(-3-2j)t})$$

$$= \frac{1}{2}e^{-3t}(e^{2jt}+e^{-2jt})$$

$$= \frac{1}{2}e^{-3t}(\cos 2t + j\sin 2t + \cos 2t - j\sin 2t)$$

$$= e^{-3t}\cos 2t$$

(b) $$\frac{2s+3}{s^2+6s+13} = \frac{2s+3}{(s-a)(s-b)} = \frac{A}{s-a}+\frac{B}{s-b}$$

where $a = -3+2j$ and $b = -3-2j$. Hence,

$$2s+3 = A(s-b)+B(s-a)$$

Put $s = a = -3+2j$,

$$-3+4j = A(4j)$$

$$A = 1+0.75j$$

Equate the coefficients of s:

$$2 = A+B$$

$$B = 1-0.75j$$

Hence,

$$\frac{2s+3}{s^2+6s+13} = \frac{1+0.75j}{s-a}+\frac{1-0.75j}{s-b}$$

Taking the inverse Laplace transform yields,

$$\mathcal{L}^{-1}\left\{\frac{2s+3}{s^2+6s+13}\right\} = (1+0.75j)e^{at}+(1-0.75j)e^{bt}$$

$$= (1+0.75j)e^{(-3+2j)t}+(1-0.75j)e^{(-3-2j)t}$$

$$= (1+0.75j)e^{-3t}(\cos 2t + j\sin 2t)$$

$$+ (1-0.75j)e^{-3t}(\cos 2t - j\sin 2t)$$

$$= e^{-3t}(2\cos 2t - 1.5\sin 2t)$$

(c) $$\frac{s-1}{2s^2+8s+11} = \frac{1}{2}\left(\frac{s-1}{s^2+4s+5.5}\right) = \frac{1}{2}\left(\frac{s-1}{(s-a)(s-b)}\right)$$

where $a = -2+\sqrt{1.5}j$, $b = -2-\sqrt{1.5}j$.

Applying the method of partial fractions produces:

$$\frac{s-1}{(s-a)(s-b)} = \frac{A}{s-a} + \frac{B}{s-b}$$

Hence,

$$s - 1 = A(s - b) + B(s - a)$$

By letting $s = a$, then $s = b$ in turn gives:

$$A = 0.5 + \sqrt{1.5}j \qquad B = 0.5 - \sqrt{1.5}j$$

Hence we may write:

$$\frac{s-1}{(s-a)(s-b)} = \frac{0.5 + \sqrt{1.5}j}{s-a} + \frac{0.5 - \sqrt{1.5}j}{s-b}$$

Taking the inverse Laplace transform yields:

$$\mathcal{L}^{-1}\left\{\frac{s-1}{(s-a)(s-b)}\right\} = (0.5 + \sqrt{1.5}j)e^{at} + (0.5 - \sqrt{1.5}j)e^{bt}$$

$$= (0.5 + \sqrt{1.5}j)e^{(-2+\sqrt{1.5}j)t}$$

$$+ (0.5 - \sqrt{1.5}j)e^{(-2-\sqrt{1.5}j)t}$$

$$= e^{-2t}\{(0.5 + \sqrt{1.5}j)e^{\sqrt{1.5}jt}$$

$$+ (0.5 - \sqrt{1.5}j)e^{-\sqrt{1.5}jt}\}$$

$$= e^{-2t}\{(0.5 + \sqrt{1.5}j)(\cos\sqrt{1.5}t + j\sin\sqrt{1.5}t)$$

$$+ (0.5 - \sqrt{1.5}j)(\cos\sqrt{1.5}t - j\sin\sqrt{1.5}t)\}$$

$$= e^{-2t}(\cos\sqrt{1.5}t - 2\sqrt{1.5}\sin\sqrt{1.5}t)$$

Hence

$$\mathcal{L}^{-1}\left\{\frac{s-1}{2s^2 + 8s + 11}\right\} = \frac{e^{-2t}}{2}(\cos\sqrt{1.5}t - 2\sqrt{1.5}\sin\sqrt{1.5}t) \quad \blacktriangle$$

As seen from Example 11.17, when complex numbers are allowed, all the factors in the denominator are linear. The unknown constants are evaluated using particular values of s or equating coefficients.

EXERCISES 11.6

Express the following expressions as partial fractions, using complex numbers if necessary. Hence find their inverse Laplace transforms.

(a) $(3s - 2)/(s^2 + 6s + 13)$

(b) $(2s + 1)/(s^2 - 2s + 2)$

(c) $\dfrac{s^2}{(s^2/2) - s + 5}$

(d) $(s^2 + s + 1)/(s^2 - 2s + 3)$

(e) $(2s + 3)/(-s^2 + 2s - 5)$

11.9 The convolution theorem

■   Let $f(t)$ and $g(t)$ be two piecewise continuous functions. The **convolution** of $f(t)$ and $g(t)$, denoted $(f * g)(t)$, is defined by

$$(f * g)(t) = \int_0^t f(t - v)g(v)\, dv$$

Example 11.18

Find the convolution of $2t$ and t^3.

Solution

$f(t) = 2t$, $g(t) = t^3$, $f(t - v) = 2(t - v)$, $g(v) = v^3$.

$$2t * t^3 = \int_0^t 2(t - v)v^3\, dv = 2\int_0^t tv^3 - v^4\, dv$$

$$= 2\left[\frac{tv^4}{4} - \frac{v^5}{5}\right]_0^t = 2\left[\frac{t^5}{4} - \frac{t^5}{5}\right]_0^t$$

$$= \frac{t^5}{10} \quad ▲$$

It can be shown that

$$f * g = g * f$$

but the proof is omitted. Instead, this property is illustrated by an example.

Example 11.19

Show that $f * g = g * f$ where $f(t) = 2t$ and $g(t) = t^3$.

Solution

$f * g = 2t * t^3 = t^5/10$ by Example 11.18. From the definition of convolution

$$(g * f)(t) = \int_0^t g(t - v)f(v)\, dv$$

We have $g(t) = t^3$, so $g(t - v) = (t - v)^3$, and $f(v) = 2v$. Therefore

$$g * f = t^3 * 2t = \int_0^t (t - v)^3 2v\, dv$$

$$= \int_0^t (t^3 - 3t^2 v + 3tv^2 - v^3)2v\, dv = 2\int_0^t t^3 v - 3t^2 v^2 + 3tv^3 - v^4\, dv$$

$$= 2\left[\frac{t^3 v^2}{2} - t^2 v^3 + \frac{3tv^4}{4} - \frac{v^5}{5}\right]_0^t = \frac{t^5}{10} \quad ▲$$

■ For any functions $f(t)$ and $g(t)$

$$f * g = g * f$$

11.9.1 *The convolution theorem*

Let $f(t)$ and $g(t)$ be piecewise continuous functions, with $\mathscr{L}\{f(t)\} = F(s)$ and $\mathscr{L}\{g(t)\} = G(s)$. The convolution theorem allows us to find the inverse Laplace transform of a product of transforms, $F(s)G(s)$.

■ $\mathscr{L}^{-1}\{F(s)G(s)\} = (f * g)(t)$

Example 11.20

Use the convolution theorem to find the inverse Laplace transforms of the following functions.

(a) $\dfrac{1}{(s + 2)(s + 3)}$ (b) $\dfrac{3}{s(s^2 + 4)}$

Solutions

(a) Let $F(s) = 1/(s + 2)$, $G(s) = 1/(s + 3)$. Then $f(t) = \mathscr{L}^{-1}\{F(s)\} = e^{-2t}$, $g(t) = \mathscr{L}^{-1}\{G(s)\} = e^{-3t}$.

$$\mathscr{L}^{-1}\left\{\frac{1}{(s + 2)(s + 3)}\right\} = \mathscr{L}^{-1}\{F(s)G(s)\} = (f * g)(t)$$

$$= \int_0^t e^{-2(t-v)} e^{-3v} \, dv = \int_0^t e^{-2t} e^{2v} e^{-3v} \, dv$$

$$= \int_0^t e^{-2t} e^{-v} \, dv$$

$$= e^{-2t}[-e^{-v}]_0^t = e^{-2t}(-e^{-t} + 1) = e^{-2t} - e^{-3t}$$

(b) Let $F(s) = 3/s$, $G(s) = 1/(s^2 + 4)$. Then $f(t) = 3$, $g(t) = \frac{1}{2}\sin 2t$. So,

$$\mathscr{L}^{-1}\left\{\frac{3}{s(s^2 + 4)}\right\} = \mathscr{L}^{-1}\{F(s)G(s)\}$$

$$= (f * g)(t)$$

$$= \int_0^t 3 \frac{\sin 2v}{2} \, dv = \frac{3}{2} \int_0^t \sin 2v \, dv$$

$$= \frac{3}{2}\left[\frac{-\cos 2v}{2}\right]_0^t = \frac{3}{4}(1 - \cos 2t) \quad \blacktriangle$$

EXERCISES 11.7

1. Find

(a) $e^{-2t} * e^{-t}$

(b) $t^2 * e^{-3t}$

2. Find $f * g$ when

(a) $f = 1, g = t$

(b) $f = t^2, g = t$

(c) $f = e^t, g = t$

(d) $f = \sin t, g = t$

In each case verify that
$\mathscr{L}\{f\} \times \mathscr{L}\{g\} = \mathscr{L}\{f * g\}$.

3. If $F(s) = 1/(s - 1)$, $G(s) = 1/s$ and $H(s) = 1/(2s + 3)$ use the convolution

theorem to find the inverse Laplace transforms of:

(a) $F(s)G(s)$

(b) $F(s)H(s)$

(c) $G(s)H(s)$

4. Use the convolution theorem to determine the inverse Laplace transforms of:

(a) $\dfrac{1}{s^2(s + 1)}$

(b) $\dfrac{1}{(s + 3)(s - 2)}$

(c) $\dfrac{1}{(s^2 + 1)^2}$

11.10 Solving linear constant coefficient differential equations using the Laplace transform

So far we have seen how to find the Laplace transform of a function of time and how to find the inverse Laplace transform. We now apply this to finding the particular solution of differential equations. The initial conditions are automatically satisfied when solving an equation using the Laplace transform. They are contained in the transform of the derivative terms.

The Laplace transform of the equation is found. This transforms the differential equation into an algebraic equation. The transform of the dependent variable is found and then the inverse transform is calculated to yield the required particular solution.

Example 11.21

Solve

$$\frac{dx}{dt} + x = 9e^{2t} \qquad x(0) = 3$$

using the Laplace transform.

Solution

The Laplace transform of both sides of the equation is found.

$$sX(s) - x(0) + X(s) = \frac{9}{s - 2}$$

$$sX(s) - 3 + X(s) = \frac{9}{s - 2}$$

$$(s+1)X(s) = \frac{9}{s-2} + 3 = \frac{3(s+1)}{s-2}$$

$$X(s) = \frac{3}{s-2}$$

Taking the inverse Laplace transform yields

$$x(t) = 3e^{2t} \quad \blacktriangle$$

Example 11.22   RL circuit with ramp input

Use the Laplace transform to solve

$$iR + L\frac{di}{dt} = t \qquad t \geq 0 \qquad i(0) = 0$$

This equation was introduced in Example 10.15.

Solution

Let $\mathcal{L}\{i(t)\} = I(s)$ and take the Laplace transform of both sides of the equation.

$$I(s)R + L(sI(s) - i(0)) = \frac{1}{s^2}$$

Now $i(0) = 0$ and so

$$I(s)(R + Ls) = \frac{1}{s^2}$$

$$I(s) = \frac{1}{s^2(R + Ls)} = \frac{A}{s} + \frac{B}{s^2} + \frac{C}{R + Ls}$$

Evaluating the constants A, B and C gives

$$I(s) = -\frac{L}{R^2 s} + \frac{1}{Rs^2} + \frac{L^2}{R^2(R + Ls)} = \frac{-L}{R^2 s} + \frac{1}{Rs^2} + \frac{L}{R^2\left(\dfrac{R}{L} + s\right)}$$

Taking the inverse Laplace transform yields

$$i(t) = -\frac{L}{R^2} + \frac{t}{R} + \frac{L}{R^2}e^{-Rt/L} = \frac{t}{R} + \frac{L}{R^2}(e^{-Rt/L} - 1) \qquad t \geq 0 \quad \blacktriangle$$

Example 11.23

Solve

$$x'' + 2x' + 2x = e^{-t} \qquad x(0) = x'(0) = 0$$

using Laplace transforms.

Solution

Taking the Laplace transform of both sides:

$$s^2X(s) - sx(0) - x'(0) + 2(sX(s) - x(0)) + 2X(s) = \frac{1}{s+1}$$

Therefore,

$$(s^2 + 2s + 2)X(s) = \frac{1}{s+1} \qquad \text{since } x(0) = x'(0) = 0$$

$$X(s) = \frac{1}{(s+1)(s^2+2s+2)} = \frac{1}{(s+1)(s-a)(s-b)}$$

where $a = -1 + j$, $b = -1 - j$. Using partial fractions gives:

$$X(s) = \frac{1}{s+1} - \frac{1}{2}\left(\frac{1}{s-a} + \frac{1}{s-b}\right)$$

Taking the inverse Laplace transform

$$x(t) = e^{-t} - \frac{1}{2}(e^{at} + e^{bt})$$

$$= e^{-t} - \frac{1}{2}(e^{(-1+j)t} + e^{(-1-j)t})$$

$$= e^{-t} - \frac{e^{-t}}{2}(e^{jt} + e^{-jt})$$

$$= e^{-t} - \frac{e^{-t}}{2}(\cos t + j \sin t + \cos t - j \sin t)$$

$$= e^{-t} - e^{-t}\cos t \quad \blacktriangle$$

Example 11.24

Solve

$$x'' - 5x' + 6x = 6t - 4 \qquad x(0) = 1 \qquad x'(0) = 2$$

Solution

The Laplace transform of both sides of the equation is found. Let $\mathscr{L}\{x\} = X(s)$.

$$s^2X(s) - sx(0) - x'(0) - 5(sX(s) - x(0)) + 6X(s) = \frac{6}{s^2} - \frac{4}{s}$$

$$(s^2 - 5s + 6)X(s) = \frac{6}{s^2} - \frac{4}{s} + s - 3 = \frac{s^3 - 3s^2 - 4s + 6}{s^2}$$

$$X(s) = \frac{s^3 - 3s^2 - 4s + 6}{s^2(s^2 - 5s + 6)} = \frac{s^3 - 3s^2 - 4s + 6}{s^2(s - 2)(s - 3)}$$

$$= \frac{A}{s} + \frac{B}{s^2} + \frac{C}{s - 2} + \frac{D}{s - 3}$$

The constants A, B, C and D are evaluated in the usual way.

$$X(s) = \frac{1}{6s} + \frac{1}{s^2} + \frac{3}{2(s - 2)} - \frac{2}{3(s - 3)}$$

Taking the inverse Laplace transform yields:

$$x(t) = \frac{1}{6} + t + \frac{3}{2} e^{2t} - \frac{2}{3} e^{3t} \quad \blacktriangle$$

Example 11.25   Discharge of a capacitor

In Example 1.24, we examined the variation in voltage across a capacitor, C, when it was switched in series with resistor, R, at time $t = 0$. We stated a relationship for the time varying voltage, v, across the capacitor. Prove this relationship. Refer to the example for details of the circuit.

Solution

First we must derive a differential equation for the circuit. Using Kirchhoff's voltage law and denoting the voltage across the resistor by v_R we obtain:

$$v + v_R = 0$$

Using Ohm's law and denoting the current in the circuit by i we obtain

$$v + iR = 0$$

For the capacitor,

$$i = C \frac{dv}{dt}$$

Combining these equations gives

$$v + RC \frac{dv}{dt} = 0$$

We now take the Laplace transform of this equation. Using $\mathcal{L}\{v\} = V(s)$ we obtain

$$V(s) + RC(sV(s) - v(0)) = 0$$

$$V(s)(1 + RCs) = RCv(0)$$

$$V(s) = \frac{RCv(0)}{1 + RCs} = \frac{v(0)}{\frac{1}{RC} + s}$$

Taking the inverse Laplace transform of the equation yields

$$v = v(0)e^{-t/(RC)} \qquad t \geq 0$$

This is equivalent to the relationship stated in Example 1.24. ▲

Example 11.26 Electronic thermometer measuring oven temperature

Many engineering systems can be modelled by a first order differential equation. The time constant is a measure of the rapidity with which these systems respond to a change in input. Suppose an electronic thermometer is used to measure the temperature of an oven. The sensing element does not respond instantly to changes in the oven temperature because it takes time for the element to heat up or cool down. Provided the electronic circuitry does not introduce further time delays then the differential equation that models the thermometer is given by:

$$\tau \frac{dv_m}{dt} + v_m = v_o$$

where v_m = measured temperature, v_o = oven temperature, τ = time constant of the sensor. For convenience the temperature is measured relative to the ambient room temperature, which forms a 'base line' for temperature measurement.

Suppose the sensing element of an electronic thermometer has a time constant of two seconds. If the temperature of the oven increases linearly at the rate of $3\,°C\,s^{-1}$ starting from an ambient room temperature of $20\,°C$ at $t = 0$, calculate the response of the thermometer to the changing oven temperature. State the maximum temperature error.

Solution

Taking Laplace transforms of the equation gives:

$$\tau(sV_m(s) - v_m(0)) + V_m(s) = V_o(s)$$

$v_m(0) = 0$ as the oven temperature and sensor temperature are identical at $t = 0$. Therefore,

$$\tau s V_m(s) + V_m(s) = V_o(s)$$

$$V_m(s) = \frac{V_o(s)}{1 + \tau s} \tag{11.1}$$

For this example, the input to the thermometer is a temperature ramp with a slope of $3\,°C\,s^{-1}$. Therefore, $v_o = 3t$ for $t \geq 0$

$$V_o(s) = \mathcal{L}\{v_o(t)\} = \frac{3}{s^2} \tag{11.2}$$

Combining Equations (11.1) and (11.2) yields:

$$V_m(s) = \frac{3}{s^2(1 + \tau s)} = \frac{3}{s^2(1 + 2s)} \qquad \text{since } \tau = 2$$

Then using partial fractions, we have:

$$V_m(s) = \frac{3}{s^2} - \frac{6}{s} + \frac{12}{1 + 2s}$$

Taking the inverse Laplace transform yields:

$$v_m = 3t - 6 + 6e^{-0.5t} \qquad t \geq 0$$

This response consists of three parts.

(1) a decaying transient which disappears with time

(2) a ramp with the same slope as the oven temperature

(3) a fixed negative temperature error.

Therefore, after the transient has decayed the measured temperature follows the oven temperature with a fixed negative error. It is instructive to obtain the temperature error by an alternative method. Given that the temperature error is v_e, then

$$v_e = v_m - v_o$$

and

$$V_e(s) = V_m(s) - V_o(s) \tag{11.3}$$

Combining Equations (11.1) and (11.3) yields,

$$V_e(s) = \frac{V_o(s)}{1 + \tau s} - V_o(s) = \frac{-\tau s V_o(s)}{1 + \tau s} \tag{11.4}$$

Combining Equations (11.2) and (11.4) yields:

$$V_e(s) = \frac{-\tau s}{1 + \tau s}\frac{3}{s^2} = \frac{-3\tau}{s(1 + \tau s)}$$

The final value theorem can be used to find the steady-state error.

$$\lim_{t \to \infty} v_e(t) = \lim_{s \to 0} sV_e(s) = \lim_{s \to 0}\left[\frac{-3\tau s}{s(1 + \tau s)}\right] = -3\tau = -6$$

that is, the steady-state temperature error is $-6\,°C$. It is important to note that the final value theorem can only be used if it is known that the time function tends to a limit as $t \to \infty$. In many cases engineers know this is the case from experience. ▲

The Laplace transform technique can also be used to solve simultaneous differential equations.

Example 11.27

Solve

$$x' + x + \frac{y'}{2} = 1 \qquad x(0) = y(0) = 0$$

$$\frac{x'}{2} + y' + y = 0$$

Solution

Take the Laplace transforms of both equations.

$$sX(s) - x(0) + X(s) + \frac{sY(s) - y(0)}{2} = \frac{1}{s}$$

$$\frac{sX(s) - x(0)}{2} + sY(s) - y(0) + Y(s) = 0$$

These are rearranged to give:

$$(s + 1)X(s) + \frac{sY(s)}{2} = \frac{1}{s}$$

$$\frac{sX(s)}{2} + (s + 1)Y(s) = 0$$

These simultaneous algebraic equations need to be solved for $X(s)$ and $Y(s)$. By Cramer's Rule

$$Y(s) = \frac{\begin{vmatrix} s+1 & 1/s \\ s/2 & 0 \end{vmatrix}}{\begin{vmatrix} s+1 & s/2 \\ s/2 & s+1 \end{vmatrix}} = \frac{-1/2}{(s+1)^2 - s^2/4}$$

$$= -\frac{1}{2} \left\{ \frac{1}{3s^2/4 + 2s + 1} \right\}$$

$$= -\frac{1}{2} \left\{ \frac{4}{3s^2 + 8s + 4} \right\}$$

$$= -\frac{1}{2} \left\{ \frac{4}{(3s + 2)(s + 2)} \right\}$$

Using partial fractions we find

$$Y(s) = -\frac{1}{2} \left\{ \frac{3}{3s + 2} - \frac{1}{s + 2} \right\}$$

$$= -\frac{1}{2} \left\{ \frac{1}{s + 2/3} - \frac{1}{s + 2} \right\}$$

and hence

$$y(t) = \tfrac{1}{2} \left(e^{-2t} - e^{-2t/3} \right)$$

Similarly,

$$X(s) = \frac{4(s+1)}{s(3s+2)(s+2)} = \frac{1}{s} - \frac{1}{2(s+2/3)} - \frac{1}{2(s+2)}$$

and so

$$x(t) = 1 - \tfrac{1}{2} \left(e^{-2t/3} + e^{-2t} \right) \quad \blacktriangle$$

EXERCISES 11.8

1. Use Laplace transforms to solve:

   (a) $x'' + x = 2t$, $x(0) = 0$, $x'(0) = 5$

   (b) $2x'' + x' - x = 27 \cos 2t + 6 \sin 2t$,
   $x(0) = -1$, $x'(0) = -2$

   (c) $x'' + x' - 2x = 1 - 2t$,
   $x(0) = 6$, $x'(0) = -11$

   (d) $x'' - 4x = 4(\cos 2t - 1)$,
   $x(0) = 1$, $x'(0) = 0$

   (e) $x' - 2x - y' + 2y = -2t^2 + 7$,
   $x'/2 + x + 3y' + y = t^2 + 6$
   $x(0) = 3$, $y(0) = 6$

   (f) $x' + x + y' + y = 6e^t$,
   $x' + 2x - y' - y = 2e^{-t}$
   $x(0) = 2$, $y(0) = 1$

2. Using Laplace transforms find the particular solution of

$$\frac{d^2 y}{dt^2} - 5\frac{dy}{dt} - 6y = 14e^{-t}$$

satisfying $y = 3$ and $dy/dt = 8$ when $t = 0$

11.11 Transfer functions

It is possible to obtain a mathematical model of an engineering system that consists of one or more differential equations. This approach was introduced in Section 10.9. We have already seen that the solution of differential equations can be found by using the Laplace transform. This leads naturally to the concept of a transfer function which will be developed in this section. Consider the differential equation:

$$\frac{dx(t)}{dt} + x(t) = f(t) \qquad x(0) = x_0 \tag{11.5}$$

and assume that it models a simple engineering system. Then $f(t)$ represents the input to the system and $x(t)$ represents the output, or response of the system to the input $f(t)$. For reasons that will be explained below it is necessary to assume that the initial conditions associated with the differential equation are zero. In Equation (11.5) this means we take x_0 to be zero. Taking the Laplace transform of Equation (11.5) yields:

$$sX(s) - x_0 + X(s) = F(s)$$

$$(1 + s)X(s) = F(s) \qquad \text{assuming } x_0 = 0$$

so that

$$\frac{X(s)}{F(s)} = \frac{1}{1+s}$$

The function, $X(s)/F(s)$, is called a **transfer function**. It is the ratio of the Laplace transform of the output to the Laplace transform of the input. It is often denoted by $G(s)$. Therefore, for Equation (11.5),

$$G(s) = \frac{1}{1+s}$$

The assumption of zero initial conditions is necessary in order to produce a simple relationship between the input and output variables. In practice, there are methods to overcome this limitation should it be necessary to do so.

Example 11.28

Find the transfer functions of the following equations assuming that $f(t)$ represents the input and $x(t)$ represents the output.

(a) $\dfrac{dx(t)}{dt} - 4x(t) = 3f(t), \qquad x(0) = 0$

(b) $\dfrac{d^2x(t)}{dt^2} + 3\dfrac{dx(t)}{dt} - x(t) = f(t), \qquad \dfrac{dx(0)}{dt} = 0, \qquad x(0) = 0$

Solutions

(a) Taking Laplace transforms of the differential equation gives:

$$sX(s) - x(0) - 4X(s) = 3F(s)$$

$$(s - 4)X(s) = 3F(s) \qquad \text{as } x(0) = 0$$

$$\frac{X(s)}{F(s)} = G(s) = \frac{3}{s - 4}$$

(b) Taking Laplace transforms of the differential equation gives:

$$s^2X(s) - sx(0) - \frac{dx(0)}{dt} + 3(sX(s) - x(0)) - X(s) = F(s)$$

$$(s^2 + 3s - 1)X(s) = F(s) \qquad \text{as } \frac{dx(0)}{dt} = 0 \text{ and } x(0) = 0$$

$$\frac{X(s)}{F(s)} = G(s) = \frac{1}{s^2 + 3s - 1} \qquad \blacktriangle$$

When creating a mathematical model of an engineering system it is often convenient to think of the variables within the system as signals and elements of the system as means by which these signals are modified. The word signal is used in a very general sense and is not restricted to, say, voltage. On this basis each of the elements of the system can be modelled by a transfer function. A

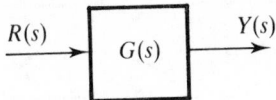

Figure 11.3 The relationship $Y(s) = G(s)R(s)$ holds for a single block.

transfer function defines the relationship between an input signal and an output signal. The relationship is defined in terms of the Laplace transforms of the signals. The advantage of this is that the rules governing the manipulation of transfer functions are then of a purely algebraic nature. Consider Figure 11.3. If,

$R(s) = \mathcal{L}\{r(t)\} =$ Laplace transform of the input signal

$Y(s) = \mathcal{L}\{y(t)\} =$ Laplace transform of the output signal

$G(s) =$ transfer function

then,

$$Y(s) = G(s)R(s)$$

Transfer functions are represented schematically by rectangular blocks, while signals are represented as arrows. Engineers often speak of the **time domain** and the **s domain** in order to distinguish between the two mathematical representations of an engineering system. However, it is important to emphasize the equivalence between the two domains.

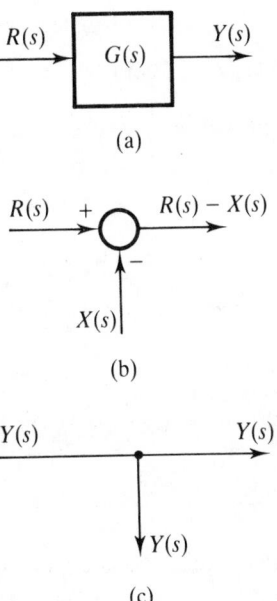

Figure 11.4 The three components of block diagrams. (a) A basic block; the block contains a transfer function which relates the input and output signals. (b) A summing point. (c) A take-off point.

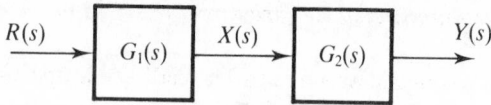

Figure 11.5 Two block diagrams are combined to form a single diagram.

Often when constructing a mathematical model of a system using transfer functions, it is convenient first to obtain transfer functions of the elements of the system and then combine them. Before the overall transfer function is calculated a block diagram is drawn which shows the relationship between the various transfer functions. Block diagrams consist of three basic components. These are shown in Figure 11.4.

We have already examined the **basic block** which is governed by the relationship $Y(s) = G(s)R(s)$. A **summing point** adds together the incoming signals to the summing point and produces an outgoing signal. The polarity of the incoming signals is denoted by means of a positive or negative sign. There can be several incoming signals but only one outgoing signal. A **take-off point** is a point where a signal is tapped. This process of tapping the signal has no effect on the signal value, that is, the tap does not load the original signal. There are several rules governing the manipulation of block diagrams. Only two will be considered here.

11.11.1 *Rule 1. Combining two transfer functions in series*

Consider Figure 11.5. The following relationships hold.

$$X(s) = G_1(s)R(s) \qquad Y(s) = G_2(s)X(s)$$

Eliminating $X(s)$ from these equations yields:

$$Y(s) = G_1(s)G_2(s)R(s)$$

$$\frac{Y(s)}{R(s)} = G_1(s)G_2(s)$$

Finally the overall transfer function, $G(s)$, is given by $G_1(s)G_2(s)$ as shown in Figure 11.6.

■ For two transfer functions in series the overall transfer function is given by:

$$G(s) = G_1(s)G_2(s)$$

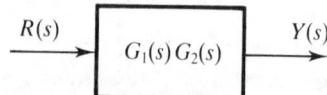

Figure 11.6 Figure 11.5 is simplified to a single block.

11.11.2 *Rule 2. Eliminating a negative feedback loop*

Consider Figure 11.7 which shows a negative feedback loop. It is so called because the output signal is 'fed back' and subtracted from the input signal. Such loops are common in a variety of engineering systems. The quantities $X_1(s)$ and $X_2(s)$ represent intermediate signals in the system. We wish to obtain an overall transfer function for this system relating $Y(s)$ and $R(s)$. For the two transfer functions the following hold.

$$Y(s) = G(s)X_2(s) \tag{11.6}$$

$$X_1(s) = H(s)Y(s) \tag{11.7}$$

For the summing point

$$X_2(s) = R(s) - X_1(s) \tag{11.8}$$

Combining Equations (11.7) and (11.8) gives:

$$X_2(s) = R(s) - H(s)Y(s) \tag{11.9}$$

Combining Equations (11.6) and (11.9) gives:

$$Y(s) = G(s)(R(s) - H(s)Y(s)) = G(s)R(s) - G(s)H(s)Y(s)$$

$$Y(s)(1 + G(s)H(s)) = G(s)R(s)$$

$$Y(s) = \frac{G(s)R(s)}{1 + G(s)H(s)}$$

■ The overall transfer function for a negative feedback loop is given by:

$$\frac{Y(s)}{R(s)} = \frac{G(s)}{1 + G(s)H(s)}$$

The simplified block diagram for a negative feedback loop is shown in Figure 11.8.

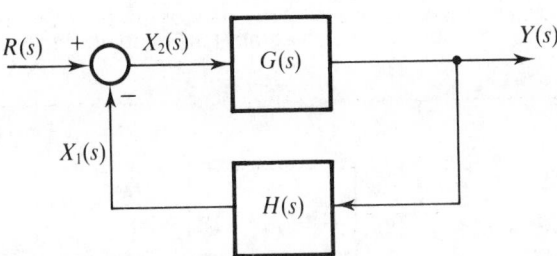

Figure 11.7 Block diagram for a negative feedback loop.

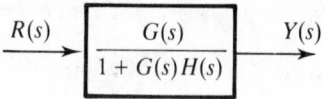

Figure 11.8 Simplified block diagram for a negative feedback loop.

A complicated engineering system may be represented by many differential equations. The output from one part of the system may form the input to another part. Consider the following example.

Example 11.29

A system is modelled by the differential equations:

$$x' + 2x = f(t) \tag{11.10}$$

$$2y' - y = x(t) \tag{11.11}$$

In Equation (11.10) the input is $f(t)$ and the output is $x(t)$. In Equation (11.11), $x(t)$ is the input and $y(t)$ is the final output of the system. Find the overall system transfer function assuming zero initial conditions.

Solution

The output from Equation (11.10) is $x(t)$; this forms the input to Equation (11.11). The block diagrams for Equations (11.10) and (11.11) are combined into a single block diagram as shown in Figure 11.9.

Using Rule 1, the overall system transfer function can then be found.

$$G(s) = \frac{Y(s)}{F(s)} = \frac{1}{(s+2)(2s-1)}$$

This transfer function relates $Y(s)$ and $F(s)$ (see Figure 11.10). ▲

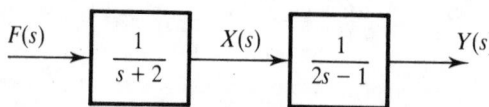

Figure 11.9 Combined block diagram for Equations 11.10 and 11.11.

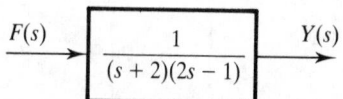

Figure 11.10 The overall system transfer function.

Example 11.30

A system is represented by the differential equations:

$$2x' - x = f(t)$$

$$y' + 3y = x(t)$$

$$z' + z = y(t)$$

The initial input is $f(t)$ and the final output is $z(t)$. Find the overall system transfer function, assuming zero initial conditions.

Solution

The transfer function for each equation is found and combined into one block diagram (see Figure 11.11). The three blocks are simplified to a single block as shown in Figure 11.12. The overall system transfer function is

$$G(s) = \frac{Z(s)}{F(s)} = \frac{1}{(2s - 1)(s + 3)(s + 1)}. \quad \blacktriangle$$

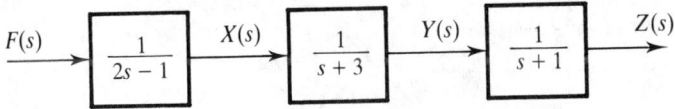

Figure 11.11   Block diagram for the system given in Example 11.30.

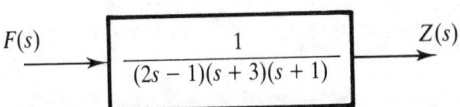

Figure 11.12   Simplified block diagram for Example 11.30.

Example 11.31   Transport lag

Transport lag is a term used to describe the time delay that may be present in certain engineering systems. A typical example would be a conveyor belt feeding a furnace with coal supplied by a hopper (see Figure 11.13). The amount of fuel supplied to the furnace can be varied by varying the opening at the base of the hopper but there is a time delay before this changed quantity of fuel reaches the furnace. The time delay depends on the speed and length of the conveyor. Mathematically, the function describing the variation in the quantity of fuel entering the furnace is a time shifted version of the function describing the variation in the quantity of fuel placed on the conveyor (see Figure 11.14).

Let
$u(t)q(t) =$ quantity of fuel placed on the conveyor, where $u(t)$ is the unit step function,

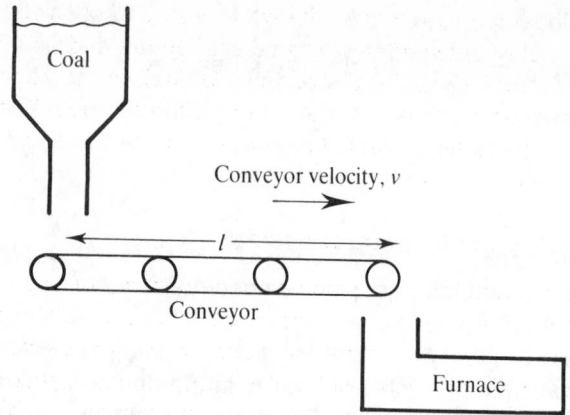

Figure 11.13 Coal is fed into the furnace via the conveyor belt.

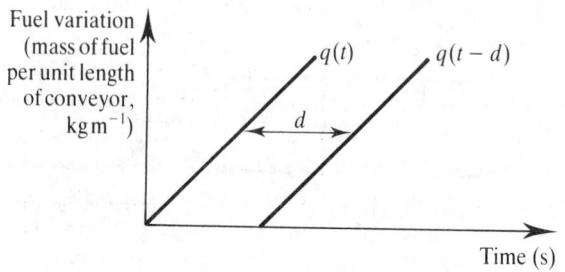

Figure 11.14 The time delay d, is introduced by the conveyor belt.

$u(t - d)q(t - d)$ = quantity of fuel entering the furnace,
d = time delay introduced by the conveyor,
l = length of conveyor (m),
v = speed of conveyor (m s^{-1}).

The input to the conveyor is $u(t)q(t)$ and the output from the conveyor is $u(t - d)q(t - d)$. If the conveyor is moving at a constant speed then $v = l/d$ and so $d = l/v$. The transfer function that models the conveyor belt can be obtained by using the second shift theorem.

$$Q_d(s) = \mathcal{L}\{u(t - d)q(t - d)\} = e^{-sd}\mathcal{L}\{q(t)\} = e^{-sd}Q(s)$$

The transfer function for the conveyor is shown in Figure 11.15.

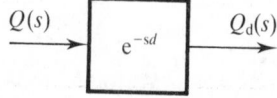

Figure 11.15 Block diagram for the conveyer belt.

Transport lags can cause difficulty when trying to control a system because of the delay between taking a control action and its effect being felt. In this example, increasing the quantity of fuel on the conveyor does not lead to an immediate increase in fuel entering the furnace. The difficulty of controlling the furnace temperature increases as the transport lag introduced by the conveyor increases. ▲

Example 11.32 Position control system

There are many examples of position control systems in engineering, for example, control of the position of a plotter pen, and control of the position of a radio telescope. The common term for these systems is **servo-systems**.

Consider the block diagram of Figure 11.16 which represents a simple servo-system. The actual position of the motor is denoted by $\Theta_a(s)$ in the s domain and $\theta_a(t)$ in the time domain. The desired position is denoted by $\Theta_d(s)$ and $\theta_d(t)$, respectively. The system is a closed loop with negative feedback. The difference between the desired and the actual position generates an error signal which is fed to a controller with gain K. The output signal from the controller is fed to a servo-motor and its associated drive circuitry. The aim of the control system is to maintain the actual position of the motor at a value corresponding to the desired position. In practice, if a new desired position is requested then the system will take some time to attain this new position. The engineer can choose a value of the controller gain to obtain the best type of response from the control system. We will examine the effect of varying K on the response of the servo-system.

We can use Rules 1 and 2 to obtain an overall transfer function for the system. The forward transfer function is:

$$G(s) = \frac{0.5K}{s(s+1)} \qquad \text{by Rule 1}$$

The overall transfer function $\Theta_a(s)/\Theta_d(s)$ is obtained by Rule 2 with $H(s) = 1$. So,

$$\frac{\Theta_a(s)}{\Theta_d(s)} = \frac{\dfrac{0.5K}{s(s+1)}}{1 + \dfrac{0.5K(1)}{s(s+1)}} = \frac{0.5K}{s(s+1) + 0.5K} = \frac{0.5K}{s^2 + s + 0.5K}$$

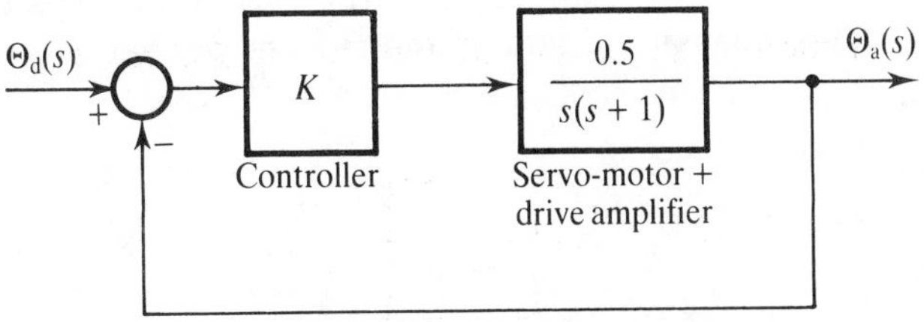

Figure 11.16 Position control system.

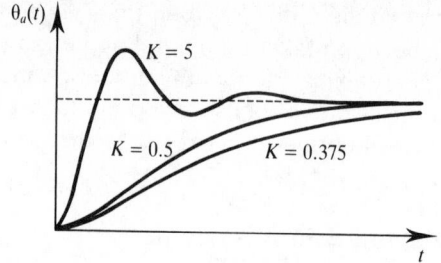

Figure 11.17   Time response for various values of *K*.

Let us now examine the effect of varying *K*. We will consider three values: *K* = 0.375, *K* = 0.5, *K* = 5 and examine the response of the system to a unit step input in each case.
For *K* = 0.375

$$\frac{\Theta_a(s)}{\Theta_d(s)} = \frac{0.1875}{s^2 + s + 0.1875}$$

With $\Theta_d(s) = 1/s$, then

$$\Theta_a(s) = \frac{0.1875}{(s^2 + s + 0.1875)s} = \frac{1}{s} + \frac{0.5}{s + 0.75} - \frac{1.5}{s + 0.25} \quad \text{using partial fractions.}$$

So,

$$\theta_a(t) = 1 + 0.5e^{-0.75t} - 1.5e^{-0.25t} \quad t \geq 0$$

This is shown in Figure 11.17. Engineers usually refer to this as an **overdamped** response. The response does not overshoot the final value.
For *K* = 0.5

$$\frac{\Theta_a(s)}{\Theta_d(s)} = \frac{0.25}{s^2 + s + 0.25}$$

$$\Theta_a(s) = \frac{0.25}{s(s^2 + s + 0.25)} = \frac{1}{s} - \frac{1}{s + 0.5} - \frac{0.5}{(s + 0.5)^2}$$

$$\theta_a(t) = 1 - e^{-0.5t} - 0.5te^{-0.5t} \quad t \geq 0$$

This is shown in Figure 11.17 and is termed a **critically damped** response. It corresponds to the fastest rise time of the system without overshooting.
For *K* = 5

$$\frac{\Theta_a(s)}{\Theta_d(s)} = \frac{2.5}{s^2 + s + 2.5}$$

$$\Theta_a(s) = \frac{2.5}{s(s^2 + s + 2.5)}$$

Rearranging to enable standard forms to be inverted gives:

$$\Theta_a(s) = \frac{1}{s} - \frac{s + 0.5}{(s + 0.5)^2 + 1.5^2} - \frac{0.5}{(s + 0.5)^2 + 1.5^2}$$

$$\theta_a(t) = 1 - e^{-0.5t} \cos 1.5t - \tfrac{1}{3}e^{-0.5t} \sin 1.5t$$

The trigonometric terms can be expressed as a single sinusoid using the techniques given in Section 1.4.8. Thus,

$$\theta_a(t) = 1 - 1.054e^{-0.5t} \sin(1.5t + 1.249) \qquad t \geq 0$$

This is shown in Figure 11.17 and is termed an **underdamped** response. The system overshoots its final value.

In a practical system it is common to design for some overshoot, provided it is not excessive, as this enables the desired value to be reached more quickly. It is interesting to compare the system response for the three cases with the nature of their respective transfer function poles. For the overdamped case the poles are real and unequal, for the critically damped case the poles are real and equal, and for the underdamped case the poles are complex. Engineers rely heavily on pole positions when designing a system to have a particular response. By varying the value of K it is possible to obtain a range of system responses and corresponding pole positions. ▲

EXERCISES 11.9

Find the transfer function for each of the following equations assuming zero initial conditions:

(a) $x'' + x = f(t)$

(b) $2x'' + x' - x = f(t)$

(c) $x'' + x' - 2x = f(t)$

(d) $x'' - 4x = f(t)$

11.12 Poles, zeros and the *s* plane

Most transfer functions for engineering systems can be written as rational functions, that is, as ratios of two polynomials in s, with a constant factor, K:

$$G(s) = K \frac{P(s)}{Q(s)}$$

$P(s)$ is of order m, and $Q(s)$ is of order n; for a physically realizable system $m < n$. Hence $G(s)$ may be written as:

$$G(s) = \frac{K(s - z_1)(s - z_2)\ldots(s - z_m)}{(s - p_1)(s - p_2)\ldots(s - p_n)}$$

The values of s that make $G(s)$ zero are known as the **system zeros** and correspond to the roots of $P(s) = 0$, that is, $s = z_1, z_2, \ldots, z_m$. The values of s that make $G(s)$

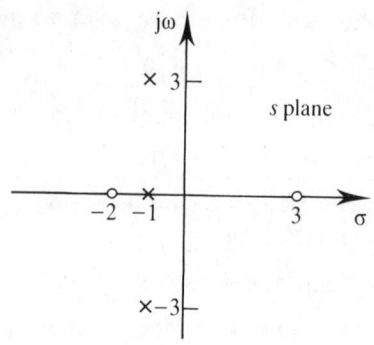

Figure 11.18   Poles and zeros plotted for the transfer function:
$$G(s) = \frac{3(s - 3)(s + 2)}{(s + 1)(s + 1 + 3j)(s + 1 - 3j)}.$$

infinite are known as the **system poles** and correspond to the roots of $Q(s) = 0$, that is, $s = p_1, p_2, \ldots, p_n$. As we have seen, poles may be real or complex. Complex poles always occur in complex conjugate pairs whenever the polynomial $Q(s)$ has real coefficients.

   Engineers find it useful to plot these poles and zeros on an s plane diagram. A complex plane plot is used with, conventionally, a real axis label of σ and an imaginary axis label of $j\omega$. Poles are marked as crosses and zeros are marked as small circles. Figure 11.18 shows an s plane plot for the transfer function:

$$G(s) = \frac{3(s - 3)(s + 2)}{(s + 1)(s + 1 + 3j)(s + 1 - 3j)}$$

The benefit of this approach is that it allows the character of a linear system to be determined by examining the s plane plot. In particular, the transient response of the system can easily be visualized by the number and positions of the system poles and zeros.

Example 11.33

Find the poles of $\dfrac{s - 2}{(s^2 + 2s + 5)(s + 1)}$.

Solution

The denominator is factorized into linear factors.

$$(s^2 + 2s + 5)(s + 1) = (s - p_1)(s - p_2)(s - p_3)$$

where $p_1 = -1 + 2j$, $p_2 = -1 - 2j$, $p_3 = -1$. The poles are $-1 + 2j$, $-1 - 2j$, -1. ▲

   If $X(s) = 1/(s - p_1)$ where the pole p_1 is given by $a + bj$, then

$$x(t) = e^{p_1 t} = e^{at} e^{btj} = e^{at}(\cos bt + j \sin bt)$$

Hence the real part of the pole, a, gives rise to an exponential term and the imaginary part, b, gives rise to an oscillatory term. If $a < 0$ the response, $x(t)$, will decrease to zero as $t \to \infty$.

Consider the Laplace transform in Example 11.23. There are three poles; -1, $-1 + j$ and $-1 - j$. The real pole is negative, and the real parts of the complex poles are also negative. This ensures the response, $x(t)$, decreases with time. The imaginary part of the complex poles gives rise to the oscillatory term, $\cos t$. The characteristics of poles and the corresponding responses are now discussed.

Given a system with transfer function, $G(s)$, input signal, $R(s)$, and output signal, $C(s)$, then

$$\frac{C(s)}{R(s)} = G(s)$$

that is,

$$C(s) = G(s)R(s)$$

and so

$$C(s) = \frac{K(s - z_1)(s - z_2)\ldots(s - z_m)R(s)}{(s - p_1)(s - p_2)\ldots(s - p_n)}$$

The poles and zeros of the system are independent of the input that is applied. All that $R(s)$ contributes to the expression for $C(s)$ is extra poles and zeros.

Consider the case where $R(s) = 1/s$, corresponding to a unit step input.

$$C(s) = \frac{K(s - z_1)(s - z_2)\ldots(s - z_m)}{(s - p_1)(s - p_2)\ldots(s - p_n)s}$$

$$= \frac{A_1}{s - p_1} + \frac{A_2}{s - p_2} + \cdots + \frac{A_n}{s - p_n} + \frac{B_1}{s}$$

where $A_1, A_2, \ldots, A_n$ and B_1 are constants. Taking inverse Laplace transforms yields:

$$c(t) = A_1 e^{p_1 t} + A_2 e^{p_2 t} + \cdots + A_n e^{p_n t} + B_1$$

If the system is stable then $p_1, p_2, \ldots, p_n$ will have negative real parts and their contribution to $c(t)$ will vanish as $t \to \infty$.

The response caused by the system poles is often called a **transient response** because it decreases with time for a stable system. The component of the transient response due to a particular pole is often termed its **transient**. Notice that the form of the transient response is independent of the system input and is determined by the nature of the system poles. It is now possible to derive a series of rules relating the transient response of the system to the positions of the system poles in the s plane.

Rule 1

The poles may be either real or complex but for a particular pole it is necessary for the real part to be negative if the transient caused by that pole is to decay with time. Otherwise the transient response will increase with time and the system will

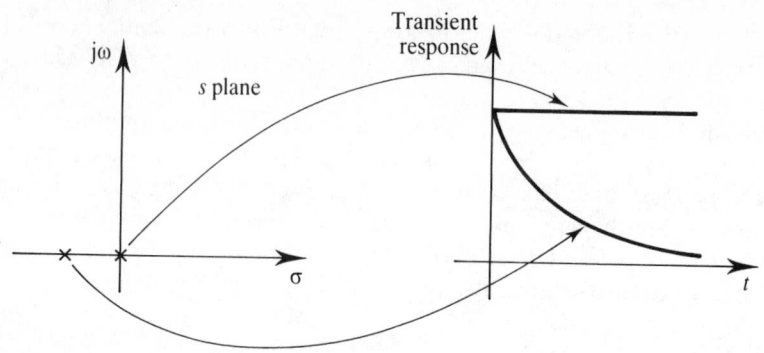

Figure 11.19 A pole with a negative real part leads to a decaying transient while a pole with a zero real part leads to a transient that does not decay with time.

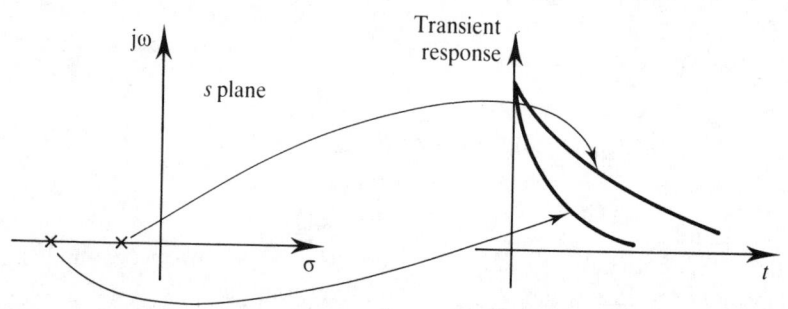

Figure 11.20 The further a pole is from the imaginary axis, the quicker the decay of its transient.

be unstable; a condition engineers usually design to avoid. In simple terms this means that the poles of a linear system must all lie in the left-hand side of the s plane for stability. Poles on the imaginary axis lead to marginal stability as the transients introduced by such poles do not grow or decay. This is illustrated in Figure 11.19.

Rule 2

The further a pole is to the left of the imaginary axis the faster its transient decays (see Figure 11.20). This is because its transient contains a larger negative exponential term. For example, e^{-5t} decays faster than e^{-2t}. The poles near to the imaginary axis are termed the **dominant poles** as their transients take the longest to decay. It is quite common for engineers to ignore the effect of poles that are more than five or six times further away from the imaginary axis than the dominant poles.

Rule 3

For a real system, poles with imaginary components occur as complex conjugate pairs. The transient resulting from this pair of poles has the form of a sinusoidal term multiplying an exponential term. For a pair of stable poles, that is, negative real

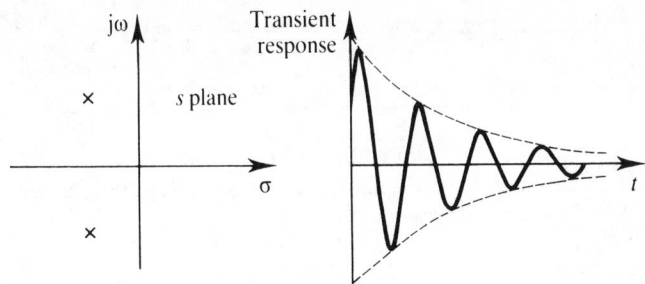

Figure 11.21 A pair of stable complex poles gives rise to a decaying sinusoid transient.

part, the transient can be sketched by drawing a sinusoid confined within a 'decaying exponential envelope' (see Figure 11.21). The reason for this is that when the sinusoidal term has a value of 1 the transient touches the decaying exponential. When the sinusoidal term has a value of -1 the transient touches the reflection in the t axis of the decaying exponential. The larger the imaginary component of the pair of poles the higher the frequency of the sinusoidal term.

It should now be clear how useful a concept the s plane plot is when analysing the response of a linear system. A complex system may have many poles and zeros but by plotting them on the s plane the engineer begins to get a feel for the character of the system. The form of the transients relating to particular poles or pairs of poles can be obtained using the above rules. The magnitude of the transients, that is, the values of the coefficients A_1, A_2,..., A_n depends on the system zeros.

It can be shown that having a zero near to a pole reduces the magnitude of the transient relating to that pole. Engineers often deliberately introduce zeros into a system to reduce the effect of unwanted transients. If a zero coincides with the pole, it cancels it and the transient corresponding to that pole is eliminated.

11.13 Laplace transforms of some special functions

In this section we apply the Laplace transform to the delta function and periodic functions. These functions were introduced in Chapter 1.

11.13.1 *The delta function, $\delta(t-d)$*

Recall the integral property of the delta function (see Example 8.48),

$$\int_{-\infty}^{\infty} f(t)\delta(t-d)\,\mathrm{d}t = f(d)$$

The Laplace transform follows from this integral property. Let $f(t) = \mathrm{e}^{-st}$ so that $f(d) = \mathrm{e}^{-sd}$. Then

$$\int_{-\infty}^{\infty} \mathrm{e}^{-st}\delta(t-d)\,\mathrm{d}t = \int_{0}^{\infty} \mathrm{e}^{-st}\delta(t-d)\,\mathrm{d}t = f(d) = \mathrm{e}^{-sd}$$

that is,

■ $\mathscr{L}\{\delta(t - d)\} = e^{-sd}$

The Laplace transform of $\delta(t)$ follows by setting $d = 0$.

■ $\mathscr{L}\{\delta(t)\} = 1$

Example 11.34 Impulse response of a linear system

Consider a linear system with input $x(t)$ and output $y(t)$. Then,

$$Y(s) = G(s)X(s)$$

where $G(s)$ is the system transfer function. If an impulse is applied to the system at $t = 0$ then $X(s) = \mathscr{L}\{\delta(t)\} = 1$. Then

$$Y(s) = G(s) \qquad y(t) = \mathscr{L}^{-1}\{G(s)\} = g(t)$$

The function $g(t)$ is termed the **impulse response** of a system. In Example 1.51 we briefly discussed how to obtain the impulse response of a system in practice. Now consider a general input signal, $x(t)$, applied to the system. It is possible to represent this signal as a series of pulses of width $\delta\lambda$ and height $x(t)$ (see Figure 11.22). If the width of these pulses is small then they can be approximated by a series of impulses of strength $x(t)\delta\lambda$. Each of these impulses will make a contribution to the output signal. At time t, the contribution to the output, $\delta y(t)$, from an impulse applied λ seconds earlier is given by the magnitude of the impulse response of the system at time λ multiplied by the strength of the impulse applied at time $t - \lambda$. So,

$$\delta y(t) = g(\lambda)x(t - \lambda)\delta\lambda$$

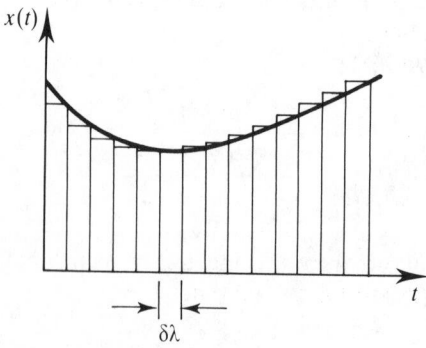

Figure 11.22 The signal is approximated by a series of pulses.

In order to obtain the total output from the system at time t we need to sum all of the contributions for each of the impulses. So,

$$y(t) = \sum_{-\infty}^{t} g(\lambda)x(t-\lambda)\delta\lambda$$

Note that the upper limit of the summation is t because the effect of any impulse applied after time t will not be felt at time t. In order to improve the accuracy of this summation $\delta\lambda$ is allowed to shrink to zero and the summation becomes an integral, that is,

$$y(t) = \int_{-\infty}^{t} g(\lambda)x(t-\lambda)\,d\lambda$$

Now for a real system $g(t)$ is zero for $t < 0$, and so the integral simplifies to:

$$y(t) = \int_{0}^{t} g(\lambda)x(t-\lambda)\,d\lambda$$

We see that this integral is the same as that discussed in Section 11.9. It is known as the **convolution integral**. It allows the output signal of a system to be calculated by **convolving** the input signal to the system with the impulse response of the system. ▲

The result of Example 11.34 can be obtained directly from the convolution theorem as follows. When $x(t) = \delta(t)$ we have:

$$Y(s) = G(s)$$

$$y(t) = \mathcal{L}^{-1}\{G(s)\} = g(t) \qquad \text{the impulse response}$$

If $x(t)$ is now a general input with Laplace transform $X(s)$, we have:

$$Y(s) = G(s)X(s)$$

To find $y(t)$ we use the convolution theorem:

$$y(t) = \mathcal{L}^{-1}\{G(s)X(s)\} = g(t) * x(t)$$

If the impulse response is known, then the response, $y(t)$, to an arbitrary input, $x(t)$, is obtained by convolution:

$$y(t) = \int_{0}^{t} g(\lambda)x(t-\lambda)\,d\lambda$$

Consider the following example.

Example 11.35

For a particular circuit it can be shown that:

$$\frac{V_o(s)}{V_i(s)} = \frac{1}{s+2}$$

where $V_i(s)$ and $V_o(s)$ are the Laplace transforms of the input and output voltages respectively.

(a) Find $v_o(t)$ when $v_i(t) = \delta(t)$

(b) Use the convolution theorem to find $v_o(t)$ when

$$v_i(t) = \begin{cases} e^{-t} & t \geq 0 \\ 0 & t < 0 \end{cases}$$

Solutions

(a) We are given the transfer function

$$G(s) = \frac{V_o(s)}{V_i(s)} = \frac{1}{s + 2}$$

When $v_i(t) = \delta(t)$, $V_i(s) = 1$ and hence:

$$V_o(s) = \frac{1}{s + 2}$$

and

$$v_o(t) = \mathcal{L}^{-1}\left\{\frac{1}{s + 2}\right\} = e^{-2t}$$

This is the impulse response of the system, $g(t)$.

(b) The response to an input, $v_i(t) = u(t)e^{-t}$ is given by:

$$v_o(t) = \int_0^t g(\lambda)v_i(t - \lambda)\,d\lambda$$

$$= \int_0^t e^{-2\lambda} e^{-(t-\lambda)}\,d\lambda$$

$$= e^{-t}\int_0^t e^{-\lambda}\,d\lambda$$

$$= e^{-t}\left[\frac{e^{-\lambda}}{-1}\right]_0^t$$

$$= e^{-t}(1 - e^{-t})$$

$$= e^{-t} - e^{-2t} \quad \blacktriangle$$

11.13.2 *Periodic functions*

Recall the definition of a periodic function, $f(t)$. Given $T > 0$, $f(t)$ is periodic if $f(t) = f(t + T)$ for all t in the domain. If $f(t)$ is periodic and we know the values of $f(t)$ over a period, then we know the values of $f(t)$ over its entire domain.

Hence, it seems reasonable that the Laplace transform of $f(t)$ can be found by studying an appropriate integral over an interval whose length is just one period. This is indeed the case and forms the basis of the following development.

Let $f(t)$ be periodic, with period T. The Laplace transform of $f(t)$ is:

$$\mathcal{L}\{f(t)\} = F(s) = \int_0^\infty e^{-st} f(t) \, dt$$

$$= \int_0^T e^{-st} f(t) \, dt + \int_T^{2T} e^{-st} f(t) \, dt + \int_{2T}^{3T} e^{-st} f(t) \, dt + \cdots$$

Let $t = x$ in the first integral, $t = x + T$ in the second, $t = x + 2T$ in the third and so on.

$$F(s) = \int_0^T e^{-sx} f(x) \, dx + \int_0^T e^{-s(x+T)} f(x+T) \, dx$$

$$+ \int_0^T e^{-s(x+2T)} f(x+2T) \, dx + \cdots$$

Since f is periodic with period T then

$$f(x) = f(x+T) = f(x+2T) = \cdots$$

So,

$$F(s) = \int_0^T e^{-sx} f(x) \, dx + \int_0^T e^{-sT} e^{-sx} f(x) \, dx + \int_0^T e^{-2sT} e^{-sx} f(x) \, dx + \cdots$$

$$= (1 + e^{-sT} + e^{-2sT} + \cdots) \int_0^T e^{-sx} f(x) \, dx$$

We recognize the terms in brackets as a geometric series whose sum to infinity is $1/(1 - e^{-sT})$. Hence,

■ $$F(s) = \frac{\int_0^T e^{-st} f(t) \, dt}{1 - e^{-sT}}$$

Example 11.36

A waveform, $f(t)$, is defined as follows:

$$f(t) = \begin{cases} 2 & 0 < t \leq 1.25 \\ 0 & 1.25 < t \leq 1.5 \end{cases}$$

and $f(t)$ is periodic with period of 1.5. Find the Laplace transform of the waveform.

Solution

$$\mathcal{L}\{f(t)\} = \frac{\int_0^{1.5} e^{-st} f(t)\, dt}{1 - e^{-1.5s}}$$

$$= \frac{\int_0^{1.25} 2e^{-st}\, dt + \int_{1.25}^{1.5} 0 e^{-st}\, dt}{1 - e^{-1.5s}}$$

$$= \frac{2(1 - e^{-1.25s})}{s(1 - e^{-1.5s})} \quad \blacktriangle$$

EXERCISES 11.10

1. Find the Laplace transforms of:

   (a) $3u(t) + \delta(t)$

   (b) $-6u(t) + 4\delta(t)$

   (c) $3u(t - 2) + \delta(t - 2)$

   (d) $u(t + 3) - \delta(t + 4)$

   (e) $\frac{1}{2}u(t - 4) + 3\delta(t - 4)$

2. Find the inverse Laplace transforms of:

   (a) $\dfrac{2}{s} - 1$

   (b) $\dfrac{2}{3s} + \dfrac{1}{2}$

   (c) $(3 - 2s)/s$

   (d) $(4s - 3)/s$

3. A periodic waveform is defined by:

   $$f(t) = \begin{cases} t & 0 \leq t \leq 1 \\ 2 - t & 1 < t \leq 2 \end{cases}$$

   and has a period of 2.

   (a) Sketch two cycles of $f(t)$

   (b) Find $\mathcal{L}\{f(t)\}$.

4. A waveform, $f(t)$, is defined by:

   $$f(t) = \begin{cases} 2 & 0 \leq t \leq 1.5 \\ -2t + 5 & 1.5 < t < 2.5 \end{cases}$$

   and has a period of 2.5.

   (a) Sketch $f(t)$ on $[0, 5]$

   (b) Find $\mathcal{L}\{f(t)\}$.

5. If the impulse response of a network is $g(t) = 10e^{-4t}$ find the output when the input is $f(t) = e^{-t} \cos 2t$, $t \geq 0$.

Miscellaneous exercises

1. Given $F(s) = (s + 1)/(s^2 + 2)$ is the Laplace transform of $f(t)$, find the Laplace transforms of the following:

   (a) $f(t)/e^{2t}$

   (b) $3e^{2t}f(t)$

   (c) $2e^{-t}(f(t) + 1)$

   (d) $u(t - 1)f(t - 1)$

   (e) $4u(t - 3)f(t - 3)$

   (f) $e^{-2t}u(t - 2)f(t - 2)$

2. Find the Laplace transforms of the following:

   (a) $2t \sin 3t$

   (b) $\frac{1}{2}(-3t \cos 2t)$

   (c) $e^{-t}u(t)$

   (d) $e^{-t}u(t - 1)$

   (e) $3t^2 u(t - 1)$

   (f) $e^{-t}\delta(t - 2)$

3. Find the inverse Laplace transforms of:

(a) $\dfrac{2s + 3}{(s + 1)(s + 2)}$

(b) $\dfrac{4s}{s^2 - 9}$

(c) $\dfrac{2(s^3 + 4s^2 + 4s + 64)}{(s^2 + 4)(s^2 + 16)}$

(d) $e^{-2s}\dfrac{6}{s^4}$

(e) $e^{-s}\dfrac{s + 1}{s^2}$.

4. Solve the following differential equations using the Laplace transform method.

(a) $x'' + 2x' - 3x = 2t$, $x(0) = 1$ $x'(0) = 2$

(b) $x'' - 2x' + 5x = \cos t$, $x(0) = 0$ $x'(0) = 1$

(c) $\dot{x} + \dot{y} + x + y = \frac{3}{2}(1 + t)$

$\dot{x} - 2\dot{y} + x + 2y = 2t$

$x(0) = 0$ $\quad y(0) = 0$

(d) $\dot{x} - \dot{y} + x = -1$, $x(0) = 0$ $\quad y(0) = 2$

$2\dot{x} - \dot{y} + \dfrac{y}{2} - x = 1$

5. The current, $i(t)$, in a series LC circuit is governed by:

$$L\frac{di}{dt} + \frac{1}{C}\int_0^t i\,dt = v(t)$$

where $v(t)$ is the applied voltage.

(a) Assuming zero initial conditions show that:

$$LsI + \frac{1}{Cs}I = V(s)$$

(b) If $v(t) = \delta(t)$ show that

$$i(t) = \frac{1}{L}\cos\frac{t}{\sqrt{LC}}$$

6. The input and output voltages, $v_i(t)$ and $v_o(t)$, of a series RC network are related by the differential equation:

$$CR\frac{dv_o}{dt} + v_o = v_i$$

(a) If $V_o(s) = \mathscr{L}\{v_o(t)\}$ and $V_i(s) = \mathscr{L}\{v_i(t)\}$, show that the transfer function, $V_o(s)/V_i(s)$, is given by $1/(sCR + 1)$.

(b) If $C = 0.1\,\mu\text{F}$, $R = 100\,\text{k}\Omega$ find, and sketch a graph of the response, $v_o(t)$, when

$$v_i(t) = \begin{cases} 5\,\text{volts} & t \ge 0 \\ 0 & t < 0 \end{cases}$$

(c) Using the component values given in (b) find the response when the input is a unit impulse, $\delta(t)$.

(d) Using the component values given in (b) use the convolution theorem to determine the response when $v_i(t) = 5e^{-100t}$

7. Express the square wave:

$$f(t) = \begin{cases} 1 & 0 < t < a \\ -1 & a < t < 2a \end{cases} \quad \text{period } 2a$$

in terms of unit step functions. Hence deduce its Laplace transform.

8. Consider the circuit shown in Figure 11.23

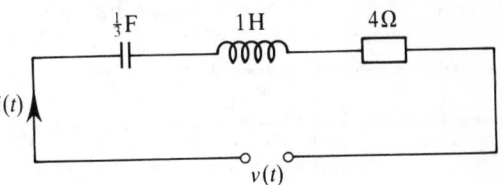

Figure 11.23 LCR circuit.

(a) Show that

$$\frac{di}{dt} + 4i + 3\int_0^t i\,dt = v(t)$$

(b) Assuming zero initial conditions, show that

$$I(s) = \frac{s}{(s + 1)(s + 3)}V(s)$$

(c) Find $i(t)$ if $v(t) = \delta(t)$

(d) Find $i(t)$ if

$$v(t) = \begin{cases} 2e^{-2t} & t \ge 0 \\ 0 & \text{otherwise} \end{cases}$$

Difference equations and 12
the z transform

12.1 Introduction

Difference equations are the discrete equivalent of differential equations. The terminology is similar and indeed the methods of solution have much in common. Difference equations arise whenever an independent variable can have only discrete values. They are of growing importance in engineering in view of their association with discrete-time systems based on the microprocessor.

In Chapter 11 the Laplace transform was shown to be a useful tool for the solution of ordinary differential equations, and for the construction of transfer functions in circuit analysis, control theory, etc. Generally, Laplace transform methods apply when the variables being measured are continuous. The z transform plays a similar role for discrete systems to that played by the Laplace transform for continuous ones. In this chapter you will be introduced to the z transform and one of its applications – the solution of linear constant coefficient difference equations. The z transform is of increasing importance as more and more engineering systems now contain a microprocessor or computer and so have one or more discrete-time components. For example, most industrial controllers now have an embedded microprocessor, and overall control of a factory is often by means of a supervisory process control computer. Many factories also have a production control computer to schedule production.

12.2 Basic definitions

Before classifying difference equations we will first derive an example of one. Suppose an algorithm for processing data is implemented on a computer. Let i be the number of units of central processing unit (c.p.u.) time used in implementing the algorithm and let n be the number of items of data. Clearly n is a non-negative integer, that is, $n \in \mathbb{N}$. Since i depends upon n we write $i = i[n]$. The square brackets notation reflects the fact that n is a discrete variable (see Chapter 3). If there are n items of data, the number of units of c.p.u. time used is $i[n]$. Suppose that if there are $n + 1$ items of data, the number of units of c.p.u. time used increases by $10n + 1$. Then,

$$i[n + 1] = i[n] + 10n + 1$$

This is an example of a **difference equation**. The dependent variable is i; the independent variable is n.

If there are no items of data, then no units of c.p.u. time are used, that is, $i[0] = 0$. Putting $n = 0$ in the difference equation leads to:

$$i[1] = i[0] + 10(0) + 1 = 1$$

Similarly, $i[2] = 12$, $i[3] = 33$, $i[4] = 64$ and so on. We see that the difference equation gives rise to a sequence of values.

There are strong similarities between difference and differential equations. The important point to note is that with difference equations, the independent variable is discrete, not continuous. In the above example, n is the number of items of data; it can have only integer values. This discrete property of the independent variable is an essential and distinguishing feature of difference equations. Much of the terminology of differential equations is applied, with identical meaning, to difference equations.

12.2.1 *Dependent and independent variables*

Consider a simple difference equation:

$$x[n + 1] - x[n] = 10$$

The dependent variable is x; the independent variable is n. In the difference equation

$$y[k + 1] - y[k] = 3k + 5$$

the dependent variable is y and the independent variable is k.

12.2.2 *The solution of a difference equation*

A **solution** is obtained when the dependent variable is known for each value of interest of the independent variable. Thus the solution takes the form of a sequence. There are frequently many different sequences which satisfy a difference equation, that is, there are many solutions. The **general solution** embraces all of these and all possible solutions can be obtained from it.

Example 12.1

Show $x[n] = A2^n$, where A is a constant, is a solution of:

$$x[n + 1] - 2x[n] = 0$$

Solution

$$x[n] = A2^n \qquad x[n + 1] = A2^{n+1} = 2A2^n$$

Hence,

$$x[n + 1] - 2x[n] = 2A2^n - 2A2^n = 0$$

Hence $x[n] = A2^n$ is a solution of the given difference equation. In fact $x[n]$ is the general solution. ▲

If additionally we are given a condition, say $x[0] = 3$, the constant A can be found. If $x[n] = A2^n$, then $x[0] = A2^0 = A$ and hence:

$$A = 3$$

The solution is thus $x[n] = 3(2^n)$. This is the **specific solution** and satisfies both the difference equation and the given condition.

12.2.3 *Linear and non-linear*

An equation is linear if the dependent variable occurs only to the first power. If an equation is not linear it is non-linear. For example:

$$3x[n + 1] - x[n] = 10$$
$$y[n + 1] - 2y[n - 1] = n^2$$
$$kz[k + 2] + z[k] = z[k - 1]$$

are all linear equations. Note that the presence of the term n^2 does not make the equation non-linear, since n is the independent variable. However,

$$(x[n + 1])^2 - x[n] = 10$$
$$y[k + 1] = \sqrt{y[k] + 1}$$

are both non-linear. Also,

$$z[n + 1]z[n] = n^2 + 100$$
$$\sin x[n] = x[n - 1]$$

are non-linear. The product term $z[n + 1]z[n]$ and the term $\sin x[n]$ are the causes of the non-linearity.

12.2.4 *Order*

The **order** is the difference between the highest and lowest arguments of the dependent variable. The equation:

$$3x[n+2] - x[n+1] - 7x[n] = n$$

is second order because the difference between $n+2$ and n is 2.

$$x[n+1]x[n-1] = 7x[n-2]$$

is third order because the difference between $n+1$ and $n-2$ is 3. In general, the higher the order of an equation, the more difficult it is to solve.

In some difference equations the dependent variable occurs only once. These are classified as zero order. Engineers refer to them as **non-recursive** difference equations because calculation of the value of the dependent variable does not require knowledge of the previous values. In contrast, difference equations of order one or greater are referred to as **recursive** difference equations because their solution requires knowledge of previous values of the dependent variable. The difference equation:

$$x[n] = n^2 + n + 1$$

has zero order and so is a non-recursive difference equation.

12.2.5 *Homogeneous and inhomogeneous*

The meanings of homogeneous and inhomogeneous as applied to linear difference equations are analogous to those meanings when applied to differential equations. To decide whether a linear equation is homogeneous or inhomogeneous it is written in standard form, with all the dependent variable terms on the left-hand side. Any remaining independent variable terms are written on the right-hand side. For example,

$$3nx[n+1] - 2n^3 = x[n-1]$$

is written as:

$$3nx[n+1] - x[n-1] = 2n^3 \qquad (12.1)$$

If the right-hand side is 0, the equation is homogeneous; otherwise it is inhomogeneous. Equation (12.1) is inhomogeneous but:

$$3nx[n+1] - x[n-1] = 0$$

is homogeneous.

Example 12.2 Signal processing using a microprocessor

In engineering, an increasing number of products contain a microprocessor or computer which is solving a difference equation. The input to the microprocessor is a sequence of signal values, in many cases formed as a result of sampling a continuous input signal. The output from the microprocessor is a sequence of signal values which may be subsequently converted into a continuous signal. For example, an inhomogeneous difference equation could be of the form:

$$y[n] - 2y[n-1] = 0.1s[n] + 0.2s[n-1] - 0.5s[n-2]$$

where $y[n]$ is the **output sequence** or dependent variable and $s[n]$ is the **input sequence**. Note that the input sequence can still be thought of as the independent variable but instead of being expressed analytically in terms of n it arises as a result of sampling. The corresponding homogeneous equation is:

$$y[n] - 2y[n-1] = 0 \quad \blacktriangle$$

12.2.6 *Coefficient*

The term **coefficient** refers to the coefficient of the dependent variable. In Equation (12.1) the coefficients are $3n$ and -1.

Example 12.3

(a) State the order of each equation

(b) State whether each equation is linear or non-linear

(c) For each linear equation, state whether it is homogeneous or inhomogeneous.

(i) $2x[n] - 3nx[n-1] + x[n-2] + n^2 = 0$

(ii) $\frac{1}{3}(x[n+1] - x[n-1]) = x[n]$

(iii) $z[n+2](2n - z[n-1]) = n+1$

(iv) $\dfrac{7x[n-1]}{x[n-2]} = \dfrac{n+1}{n-1}$

(v) $w[n+3]w[n+1] = n^3 - 1$

(vi) $y[n+2] + 2y[n+1] = 6s[n+2] - 2s[n+1] + s[n]$ where y is the dependent variable

(vii) $x[k+3] - 2x[k+2] + x[k] = e[k+2] - e[k]$ where x is the dependent variable.

Solutions

(a) (i)   Second order

(ii)   Second order

(iii)   Third order

(iv)   First order

(v)   Second order

(vi) First order

(vii) Third order

(b) (i)   Linear

(ii) Linear

(iii) Non-linear

(iv) Linear

(v)   Non-linear

(vi) Linear

(vii) Linear

(c) Equations (iii) and (v) are non-linear. The linear equations are written in standard form

(i) $2x[n] - 3nx[n-1] + x[n-2] = -n^2$

(ii) $x[n+1] - 3x[n] - x[n-1] = 0$

(iv) $7(n-1)x[n-1] - (n+1)x[n-2] = 0$

(vi) and (vii) are already in standard form.

Hence we find the following:

(i)   Inhomogeneous

(ii) Homogeneous

(iv) Homogeneous

(vi) Inhomogeneous

(vii) Inhomogeneous ▲

EXERCISE 12.1

(a) State the order of the equation

(b) State whether each equation is linear or non-linear

(c) For each linear equation, state whether it is homogeneous or inhomogeneous

(i) $n(3n + x[n]) = x[n-1]$

(ii) $2z[k-4]/z[k-3] = z[k-2]$

(iii) $y[n-2] + y[n-1] + y[n] = n^2$

(iv) $\sqrt{n + x[n]} = x[n-2] + e[n]$, where x is the dependent variable

(v) $(2w[n-1] + 1)^2 = w[n-2] + s[n-1] - s[n-2]$, where w is the dependent variable.

12.3 Rewriting difference equations

Sometimes an equation or expression can be written in different ways. At first sight, it may appear there are two independent equations when in fact there is only one. Thus we need to be able to rewrite equations so that comparisons can be made. When general solutions of equations are to be found, usually the equation is first written in a standard form. So once again there is a need to rewrite equations.

Example 12.4

Rewrite the equation so that the highest argument of the dependent variable is $n + 1$.

$$x[n + 3] - x[n + 2] = 2n \qquad x[2] = 7$$

Solution

The highest argument in the given equation is $n + 3$; this must be reduced by 2 to $n + 1$. To do this n is replaced by $n - 2$. The equation becomes:

$$x[n + 1] - x[n] = 2(n - 2) \qquad x[2] = 7$$

Note, however, that the initial condition, $x[2] = 7$, is not changed. This is simply stating that x has a value of 7 when the independent variable has a value of 2. ▲

Example 12.5

Write the following equations so that the highest argument of the dependent variable is $n + 2$.

(a) $3x[n + 4] - 2nx[n + 2] = (n - 1)^2 \qquad x[3] = 6 \qquad x[4] = -7$

(b) $z[n - 2] + z[n - 1] + z[n] = 1 + n \qquad z[0] = 1 \qquad z[1] = 0$

Solutions

(a) The highest argument, $n + 4$, must be replaced by $n + 2$, that is, n is replaced by $n - 2$ throughout the equation.

$$3x[n + 2] - 2(n - 2)x[n] = (n - 3)^2 \qquad x[3] = 6 \qquad x[4] = -7$$

(b) The highest argument, n, is increased to $n + 2$, that is, n is replaced by $n + 2$.

$$z[n] + z[n + 1] + z[n + 2] = n + 3 \qquad z[0] = 1 \qquad z[1] = 0 \quad ▲$$

Example 12.6

Write the following equations so that the highest argument of the dependent variable is k.

(a) $\quad a[k + 2] = \dfrac{s[k + 2] - 2s[k + 1] + s[k]}{15}$

where a is the dependent variable.

(b) $\quad a[k + 3] = \dfrac{l[k + 3] + l[k + 2] + l[k + 1] + l[k] + l[k - 1]}{5}$

where a is the dependent variable.

Solutions

(a) The highest argument, $k + 2$, must be replaced by k, that is, k is replaced by $k - 2$ throughout the equation.

$$a[k] = \frac{s[k] - 2s[k-1] + s[k-2]}{15}$$

(b) The highest argument, $k + 3$, must be replaced by k, that is, k is replaced by $k - 3$ throughout the equation.

$$a[k] = \frac{l[k] + l[k-1] + l[k-2] + l[k-3] + l[k-4]}{5} \quad \blacktriangle$$

EXERCISE 12.2

Write each equation so that the highest argument of the dependent variable is as specified.

(a) $p[k] - 3p[k+1] = p[k-2]$, highest argument of the dependent variable is to be $k + 2$.

(b) $R[n-1] - R[n-2] - R[n-3] = n$, $R[0] = 1$, $R[1] = -2$, highest argument of the dependent variable is to be n.

(c) $q[t] + tq[t-1] = 3q[t+1]$, $q[1] = 0$, $q[2] = -2$, highest argument of the dependent variable is to be $t - 1$.

(d) $T[m] + (m-1)T[m-2] = m^2$, $T[0] = T[1] = 1$, highest argument of the dependent variable is to be $m + 2$.

(e) $y[k+1] - y[k+3] = (s[k+2] - s[k+4])/2$, where y is the dependent variable and the highest argument of the dependent variable is to be k.

12.4 Block diagram representation of difference equations

Many engineering systems can be modelled by means of difference equations. It is possible to represent a difference equation pictorially by means of a block diagram. The use of a block diagram representation helps an engineer to visualize a system and may often be helpful in suggesting the required hardware or software to implement a particular difference equation. This is particularly important in areas such as digital signal processing and digital control engineering.

Before discussing block diagrams it is necessary to review the topic of sampling. Difference equations operate on discrete-time data and therefore a continuous signal needs to be sampled before use. In the most common form of sampling, a sample is taken at regular intervals, T. A continuous signal and the sequence produced by sampling it are shown in Figure 12.1. Some authors write the sequence as $x[nT]$ to indicate that the sequence has been obtained by sampling a continuous waveform at intervals T. We will not use this convention but simply refer to the sampled sequence as $x[n]$.

Several components are used in a block diagram. The **delay block** is shown in Figure 12.2. The effect of this element is to delay the sequence by one sampling

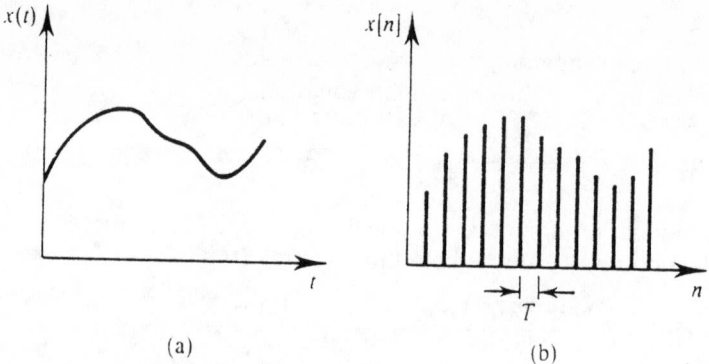

Figure 12.1   (a) Continuous signal; (b) sequence produced as a result of sampling.

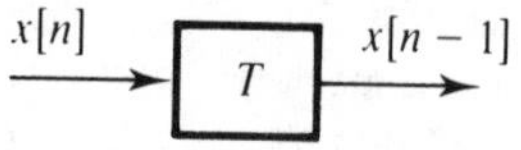

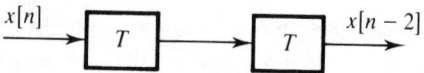

Figure 12.2   A delay block delays a sequence by a time interval, *T*.

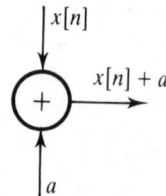

Figure 12.3   Two delay blocks in series.

Figure 12.4   Adding a constant to a sequence.

interval, *T*. For example, if

$$x[n] = 6, 4, 3, -2, 0, 2, 5, 0, \ldots \qquad n \in \mathbb{N}$$

We can write this as $x[0] = 6$, $x[1] = 4$, $x[2] = 3$, $x[3] = -2, \ldots$, and then,

$$x[n - 1] = 0, 6, 4, 3, -2, 0, 2, 5, 0, \ldots$$

Note that $x[n - 1]$ is undefined when $n = 0$ and so this is assigned a value of 0. A delay of two sampling intervals results in the sequence $x[n - 2]$ as shown in Figure 12.3.

Another block diagram element represents the addition of a constant to a sequence and is shown in Figure 12.4. A sequence can be scaled by a constant.

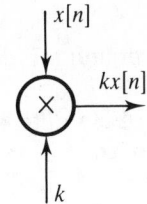

Figure 12.5   Scaling a sequence by a constant.

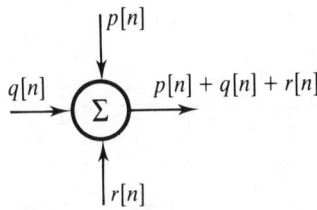

Figure 12.6   Adding sequences together using a summer.

This is shown in Figure 12.5. Finally, sequences can be added together using a **summer**. This is shown in Figure 12.6.

Example 12.7 Discrete-time filter

A simple example of a discrete-time filter is one described by the difference equation:

$$y[n] - ay[n-1] = x[n]$$

where $x[n]$ is the input sequence, $y[n]$ is the output sequence and a is a constant. If a is positive then the filter behaves as a low pass filter which rejects high frequencies but allows low frequencies to pass. If a is negative then the filter behaves as a high pass filter. A block diagram for the filter is shown in Figure 12.7. Note that this is a recursive filter because calculation of $y[n]$ requires knowledge of previous values of the output sequence. Note also that the block diagram contains a feedback path. This is a feature of recursive difference equations. ▲

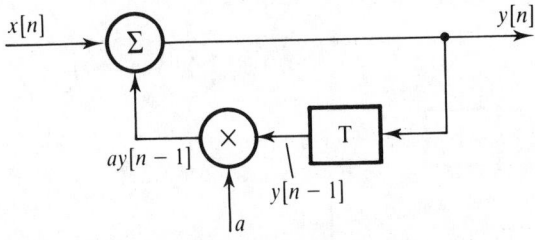

Figure 12.7   Discrete-time filter.

Example 12.8

A computer is fed a signal representing the position of an object as a function of time. Prior to entering the computer the signal is sampled using an analog-to-digital converter. Derive a difference equation and associated block diagram to obtain the acceleration of the object as a function of time.

Solution

Let s = position, v = speed and a = acceleration.

$$v = \frac{ds}{dt} \qquad a = \frac{dv}{dt} = \frac{d^2s}{dt^2}$$

Therefore, in order to obtain an acceleration signal the position signal must be differentiated twice. For a small time interval T, the derivative, $y(t)$, of a signal $x(t)$ can be approximated by:

$$y(t) \approx \frac{x(t) - x(t - T)}{T}$$

This follows directly from the definition of differentiation. If the signal $x(t)$ is sampled to give $x[n]$ then the process of differentiation is represented by the difference equation:

$$y[n] = \frac{x[n] - x[n - 1]}{T}$$

Figure 12.8 shows a block diagram for the differentiator. It is important to note that this difference equation is only an approximation to the process of differentiation. This could be implemented using special purpose hardware or by software on a microprocessor. It follows that the speed of the object is given by:

$$v[n] = \frac{s[n] - s[n - 1]}{T}$$

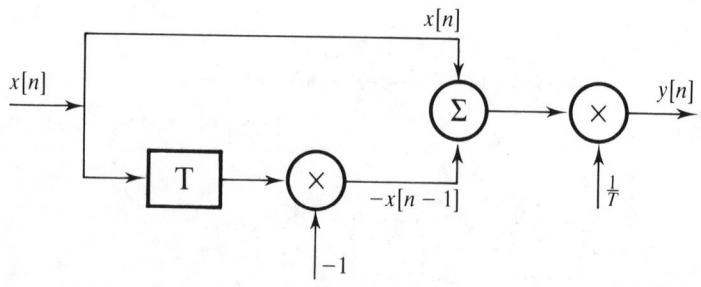

Figure 12.8   Block diagram of a differentiator.

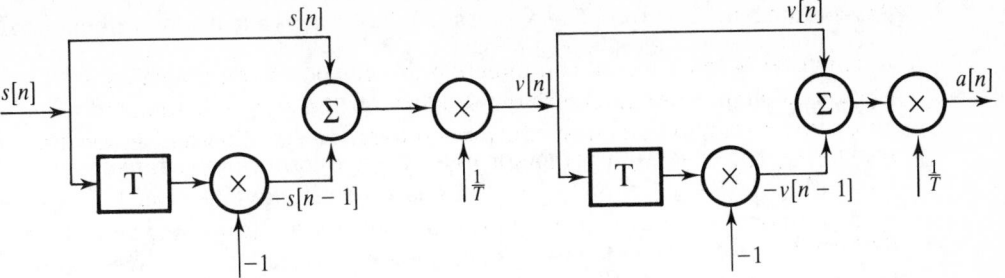

Figure 12.9   Two differentiators in series.

The problem of finding the acceleration, $a[n]$, can now be solved by coupling two differentiators together as shown in Figure 12.9. An alternative approach to this problem is to obtain a difference equation for the process of finding a second derivative. Given that:

$$v[n] = \frac{s[n] - s[n-1]}{T} \tag{12.2}$$

and

$$a[n] = \frac{v[n] - v[n-1]}{T} \tag{12.3}$$

then substituting Equation (12.2) into Equation (12.3) gives:

$$a[n] = \frac{(s[n] - s[n-1]) - (s[n-1] - s[n-2])}{T^2}$$

$$a[n] = \frac{s[n] - 2s[n-1] + s[n-2]}{T^2}$$

The block diagram for this difference equation is shown in Figure 12.10. Note that the difference equation is non-recursive and so there are no feedback

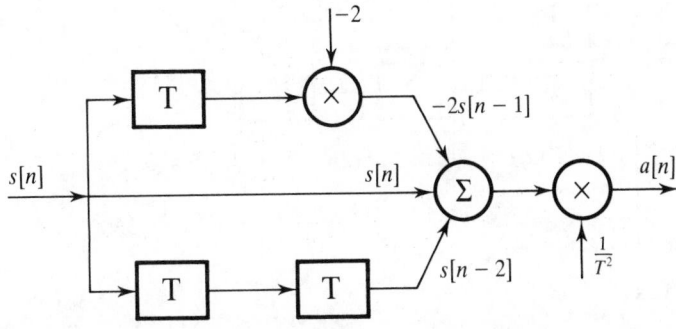

Figure 12.10 $a[n]$ is obtained by calculating the second derivative of $s[n]$.

paths in the block diagram. The output sequence is $a[n]$ and the input sequence is $s[n]$. The independent variable is n. ▲

Example 12.9

A signal is received by a computer in a sampled form from a transducer measuring the height of acetic acid in a large chemical tank. The measurements are known to fluctuate as a result of the acid swilling about in the tank. It is therefore decided to smooth out these fluctuations by averaging the five most recently measured values of the level and to use this **moving average** as a measure of the height of the acid in the tank. Formulate a difference equation to carry out this averaging and draw a block diagram of the difference equation.

Solution

Let $a[n]$ represent the average level of the acid in the tank and let $l[n]$ represent the sampled values of the level measurements received from the transducer. Then,

$$a[n] = \frac{l[n] + l[n-1] + l[n-2] + l[n-3] + l[n-4]}{5}$$

The block diagram for this difference equation is shown in Figure 12.11. The action of taking a moving average of sampled values is equivalent to passing the sampled values through a low pass filter because it filters out high frequency variations in the sampled values. This process is termed **digital filtering** or **digital signal processing**. ▲

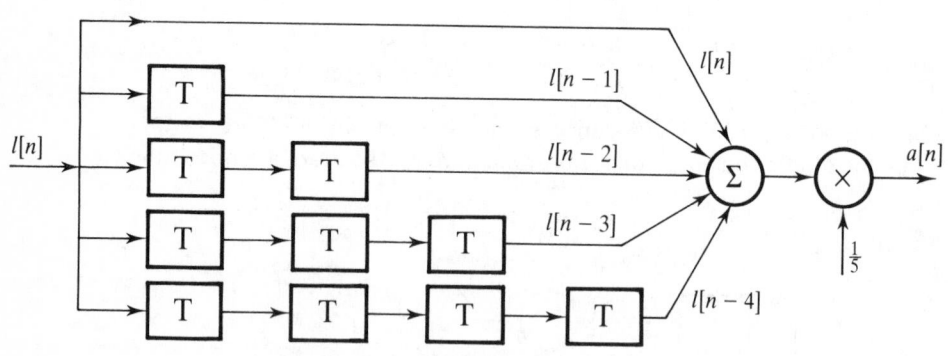

Figure 12.11   Block diagram of a moving averager.

EXERCISE 12.3

Design a digital filter based on taking a moving average of the last three values of a sampled signal.

12.5 Design of a discrete-time controller

Figure 12.12 shows a block diagram of a single loop industrial control system. The control system employs feedback to compare the desired value of a process variable with the actual value. Any difference between the two generates an error signal $e(t)$. This signal is processed by the controller to produce a controller signal $m(t)$. An amplifier is often present to magnify this signal to make it suitable for driving the plant that is being controlled.

The controller can be implemented by means of an analog electronic circuit. However, digital computers are being used increasingly as controllers. The signals $e(t)$ and $m(t)$ are both continuous in time and so it is necessary to sample $e(t)$ before it can be used by the computer and to reconstruct the signal generated by the computer to produce the controller output, $m(t)$. The arrangement for implementing a digital controller is shown in Figure 12.13.

The most common type of controller used in industry is the proportional/integral/derivative (p.i.d.) controller. It can be shown that the analog form of this controller is modelled by the equation:

$$m(t) = K_p e(t) + K_i \int_0^t e(t)\, dt + K_d \frac{de(t)}{dt} \tag{12.4}$$

where K_p, K_i and K_d are constants. In order to implement a discrete-time (digital) controller it is necessary to convert this equation into an equivalent difference equation. The approximation for the process of differentiation has already been examined in Example 12.8, and is given by:

$$\frac{de(t)}{dt} \approx \frac{e[n] - e[n-1]}{T}$$

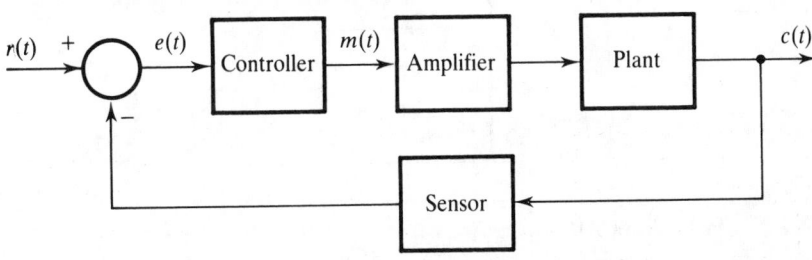

Figure 12.12 A single loop industrial control system.

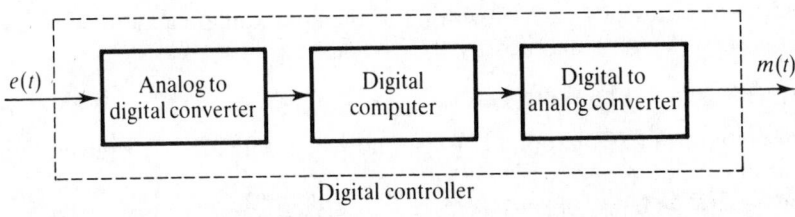

Figure 12.13 Block diagram of a digital controller.

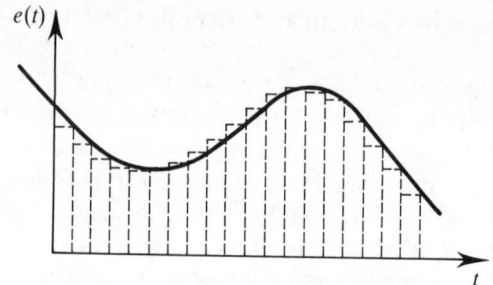

Figure 12.14 Approximating the area under the curve by a series of rectangles.

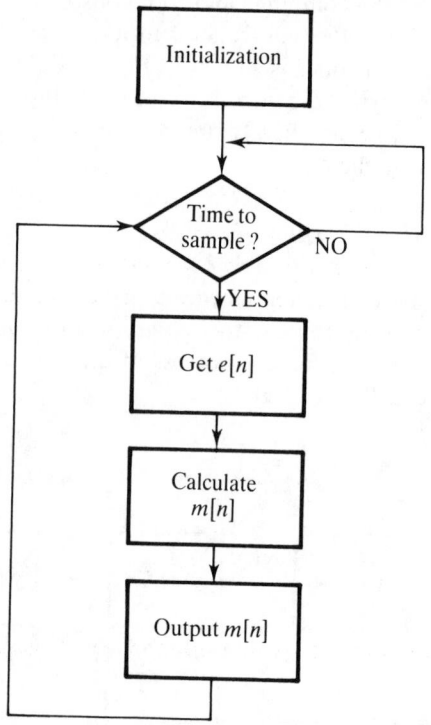

Figure 12.15 Flow chart for a p.i.d. controller.

There are several possible ways of approximating the process of integration. One method is illustrated in Figure 12.14. Here the area under the curve is approximated by a series of rectangles, each of width T. If the approximate area under the curve from $t = 0$ to $t = nT$ is denoted by $x[n]$, then

$$x[n] = x[n-1] + Te[n] \tag{12.5}$$

The discrete form of Equation (12.4) can now be formulated. It is given by:

$$m[n] = K_p e[n] + K_i x[n] + K_d \frac{e[n] - e[n-1]}{T} \tag{12.6}$$

Equations (12.5) and (12.6) form a set of equations to implement a discrete form of the p.i.d. controller on a digital computer or microprocessor. These two equations are termed coupled difference equations because both are needed to calculate $m[n]$. In addition, Equation (12.5) is recursive. A flow chart for implementing these equations is shown in Figure 12.15.

EXERCISES 12.4

1. A transducer is used to measure the speed of a motor car. Design a digital filter to calculate the distance travelled by the car.

2. Draw a block diagram for Equations (12.5) and (12.6).

12.6 Numerical solution of difference equations

Having seen how difference equations are formulated we now proceed to methods of solution. The numerical method illustrated may be applied to all classes of difference equation.

Example 12.10

Given

$$x[n + 1] - x[n] = n \qquad x[0] = 1$$

Determine $x[1]$, $x[2]$ and $x[3]$.

Solution

The terms in the equation are evaluated for various values of n.

$n = 0$

$$x[1] - x[0] = 0$$
$$x[1] = 1$$

$n = 1$

$$x[2] - x[1] = 1$$
$$x[2] = 2$$

$n = 2$

$$x[3] - x[2] = 2$$
$$x[3] = 4 \quad \blacktriangle$$

Example 12.11

Determine $x[4]$ given

$$2x[k + 2] - x[k + 1] + x[k] = -k^2 \qquad x[0] = 1 \qquad x[1] = 3$$

Solution

$k = 0$

$$2x[2] - x[1] + x[0] = 0$$

$$x[2] = 1$$

$k = 1$

$$2x[3] - x[2] + x[1] = -1$$

$$x[3] = -\tfrac{3}{2}$$

$k = 2$

$$2x[4] - x[3] + x[2] = -4$$

$$x[4] = -\tfrac{13}{4} \quad \blacktriangle$$

As the previous examples illustrate, to determine a unique solution to a first order equation requires one initial condition; for a second order equation, two initial values are required.

Example 12.12 Low pass filter

Recall from Example 12.7, the formula for a simple discrete-time filter

$$y[n] - ay[n - 1] = x[n]$$

If a is positive then the filter is a low pass filter. We will choose $a = 0.5$ and so

$$y[n] = 0.5y[n - 1] + x[n]$$

Let us examine the response of this filter to a unit step input applied at $n = 0$, that is,

$$x[n] = \begin{cases} 0 & n < 0 \\ 1 & \text{otherwise} \end{cases} \qquad n \in \mathbb{Z}$$

Assume the output of the filter is zero prior to the application of the step input, that is, $y[n] = 0$ for $n \leq -1$. From the difference equation, we find:

$$y[0] = 0.5y[-1] + x[0]$$

$$= 0.5(0) + 1$$

$$= 1$$

Table 12.1 Numerical solution of a difference equation.

n	$x[n]$	$y[n-1]$	$y[n]$
-1	0	0	0
0	1	0	1
1	1	1	1.5
2	1	1.5	1.75
3	1	1.75	1.88
4	1	1.88	1.94
5	1	1.94	1.97
6	1	1.97	1.99
7	1	1.99	2.00
8	1	2.00	2.00
9	1	2.00	2.00
10	1	2.00	2.00

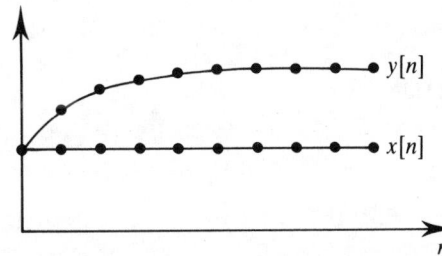

Figure 12.16   Input and output sequences for the low pass filter.

Similarly,

$$y[1] = 0.5y[0] + x[1]$$

$$= 0.5(1) + 1$$

$$= 1.5$$

and so on. When numerically solving a difference equation, it is often useful to form a table with intermediate results. Table 12.1 shows such a table.

Figure 12.16 shows the input and output sequences superimposed on the same graph. The sequence values have been joined to illustrate their trends. Note that the input signal reaches its final value immediately, whereas the output signal takes several sample intervals to reach its final value. Engineers often refer to this process as 'smoothing' the input signal. We will see in Chapter 13 that rapidly changing signals tend to be richer in high frequencies than those that change more slowly. The effect of the low pass filter is to filter out these high frequencies and so the output from the filter changes more slowly than the input. In Chapter 13 we will also examine a continuous low pass filter which does the same job for continuous signals. ▲

Numerical solution of difference equations is often the only feasible method of obtaining a solution for many practical engineering systems. This is because the input terms are usually obtained as a result of sampling an input signal and so cannot be expressed analytically. However, in some cases it is possible to express the input analytically and then an analytical solution to the difference equation may be feasible. While analytical methods analogous to those applied to differential equations are available, z transform techniques are more popular with engineers and so these are introduced in the following sections.

EXERCISES 12.5

1. Given

$$x[n + 2] + x[n + 1] - x[n] = 2$$

$$x[0] = 3 \qquad x[1] = 5$$

find $x[2]$, $x[3]$, $x[4]$ and $x[5]$.

2. If

$$z[n]z[n - 1] = n^2 \qquad z[1] = 7$$

find $z[2]$, $z[3]$ and $z[4]$.

3. Determine $x[2]$ and $x[3]$ given

(a) $2x[n + 2] - 5x[n + 1] = 4n$, $x[1] = 2$
(b) $6x[n] - x[n - 1] + 2x[n - 2]$
    $= n^2 - n$, $x[0] = 1$, $x[1] = 2$
(c) $3x[n - 1] + x[n - 2] - 9x[n - 3]$
    $= (n - 1)^2$, $x[0] = 3$, $x[1] = 2$.

12.7 Definition of the z transform

Suppose we have a sequence $f[k]$, $k \in \mathbb{N}$. Such a sequence may have arisen by sampling a continuous signal. We define its z transform to be:

■ $\quad F(z) = \mathscr{Z}\{f[k]\} = \displaystyle\sum_{k=0}^{\infty} f[k]z^{-k}$ $\hfill$ (12.7)

We see from the definition that the z transform is an infinite series formed from the terms of the sequence. Explicitly, we have:

$$\mathscr{Z}\{f[k]\} = f[0] + \frac{f[1]}{z} + \frac{f[2]}{z^2} + \frac{f[3]}{z^3} + \cdots$$

In most engineering applications we do not actually need to work with the infinite series since it is often possible to express this in a closed form. The closed form is generally valid for values of z within a region known as the **radius of absolute convergence** as will become apparent from the following examples.

Example 12.13

Find the z transform of the sequence defined by:

$$f[k] = \begin{cases} 1 & k = 0 \\ 0 & k \neq 0 \end{cases}$$

This sequence is sometimes called the **Kronecker delta sequence**, often denoted by $\delta[k]$.

Solution

$$\mathscr{Z}\{f[k]\} = \sum_{k=0}^{\infty} f[k]z^{-k}$$

$$= f[0] + \frac{f[1]}{z} + \frac{f[2]}{z^2} + \frac{f[3]}{z^3} + \cdots$$

$$= 1 + \frac{0}{z} + \frac{0}{z^2} + \frac{0}{z^3} + \cdots$$

$$= 1$$

Hence $F(z) = 1$. ▲

Example 12.14

Find the z transform of the sequence defined by:

$$f[k] = 1 \qquad k \in \mathbb{N}$$

This is the **unit step sequence**, often denoted by $u[k]$.

Solution

$$\mathscr{Z}\{f[k]\} = \sum_{k=0}^{\infty} f[k]z^{-k}$$

$$= 1 + \frac{1}{z} + \frac{1}{z^2} + \frac{1}{z^3} + \cdots$$

This is a geometric progression with first term 1 and common ratio $1/z$. The progression converges if $|z| > 1$ in which case the sum to infinity is:

$$\frac{1}{1 - 1/z} = \frac{z}{z - 1}$$

We see that $F(z)$ has the convenient closed-form solution:

$$F(z) = \frac{z}{z - 1}$$

for $|z| > 1$. ▲

Note that the process of taking the z transform converts the sequence $f[k]$ into the continuous function $F(z)$.

Example 12.15

Find the z transform of the sequence defined by $f[k] = k$, $k \in \mathbb{N}$. This sequence is called the **unit ramp sequence**.

Solution

$$F(z) = \mathcal{Z}\{f[k]\} = \sum_{k=0}^{\infty} kz^{-k}$$

$$= \frac{1}{z} + \frac{2}{z^2} + \frac{3}{z^3} + \cdots$$

$$= \frac{1}{z}\left\{1 + \frac{2}{z} + \frac{3}{z^2} + \cdots\right\}$$

If we use the binomial theorem (Section 3.4) to express $(1 - 1/z)^{-2}$ as an infinite series, we find that

$$\left(1 - \frac{1}{z}\right)^{-2} = 1 + \frac{2}{z} + \frac{3}{z^2} + \cdots \qquad \text{provided } \left|\frac{1}{z}\right| < 1, \text{ that is, } |z| > 1$$

Using this result we see that $\mathcal{Z}\{f[k]\}$ can be written as

$$F(z) = \frac{1}{z}\left(1 - \frac{1}{z}\right)^{-2}$$

and so,

$$F(z) = \frac{1}{z}\frac{1}{(1 - 1/z)^2} = \frac{z}{(z - 1)^2} \text{ for } |z| > 1 \quad \blacktriangle$$

Example 12.16

Find the z transform of the sequence defined by:

$$f[k] = Ak \qquad A \text{ constant}$$

Solution

We find:

$$F(z) = \mathcal{Z}\{f[k]\} = \sum_{k=0}^{\infty} Akz^{-k}$$

$$= A \sum_{k=0}^{\infty} kz^{-k}$$

$$= \frac{Az}{(z - 1)^2} \qquad \text{using Example 12.15} \quad \blacktriangle$$

In the same way as has been done for Laplace transforms, we can build up a library of sequences and their z transforms. Some common examples appear in Table 12.2. Note that in Table 12.2 a and b are constants.

Table 12.2 The z transforms of some common functions.

$f[k]$	$F(z)$
$\delta[k] = \begin{cases} 1 & k = 0 \\ 0 & k \neq 0 \end{cases}$	1
$u[k] = \begin{cases} 1 & k \geq 0 \\ 0 & k < 0 \end{cases}$	$\dfrac{z}{z - 1}$
k	$\dfrac{z}{(z - 1)^2}$
e^{-ak}	$\dfrac{z}{z - e^{-a}}$
a^k	$\dfrac{z}{z - a}$
ka^k	$\dfrac{az}{(z - a)^2}$
$k^2 a^k$	$\dfrac{az(z + a)}{(z - a)^3}$
k^2	$\dfrac{z(z + 1)}{(z - 1)^3}$
k^3	$\dfrac{z(z^2 + 4z + 1)}{(z - 1)^4}$
$\sin ak$	$\dfrac{z \sin a}{z^2 - 2z \cos a + 1}$
$\cos ak$	$\dfrac{z(z - \cos a)}{z^2 - 2z \cos a + 1}$
$e^{-ak} \sin bk$	$\dfrac{ze^{-a} \sin b}{z^2 - 2ze^{-a} \cos b + e^{-2a}}$
$e^{-ak} \cos bk$	$\dfrac{z^2 - ze^{-a} \cos b}{z^2 - 2ze^{-a} \cos b + e^{-2a}}$

Example 12.17

Use Table 12.2 to find the z transforms of

(a) $\sin \frac{1}{2}k$

(b) $e^{3k} \cos 2k$

Solutions

Directly from Table 12.2 we find:

(a) $\mathcal{Z}\{\sin\frac{1}{2}k\} = \dfrac{z\sin\frac{1}{2}}{z^2 - 2z\cos\frac{1}{2} + 1}$

(b) $\mathcal{Z}\{e^{3k}\cos 2k\} = \dfrac{z^2 - ze^3\cos 2}{z^2 - 2ze^3\cos 2 + e^6}$ ▲

EXERCISES 12.6

1. Using the definition of the z transform, find closed form expressions for the z transforms of the following sequences $f[k]$ where:

    (a) $f[0] = 0$, $f[1] = 0$, $f[k] = 1$ for $k \geq 2$

    (b) $f[k] = \begin{cases} 0 & k = 0, 1, \ldots, 5 \\ 4 & k > 5 \end{cases}$

    (c) $f[k] = 3k$, $k \geq 0$

    (d) $f[k] = e^{-k}$, $k = 0, 1, 2, \ldots$

    (e) $f[0] = 1$, $f[1] = 2$, $f[2] = 3$, $f[k] = 0$, $k \geq 3$

    (f) $f[0] = 3$, $f[k] = 0$, $k \neq 0$

    (g) $f[k] = \begin{cases} 2 & k \geq 0 \\ 0 & k < 0 \end{cases}$

2. Use Table 12.2 to find the z transforms of

    (a) $\cos 3k$

    (b) e^k

    (c) $e^{-2k}\cos k$

    (d) $e^{4k}\sin 2k$

    (e) 4^k

    (f) $(-3)^k$

    (g) $\sin(k\pi/2)$

    (h) $\cos(k\pi/2)$.

3. Find, from Table 12.2, the sequences which have the following z transforms:

    (a) $z/(z + 4)$,

    (b) $2z/(2z - 1)$

    (c) $3z/(3z + 1)$

    (d) $z/(z - e^3)$

    (e) $z/(z^2 + 1)$

4. By considering the Taylor series expansion of $e^{1/z}$, find the z transform of the sequence $f[k] = 1/k!$, $k \geq 0$.

12.8 Sampling a continuous signal

We have already introduced sampling in Section 12.4. We now return to the topic. Most of the signals that are encountered in the physical world are **continuous** in time. This means that they have a signal level for every value of time over a particular time interval of interest. An example is the measured value of the temperature of an oven obtained using an electronic thermometer. This type of signal can be modelled using a continuous mathematical function in which for each value of t there is a continuous signal level, $f(t)$. Several engineering systems contain signals whose value is important only at particular points in time. These points are usually equally spaced and separated by a time interval, T. Such signals are referred to as discrete time, or more compactly, **discrete** signals. They are modelled by a mathematical function that is only defined at certain points in time.

An example of a discrete system is a digital computer. It carries out calculations at fixed intervals governed by an electronic clock.

Suppose we have a continuous signal $f(t)$, defined for $t \geq 0$ which we **sample**, that is, measure, at intervals of time, T. We obtain a sequence of sampled values of $f(t)$, that is, $f[0], f[1], f[2],\ldots,f[k],\ldots$. Returning to the example of the oven temperature signal, a discrete signal with a time interval of five seconds can be obtained by noting the value of the electronic thermometer display every five seconds. Some textbooks use the notation $f[kT]$ as a reminder that the sequence has been obtained by sampling at an interval T. We will not use this notation as it can become clumsy. However, it is important to note that changing the value of T changes the z transform as we shall see in Example 12.18. It can be shown that sampling a continuous signal does not lose the essence of the signal provided the sampling rate is sufficiently high, and it is in fact possible to recreate the original continuous signal from the discrete signal, if required. It is often convenient to represent a discrete signal as a series of weighted impulses. The strength of each impulse is the level of the signal at the corresponding point in time. We write:

$$■ \quad f^*(t) = \sum_{k=0}^{\infty} f[k]\, \delta(t - kT) \tag{12.8}$$

the $*$ indicating that $f(t)$ has been sampled. This representation is discussed in Appendix V. This is a useful mathematical way of representing a discrete signal as the properties of the impulse function lend themselves to a value that only exists for a short interval of time. In practice, no sampling method has zero sampling time but provided the sampling time is much smaller than the sampling interval, then this is a valid mathematical model of a discrete signal.

We can apply the z transform directly to a continuous function, $f(t)$, if we regard the function as having been sampled at discrete intervals of time. Consider the following example.

Example 12.18

(a) Find the z transform of $f(t) = e^{-t}$ sampled at $t = 0, 0.1, 0.2, \ldots$

(b) Find the z transform of $f(t) = e^{-t}$ sampled at $t = 0, 0.01, 0.02, \ldots$

(c) Express the sequences obtained in (a) and (b) as series of weighted impulses.

Solutions

(a) The sequence of sampled values is:

$$1, e^{-0.1}, e^{-0.2}, \ldots$$

that is,

$$e^{-0.1k} \qquad k \in \mathbb{N} \text{ and } T = 0.1$$

From Table 12.2 we find the z transform of this sequence is $z/(z - e^{-0.1})$.

(b) The sequence of sampled values is:

$$1, e^{-0.01}, e^{-0.02}, \ldots$$

that is,

$$e^{-0.01k} \qquad k \in \mathbb{N} \qquad \text{and} \qquad T = 0.01$$

From Table 12.2 we find the z transform of this sequence is $z/(z - e^{-0.01})$. We note that modifying the sampling interval, T, alters the z transform even though we are dealing with the same function $f(t)$.

(c) When $T = 0.1$ we have, from Equation (12.8),

$$f^*(t) = \sum_{k=0}^{\infty} e^{-0.1k} \delta(t - 0.1k)$$

$$= 1\delta(t) + e^{-0.1} \delta(t - 0.1) + e^{-0.2} \delta(t - 0.2) + \cdots$$

Note that an advantage of expressing the sequence as a series of weighted impulses is that information concerning the time of occurrence of a particular value is contained in the corresponding δ term.
When $T = 0.01$, we have

$$f^*(t) = \sum_{k=0}^{\infty} e^{-0.01k} \delta(t - 0.01k)$$

$$= 1\delta(t) + e^{-0.01} \delta(t - 0.01) + e^{-0.02} \delta(t - 0.02) + \cdots \quad \blacktriangle$$

Example 12.19

(a) Find the z transform of the continuous function $f(t) = \cos 3t$ sampled at $t = kT, k \in \mathbb{N}$

(b) Write down the first four terms of the sampled sequence when $T = 0.2$, and express the sequence as a series of weighted impulses.

Solutions

(a) The sampled sequence is:

$$f[k] = \cos 3kT = \cos((3T)k)$$

The z transform of this sequence can be obtained directly from Table 12.2 from which we have

$$\mathcal{L}\{\cos ak\} = \frac{z(z - \cos a)}{z^2 - 2z \cos a + 1}$$

Writing $a = 3T$ we find

$$\mathcal{L}\{\cos 3kT\} = \frac{z(z - \cos 3T)}{z^2 - 2z \cos 3T + 1}$$

(b) When $T = 0.2$, the first four terms are:

$$1 \qquad \cos 0.6 \qquad \cos 1.2 \qquad \cos 1.8$$

From Equation (12.8), the sequence of sampled values can be represented as the following series of weighted impulses:

$$f^*(t) = \sum_{k=0}^{\infty} \cos 3kT\, \delta(t - kT)$$

$$= \sum_{k=0}^{\infty} \cos 0.6k\, \delta(t - 0.2k)$$

$$= \delta(t) + \cos 0.6\, \delta(t - 0.2) + \cos 1.2\, \delta(t - 0.4)$$

$$+ \cos 1.8\, \delta(t - 0.6) + \cdots \quad \blacktriangle$$

12.9 The relationship between the z transform and the Laplace transform

We have defined the z transform quite independently of any other transform. However, there is a close relationship between the z transform and the Laplace transform, the z being regarded as the discrete equivalent of the Laplace. This can be seen from the following argument.

If the continuous signal $f(t)$ is sampled at intervals of time, T, we obtain a sequence of sampled values $f[k]$, $k \in \mathbb{N}$. From Section 12.8 we note that this sequence can be regarded as a train of impulses.

$$f^*(t) = \sum_{k=0}^{\infty} f[k]\, \delta(t - kT)$$

Taking the Laplace transform, we have:

$$\mathscr{L}\{f^*(t)\} = \int_0^{\infty} e^{-st} \sum_{k=0}^{\infty} f[k]\, \delta(t - kT)\, dt$$

$$= \sum_{k=0}^{\infty} f[k] \int_0^{\infty} e^{-st}\, \delta(t - kT)\, dt$$

assuming that it is permissible to interchange the order of summation and integration. Noting from Table 11.1 that the Laplace transform of the function $\delta(t - kT)$ is e^{-skT}, we can write:

$$\mathscr{L}\{f^*(t)\} = \sum_{k=0}^{\infty} f[k]e^{-skT} \tag{12.9}$$

Now, making the change of variable $z = e^{sT}$, we have:

$$\mathscr{L}\{f^*(t)\} = \sum_{k=0}^{\infty} f[k]z^{-k}$$

which is the definition of the z transform. The expression $\mathscr{L}\{f^*(t)\}$ is commonly written as $F^*(s)$.

12.9.1 *Mapping the s plane to the z plane*

When designing an engineering system it is often useful to consider an s plane representation of the system. The characteristics of a system can be quickly identified by the positions of the poles and zeros as we saw in Chapter 11. Engineers will often modify the system characteristics by introducing new poles and zeros or by changing the positions of existing ones. Unfortunately, it is not convenient to use the s plane to analyse discrete systems. For a sampled signal Equation (12.9) yields:

$$F^*(s) = \mathscr{L}\{f^*(t)\} = \sum_{k=0}^{\infty} f[k]e^{-skT}$$

The continuous signals and systems that were analysed in Chapter 11 had Laplace transforms that were simple ratios of polynomials in s. This was one of the main reasons for using Laplace transforms to solve differential equations; the problem was reduced to one of reasonably straightforward algebraic manipulation. Here we have a Laplace transform that is very complicated. In fact it can have an infinite number of poles and zeros. To see this consider the following example.

Example 12.20

The continuous signal $f(t) = \cos(\pi t/2)$ is sampled at one second intervals starting from $t = 0$

(a) Find the Laplace transform of the sampled signal $f^*(t)$.

(b) Show that $F^*(s)$ has an infinity of poles.

(c) Find the z transform of the sampled signal and show that this has just two poles.

Solutions

(a) The continuous signal $f(t) = \cos(\pi t/2)$ sampled at one second intervals gives rise to the sequence $1, 0, -1, 0, 1, 0, -1, \ldots$. Consequently, from Equation (12.9)

$$F^*(s) = \mathscr{L}\{f^*(t)\} = \sum_{k=0}^{\infty} f[k]e^{-skT} = \sum_{k=0}^{\infty} f[k]e^{-sk} \qquad \text{since } T = 1$$

that is,

$$F^*(s) = 1 + 0 - e^{-2s} + 0 + e^{-4s} + 0 - e^{-6s} + \cdots$$

This is a geometric progression with common ratio $-e^{-2s}$ and hence its sum to infinity is

$$\frac{1}{1 - (-e^{-2s})}$$

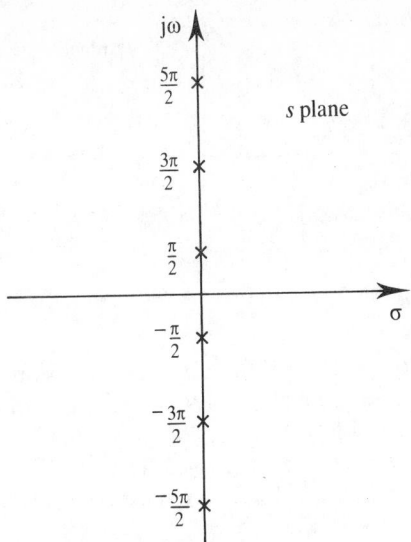

Figure 12.17   The sampled signal has an infinite number of poles.

that is,

$$F^*(s) = \frac{1}{1 + e^{-2s}}$$

(b) Poles of $F^*(s)$ will occur when $1 + e^{-2s} = 0$. Writing $s = \sigma + j\omega$ we see that poles will occur when $e^{-2(\sigma + j\omega)} = -1$. Since -1 can be written as $e^{j(2n-1)\pi}$, $n \in \mathbb{Z}$ (see Chapter 6) we see that poles will occur when:

$$e^{-2\sigma - 2j\omega} = e^{j(2n-1)\pi}$$

that is, when $\sigma = 0$ and $\omega = -(2n-1)\pi/2$. Thus there exist an infinite number of poles occuring when

$$s = -(2n-1)\frac{\pi}{2}j \qquad n \in \mathbb{Z}$$

Some of these are illustrated in Figure 12.17.

(c) The z transform of the sampled signal is:

$$\mathscr{Z}\{f^*(t)\} = 1 + \frac{0}{z} - \frac{1}{z^2} + \frac{0}{z^3} + \frac{1}{z^4} + \cdots$$

$$= \frac{1}{1 - (-1/z^2)}$$

$$= \frac{z^2}{z^2 + 1}$$

which has just two poles at $z = \pm j$ as shown in Figure 12.18.   ▲

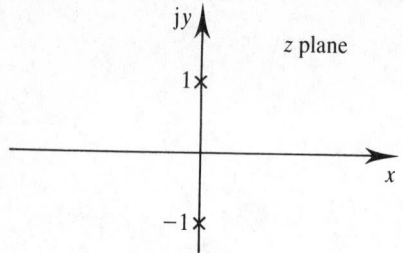

Figure 12.18   There are two poles at $z = \pm j$.

It is possible to show in general that the result of sampling a continuous signal is to convert each simple pole of the Laplace transform into an infinite set of poles. Suppose that the Laplace transform of the signal $f(t)$ can be broken down by using partial fractions into a series of $n + 1$ terms with simple poles $a_0, a_1, \ldots, a_n$. For simplicity, repeating poles will not be considered but the proof for such a case is similar. So,

$$F(s) = \frac{A_0}{s - a_0} + \frac{A_1}{s - a_1} + \frac{A_2}{s - a_2} + \cdots + \frac{A_n}{s - a_n}$$

In the time domain this corresponds to:

$$f(t) = A_0 e^{a_0 t} + A_1 e^{a_1 t} + \cdots + A_n e^{a_n t}$$

Now we consider the Laplace transform of the sampled signal $f^*(t)$.

$$F^*(s) = \mathscr{L}\{f^*(t)\} = \sum_{k=0}^{\infty} f[k] e^{-skT}$$

$$= \sum_{k=0}^{\infty} (A_0 e^{a_0 kT} + A_1 e^{a_1 kT} + \cdots + A_n e^{a_n kT}) e^{-skT}$$

$$= A_0 \sum_{k=0}^{\infty} e^{-kT(s - a_0)} + A_1 \sum_{k=0}^{\infty} e^{-kT(s - a_1)} + \cdots$$

$$+ A_n \sum_{k=0}^{\infty} e^{-kT(s - a_n)}$$

Now each of the summations can be converted into a closed form. For example,

$$\sum_{k=0}^{\infty} e^{-kT(s - a_0)} = 1 + e^{-T(s - a_0)} + e^{-2T(s - a_0)} + e^{-3T(s - a_0)} + \cdots$$

$$= \frac{1}{1 - e^{-T(s - a_0)}}$$

Therefore,

$$F^*(s) = \frac{A_0}{1 - e^{-T(s - a_0)}} + \frac{A_1}{1 - e^{-T(s - a_1)}} + \cdots + \frac{A_n}{1 - e^{-T(s - a_n)}}$$

It is possible to show that for each simple pole in $F(s)$ there is now an infinite set of poles. Consider the pole at $s = a_0$. This contributes the term:

$$\frac{A_0}{1 - e^{-T(s - a_0)}}$$

to $F^*(s)$. This term has poles whenever $1 - e^{-T(s - a_0)} = 0$, that is, $e^{-T(s - a_0)} = 1$. This corresponds to $T(s - a_0) = 2\pi m\mathrm{j}$, $m \in \mathbb{Z}$. Therefore,

$$T(s - a_0) = 2\pi m\mathrm{j}$$

$$s - a_0 = \frac{2\pi m}{T}\mathrm{j}$$

$$s = a_0 + \frac{2\pi m}{T}\mathrm{j} \qquad m \in \mathbb{Z}$$

The effect of sampling is to introduce an infinite set of poles. Each one is equal to the pole of the original continuous signal but displaced by an imaginary component. This is illustrated in Figure 12.19 for a real pole a_0. However, the proof is equally valid for a complex conjugate pair of poles but the diagram is more cluttered and has, therefore, not been shown.

Clearly discrete systems are not amenable to s plane design techniques. Fortunately, the z plane can be used for analysing discrete systems in the same way that the s plane can be used when analysing continuous systems. It is possible to map points from the s plane to the z plane using the relation $z = e^{sT}$ which gives rise to the definition of the z transform as described previously.

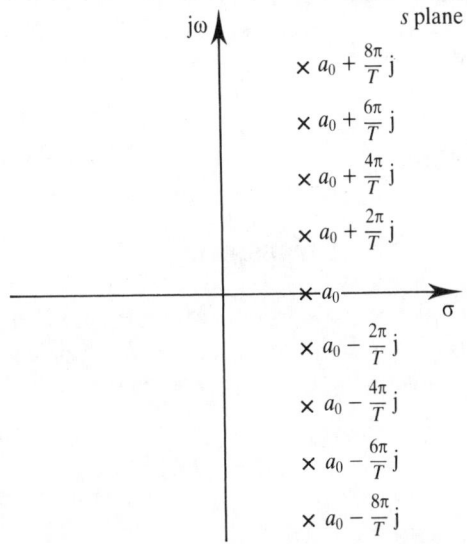

Figure 12.19 The effect of sampling is to introduce an infinite set of poles.

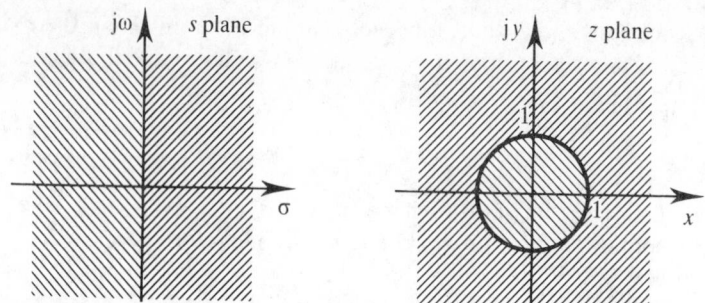

Figure 12.20   The left-hand side of the *s* plane maps to the inside of the unit circle of the *z* plane.

The great advantage of the *z* plane is that it eliminates the problem of infinitely repeating poles and zeros when analysing discrete systems. This can be illustrated by considering how the imaginary axis of the *s* plane maps to the *z* plane.

Referring to Figure 12.20, we see that $s = \sigma + j\omega$. On the imaginary axis $\sigma = 0$ and therefore $z = e^{sT} = e^{j\omega T}$. As ω varies between $-\pi/T$ and π/T, the locus of *z* is a circle of radius 1, centred at the origin (see Section 6.9). As ω is increased from 0 to π/T the upper half of the unit circle is traced out, while as ω is decreased from 0 to $-\pi/T$, the lower half of the unit circle is traced out. Increasing ω above π/T or decreasing it below $-\pi/T$ leads to a retracing of the unit circle. In other words, the repeated *s* plane points are superimposed on top of each other. This is the reason why the *z* plane approach is much simpler than the *s* plane approach when analysing discrete systems.

The *z* transforms of discrete signals and systems are, in many cases, simple ratios of polynomials. We shall see shortly that this means the process of analysing difference equations which model these signals and systems, is reduced to relatively simple algebraic manipulations.

12.10 Properties of the *z* transform

Because of the relationship between the two transforms we would expect that many of the properties of the Laplace transform would be mirrored by properties of the *z* transform. This is indeed the case and some of these properties are given now. These are:

(1) Linearity
(2) Shift theorems
(3) The complex translation theorem

12.10.1 *Linearity*

If $f[k]$ and $g[k]$ are two sequences then:

■ $\mathscr{L}\{f[k] + g[k]\} = \mathscr{L}\{f[k]\} + \mathscr{L}\{g[k]\}$

This statement simply says that to find the z transform of the sum of two sequences we can add the z transforms of the two sequences. If c is a constant, which may be negative, and $f[k]$ is a sequence, then:

■ $\mathscr{L}\{cf[k]\} = c\mathscr{L}\{f[k]\}$

Together these two properties mean that the z transform is a linear operator.

Example 12.21

Find the z transform of $e^{-k} + k$.

Solution

From Table 12.2 we have:

$$\mathscr{L}\{e^{-k}\} = \frac{z}{z - e^{-1}}$$

and

$$\mathscr{L}\{k\} = \frac{z}{(z - 1)^2}$$

Therefore,

$$\mathscr{L}\{e^{-k} + k\} = \frac{z}{z - e^{-1}} + \frac{z}{(z - 1)^2} \quad \blacktriangle$$

Example 12.22

Find the z transform of $3k$.

Solution

From Table 12.2 we have:

$$\mathscr{L}\{k\} = \frac{z}{(z - 1)^2}$$

Therefore,

$$\mathscr{L}\{3k\} = 3\mathscr{L}\{k\} = 3 \times \frac{z}{(z-1)^2} = \frac{3z}{(z-1)^2} \quad \blacktriangle$$

Example 12.23

Find the z transform of the function $f(t) = 2t^2$ sampled at $t = kT$, $k \in \mathbb{N}$.

Solution

The sequence of sampled values is:

$$f[k] = 2(kT)^2 = 2T^2k^2$$

The z transform of this sequence can be read directly from Table 12.2 using the linearity properties. We have:

$$\mathscr{L}\{2T^2k^2\} = 2T^2\mathscr{L}\{k^2\} = \frac{2T^2z(z+1)}{(z-1)^3} \quad \blacktriangle$$

12.10.2 *First shift theorem*

■ If $f[k]$ is a sequence and $F(z)$ is its z transform, then:

$$\mathscr{L}\{f[k+i]\} = z^iF(z) - (z^if[0] + z^{i-1}f[1] + \cdots + zf[i-1]) \qquad i \in \mathbb{N}^+$$
$$(12.10)$$

In particular, if $i = 1$ we have

$$\mathscr{L}\{f[k+1]\} = zF(z) - zf[0]$$
$$(12.11)$$

If $i = 2$ we have

$$\mathscr{L}\{f[k+2]\} = z^2F(z) - z^2f[0] - zf[1]$$

Example 12.24

The sequence $f[k]$ is defined by:

$$f[k] = \begin{cases} 0 & k = 0, 1, 2, 3 \\ 1 & k = 4, 5, 6, \ldots \end{cases}$$

Write down the sequence $f[k + 1]$ and verify that:

$$\mathscr{L}\{f[k+1]\} = zF(z) - zf[0]$$

where $F(z)$ is the z transform of $f[k]$.

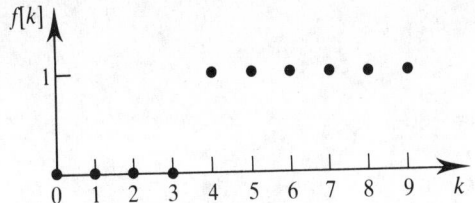

Figure 12.21   The sequence of Example 12.24.

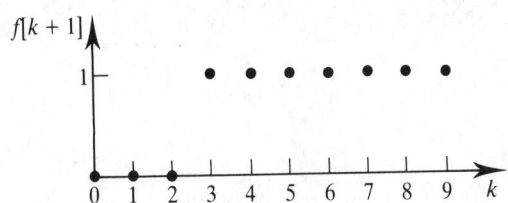

Figure 12.22   A shifted version of the sequence of Figure 12.21.

Solution

The graph of $f[k]$ is illustrated in Figure 12.21. The sequence $f[k + 1]$ is defined as follows:

When $k = 0$, $f[k + 1] = f[1]$ which is 0.

When $k = 1$, $f[k + 1] = f[2]$ which is 0.

When $k = 2$, $f[k + 1] = f[3]$ which is 0.

When $k = 3$, $f[k + 1] = f[4]$ which is 1, and so on.

Consequently,

$$f[k + 1] = \begin{cases} 0 & k = 0, 1, 2 \\ 1 & k = 3, 4, 5, \ldots \end{cases}$$

as illustrated in Figure 12.22. We see that the graph of $f[k + 1]$ is simply that of $f[k]$ shifted one place to the left. More generally, $f[k + i]$ is the sequence $f[k]$ shifted i places to the left. Now

$$F(z) = \mathscr{Z}\{f[k]\} = \sum_{k=0}^{\infty} f[k]z^{-k}$$

$$= \sum_{k=4}^{\infty} z^{-k}$$

$$= \frac{1}{z^4} + \frac{1}{z^5} + \frac{1}{z^6} + \cdots$$

$$= \frac{1}{z^4}\left(1 + \frac{1}{z} + \frac{1}{z^2} + \cdots\right)$$

$$= \frac{1}{z^4}\frac{1}{1 - 1/z}$$

$$= \frac{1}{z^4}\frac{z}{z - 1}$$

$$= \frac{1}{z^3}\frac{1}{z - 1}$$

The same argument shows that:

$$\mathscr{Z}\{f[k + 1]\} = \frac{1}{z^2}\frac{1}{z - 1}$$

It then follows that:

$$\mathscr{Z}\{f[k + 1]\} = \frac{1}{z^2}\frac{1}{z - 1} = zF(z) - zf[0]$$

since $f[0] = 0$. This illustrates the first shift theorem. ▲

12.10.3 *Second shift theorem*

The function $f(t)u(t)$ is defined by:

$$f(t)u(t) = \begin{cases} f(t) & t \geq 0 \\ 0 & t < 0 \end{cases}$$

where $u(t)$ is the unit step function. The function $f(t - iT)u(t - iT)$, where i is a positive integer, represents a shift to the right of i sample intervals. Suppose this shifted function is sampled, then we obtain:

$$f[k - i]u[k - i] \qquad k \in \mathbb{N}$$

The second shift theorem states:

■ $\mathscr{Z}\{f[k - i]u[k - i]\} = z^{-i}F(z) \qquad i \in \mathbb{N}^+$

where $F(z)$ is the z transform of $f[k]$.

Example 12.25

The function $t\,u(t)$ is sampled at intervals $T = 1$ to give $k\,u[k]$. This sample is then shifted to the right by one sampling interval to give:

$$(k - 1)u[k - 1]$$

Find its z transform.

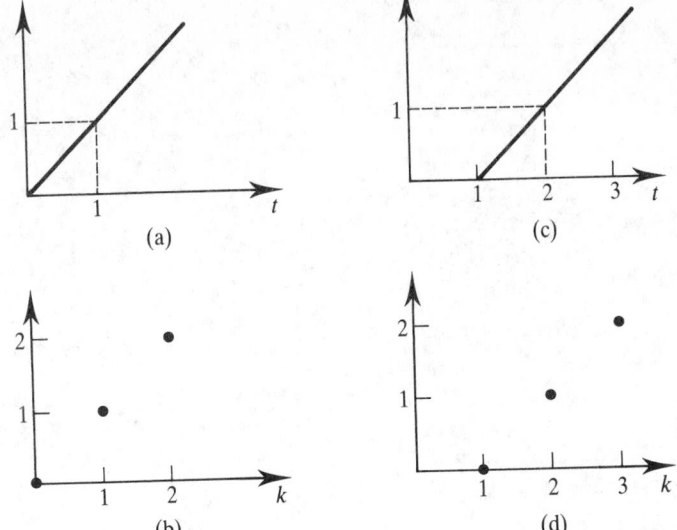

Figure 12.23 Graphs for Example 12.25: (a) $tu(t)$; (b) $ku[k]$; (c) $(t-1)u(t-1)$; (d) $(k-1)u[k-1]$.

Solution

Figure 12.23(a) shows $tu(t)$ and Figure 12.23(b) shows the sampled function. Figures 12.23(c) and (d) show $(t-1)u(t-1)$ and $(k-1)u[k-1]$, respectively. From Table 12.2, we have

$$\mathcal{Z}\{k\} = \frac{z}{(z-1)^2}$$

and so, from the second shift theorem with $i = 1$ we have:

$$\mathcal{Z}\{(k-1)u[k-1]\} = z^{-1}\frac{z}{(z-1)^2} = \frac{1}{(z-1)^2} \quad \blacktriangle$$

Example 12.26

Find the z transform of the unit step function $u(t)$ and the shifted unit step $u(t-2T)$, sampled at intervals of T seconds.

Solution

If the function $u(t)$ is sampled at intervals T then we are concerned with finding the z transform of the sequence $u[k]$. This has been derived earlier: $\mathcal{Z}\{u[k]\} = z/(z-1)$. If $u(t-2T)$ is sampled, we have:

$$u[k-2] = \begin{cases} 1 & k = 2, 3, 4, \ldots \\ 0 & \text{otherwise} \end{cases}$$

Therefore, by the second shift theorem,

$$\mathscr{L}\{u[k-2]\} = z^{-2}\mathscr{L}\{u[k]\}$$

$$= z^{-2}\frac{z}{z-1}$$

$$= \frac{1}{z(z-1)} \quad \blacktriangle$$

Example 12.27

Find the sequence whose z transform is $1/(z-1)$.

Solution

$$\frac{1}{z-1} = \frac{1}{z}\frac{z}{z-1} = z^{-1}\frac{z}{z-1}$$

From Table 12.2 we have

$$\mathscr{L}\{u[k]\} = \frac{z}{z-1}$$

So from the second shift property, we have:

$$\mathscr{L}\{u[k-1]\} = z^{-1}\frac{z}{z-1}$$

The required sequence is therefore $u[k-1]$. $\quad \blacktriangle$

Example 12.28

Find the sequence whose z transform is $\dfrac{1}{z^2(z-1)^2}$.

Solution

The expression $\dfrac{1}{z^2(z-1)^2}$ does not appear in the table of transforms, but we observe that:

$$\frac{1}{z^2(z-1)^2} = \frac{1}{z^3}\frac{z}{(z-1)^2}$$

and $z/(z-1)^2$ does appear. It follows from Table 12.2 that

$$\mathscr{L}\{k\} = \frac{z}{(z-1)^2}$$

From the second shift property, $z^{-3}z/(z-1)^2$ is the z transform of $(k-3)u[k-3]$. $\quad \blacktriangle$

12.10.4 *The complex translation theorem*

■ $\mathscr{L}\{e^{-bk}f[k]\} = F(e^bz)$ where $F(z)$ is the z transform of $f[k]$

Example 12.29

Given that the z transform of $\cos(ak)$ is:

$$\frac{z(z - \cos a)}{z^2 - 2z \cos a + 1}$$

find the z transform of $e^{-2k}\cos(ak)$.

Solution

Since $b = 2$, the complex translation theorem states that we replace z by e^2z in the z transform $F(z)$.

$$F(e^2z) = \frac{e^2z(e^2z - \cos a)}{e^4z^2 - 2e^2z \cos a + 1}$$

is therefore the required transform. ▲

EXERCISES 12.7

1. Use Table 12.2 to find the z transforms of

   (a) $3(4)^k + 7k^2$, $k \geq 0$

   (b) $3e^{-k}\sin 4k - k$, $k \geq 0$

2. Find the z transforms of the following continuous functions sampled at $t = kT$, $k \in \mathbb{N}$.

   (a) t^2

   (b) $4t$

   (c) $\sin 2t$

   (d) $u(t - 4T)$

   (e) e^{3t}

3. Find the z transform of $(k - 3)u[k - 3]$ by direct use of the definition of the z transform. Hence verify the result of Example 12.28.

4. Prove that the z transform of $e^{-at}f(t)$ is $F(e^{aT}z)$.

5. Prove the first and second shift theorems.

6. Use the complex translation theorem to find the z transforms of

   (a) ke^{-bk}

   (b) $e^{-k}\sin k$

7. If $f[k] = 4(3)^k$ find $\mathscr{L}\{f[k]\}$. Use the first shift theorem to deduce $\mathscr{L}\{f[k + 1]\}$. Show that $\mathscr{L}\{f[k + 1]\} - 3\mathscr{L}\{f[k]\} = 0$.

8. Write down the first five terms of the sequence defined by
   $f[k] = 4.2^{k-1}u[k - 1]$, $k \geq 0$. Find its z transform directly, and also by using the second shift theorem.

12.11 Inversion of z transforms

Just as it is necessary to invert Laplace transforms we need to be able to invert z transforms and as before we can make use of tables of transforms, partial fractions and the shift theorems. In very complicated cases more advanced techniques are required.

Example 12.30

If $F(z) = (z + 3)/(z - 2)$, find $f[k]$.

Solution

Instead of expressing $F(z)$ in partial fractions in the usual way we shall write it as:

$$\frac{z + 3}{z - 2} = \frac{z}{z - 2} + \frac{3}{z - 2}$$

The reason for this choice is that quantities like the first on the right-hand side appear in Table 12.2. From this table we find:

$$\mathscr{L}\{2^k\} = \frac{z}{z - 2}$$

and write

$$\mathscr{L}^{-1}\left\{\frac{z}{z - 2}\right\} = 2^k$$

where $\mathscr{L}^{-1}$ denotes the inverse z transform. Also,

$$\frac{3}{z - 2} = \frac{3}{z}\frac{z}{z - 2}$$

$$= 3z^{-1}\frac{z}{z - 2}$$

Using the second shift property we see that:

$$\mathscr{L}\{3.2^{k-1}u[k - 1]\} = \frac{3}{z}\frac{z}{z - 2}$$

so that

$$\mathscr{L}^{-1}\left\{\frac{z + 3}{z - 2}\right\} = 2^k + 3.2^{k-1}u[k - 1]$$

$$= \begin{cases} 1 & k = 0 \\ 2^k + 3.2^{k-1} & k = 1, 2, \ldots \end{cases}$$

$$= \begin{cases} 1 & k = 0 \\ 5.2^{k-1} & k = 1, 2, \ldots \end{cases} \quad \blacktriangle$$

Example 12.31

Find the sequence whose z transform is:

$$\frac{2z^2 - z}{(z - 5)(z + 4)}$$

Solution

Instead of first dividing out to obtain partial fractions in the usual form we try to write $F(z)$ in the form:

$$\frac{Az}{z - 5} + \frac{Bz}{z + 4}$$

if possible. The reason for selecting this form of partial fractions is that the form of such terms appears in Table 12.2. We write:

$$\frac{2z^2 - z}{(z - 5)(z + 4)} = \frac{Az}{z - 5} + \frac{Bz}{z + 4}$$

$$= \frac{Az(z + 4) + Bz(z - 5)}{(z - 5)(z + 4)}$$

Hence,

$$2z^2 - z = (A + B)z^2 + (4A - 5B)z$$

Equating coefficients we find that $B = 1$ and $A = 1$, so that:

$$\frac{2z^2 - z}{(z - 5)(z + 4)} = \frac{z}{z - 5} + \frac{z}{z + 4}$$

Therefore, from Table 12.2:

$$f[k] = 5^k + (-4)^k \quad \blacktriangle$$

12.11.1 *Direct inversion*

Sometimes it is possible to invert a transform $F(z)$ directly by reading off the coefficients.

Example 12.32

Find $f[k]$ if

$$F(z) = 1 + z^{-1} + z^{-2} + z^{-3} + \cdots$$

Solution

Using the definition of the z transform we see that:

$$f[k] = 1, 1, 1, \ldots$$

that is, $f[k] = 1 \quad k \geq 0. \quad \blacktriangle$

Occasionally it is possible to rewrite $F(z)$ to obtain the required form.

Example 12.33

Use the binomial theorem to expand $(1 - 1/z)^{-3}$ up to the term in $1/z^4$. Hence find the sequence with z transform $F(z) = z^3/(z - 1)^3$.

Solution

Using the binomial theorem, we have:

$$\left(1 - \frac{1}{z}\right)^{-3} = 1 + (-3)\left(-\frac{1}{z}\right) + \frac{(-3)(-4)}{2!}\left(-\frac{1}{z}\right)^2$$

$$+ \frac{(-3)(-4)(-5)}{3!}\left(-\frac{1}{z}\right)^3 + \frac{(-3)(-4)(-5)(-6)}{4!}\left(-\frac{1}{z}\right)^4 + \cdots$$

$$= 1 + \frac{3}{z} + \frac{6}{z^2} + \frac{10}{z^3} + \frac{15}{z^4} + \cdots$$

provided $|z| > 1$. Since

$$F(z) = \frac{z^3}{(z-1)^3} = \left(\frac{z-1}{z}\right)^{-3} = \left(1 - \frac{1}{z}\right)^{-3}$$

we have

$$F(z) = 1 + \frac{3}{z} + \frac{6}{z^2} + \frac{10}{z^3} + \frac{15}{z^4} + \cdots$$

Thus $F(z)$ can be inverted directly to give:

$$f[k] = 1, 3, 6, 10, 15, \ldots \qquad \text{that is, } f[k] = \frac{(k+2)(k+1)}{2} \qquad k \geq 0 \quad \blacktriangle$$

EXERCISES 12.8

1. Find the inverse z transforms of the following:

   (a) $\dfrac{4z}{z - 4}$

   (b) $\dfrac{z^2 + 2z}{3z^2 - 4z - 7}$

   (c) $\dfrac{z + 1}{(z - 3)z^2}$

   (d) $\dfrac{2z^3 + z}{(z - 3)^2(z - 1)}$

2. Find the inverse z transform of

   (a) $\dfrac{2z}{(z - 2)(z - 3)}$

   (b) $\dfrac{ez}{(ez - 1)^2}$

   (c) $1 - \dfrac{2}{z} + \dfrac{z}{(z - 3)(z - 4)}$

   (d) $\dfrac{z^2}{(z^2 - \frac{1}{9})}$

   (e) $\dfrac{2z^2}{(z - 1)(z - 0.905)}$

3. Express

   $$F(z) = \frac{(z + 1)(2z - 3)(z - 2)}{z^3}$$

   in partial fractions and hence obtain its inverse z transform.

4. If

   $$F(z) = \frac{10z}{(z - 1)(z - 2)}$$

   find $f[k]$.

12.12 The *z* transform and difference equations

In Chapter 11 we saw how useful the Laplace transform can be in the solution of linear, constant coefficient, ordinary differential equations. Similarly the *z* transform has a role to play in the solution of difference equations.

Example 12.34

Solve the difference equation $y[k + 1] - 3y[k] = 0$ $y[0] = 4$.

Solution

Taking the *z* transform of both sides of the equation we have:

$$\mathscr{Z}\{y[k + 1] - 3y[k]\} = \mathscr{Z}\{0\} = 0$$

since $\mathscr{Z}\{0\} = 0$. Using the properties of linearity we find

$$\mathscr{Z}\{y[k + 1]\} - 3\mathscr{Z}\{y[k]\} = 0$$

Using the first shift theorem on the first of the terms on the left-hand side, we obtain:

$$z\mathscr{Z}\{y[k]\} - 4z - 3\mathscr{Z}\{y[k]\} = 0$$

Writing $\mathscr{Z}\{y[k]\} = Y(z)$, this becomes

$$(z - 3)Y(z) = 4z$$

so that

$$Y(z) = \frac{4z}{z - 3}$$

The function on the right-hand side is the *z* transform of the required solution. Inverting this from Table 12.2 we find $y[k] = 4.3^k$. ▲

Higher order equations are treated in the same way:

Example 12.35

Solve the second order difference equation:

$$y[k + 2] - 5y[k + 1] + 6y[k] = 0 \qquad y[0] = 0 \qquad y[1] = 2$$

Solution

Taking the *z* transform of both sides of the equation and using the properties of linearity we have:

$$\mathscr{Z}\{y[k + 2]\} - 5\mathscr{Z}\{y[k + 1]\} + 6\mathscr{Z}\{y[k]\} = 0$$

From the first shift theorem we have:

$$z^2Y(z) - z^2y[0] - zy[1] - 5(zY(z) - zy[0]) + 6Y(z) = 0$$

where

$$Y(z) = \mathscr{L}\{y[k]\}$$

Hence

$$(z^2 - 5z + 6)Y(z) = 2z$$

so that

$$Y(z) = \mathscr{L}\{y[k]\} = \frac{2z}{(z-2)(z-3)}$$

The function on the right-hand side is the z transform of the required solution. We invert this by first expressing it in the special sort of partial fractions discussed in Example 12.31 and then using Table 12.2. We have:

$$\frac{2z}{(z-2)(z-3)} = \frac{Az}{z-2} + \frac{Bz}{z-3}$$

$$= \frac{Az(z-3) + Bz(z-2)}{(z-2)(z-3)}$$

so that

$$2z = (A+B)z^2 - (3A+2B)z$$

Equating coefficients shows that $A = -2$ and $B = 2$. Hence:

$$Y(z) = \frac{2z}{(z-2)(z-3)} = \frac{-2z}{z-2} + \frac{2z}{z-3}$$

Finally, from Table 12.2 we have:

$$y[k] = -2(2^k) + 2(3^k)$$

It is worth pointing out that if the function to be inverted is expressed in partial fractions in the more usual form, progress is still possible although the second shift theorem is required. In this case we have:

$$\frac{2z}{(z-2)(z-3)} = \frac{-4}{(z-2)} + \frac{6}{(z-3)}$$

Now,

$$\frac{4}{z-2} = \frac{4}{z}\frac{z}{z-2}$$

$$= 4z^{-1}\frac{z}{z-2}$$

From Table 12.2, $z/(z-2)$ is the transform of 2^k so using the second shift property we have that $4z^{-1}\left(\dfrac{z}{z-2}\right)$ will be the z transform of

$4(2^{k-1})u[k-1]$. Similarly,

$$\frac{6}{z-3} = \frac{6}{z}\frac{z}{z-3}$$

$$= 6z^{-1}\frac{z}{z-3}$$

Now $z/(z-3)$ is the z transform of 3^k, and so $6z^{-1}\left(\dfrac{z}{z-3}\right)$ is the transform of $6(3^{k-1})u[k-1]$. Finally the solution of the difference equation is:

$$y[k] = -4(2^{k-1})u[k-1] + 6(3^{k-1})u[k-1]$$

For $k \geq 1$, $u[k-1]$ is, of course, equal to 1 so that this solution reduces to that obtained earlier. ▲

EXERCISES 12.9

1. Use z transforms to solve the following difference equations.

   (a) $x[k+1] - 3x[k] = -6$, $x[0] = 1$
   (b) $2x[k+1] - x[k] = 2^k$, $x[0] = 2$
   (c) $x[k+1] + x[k] = 2k + 1$, $x[0] = 0$
   (d) $x[k+2] - 8x[k+1] + 16x[k] = 0$,
        $x[0] = 10$, $x[1] = 20$
   (e) $x[k+2] - x[k] = 0$, $x[0] = 0$, $x[1] = 1$

2. Solve the difference equation

   $$x[k+2] - 3x[k+1] + 2x[k] = \delta[k]$$

   subject to the conditions $x[0] = x[1] = 0$.

3. Solve the difference equation

   $$y[k+2] + 3y[k+1] + 2y[k] = 0$$

   subject to the conditions $y[0] = 0$, $y[1] = 1$.

Miscellaneous exercises

1. State:

   (a) the order
   (b) the independent variable
   (c) the dependent variable
   (d) whether linear or non-linear

   for each of the following equations:

   (i) $x[n] + x[n-2] = 6$
   (ii) $y[k+1] + ky[k-1] - k = 0$
   (iii) $(y[z] + 1)y[z+1] = z^2$
   (iv) $z[n] - z[n-1] = n^2z[n-2]$
   (v) $q[k+3] + \sqrt{q[k+2]} = q[k] - 1$

   For each linear equation state whether it is homogeneous or inhomogeneous.

2. Given,

   $$3(x[n+1])^2 - 2x[n] = n^2 \qquad x[0] = 2$$

   find $x[1]$, $x[2]$ and $x[3]$.

3. Rewrite each equation so that the highest argument of the dependent variable is as specified.

   (a) $3ny[n+1] - y[n-1] = n^2$, highest argument of the dependent variable is to be n.

(b) $z[k + 2] + (3 + k/2)z[k] = \sqrt{kz[k - 1]}$, highest argument of the dependent variable is to be $k + 1$.

(c) $x[3]x[n] - x[2]x[n - 1] = (n + 1)^2$, highest argument of the dependent variable is to be $n + 1$.

4. Find $f[k]$ if

$$F(z) = \frac{z(1 - a)}{(z - 1)(z - a)}.$$

5. Find the inverse z transform of:

(a) $\dfrac{3z(z + 2)}{(z - 2)(z - 3)^2}$

(b) $\dfrac{z^2 + 3z}{3z^2 + 2z - 5}.$

6. The sequence $\delta[k - i]$ is the Kronecker delta sequence shifted i units to the right. Find its z transform.

7. The series LR circuit with $L = R = 1$ is subject to an applied voltage $v_i(t)$ and produces an output voltage across the resistor $v_o(t)$. These voltages are sampled at one second intervals to give an input signal $v_i[k]$ and an output signal $v_o[k]$. It can be shown that these are related by:

$$2v_o[k] - v_o[k - 1] = v_i[k]$$
$$v_o[0] = 0$$

By taking z transforms obtain the transfer function $V_o(z)/V_i(z)$. Show that the response to an input $\delta[k]$ is $\frac{1}{2}(\frac{1}{2})^k$, $k \geq 0$.

8. Show that $\sin ak$ can be written as

$$\frac{e^{akj} - e^{-akj}}{2j}$$

Given that

$$\mathscr{L}\{e^{-ak}\} = \frac{z}{z - e^{-a}}$$

show that

$$\mathscr{L}\{\sin ak\} = \frac{z \sin a}{z^2 - 2z \cos a + 1}$$

Fourier series and the Fourier transform

13

13.1 Introduction

The ability to analyse waveforms of various types is an important engineering skill. Fourier analysis provides a set of mathematical tools which enable the engineer to break down a wave into its various frequency components. It is then possible to predict the effect a particular waveform may have from knowledge of the effects of its individual frequency components. Often an engineer finds it useful to think of a signal in terms of its frequency components rather than in terms of its time domain representation. This alternative view is called a frequency domain representation. It is particularly useful when trying to understand the effect of a filter on a signal. Filters are used extensively in many areas of engineering. In particular, communication engineers use them in signal reception equipment for filtering out unwanted frequencies in the received signal, that is, removing the transmission signal to leave the audio signal. We shall begin this chapter by reviewing the essential properties of waves before describing how breaking down into frequency components is achieved.

13.2 Periodic waveforms

In this chapter we shall be concerned with periodic functions, especially sine and cosine functions. Let us recall some important definitions and properties already discussed in Section 1.4.8. The function $f(t) = A \sin(\omega t + \phi) = A \sin \omega(t + \phi/\omega)$ is a sine wave of amplitude A, angular frequency ω, frequency $\omega/2\pi$, period $T = 2\pi/\omega$ and phase angle ϕ. The time displacement is defined to be ϕ/ω. These quantities are shown in Figure 13.1.

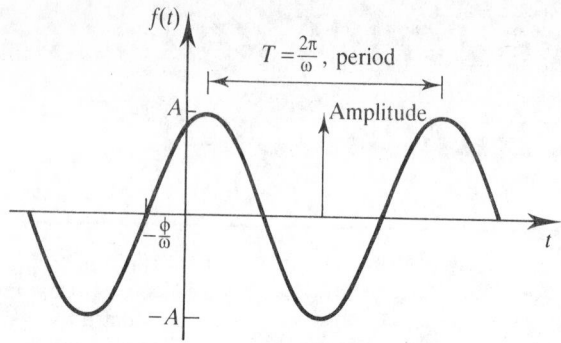

Figure 13.1   The function: $f(t) = A \sin(\omega t + \phi)$.

Similar remarks can be made about the function $A \cos(\omega t + \phi)$ and together the sine and cosine functions form a class of functions known as **sinusoids** or **harmonics**. It will be particularly important for what follows that you have mastered the skills of integrating these functions. The following results can be found in Table 8.1.

$$\int \sin n\omega t \, dt = -\frac{\cos n\omega t}{n\omega} + c \qquad \int \cos n\omega t \, dt = \frac{\sin n\omega t}{n\omega} + c$$

$$\text{for } n = \pm 1, \pm 2, \ldots$$

Sometimes a function occurs as the sum of a number of different sine or cosine components such as

$$f(t) = 2 \sin \omega_1 t + 0.8 \sin 2\omega_1 t + 0.7 \sin 4\omega_1 t \tag{13.1}$$

The right hand side of Equation (13.1) is a **linear combination** of sinusoids.

Note in particular that the angular frequencies of all components in Equation (13.1) are integer multiples of the angular frequency ω_1. Functions like these can easily be plotted using a graphics calculator or computer graph plotting package. The component with the lowest frequency, or largest period, is $2 \sin \omega_1 t$. The quantity ω_1 is called the **fundamental angular frequency** and this component is called the **fundamental** or **first harmonic**. The component with angular frequency $2\omega_1$ is called the second harmonic and so on. In what follows all angular frequencies are integer multiples of the fundamental angular frequency as in Equation (13.1). A consequence of this is that the resulting function, $f(t)$, is periodic and has the same frequency as the fundamental. Some harmonics may be missing. For example, in Equation (13.1) the third harmonic is missing. In some cases the first harmonic may be missing. For example, if

$$f(t) = \cos 2\omega_1 t + 0.5 \cos 3\omega_1 t + 0.4 \cos 4\omega_1 t + \cdots$$

all angular frequencies are integer multiples of the fundamental angular frequency ω_1, which is missing. Nevertheless, $f(t)$ has the same angular frequency as the fundamental. A common value for ω_1 is 100π as this corresponds to a frequency of 50 Hz, the frequency of the UK mains supply.

Example 13.1

Describe the frequency and amplitude characteristics of the different harmonic components of the function:

$$f(t) = \cos 20\pi t + 0.6 \cos 60\pi t - 0.2 \sin 140\pi t$$

Solution

The fundamental angular frequency is 20π arising through the term $\cos 20\pi t$. This corresponds to a frequency of 10 Hz. This term has amplitude 1. The second, fourth, fifth and sixth harmonics are missing, while the third and seventh have amplitudes 0.6 and 0.2, respectively. ▲

Example 13.2

If $f(t) = 2 \sin t + 3 \cos t$, express $f(t)$ as a single sinusoid and hence determine its amplitude and phase.

Solution

Both terms have angular frequency $\omega = 1$. Recalling the trigonometric identity (Section 1.4.8),

$$R \cos(\omega t - \theta) = a \cos \omega t + b \sin \omega t$$

where $R = \sqrt{a^2 + b^2}$, $\tan \theta = b/a$ we see that in this case $R = \sqrt{3^2 + 2^2} = \sqrt{13}$ and $\tan \theta = 2/3$, that is, $\theta = 0.59$ radians. Therefore we can express $f(t)$ in the form

$$f(t) = \sqrt{13} \cos(t - 0.59)$$

We see immediately that this is a sinusoid of amplitude $\sqrt{13}$ and phase angle -0.59 radians. ▲

Example 13.3

Find the amplitude and phase of the fundamental component of the function

$$f(t) = 0.5 \sin \omega_1 t + 1.5 \cos \omega_1 t + 3.5 \sin 2\omega_1 t - 3 \cos 3\omega_1 t$$

Solution

Contributions to the fundamental component, that is, that with the lowest frequency, come from the terms $0.5 \sin \omega_1 t$ and $1.5 \cos \omega_1 t$ only. To find the amplitude and phase we must express these as a single component. Using the trigonometric identity:

$$R \cos(\omega t - \theta) = a \cos \omega t + b \sin \omega t$$

where $R = \sqrt{a^2 + b^2}$, $\tan\theta = b/a$, we find

$$R = \sqrt{1.5^2 + 0.5^2} = 1.58$$

$$\tan\theta = \frac{0.5}{1.5} = \frac{1}{3} \qquad \text{that is, } \theta = 0.32 \text{ radians}$$

Therefore the fundamental can be written $1.58\cos(\omega_1 t - 0.32)$ and has amplitude 1.58 and phase angle -0.32 radians. ▲

Many other periodic functions arise in engineering applications as well as the more familiar harmonic waves. Remember, to be periodic the function values must repeat at regular intervals known as the period, T. The angular frequency ω is given by $\omega = 2\pi/T$. To describe a periodic function mathematically it is sufficient to give its equation over one full period and indicate that period. From this information the complete graph can be drawn as Examples 13.4 and 13.5 show.

Example 13.4

Sketch the graph of the periodic function defined by:

$$f(t) = t \qquad 0 \le t < 1 \qquad \text{period } 1$$

Solution

To proceed we first sketch the graph on the given interval $0 \le t < 1$ (Figure 13.2), and then use the fact that the function repeats regularly with period 1 to complete the picture. ▲

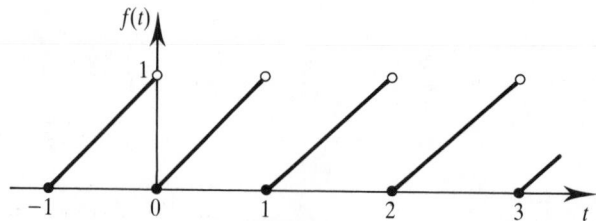

Figure 13.2 Graph for Example 13.4.

Example 13.5

Write down a mathematical expression for the function whose graph is shown in Figure 13.3.

Solution

We first note that the interval over which the function repeats itself is 2, that is, period = 2. It is then sufficient to describe the function over any interval

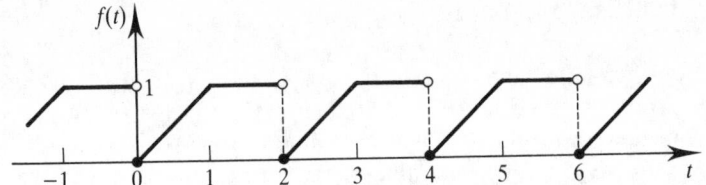

Figure 13.3 Graph for Example 13.5.

of length 2. The simplest interval to take is $0 \leq t < 2$. We note in this
example that a single formula is insufficient to describe the function for
$0 \leq t < 2$ since different behaviour is exhibited in the two intervals
$0 \leq t < 1$ and $1 \leq t < 2$. For $0 \leq t < 1$ the function is a ramp with slope 1 and
passes through the origin, that is, it has equation $f(t) = t$. For $1 \leq t < 2$ the
function value remains constant at 1. Therefore this periodic function can be
described by the expression

$$f(t) = \begin{cases} t & 0 \leq t < 1 \\ 1 & 1 \leq t < 2 \end{cases} \quad \text{period 2} \quad \blacktriangle$$

EXERCISES 13.1

1. Describe the frequency and amplitude
   characteristics of the different harmonic
   components of the following waveforms:

   (a) $f(t) = 3 \sin 100\pi t - 4 \sin 200\pi t$
   $\qquad + 0.7 \sin 300\pi t$

   (b) $f(t) = \sin 40t - 0.5 \cos 120t$
   $\qquad + 0.3 \cos 240t$

   Use a graph plotting computer package or
   graphics calculator to graph these
   waveforms.

2. Express each of the following functions as
   a single sinusoid and hence find their
   amplitudes and phases.

   (a) $f(t) = 2 \cos t - 3 \sin t$

   (b) $f(t) = 0.5 \cos t + 3.2 \sin t$
   (c) $f(t) = 3 \cos 3t$
   (d) $f(t) = 2 \cos 2t + 3 \sin 2t$

3. Sketch the graphs of the following functions.

   (a) $f(t) = t^2$, $-1 \leq t \leq 1$, period 2

   (b) $f(t) = \begin{cases} 0 & 0 \leq t < \pi/2 \\ \sin t & \pi/2 \leq t \leq \pi \end{cases}$ period π

   (c) $f(t) = \begin{cases} -t & -2 \leq t < 0 \\ t & 0 \leq t < 1 \end{cases}$ period 3

4. Write down mathematical expressions to
   describe the functions whose graphs are
   shown in Figure 13.4.

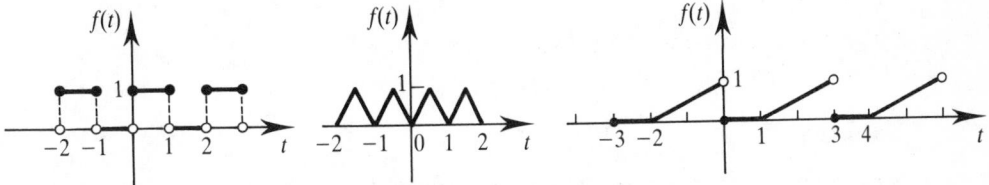

Figure 13.4 Graphs for Exercise 13.1.4.

13.3 Odd and even functions

The functions $\sin t$ and $\cos t$ each possess certain properties which can be generalized to other functions. Figure 13.5 shows the graph of $f(t) = \cos t$. It is obvious from the graph that the function value at a negative t value, say $-\pi/4$, will be the same as the function value at the corresponding positive t value, in this case $+\pi/4$. This is true because the graph is symmetrical about the vertical axis. We can therefore state that for any value of t, $\cos(-t) = \cos t$.

More generally, any function with the property that $f(t) = f(-t)$ for any value of its argument, t, is said to be an **even** function.

■ If $f(t) = f(-t)$ then f is an even function

In particular, the set of functions $\cos n\omega t$, $n \in \mathbb{Z}$, is even. The graphs of all even functions are symmetrical about the vertical axis – or equivalently, the graph on the left-hand side of the origin can be obtained by reflecting in the vertical axis that on the right. Some other examples of even functions are shown in Figure 13.6.

Sketching a graph immediately shows up the required symmetry. However, even functions can be identified by an algebraic approach as shown in Example 13.6.

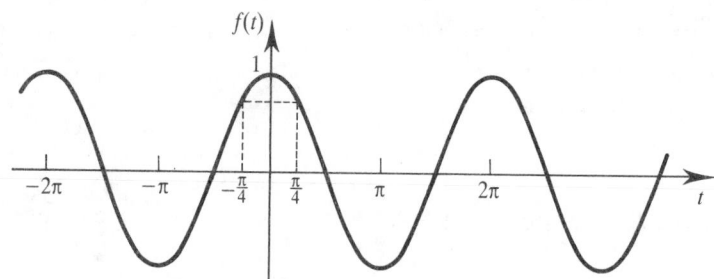

Figure 13.5   The function: $f(t) = \cos t$.

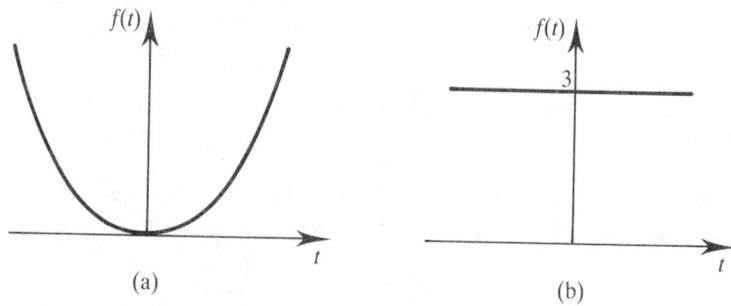

Figure 13.6   Examples of even functions. (a) $f(t) = t^2$; (b) $f(t) = 3$.

Example 13.6

Show that $f(t) = t^2$ is even.

Solution

We can argue as follows. If:

$$f(t) = t^2$$

then

$$f(-t) = (-t)^2$$
$$= t^2$$
$$= f(t)$$

so that $f(t)$ is even, by definition. ▲

Example 13.7

Test whether or not the function $f(t) = 4t^3$ is even.

Solution

If

$$f(t) = 4t^3$$

then

$$f(-t) = 4(-t)^3$$
$$= -4t^3$$
$$= -f(t)$$

so that $f(-t)$ is not equal to $f(t)$ and therefore the given function is not even. ▲

Let us turn now to the graph of $f(t) = \sin t$ in Figure 13.7. It is obvious from the graph that the function value at a negative t value, say $-\pi/4$, will not be the same as the function value at the corresponding positive t value, in this case $+\pi/4$. This graph is not symmetrical about the vertical axis. However, we can state

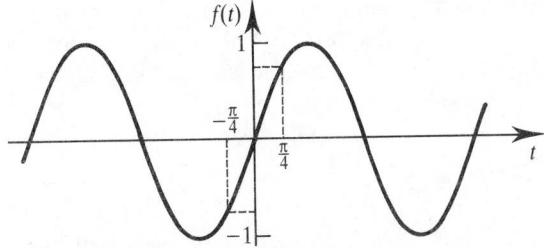

Figure 13.7   The function: $f(t) = \sin t$.

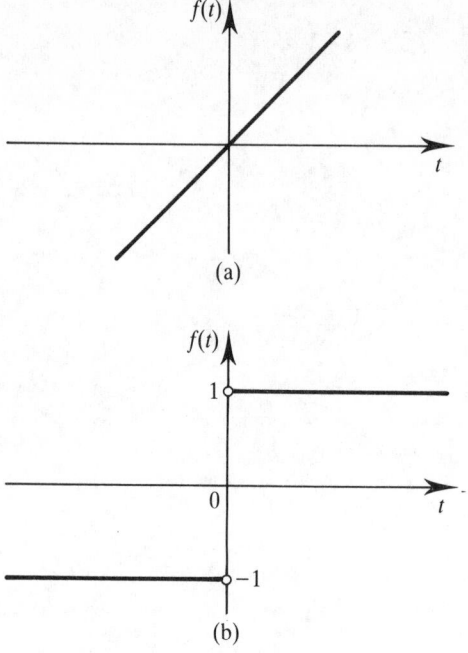

Figure 13.8 Examples of odd functions. (a) $f(t) = t$; (b) $f(t) = \begin{cases} 1 & t > 0 \\ -1 & t < 0. \end{cases}$

something else. The function value at a negative t value is minus the function value at the corresponding positive t value. For example, $\sin(-\pi/4) = -\sin(\pi/4)$. In fact, for all values of t we can state that $\sin(-t) = -\sin t$. More generally, any function with the property that $f(-t) = -f(t)$ for all values of its argument, t, is said to be an **odd** function.

■ If $f(-t) = -f(t)$ then f is an odd function

In particular, the set of functions $\sin n\omega t$, $n \in \mathbb{Z}$, is odd. In Example 13.7, $f(t) = 4t^3$, we found that $f(-t) = -f(t)$ so this function is odd. The graph of an odd function can be obtained by first reflecting in the horizontal axis and then the vertical axis. Some more examples of odd functions are shown in Figure 13.8.

There are some functions that are neither odd nor even – for example, the exponential function (see Chapter 1).

Example 13.8

Show that any function, $f(t)$, can be expressed as the sum of an odd component and an even component.

Solution

We can write

$$f(t) = f(t) + \frac{f(-t)}{2} - \frac{f(-t)}{2}$$

$$= \frac{f(t)}{2} + \frac{f(t)}{2} + \frac{f(-t)}{2} - \frac{f(-t)}{2}$$

Rearranging gives

$$f(t) = \frac{f(t) + f(-t)}{2} + \frac{f(t) - f(-t)}{2}$$

Now it is easy to check that the first term on the right-hand side is even and the second term is odd, so that we have expressed $f(t)$ as the sum of an even and an odd component as required. ▲

Example 13.9

Show that the product of two even functions is itself an even function. Determine whether the product of two odd functions is even or odd. Is the product of an even function and an odd function even or odd?

Solution

If $f(t)$ and $g(t)$ are even then $f(-t) = f(t)$ and $g(-t) = g(t)$. Let $P(t) = f(t)g(t)$ be the product of f and g. Then

$$P(-t) = f(-t)g(-t) \qquad \text{by definition}$$

$$= f(t)g(t) \qquad \text{since } f \text{ and } g \text{ are even}$$

$$= P(t)$$

Therefore, $P(-t) = P(t)$ and so the product $f(t)g(t)$ is itself an even function. On the other hand, if $f(t)$ and $g(t)$ are both odd we find:

$$P(-t) = f(-t)g(-t)$$

$$= (-f(t))(-g(t))$$

$$= f(t)g(t)$$

$$= P(t)$$

so that the product $f(t)g(t)$ is even.

If $f(t)$ is even and $g(t)$ is odd, we find:

$$P(-t) = f(-t)g(-t)$$

$$= f(t)(-g(t))$$

$$= -f(t)g(t)$$

$$= -P(t)$$

so the product is an odd function. These rules are obviously analogous to the rules for multiplying positive and negative numbers. ▲

The results of Example 13.9 are summarized thus:

■   (even) × (even) = even

(odd) × (odd) = even

(even) × (odd) = odd

13.3.1 *Integral properties of even and odd functions*

Consider a typical odd function, $f(t)$, such as that shown in Figure 13.9. Suppose we wish to evaluate $\int_{-a}^{a} f(t)\, dt$ where the interval of integration $[-a, a]$ is symmetrical about the vertical axis. Recall from Chapter 8 that a definite integral can be regarded as the area bounded by the graph of the integrand and the horizontal axis. Areas above the horizontal axis are positive while those below are negative. We see that because positive and negative contributions cancel, the integral of an odd function over an interval which is symmetrical about the vertical axis will be 0.

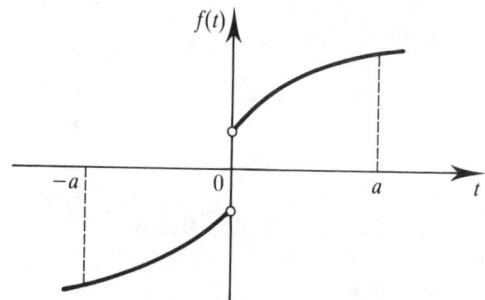

Figure 13.9   A typical odd function, $f(t)$.

Example 13.10

Evaluate $\int_{-\pi}^{\pi} t \cos n\omega t\, dt$.

Solution

The function t is odd. The function $\cos n\omega t$ is even and hence the product $t \cos n\omega t$ is odd. The interval $[-\pi, \pi]$ is symmetrical about the vertical axis and hence the required integral is zero. ▲

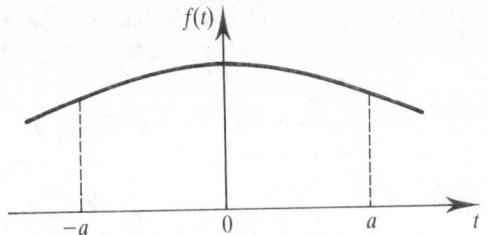

Figure 13.10 A typical even function, $f(t)$.

Consider now a typical even function, $f(t)$, such as that shown in Figure 13.10. Suppose we wish to evaluate $\int_{-a}^{a} f(t)\,\mathrm{d}t$. Clearly the area bounded by the graph and the t axis in the interval $[-a, 0]$ is the same as the corresponding area in the interval $[0, a]$. Hence we can write:

$$\int_{-a}^{a} f(t)\,\mathrm{d}t = 2 \int_{0}^{a} f(t)\,\mathrm{d}t$$

Example 13.11

Evaluate $\int_{-\pi}^{\pi} t \sin t\,\mathrm{d}t$.

Solution

The functions t and $\sin t$ are both odd, and hence their product is even. Therefore, using integration by parts:

$$\int_{-\pi}^{\pi} t \sin t\,\mathrm{d}t = 2 \int_{0}^{\pi} t \sin t\,\mathrm{d}t$$

$$= 2\left([-t \cos t]_0^{\pi} + \int_{0}^{\pi} \cos t\,\mathrm{d}t \right)$$

$$= 2((-\pi \cos \pi) - (0) + [\sin t]_0^{\pi})$$

$$= 2\pi \quad \blacktriangle$$

EXERCISES 13.2

1. Determine by inspection whether each of the functions in Figure 13.11 is odd, even or neither.

2. By using the properties of odd and even functions developed in Example 13.9 state whether the following are odd, even or

neither.

(a) $t^3 \sin \omega t$

(b) $t \cos 2t$

(c) $\sin t \sin 4t$

(d) $\cos \omega t \sin 2\omega t$

(e) $e^t \sin t$

3. Evaluate the following integrals using the integral properties of odd and even functions where appropriate.

(a) $\int_{-5}^5 t^3 \, dt$

(b) $\int_{-5}^5 t^3 \cos 3t \, dt$

(c) $\int_{-\pi}^\pi t^2 \sin t \, dt$

(d) $\int_{-2}^2 t \cosh 3t \, dt$

(e) $\int_{-1}^1 |t| \, dt$

(f) $\int_{-1}^1 t|t| \, dt$

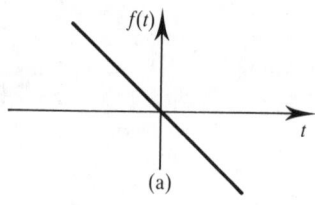

(a)

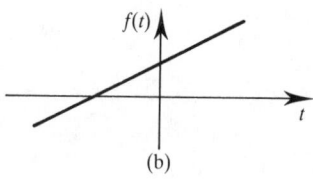

(b)

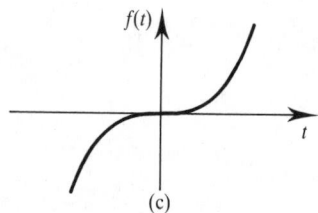

(c)

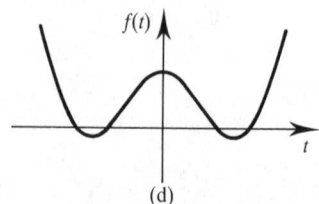

(d)

Figure 13.11   (a) The function $f(t) = -t$; (b) $f(t) = \dfrac{t}{2} + 1$; (c) $f(t) = t^3$;

(d) $f(t) = \cos t + 0.1t^2$.

13.4 Orthogonality relations and other useful identities

Recall from Chapter 8 that two functions $f(t)$ and $g(t)$ are said to be orthogonal on the interval $a \leq t \leq b$ if

$$\int_a^b f(t)g(t)\, dt = 0$$

Example 13.12

Show that the functions $\cos m\omega t$ and $\cos n\omega t$ with m, n positive integers and $m \neq n$, are orthogonal on the interval $-\pi/\omega \leq t \leq \pi/\omega$.

Solution

We must evaluate

$$\int_{-\pi/\omega}^{\pi/\omega} \cos m\omega t \cos n\omega t\, dt$$

Using the trigonometrical identity $2 \cos A \cos B = \cos(A + B) + \cos(A - B)$, we find the integral becomes:

$$\frac{1}{2} \int_{-\pi/\omega}^{\pi/\omega} \cos(m + n)\omega t + \cos(m - n)\omega t\, dt$$

$$= \frac{1}{2} \left[\frac{\sin(m + n)\omega t}{(m + n)\omega} + \frac{\sin(m - n)\omega t}{(m - n)\omega} \right]_{-\pi/\omega}^{\pi/\omega}$$

$$= 0$$

since $\sin(m \pm n)\pi = 0$ for all integers m, n. It was necessary to require $m \neq n$ since otherwise the second quantity in brackets becomes undefined. ▲

A number of other functions regularly appearing in work connected with Fourier analysis are orthogonal. The main results together with some other useful integral identities are given in Table 13.1. In this table m and n are non-negative integers.

13.5 Fourier series

We have seen that the functions $\sin \omega t$, $\sin 2\omega t$, $\sin 3\omega t, \ldots$, $\cos \omega t$, $\cos 2\omega t, \ldots$ are periodic. Furthermore, linear combinations of them are also periodic. They are also convenient functions to deal with because they can be easily differentiated, integrated, etc. They also possess another very useful property – that of **completeness**. This means that almost any periodic function can be expressed as a linear combination of them and no additional functions are required to do this. In other words, they can be used as building blocks to construct periodic functions simply by adding particular multiples of them together.

Table 13.1 Some useful integral identities.

$$\int_0^T \sin \frac{2n\pi t}{T} \, dt = 0$$

$$\int_0^T \cos \frac{2n\pi t}{T} \, dt = 0 \qquad n = 1, 2, 3, \ldots$$

$$\int_0^T \cos \frac{2n\pi t}{T} \, dt = T \qquad n = 0$$

$$\int_0^T \cos \frac{2m\pi t}{T} \cos \frac{2n\pi t}{T} \, dt = \begin{cases} 0 & m \neq n \\ T/2 & m = n \neq 0 \end{cases}$$

$$\int_0^T \sin \frac{2m\pi t}{T} \sin \frac{2n\pi t}{T} \, dt = \begin{cases} 0 & m \neq n \\ T/2 & m = n \neq 0 \end{cases}$$

$$\int_0^T \sin \frac{2m\pi t}{T} \cos \frac{2n\pi t}{T} \, dt = 0 \qquad \text{for all integers } m \text{ and } n$$

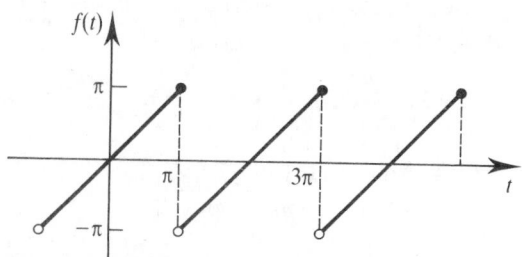

Figure 13.12 Saw-tooth waveform.

We shall see, for example, that the saw-tooth waveform with period 2π, shown in Figure 13.12, is given by the particular combination:

$$f(t) = 2(\sin t - \tfrac{1}{2} \sin 2t + \tfrac{1}{3} \sin 3t - \tfrac{1}{4} \sin 4t + \tfrac{1}{5} \sin 5t - \cdots)$$

This is an infinite series which can be shown to converge for almost all values of t to the function f. This means that if any value of t is substituted into the infinite series and the series is summed, the result will be the same as the value of the saw-tooth function at that value of t. There is an exception: if t is one of the points of discontinuity the infinite series will converge to the mean of the values to its left and right, that is, 0.

To obtain a feel for what is happening consider Figure 13.13. As more terms are taken we see that the series approaches the desired saw-tooth waveform. The process of adding together sinusoids to form a new periodic function is called **Fourier synthesis**. We see that the saw-tooth waveform has been expressed as an infinite series of harmonic waves, $\sin t$ being the fundamental or first harmonic, and the rest being waves with frequencies that are integer multiples of the

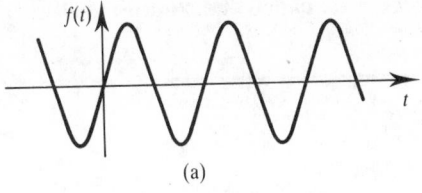

(a)

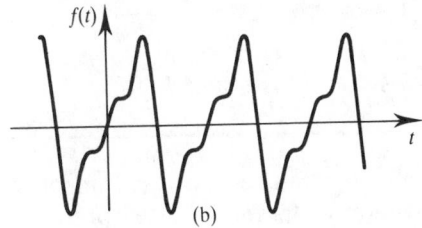

(b)

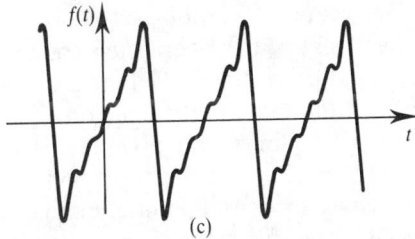

(c)

Figure 13.13 Fourier synthesis of a saw-tooth waveform. (a) $f(t) = 2\sin t$;
(b) $f(t) = 2(\sin t - \frac{1}{2}\sin 2t + \frac{1}{3}\sin 3t)$;
(c) $f(t) = 2(\sin t - \frac{1}{2}\sin 2t + \frac{1}{3}\sin 3t - \frac{1}{4}\sin 4t + \frac{1}{5}\sin 5t)$.

fundamental frequency. This infinite series is called the **Fourier series** representation of $f(t)$ and what we have succeeded in doing is to break down $f(t)$ into its component harmonic waveforms. In this example, only sine waves were required to construct the function. More generally we shall need both sine and cosine waves.

Suppose the function $f(t)$ is defined in the interval $0 < t < T$ and is periodic with period T. Then, under certain conditions, its Fourier series is given by:

$$■ \quad f(t) = \frac{a_0}{2} + \sum_{n=1}^{\infty} \left(a_n \cos \frac{2n\pi t}{T} + b_n \sin \frac{2n\pi t}{T} \right) \qquad (13.2)$$

or equivalently

$$f(t) = \frac{a_0}{2} + \sum_{n=1}^{\infty} (a_n \cos n\omega t + b_n \sin n\omega t)$$

where a_n and b_n are constants called the **Fourier coefficients**. These are given by the formulae:

$$\blacksquare \quad a_0 = \frac{2}{T}\int_0^T f(t)\,dt \tag{13.3}$$

$$a_n = \frac{2}{T}\int_0^T f(t)\cos\frac{2n\pi t}{T}\,dt \qquad \text{for } n \in \mathbb{N}^+ \tag{13.4}$$

$$b_n = \frac{2}{T}\int_0^T f(t)\sin\frac{2n\pi t}{T}\,dt \qquad \text{for } n \in \mathbb{N}^+ \tag{13.5}$$

The term $a_0/2$ represents the mean value or d.c. component of the waveform (see Section 8.5). The derivation of these formulae appears in Example 13.16. It is important to point out that the integrals in Equations (13.3), (13.4) and (13.5) can be evaluated over any complete period, for example, from $t = -T/2$ to $t = T/2$. Prudent choice of the interval of integration can often save effort. The expression appearing in the right-hand side of the Fourier representation, Equation (13.2), is an infinite series. We list conditions, often called the Dirichlet conditions, sufficient for the series to converge to the value of the function $f(t)$. The integral $\int |f(t)|\,dt$ over a complete period must be finite, and $f(t)$ may have no more than a finite number of discontinuities in any finite interval. Fortunately, most signals of interest to engineers satisfy these conditions. At a point of discontinuity the Fourier series converges to the average of the two function values at either side of the discontinuity.

Example 13.13

Find the Fourier series representation of the function with period $T = 1/50$ given by

$$f(t) = \begin{cases} 1 & 0 \le t < 0.01 \\ 0 & 0.01 \le t < 0.02 \end{cases}$$

Solution

The function $f(t)$ is shown in Figure 13.14. Using Equations (13.3)–(13.5) we find:

$$a_0 = 100\int_0^{0.02} f(t)\,dt = 100\int_0^{0.01} 1\,dt + 100\int_{0.01}^{0.02} 0\,dt$$

$$= 100[t]_0^{0.01} = 1$$

$$a_n = 100\int_0^{0.01}\cos 100n\pi t\,dt = 100\left[\frac{\sin 100n\pi t}{100n\pi}\right]_0^{0.01}$$

$$= 0$$

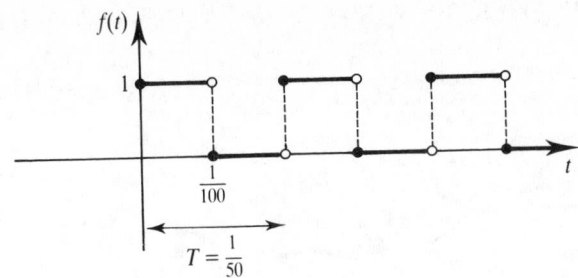

Figure 13.14 Graph for Example 13.13.

$$b_n = 100 \int_0^{0.01} \sin 100 n \pi t \, dt = 100 \left[\frac{-\cos 100 n \pi t}{100 n \pi} \right]_0^{0.01}$$

$$= -\frac{1}{n \pi} (\cos n \pi - \cos 0)$$

Noting that $\cos n \pi = (-1)^n$ we find

$$b_n = \frac{1}{n \pi} (1 - (-1)^n)$$

If n is even $b_n = 0$. If n is odd $b_n = 2/(n \pi)$. Therefore the Fourier series representation of $f(t)$ is:

$$f(t) = \frac{1}{2} + \frac{2}{\pi} \left(\sin 100 \pi t + \frac{\sin 300 \pi t}{3} + \frac{\sin 500 \pi t}{5} + \cdots \right)$$

The average value of the waveform is $1/2$. This is the zero frequency component or d.c. value. We note that in this example only odd harmonics are present. ▲

Example 13.14

Find the Fourier series representation of $f(t) = 1 + t$, $-\pi < t \le \pi$, period 2π.

Solution

As usual we sketch $f(t)$ first as this often provides insight into what follows (see Figure 13.15). Here $T = 2\pi$, $\omega = 1$, and for convenience we shall

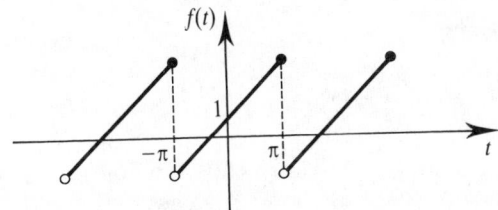

Figure 13.15 Graph for Example 13.14.

consider the period of integration to be $[-\pi, \pi]$. Using Equation (13.3) we find

$$a_0 = \frac{1}{\pi} \int_{-\pi}^{\pi} 1 + t \, dt = \frac{1}{\pi} \left[t + \frac{t^2}{2} \right]_{-\pi}^{\pi}$$

$$= \frac{1}{\pi} \left(\left(\pi + \frac{\pi^2}{2} \right) - \left(-\pi + \frac{\pi^2}{2} \right) \right)$$

$$= \frac{1}{\pi} (2\pi)$$

$$= 2$$

Similarly, using Equation (13.4) we find

$$a_n = \frac{1}{\pi} \int_{-\pi}^{\pi} (1 + t) \cos nt \, dt$$

Integrating by parts gives:

$$a_n = \frac{1}{\pi} \left(\left[(1 + t) \frac{\sin nt}{n} \right]_{-\pi}^{\pi} - \int_{-\pi}^{\pi} \frac{\sin nt}{n} \, dt \right)$$

$$= \frac{1}{\pi} \left(0 + \left[\frac{\cos nt}{n^2} \right]_{-\pi}^{\pi} \right) \qquad \text{since } \sin \pm n\pi = 0$$

$$= \frac{1}{\pi n^2} (\cos n\pi - \cos(-n\pi))$$

but $\cos(-n\pi) = \cos n\pi$ and hence $a_n = 0$, $n \in \mathbb{N}^+$. Using Equation (13.5) we find:

$$b_n = \frac{1}{\pi} \int_{-\pi}^{\pi} (1 + t) \sin nt \, dt$$

$$= \frac{1}{\pi} \left(\left[-(1 + t) \frac{\cos nt}{n} \right]_{-\pi}^{\pi} + \int_{-\pi}^{\pi} \frac{\cos nt}{n} \, dt \right)$$

$$= \frac{1}{\pi} \left(-(1 + \pi) \frac{\cos n\pi}{n} + (1 - \pi) \frac{\cos(-n\pi)}{n} + \left[\frac{\sin nt}{n^2} \right]_{-\pi}^{\pi} \right)$$

$$= \frac{1}{\pi n} (-2\pi \cos n\pi)$$

since $\sin \pm n\pi = 0$. Hence,

$$b_n = -\frac{2}{n} \cos n\pi = -\frac{2}{n} (-1)^n$$

We find $b_1 = 2$, $b_2 = -1$, $b_3 = 2/3, \dots$. Thus the Fourier series representation is given from Equation (13.2) as:

$$f(t) = 1 + 2 \sin t - \sin 2t + \tfrac{2}{3} \sin 3t + \cdots$$

which we can write concisely as:

$$f(t) = 1 - \sum_{n=1}^{\infty} \frac{2}{n}(-1)^n \sin nt \quad \blacktriangle$$

Example 13.15

Find the Fourier series representation of the function with period 2π defined by $f(t) = t^2$, $0 < t \leq 2\pi$.

Solution

As usual we sketch $f(t)$, as shown in Figure 13.16. Here $T = 2\pi$ and we shall integrate, for convenience, over the interval $[0, 2\pi]$. Using Equation (13.3) we find:

$$a_0 = \frac{1}{\pi}\int_0^{2\pi} t^2 \, dt = \frac{1}{\pi}\left[\frac{t^3}{3}\right]_0^{2\pi} = \frac{8\pi^2}{3}$$

Using Equation (13.4) we have:

$$a_n = \frac{1}{\pi}\int_0^{2\pi} t^2 \cos nt \, dt$$

Integrating by parts, we find:

$$a_n = \frac{1}{\pi}\left(\left[t^2\frac{\sin nt}{n}\right]_0^{2\pi} - \int_0^{2\pi} 2t\frac{\sin nt}{n} \, dt\right)$$

$$= -\frac{2}{n\pi}\int_0^{2\pi} t \sin nt \, dt$$

$$= -\frac{2}{n\pi}\left(\left[-t\frac{\cos nt}{n}\right]_0^{2\pi} + \int_0^{2\pi}\frac{\cos nt}{n} \, dt\right)$$

$$= -\frac{2}{n\pi}\left(\frac{-2\pi \cos 2n\pi}{n} + \left[\frac{\sin nt}{n^2}\right]_0^{2\pi}\right)$$

$$= \frac{4}{n^2}$$

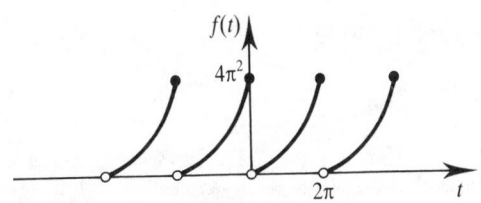

Figure 13.16 Graph for Example 13.15.

Hence $a_1 = 4$, $a_2 = 1$, $a_3 = 4/9, \ldots$ Similarly,

$$b_n = \frac{1}{\pi} \int_0^{2\pi} t^2 \sin nt \, dt$$

$$= \frac{1}{\pi} \left(\left[-t^2 \frac{\cos nt}{n} \right]_0^{2\pi} + \int_0^{2\pi} 2t \frac{\cos nt}{n} \, dt \right)$$

$$= \frac{1}{\pi} \left(-\frac{4\pi^2}{n} \cos 2n\pi + \frac{2}{n} \int_0^{2\pi} t \cos nt \, dt \right)$$

$$= \frac{1}{\pi} \left(-\frac{4\pi^2}{n} + \frac{2}{n} \left(\left[\frac{t \sin nt}{n} \right]_0^{2\pi} - \int_0^{2\pi} \frac{\sin nt}{n} \, dt \right) \right)$$

$$= \frac{1}{\pi} \left(-\frac{4\pi^2}{n} - \frac{2}{n^2} \left[-\frac{\cos nt}{n} \right]_0^{2\pi} \right)$$

$$= -\frac{4\pi}{n}$$

Thus $b_1 = -4\pi$, $b_2 = -2\pi, \ldots$ Finally, the required Fourier series representation is given by:

$$f(t) = \frac{4\pi^2}{3} + \left(4 \cos t + \cos 2t + \frac{4}{9} \cos 3t + \cdots \right)$$

$$- \pi \left(4 \sin t + 2 \sin 2t + \frac{4 \sin 3t}{3} + \cdots \right) \quad \blacktriangle$$

Example 13.16

Obtain the expressions for the Fourier coefficients a_0, a_n and b_n in Equations (13.3), (13.4) and (13.5).

Solution

Assume that $f(t)$ can be expressed in the form:

$$f(t) = \frac{a_0}{2} + \sum_{n=1}^{\infty} \left(a_n \cos \frac{2n\pi t}{T} + b_n \sin \frac{2n\pi t}{T} \right) \tag{13.6}$$

Multiplying Equation (13.6) through by $\cos (2m\pi t/T)$ and integrating from 0 to T we find

$$\int_0^T f(t) \cos \frac{2m\pi t}{T} \, dt = \int_0^T \frac{a_0}{2} \cos \frac{2m\pi t}{T} \, dt$$

$$+ \int_0^T \sum_{n=1}^{\infty} \left(a_n \cos \frac{2n\pi t}{T} + b_n \sin \frac{2n\pi t}{T} \right) \cos \frac{2m\pi t}{T} \, dt$$

If we now assume that it is legitimate to interchange the order of integration and summation we obtain:

$$\int_0^T f(t)\cos\frac{2m\pi t}{T}\,dt = \int_0^T \frac{a_0}{2}\cos\frac{2m\pi t}{T}\,dt$$

$$+ \sum_{n=1}^{\infty}\int_0^T\left(a_n\cos\frac{2n\pi t}{T} + b_n\sin\frac{2n\pi t}{T}\right)\cos\frac{2m\pi t}{T}\,dt$$

The first integral on the right-hand side is easily shown to be zero unless $m = 0$. Furthermore, we can use the previously found orthogonality properties (Table 13.1) to show that the rest of the integrals on the right-hand side vanish except for the case when $n = m$ in which case the right-hand side reduces to $a_m T/2$. Consequently,

$$a_m = \frac{2}{T}\int_0^T f(t)\cos\frac{2m\pi t}{T}\,dt \qquad m\in\mathbb{N}^+$$

as required. When $m = 0$ all terms on the right-hand side except the first vanish and we obtain

$$\int_0^T f(t)\,dt = \int_0^T \frac{a_0}{2}\,dt$$

$$= \frac{a_0 T}{2}$$

so that

$$a_0 = \frac{2}{T}\int_0^T f(t)\,dt$$

To obtain the formula for the b_n multiply Equation (13.6) through by $\sin(2m\pi t/T)$ and integrate from 0 to T.

$$\int_0^T f(t)\sin\frac{2m\pi t}{T}\,dt = \int_0^T \frac{a_0}{2}\sin\frac{2m\pi t}{T}\,dt$$

$$+ \int_0^T \sum_{n=1}^{\infty}\left(a_n\cos\frac{2n\pi t}{T} + b_n\sin\frac{2n\pi t}{T}\right)\sin\frac{2m\pi t}{T}\,dt$$

Again assuming that it is legitimate to interchange the order of integration and summation, we obtain

$$\int_0^T f(t)\sin\frac{2m\pi t}{T}\,dt = \int_0^T \frac{a_0}{2}\sin\frac{2m\pi t}{T}\,dt$$

$$+ \sum_{n=1}^{\infty}\int_0^T\left(a_n\cos\frac{2n\pi t}{T} + b_n\sin\frac{2n\pi t}{T}\right)\sin\frac{2m\pi t}{T}\,dt$$

The first integral on the right-hand side is easily shown to be zero. Furthermore, we can use the properties given in Table 13.1 to show that the rest of the integrals on the right-hand side vanish except for the case when $n = m$, in which case the right-hand side reduces to $b_m T/2$. Hence we find:

$$b_m = \frac{2}{T} \int_0^T f(t) \sin \frac{2m\pi t}{T} \, dt$$

as required. ▲

13.5.1 *Fourier series of odd and even functions*

Let us now consider what happens when we determine Fourier series of functions which are either odd or even.

Example 13.17

Find the Fourier series for the function with period 2π defined by:

$$f(t) = \begin{cases} 0 & -\pi < t < -\pi/2 \\ 4 & -\pi/2 \le t \le \pi/2 \\ 0 & \pi/2 < t < \pi \end{cases}$$

Solution

As usual we sketch the function first (Figure 13.17). Inspection of Figure 13.17 shows that the Dirichlet conditions are satisfied. We note from the graph that the function is symmetrical about the vertical axis, that is, it is an even function. We shall see shortly that this fact has important implications for the Fourier series representation. For convenience we consider the period of integration to be $[-T/2, T/2]$. Hence Equation (13.4) becomes

$$a_n = \frac{2}{T} \int_{-T/2}^{T/2} f(t) \cos \frac{2n\pi t}{T} \, dt \qquad n \in \mathbb{N}^+$$

In this example the period T equals 2π. The formula for a_n then simplifies to:

$$a_n = \frac{1}{\pi} \int_{-\pi}^{\pi} f(t) \cos nt \, dt$$

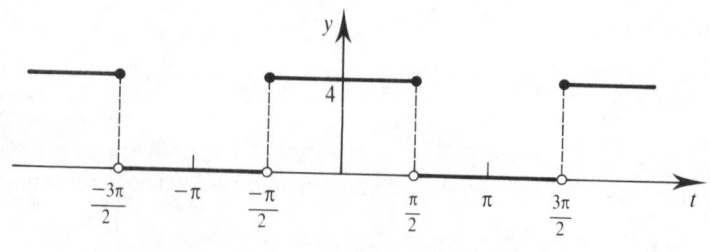

Figure 13.17   Graph for Example 13.17.

The interval of integration is from $t = -\pi$ to $t = \pi$. However, a glance at the graph shows that the function is zero outside the interval $-\pi/2 \leq t \leq \pi/2$, and takes the value 4 inside. The integral thus reduces to:

$$a_n = \frac{1}{\pi} \int_{-\pi/2}^{\pi/2} 4 \cos nt \, dt$$

$$= \frac{4}{\pi} \int_{-\pi/2}^{\pi/2} \cos nt \, dt$$

$$= \frac{4}{\pi} \left[\frac{\sin nt}{n} \right]_{-\pi/2}^{\pi/2}$$

$$= \frac{4}{n\pi} [\sin(n\pi/2) - \sin(-n\pi/2)]$$

$$= \frac{8}{n\pi} \sin \frac{n\pi}{2}$$

We obtain $a_1 = 8/\pi$, $a_2 = 0$, $a_3 = -8/3\pi$, etc. We find a_0 using Equation (13.3), again integrating over $[-T/2, T/2]$

$$a_0 = \frac{2}{T} \int_{-T/2}^{T/2} f(t) \, dt$$

$$= \frac{1}{\pi} \int_{-\pi/2}^{\pi/2} 4 \, dt = \frac{4}{\pi} [t]_{-\pi/2}^{\pi/2}$$

$$= 4$$

Similarly to find the Fourier coefficients, b_n, we use Equation (13.5)

$$b_n = \frac{2}{T} \int_{-T/2}^{T/2} f(t) \sin \frac{2n\pi t}{T} \, dt$$

which reduces to

$$b_n = \frac{1}{\pi} \int_{-\pi/2}^{\pi/2} 4 \sin nt \, dt$$

$$= \frac{4}{\pi} \left[\frac{-\cos nt}{n} \right]_{-\pi/2}^{\pi/2}$$

$$= \frac{4}{n\pi} [-\cos(n\pi/2) + \cos(n\pi/2)]$$

$$= 0$$

that is, all the Fourier coefficients, b_n, are zero. Finally we can gather together all our results and write down the Fourier series representation of $f(t)$:

$$f(t) = 2 + \frac{8}{\pi} \cos t - \frac{8}{3\pi} \cos 3t + \frac{8}{5\pi} \cos 5t - \cdots \quad \blacktriangle$$

In this example we see that there are no sine terms at all. In fact, whenever a function is even its Fourier series will possess no sine terms. To see this we note that b_n can be found from

$$b_n = \frac{2}{T} \int_{-T/2}^{T/2} f(t) \sin \frac{2n\pi t}{T} \, dt$$

Since $f(t)$ is even and $\sin(2n\pi t/T)$ is odd, the product $f(t) \sin(2n\pi t/T)$ is odd also. Now the integral of an odd function on an interval which is symmetrical about the vertical axis was shown in Section 13.3 to be zero. Hence whenever $f(t)$ is even we can immediately assume $b_n = 0$ for all n.

Correspondingly, when a function is odd its Fourier series will contain no cosine or constant terms. This is because the product:

$$f(t) \cos \frac{2n\pi t}{T}$$

is odd also and so the integral:

$$a_n = \frac{2}{T} \int_{-T/2}^{T/2} f(t) \cos \frac{2n\pi t}{T} \, dt$$

will equal zero. We conclude that when $f(t)$ is odd, $a_n = 0$ for all n. These facts can often be used to save time and effort. Knowing the function in Example 13.17 was even before we started the Fourier analysis, we could have assumed that the b_n would all be zero.

Example 13.18

Find the Fourier series representation of the saw-tooth waveform described at the beginning of this section (see Figure 13.12).

Solution

This function is defined by $f(t) = t$, $-\pi < t < \pi$, and has period $T = 2\pi$. It is an odd function and hence $a_n = 0$ for all n. To find the b_n we must evaluate

$$b_n = \frac{1}{\pi} \int_{-\pi}^{\pi} t \sin nt \, dt$$

$$= \frac{1}{\pi} \left\{ \left[\frac{-t \cos nt}{n} \right]_{-\pi}^{\pi} + \int_{-\pi}^{\pi} \frac{\cos nt}{n} \, dt \right\}$$

$$= \frac{1}{n\pi} \{ -\pi \cos n\pi - \pi \cos n\pi \}$$

since the last integral vanishes. Therefore:

$$b_n = -\frac{2}{n} \cos n\pi = -\frac{2}{n}(-1)^n$$

We conclude that $b_1 = 2$, $b_2 = -1$, $b_3 = 2/3, \ldots$ Therefore $f(t)$ has Fourier series

$$f(t) = 2\{\sin t - \tfrac{1}{2} \sin 2t + \tfrac{1}{3} \sin 3t - \tfrac{1}{4} \sin 4t + \tfrac{1}{5} \sin 5t - \cdots\} \quad \blacktriangle$$

EXERCISES 13.3

1. Find the Fourier series representation of the function:

$$f(t) = \begin{cases} 0 & -5 < t < 0 \\ 1 & 0 < t < 5 \end{cases} \quad \text{period 10}$$

2. Find the Fourier series representation of the function:

$$f(t) = \begin{cases} -t & -\pi < t < 0 \\ 0 & 0 < t < \pi \end{cases} \quad \text{period } 2\pi$$

3. Find the Fourier series representation of the function

$$f(t) = t^2 + \pi t \quad -\pi < t < \pi \quad \text{period } 2\pi$$

4. Find the Fourier series representation of the function:

$$f(t) = \begin{cases} -4 & -\pi < t \leq 0 \\ 4 & 0 < t < \pi \end{cases} \quad \text{period } 2\pi$$

5. Find the Fourier series representation of the function:

$$f(t) = \begin{cases} 2(1+t) & -1 < t \leq 0 \\ 0 & 0 < t < 1 \end{cases} \quad \text{period 2}$$

6. Find the Fourier series representation of the function with period 2π given by:

$$f(t) = \begin{cases} t^2 & 0 \leq t < \pi \\ 0 & \pi \leq t < 2\pi \end{cases}$$

7. Find the Fourier series representation of the function

$$f(t) = 2 \sin t \quad 0 < t < 2\pi \quad \text{period } 2\pi$$

13.6 Half-range series

Sometimes an engineering function is not periodic but is only defined over a finite interval, $0 < t < T/2$ say, as shown in Figure 13.18. In cases like this Fourier analysis can still be useful. Because the region of interest is only that between $t = 0$ and $t = T/2$ we may choose to define the function arbitrarily outside the interval. In particular, we can make our choice so that the resulting function is

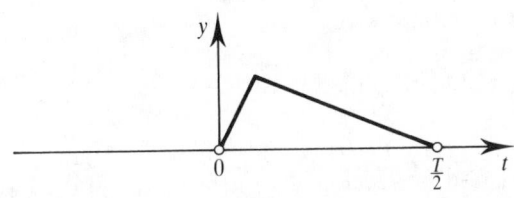

Figure 13.18 Function defined over interval $0 < t < \dfrac{T}{2}$.

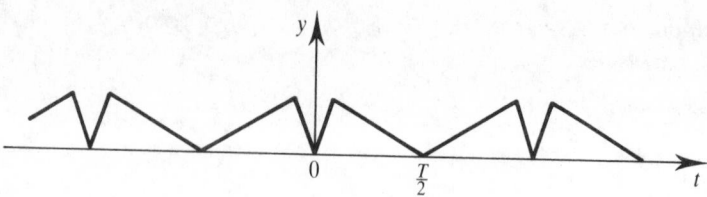

Figure 13.19 An even periodic extension.

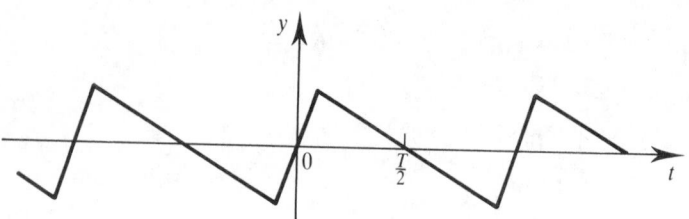

Figure 13.20 An odd periodic extension.

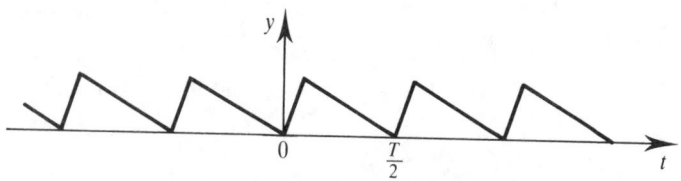

Figure 13.21 A periodic extension that is neither even nor odd.

periodic, with period T. There is more than one way to proceed. For example, we can reflect the above function in the vertical axis and then repeat it periodically so that the result is the periodic even function shown in Figure 13.19. We have performed what is called a **periodic extension** of the given function. Note that within the interval of interest nothing has altered but we have now achieved our objective of finding a periodic function. We can find the Fourier series of this periodic function and within the interval of interest this will converge to the required function. What happens outside this interval is not important. Moreover, since the periodic function is even the Fourier series will contain no sine terms.

An alternative periodic extension is that shown in Figure 13.20, which has been obtained by reflecting in both the vertical and t axes before repeating it periodically to give a periodic odd function. Its Fourier series will contain no cosine terms and within the interval of interest will converge to the function required.

A third alternative periodic extension is shown in Figure 13.21. However, this extension is neither odd nor even and so it has none of the desirable properties of the other two. Whichever extension we choose, the resulting Fourier series only gives a representation of the original function in the interval $0 < t < T/2$ and as such is termed a **half-range** Fourier series. Similarly we have the terminology **half-range sine series** for a series containing only sine terms and **half-range cosine series** for a series containing only cosine terms. The Fourier series formulae then simplify to give the following half-range formulae:

■ Half-range sine series:

$$a_n = 0 \qquad n \in \mathbb{N}$$

$$b_n = \frac{4}{T} \int_0^{T/2} f(t) \sin \frac{2n\pi t}{T} \, dt \qquad n \in \mathbb{N}^+ \tag{13.7}$$

and $f(t)$ is given by:

$$f(t) = \sum_{n=1}^{\infty} b_n \sin \frac{2n\pi t}{T}$$

■ Half-range cosine series:

$$a_0 = \frac{4}{T} \int_0^{T/2} f(t) \, dt \tag{13.8}$$

$$a_n = \frac{4}{T} \int_0^{T/2} f(t) \cos \frac{2n\pi t}{T} \, dt \qquad n \in \mathbb{N}^+ \tag{13.9}$$

$$b_n = 0 \qquad n \in \mathbb{N}^+$$

and then $f(t)$ is given by:

$$f(t) = \frac{a_0}{2} + \sum_{n=1}^{\infty} a_n \cos \frac{2n\pi t}{T}$$

Example 13.19

By defining an appropriate periodic extension of the function illustrated in Figure 13.22, find the half-range cosine series representation.

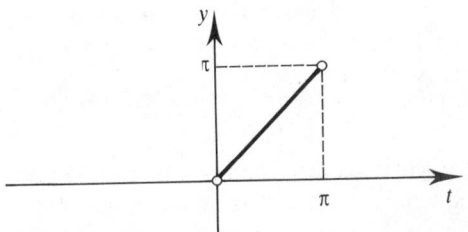

Figure 13.22 Graph for Example 13.19.

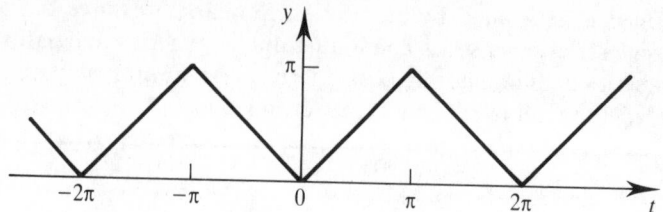

Figure 13.23 Graph for Example 13.19.

Solution

The function illustrated in Figure 13.22 is given by the formula $f(t) = t$ for $0 < t < \pi$ and is undefined outside this interval. Since the cosine series is required an even periodic extension must be formed. This is illustrated in Figure 13.23. Taking $T = 2\pi$ in Equations (13.8) and (13.9), we find a_0 and a_n.

$$a_0 = \frac{2}{\pi} \int_0^\pi t \, dt$$

$$= \frac{2}{\pi} \left[\frac{t^2}{2} \right]_0^\pi$$

$$= \pi$$

$$a_n = \frac{2}{\pi} \int_0^\pi t \cos nt \, dt$$

$$= \frac{2}{\pi} \left\{ \left[\frac{t \sin nt}{n} \right]_0^\pi - \int_0^\pi \frac{\sin nt}{n} \, dt \right\}$$

$$= \frac{2}{\pi} \left[\frac{\cos nt}{n^2} \right]_0^\pi$$

Now $\cos n\pi = (-1)^n$, so that:

$$a_n = \frac{2}{\pi} \left(\frac{(-1)^n}{n^2} - \frac{1}{n^2} \right) \qquad n = 1, 2, \ldots$$

Of course, all the b_n are zero. Therefore the half-range cosine series is:

$$f(t) = \frac{\pi}{2} - \frac{4}{\pi} \cos t - \frac{4}{9\pi} \cos 3t \ldots$$

and this series converges to the given function within the interval $0 < t < \pi$. ▲

EXERCISES 13.4

1. Graph an appropriate periodic extension of:

    $f(t) = 3t \qquad 0 < t < \pi$

    and hence find its half-range cosine series representation.

2. Find the half-range sine series representation of the function given in Example 13.19.

3. Find the half-range cosine series representing the function:

    $f(t) = \sin t \qquad 0 < t < \pi$

4. Graph an appropriate periodic extension of:

    $f(t) = e^t \qquad 0 < t < 1$

    and find its half-range cosine series.

5. Find the half-range sine series representation of $f(t) = 2 - t, \, 0 \le t \le 2$.

13.7 Parseval's theorem

If the function $f(t)$ is periodic with period T and has Fourier coefficients a_n and b_n, then Parseval's theorem states:

■ $\dfrac{2}{T} \displaystyle\int_0^T (f(t))^2 \, dt = \tfrac{1}{2}a_0^2 + \sum_{n=1}^{\infty} (a_n^2 + b_n^2)$

It is frequently useful in power calculations as the following example shows.

Example 13.20 Average power of a signal

Find the average power developed across a 1 Ω resistor by a voltage signal with period 2π given by:

$$v(t) = \cos t - \tfrac{1}{3}\sin 2t + \tfrac{1}{2}\cos 3t$$

Solution

We note that $v(t)$ is periodic with period $T = 2\pi$; $v(t)$ is already expressed as a Fourier series with $a_1 = 1$, $a_3 = 1/2$ and $b_2 = -1/3$. All other Fourier coefficients are 0. The instantaneous power is $(v(t))^2$ and hence the average power over one period is given by:

$$P_{av} = \frac{1}{2\pi} \int_0^{2\pi} (v(t))^2 \, dt$$

Therefore, using Parseval's theorem we find:

$$P_{av} = \tfrac{1}{2}(1^2 + (-\tfrac{1}{3})^2 + (\tfrac{1}{2})^2) = 0.68 \text{ W} \quad \blacktriangle$$

13.8 Complex notation

An alternative notation for Fourier series involving complex numbers is available which leads naturally into the more general topic of Fourier transforms. Recall from Chapter 6 the Euler relations

$$e^{\pm j\theta} = \cos\theta \pm j\sin\theta$$

from which we can obtain expressions for $\cos\theta$ and $\sin\theta$:

$$\cos\theta = \frac{e^{j\theta} + e^{-j\theta}}{2} \qquad \sin\theta = \frac{e^{j\theta} - e^{-j\theta}}{2j}$$

which enable us to rewrite the Fourier representation

$$f(t) = \frac{a_0}{2} + \sum_{n=1}^{\infty}\left(a_n\cos\frac{2n\pi t}{T} + b_n\sin\frac{2n\pi t}{T}\right)$$

as

$$f(t) = \frac{a_0}{2} + \sum_{n=1}^{\infty}\left(a_n\frac{e^{j2n\pi t/T} + e^{-j2n\pi t/T}}{2} + b_n\frac{e^{j2n\pi t/T} - e^{-j2n\pi t/T}}{2j}\right)$$

$$= \frac{a_0}{2} + \sum_{n=1}^{\infty}\left(\frac{a_n - jb_n}{2}e^{j2n\pi t/T} + \frac{a_n + jb_n}{2}e^{-j2n\pi t/T}\right)$$

which we can write equivalently as:

$$\blacksquare \quad f(t) = \sum_{-\infty}^{\infty} c_n e^{j2n\pi t/T}$$

---

where

$$c_n = \frac{a_n - jb_n}{2} \qquad c_{-n} = \frac{a_n + jb_n}{2} \qquad n = 1, 2, \ldots$$

and $c_0 = a_0/2$. It can be shown that the Fourier coefficients, c_n, are then given by:

$$\blacksquare \quad c_n = \frac{1}{T}\int_{-T/2}^{T/2} f(t)e^{-j2n\pi t/T}\,dt$$

The integral can also be evaluated over any complete period as convenient. Further, if we write $T = 2\pi/\omega_1$ then this complex form can be expressed as:

$$f(t) = \sum_{-\infty}^{\infty} c_n e^{jn\omega_1 t}$$

where

$$c_n = \frac{\omega_1}{2\pi} \int_{-\pi/\omega_1}^{\pi/\omega_1} f(t)e^{-jn\omega_1 t} \, dt$$

Example 13.21

Find the complex Fourier series representation of the function with period T defined by:

$$f(t) = \begin{cases} 1 & |t| < T/4 \\ 0 & \text{otherwise} \end{cases}$$

Solution

We find

$$c_n = \frac{1}{T} \int_{-T/4}^{T/4} 1 e^{-j2n\pi t/T} \, dt$$

$$= \frac{1}{T} \left[\frac{e^{-j2n\pi t/T}}{-j2n\pi/T} \right]_{-T/4}^{T/4}$$

$$= \frac{-1}{2n\pi j} (e^{-jn\pi/2} - e^{jn\pi/2})$$

$$= \frac{1}{n\pi} \left(\frac{e^{jn\pi/2} - e^{-jn\pi/2}}{2j} \right)$$

$$= \frac{1}{n\pi} \sin \frac{n\pi}{2}$$

Therefore,

$$f(t) = \sum_{-\infty}^{\infty} \frac{1}{n\pi} \sin \frac{n\pi}{2} e^{j2n\pi t/T}$$

The observant reader will note that the expressions for c_n appear invalid when $n = 0$, since the denominator is then zero. We can compute c_0 in either of two ways: using an integral expression or evaluating

$$\frac{1}{T} \left[\frac{e^{-j2n\pi t/T}}{-j2n\pi/T} \right]_{-T/4}^{T/4} \quad \text{as } n \to 0. \text{ We see}$$

$$c_0 = \frac{1}{T} \int_{-T/4}^{T/4} 1 \, dt = \frac{1}{2}$$

Also using a Taylor series expansion it is possible to show

$$\lim_{n \to 0} \frac{1}{T} \left[\frac{e^{-j2n\pi t/T}}{-j2n\pi/T} \right]_{-T/4}^{T/4} = \frac{1}{2}$$

giving a consistent result. ▲

EXERCISES 13.5

1. Find the complex Fourier series representation of:

(a) $f(t) = \begin{cases} 1 & 0 < t < 2 \\ 0 & 2 < t < 4 \end{cases}$ period 4

(b) $f(t) = e^t$ $-1 < t < 1$ period 2

(c) $f(t) = \begin{cases} A \sin \omega t & 0 < t < \pi/\omega \\ 0 & \pi/\omega < t < 2\pi/\omega \end{cases}$

period $2\pi/\omega$

2. If $f(t) = \sum_{-\infty}^{\infty} c_n e^{j2n\pi t/T}$, show that the coefficients, c_n, are given by:

$$c_n = \frac{1}{T} \int_0^T f(t) e^{-j2n\pi t/T} \, dt$$

Hint: multiply both sides by $e^{-j2m\pi t/T}$ and integrate over $[0, T]$.

13.9 Frequency response of a linear system

We have already examined some of the basic features of linear systems in Section 9.2. Linear systems have the property that the response to several inputs being applied to the system can be obtained by adding the effects of the individual inputs. Another useful property of linear systems is that if a sinusoidal input is applied to the system then the output will also be a sinusoid of the same frequency but with modified amplitude and phase. This is illustrated in Figure 13.24.

In Section 6.7 we saw that sinusoidal signals can be represented by complex numbers and that an a.c. electrical circuit can be analysed using complex numbers. This is true for linear systems in general. It is possible to define a **complex frequency function**, $G(j\omega)$, where ω is the frequency of the input; G relates the output and the input of a linear system.

If a sine wave of amplitude A_i is applied to the system then the amplitude, A_o, of the output is given by

$$A_o = |G(j\omega)| A_i$$

The phase shift, ϕ, is given by:

$$\phi = \angle G(j\omega)$$

Note that A_o and ϕ depend upon ω. It is important to note that $G(j\omega)$ is a

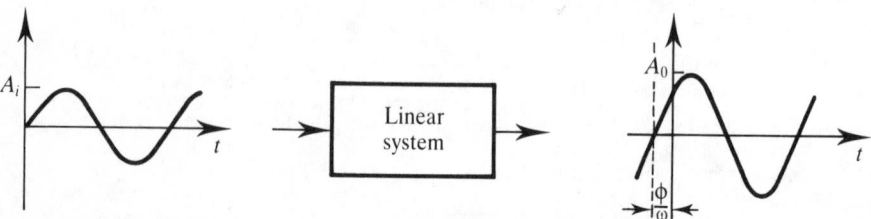

Figure 13.24 The response of a linear system to a sinusoidal input is also sinusoidal.

frequency-dependent function. Although the notation for $G(j\omega)$ may seem slightly odd it arises because one method of obtaining the frequency function for a linear system is to substitute $s = j\omega$ in the Laplace transform transfer function, $G(s)$, of the system.

It is now possible to analyse the effect of applying a generalized periodic waveform to a linear system. The first stage is to calculate the Fourier components of the input waveform. The amplitude and phase shift of each of the output components is then calculated using $G(j\omega)$. Finally, the output components are added to obtain the output waveform. This is only possible because of the additive nature of linear systems. An example will help to clarify these points.

Example 13.22 Low pass filter

Consider the circuit of Figure 13.25. Using Kirchhoff's voltage law and Ohm's law we obtain:

$$v_i = iR + v_o$$

For the capacitor,

$$v_o = \frac{i}{j\omega C}$$

Eliminating i yields:

$$v_i = v_o j\omega CR + v_o = v_o(1 + j\omega RC)$$

$$\frac{v_o}{v_i} = \frac{1}{1 + j\omega RC} \qquad (13.10)$$

Equation (13.10) relates the output of the system to the input of the system. Therefore,

$$G(j\omega) = \frac{1}{1 + j\omega RC}$$

It is convenient to convert $G(j\omega)$ into polar form:

$$G(j\omega) = \frac{1 \angle 0}{\sqrt{1 + (\omega RC)^2} \angle \tan^{-1} \omega RC}$$

$$= \frac{1}{\sqrt{1 + (\omega RC)^2}} \angle -\tan^{-1} \omega RC$$

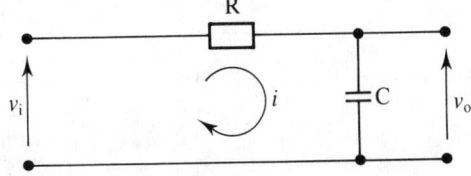

Figure 13.25 Circuit for Example 13.22.

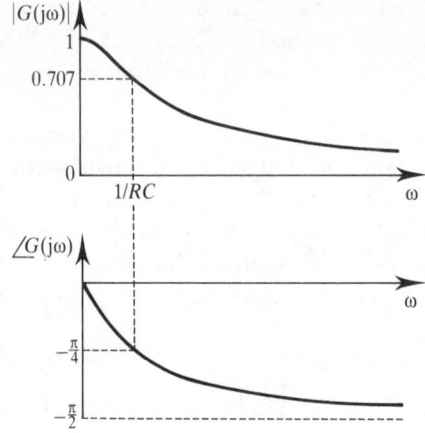

Figure 13.26 Amplitude and phase characteristics for the circuit of Figure 13.25.

Therefore,

$$|G(j\omega)| = \frac{1}{\sqrt{1 + (\omega RC)^2}} \tag{13.11}$$

$$\angle G(j\omega) = -\tan^{-1} \omega RC \tag{13.12}$$

The amplitude and phase characteristics for the circuit of Figure 13.25 are shown in Figure 13.26. These show the variation of $|G(j\omega)|$ and $\angle G(j\omega)$ with angular frequency ω.

Note that the circuit is a low pass filter; it allows low frequencies to pass easily and rejects high frequencies. The cut-off point of the filter, that is, the point at which significant frequency attenuation begins to occur, can be varied by changing the values of R and C. The quantity RC is usually known as the time constant for the system. Consider the case when $RC = 0.3$. Equations (13.11) and (13.12) reduce to:

$$|G(j\omega)| = \frac{1}{\sqrt{1 + 0.09\omega^2}} \tag{13.13}$$

$$\angle G(j\omega) = -\tan^{-1} 0.3\omega \tag{13.14}$$

Let us examine the response of this system to a square wave input with fundamental angular frequency 1 and amplitude 1. This waveform is shown in Figure 13.27(a). We note that $T = 2\pi$. The waveform function is odd and so will not contain any cosine Fourier components. It has an average value of 0 and so will not have a zero frequency component, that is, there will be

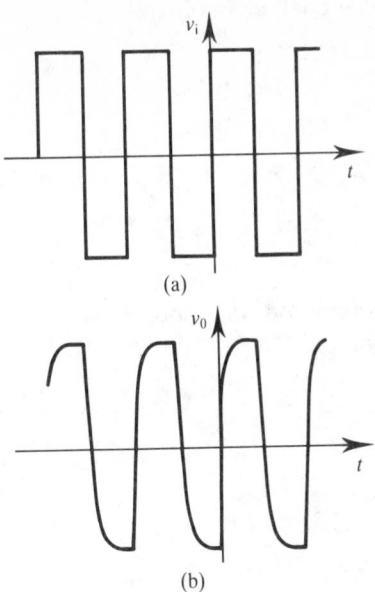

(a)

(b)

Figure 13.27   (a) Input to low pass filter; (b) output from low pass filter.

no d.c. component. Therefore calculating the Fourier components reduces to evaluating:

$$f(t) = \sum_{n=1}^{\infty} b_n \sin \frac{2\pi nt}{T}$$

$$b_n = \frac{2}{T} \int_{-T/2}^{T/2} f(t) \sin \frac{2\pi nt}{T} dt \qquad n \in \mathbb{N}^+$$

Since $T = 2\pi$, we find:

$$b_n = \frac{1}{\pi} \left(\int_{-\pi}^{0} -1 \sin nt \, dt + \int_{0}^{\pi} \sin nt \, dt \right)$$

$$= \frac{1}{\pi} \left(\left[\frac{\cos nt}{n} \right]_{-\pi}^{0} + \left[\frac{-\cos nt}{n} \right]_{0}^{\pi} \right)$$

$$= \frac{1}{n\pi} (\cos 0 - \cos \pi n - \cos \pi n + \cos 0)$$

$$= \frac{1}{n\pi} (2 - 2 \cos \pi n)$$

$$= \frac{2}{n\pi} (1 - \cos \pi n)$$

The values of the first few coefficients are:

$$b_1 = \frac{2}{\pi}(1 - \cos \pi) = \frac{4}{\pi}$$

$$b_2 = \frac{2}{2\pi}(1 - \cos 2\pi) = 0$$

$$b_3 = \frac{2}{3\pi}(1 - \cos 3\pi) = \frac{4}{3\pi}$$

The next stage is to evaluate the gain and phase changes of the Fourier components. Using Equations (13.13) and (13.14):

$n = 1$

$$\omega_1 = 1$$

$$|G(j\omega_1)| = \frac{1}{\sqrt{1 + 0.09 \times 1}} = 0.96$$

$$\angle G(j\omega_1) = -\tan^{-1} 0.3 = -16.7°$$

$n = 3$

$$\omega_3 = 3$$

$$|G(j\omega_3)| = \frac{1}{\sqrt{1 + 0.09 \times 9}} = 0.74$$

$$\angle G(j\omega_3) = -\tan^{-1} 0.9 = -42.0°$$

$n = 5$

$$\omega_5 = 5$$

$$|G(j\omega_5)| = \frac{1}{\sqrt{1 + 0.09 \times 25}} = 0.55$$

$$\angle G(j\omega_5) = -\tan^{-1} 1.5 = -56.3°$$

It is clear that high frequency Fourier components are attenuated and phase-shifted more than low frequency Fourier components. The effect is to produce a rounding of the rising and falling edges of the square wave input signal. This is illustrated in Figure 13.27(b). The output signal has been obtained by adding together the attenuated and phase-shifted output Fourier components. This is possible because the system is linear. ▲

13.10 The Fourier transform – definitions

We have seen that almost any periodic signal can be represented as a linear combination of sine and cosine waves of various frequencies and amplitudes. All frequencies are integer multiples of the fundamental. However, many practical

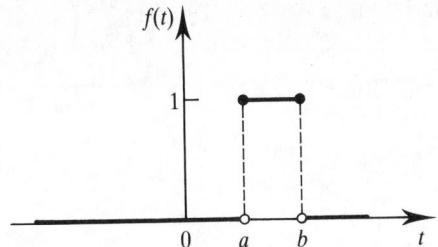

Figure 13.28 A non-periodic function.

waveforms are not periodic. Examples are pulse signals and noise signals. The function shown in Figure 13.28 is an example of a non-periodic signal.

We shall now see how Fourier techniques can still be useful by introducing the Fourier transform which is used extensively in communications engineering and signal processing. For example, it can be used to analyse the processes of modulation, which involves superimposing an audio signal on to a carrier signal, and demodulation, which involves removing the carrier signal to leave the audio signal.

Under certain conditions it can be shown that a non-periodic function, $f(t)$, can be expressed not as the sum of sine and cosine waves but as an integral. In particular,

$$f(t) = \int_0^\infty A(\omega) \cos \omega t + B(\omega) \sin \omega t \, d\omega \tag{13.15}$$

where:

$$A(\omega) = \frac{1}{\pi} \int_{-\infty}^\infty f(t) \cos \omega t \, dt \quad \text{and} \quad B(\omega) = \frac{1}{\pi} \int_{-\infty}^\infty f(t) \sin \omega t \, dt \tag{13.16}$$

The conditions required are those necessary for the integrals to exist. Because an integral can be regarded as the area under the curve of the integrand, $f(t)$ must be such that $f(t) \cos \omega t$ and $f(t) \sin \omega t$ both have finite areas under them for $-\infty < t < \infty$. Clearly this imposes rather severe conditions on $f(t)$ because $\sin \omega t$ and $\cos \omega t$ continue to oscillate as $t \to \pm\infty$. However, provided

(1) $f(t)$ and $f'(t)$ are piecewise continuous in every finite interval, and

(2) $\int_{-\infty}^\infty |f(t)| \, dt$ exists

then the above Fourier integral representation of $f(t)$ holds. At a point of discontinuity of $f(t)$ the integral representation converges to the average value of the right and left-hand limits. As with Fourier series, an equivalent complex representation exists which is, in fact, more commonly used:

$$■ \quad f(t) = \frac{1}{2\pi} \int_{-\infty}^{\infty} F(\omega) e^{j\omega t} \, d\omega \tag{13.17}$$

where

$$F(\omega) = \int_{-\infty}^{\infty} f(t) e^{-j\omega t} \, dt \tag{13.18}$$

There is no universal convention concerning the definition of these integrals and a number of variants are still correct. For instance, some authors write the factor $1/2\pi$ in the second integral rather than the first while others place a factor $1/\sqrt{2\pi}$ in both giving some symmetry to the equations. There is also variation in the location of the factors $e^{-j\omega t}$ and $e^{j\omega t}$. We shall use definitions (13.17) and (13.18) throughout but it is important to be aware of possible differences when consulting other texts.

Equations (13.17) and (13.18) form what is called a **Fourier transform pair**. The **Fourier transform** of $f(t)$ is $F(\omega)$ which is sometimes written $\mathscr{F}\{f(t)\}$. Similarly $f(t)$ in Equation (13.17) is the **inverse Fourier transform** of $F(\omega)$, usually denoted $\mathscr{F}^{-1}\{F(\omega)\}$.

■ The Fourier transform of $f(t)$ is defined to be:

$$\mathscr{F}\{f(t)\} = F(\omega) = \int_{-\infty}^{\infty} f(t) e^{-j\omega t} \, dt$$

You will also note the similarity between Equation (13.18) and the definition of the Laplace transform of $f(t)$:

$$\mathscr{L}\{f(t)\} = \int_{0}^{\infty} f(t) e^{-st} \, dt \tag{13.19}$$

We see that, apart from the limits of integration, the substitution $j\omega = s$ in Equation (13.18) results in the Laplace transform of Equation (13.19). There is indeed an important relationship between the two transforms which we shall discuss in Section 13.15. We note that Equation (13.17) provides a formula for the inverse Fourier transform of $F(\omega)$, although the integral is frequently difficult to evaluate.

Example 13.23

Find the Fourier transform of the function $f(t) = u(t) e^{-t}$, where $u(t)$ is the unit step function.

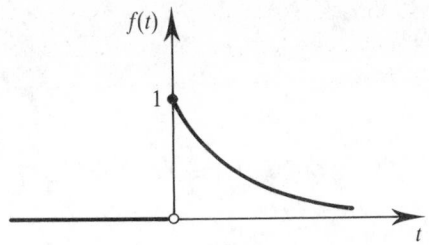

Figure 13.29   Graph of $u(t)e^{-t}$.

Solution

The function $u(t)e^{-t}$ is shown in Figure 13.29. Using Equation (13.18), its Fourier transform is given by:

$$F(\omega) = \int_{-\infty}^{\infty} f(t)e^{-j\omega t}\,dt$$

$$= \int_{0}^{\infty} e^{-t}e^{-j\omega t}\,dt \qquad \text{since } f(t) = 0 \text{ for } t < 0$$

$$= \int_{0}^{\infty} e^{-(1+j\omega)t}\,dt$$

$$= \left[\frac{e^{-(1+j\omega)t}}{-(1+j\omega)} \right]_{0}^{\infty}$$

$$= \frac{1}{1+j\omega} \qquad \text{since } e^{-(1+j\omega)t} \to 0 \text{ as } t \to \infty$$

that is,

$$F(\omega) = \frac{1}{1+j\omega} \quad \blacktriangle$$

Example 13.24

Use Equation (13.17) to find the Fourier integral representation of the function defined by:

$$f(t) = \begin{cases} 1 & -1 \le t \le 1 \\ 0 & |t| > 1 \end{cases}$$

Solution

Using Equation (13.17) we find:

$$f(t) = \frac{1}{2\pi} \int_{-\infty}^{\infty} F(\omega)e^{j\omega t}\,d\omega$$

where

$$F(\omega) = \int_{-\infty}^{\infty} f(t)e^{-j\omega t}\,dt$$

$$= \int_{-1}^{1} 1e^{-j\omega t}\,dt \qquad \text{since } f(t) \text{ is zero outside } [-1, 1]$$

$$= \left[\frac{e^{-j\omega t}}{-j\omega}\right]_{-1}^{1}$$

$$= \frac{e^{-j\omega} - e^{j\omega}}{-j\omega}$$

$$= \frac{e^{j\omega} - e^{-j\omega}}{j\omega}$$

Using Euler's relation (Section 6.6)

$$\sin\theta = \frac{e^{j\theta} - e^{-j\theta}}{2j}$$

we find

$$F(\omega) = \frac{2\sin\omega}{\omega}$$

so that

$$f(t) = \frac{1}{2\pi}\int_{-\infty}^{\infty} \frac{2\sin\omega}{\omega} e^{j\omega t}\,d\omega$$

Table 13.2   Common Fourier transforms.

$f(t)$	$F(\omega)$
$f(t) = Au(t)e^{-\alpha t},\ \alpha > 0$	$\dfrac{A}{\alpha + j\omega}$
$f(t) = \begin{cases} 1 & -\alpha \le t \le \alpha \\ 0 & \text{otherwise} \end{cases}$	$\dfrac{2\sin\omega\alpha}{\omega}$
$f(t) = A \qquad$ constant	$2\pi A\delta(\omega)$
$f(t) = u(t)A$	$A\left(\pi\delta(\omega) - \dfrac{j}{\omega}\right)$
$f(t) = \delta(t)$	1
$f(t) = \delta(t - a)$	$e^{-j\omega a}$
$f(t) = \cos at$	$\pi(\delta(\omega + a) + \delta(\omega - a))$
$f(t) = \sin at$	$\dfrac{\pi}{j}(\delta(\omega - a) - \delta(\omega + a))$

is the required integral representation. Note that $F(\omega) = (2 \sin \omega)/\omega$ is the Fourier transform of $f(t)$. The function $\sin \omega/\omega$ occurs frequently and is often referred to as the sinc function. ▲

As with Laplace transforms, tables have been compiled for reference. Such a table of common transforms appears in Table 13.2.

EXERCISES 13.6

1. Find the Fourier transforms of

(a) $f(t) = \begin{cases} 1/4 & |t| \le 3 \\ 0 & |t| > 3 \end{cases}$

(b) $f(t) = \begin{cases} 1 - t/2 & 0 \le t \le 2 \\ 1 + t/2 & -2 \le t \le 0 \\ 0 & \text{otherwise} \end{cases}$

(c) $f(t) = \begin{cases} e^{-\alpha t} & t \ge 0 \quad \alpha > 0 \\ e^{\alpha t} & t < 0 \end{cases}$

(d) $f(t) = \begin{cases} e^{-t} \cos t & t \ge 0 \\ 0 & t < 0 \end{cases}$

(e) $f(t) = u(t)e^{-t/\tau}$ where τ is a constant

2. Find

(a) the Fourier transform, and

(b) the Laplace transform of

$$f(t) = u(t)e^{-\alpha t} \qquad \alpha > 0$$

Show that making the substitution $s = j\omega$ in the Laplace transform of f results in the Fourier transform.

3. If $f(t) = \begin{cases} 1 & |t| \le 2 \\ 0 & \text{otherwise} \end{cases}$ and $g(t) = e^{jt}$ find $\mathcal{F}\{f(t)g(t)\}$.

13.11 Some properties of the Fourier transform

A number of the properties of Laplace transforms that we have already discussed hold for Fourier transforms. We consider linearity and two shift theorems.

13.11.1 *Linearity*

■ If f and g are functions of t and k is a constant, then:

$$\mathcal{F}\{f + g\} = \mathcal{F}\{f\} + \mathcal{F}\{g\}$$
$$\mathcal{F}\{kf\} = k\mathcal{F}\{f\}$$

Both of these properties follow directly from the definition and linearity properties of integrals, and mean that $\mathcal{F}$ is a linear operator.

Example 13.25

Find $\mathscr{F}\{u(t)e^{-t} + u(t)e^{-2t}\}$.

Solution

We saw in Example 13.23 that

$$\mathscr{F}\{u(t)e^{-t}\} = \frac{1}{1 + j\omega}$$

Furthermore,

$$\mathscr{F}\{u(t)e^{-2t}\} = \int_{-\infty}^{\infty} u(t)e^{-2t}e^{-j\omega t}\,dt$$

$$= \int_{0}^{\infty} e^{-(2+j\omega)t}\,dt$$

$$= \left[\frac{e^{-(2+j\omega)t}}{-(2+j\omega)}\right]_{0}^{\infty}$$

$$= \frac{1}{2 + j\omega}$$

Therefore,

$$\mathscr{F}\{u(t)e^{-t} + u(t)e^{-2t}\} = \frac{1}{1 + j\omega} + \frac{1}{2 + j\omega} \qquad \text{by linearity}$$

$$= \frac{2 + j\omega + 1 + j\omega}{(1 + j\omega)(2 + j\omega)}$$

$$= \frac{3 + 2j\omega}{2 - \omega^2 + 3j\omega} \qquad \blacktriangle$$

13.11.2 *First shift theorem*

■ If $F(\omega)$ is the Fourier transform of $f(t)$, then:
$$\mathscr{F}\{e^{jat}f(t)\} = F(\omega - a) \qquad \text{where } a \text{ is a constant}$$

Example 13.26

(a) Show that the Fourier transform of

$$f(t) = \begin{cases} 3 & -2 \leq t \leq 2 \\ 0 & \text{otherwise} \end{cases}$$

is given by $F(\omega) = \dfrac{6\sin 2\omega}{\omega}$

(b) Use the first shift theorem to find the Fourier transform of $e^{-jt}f(t)$.

(c) Verify the first shift theorem by obtaining the Fourier transform of $e^{-jt}f(t)$ directly.

Solutions

(a) $F(\omega) = 3 \displaystyle\int_{-2}^{2} e^{-j\omega t}\, dt = 3\left[\dfrac{e^{-j\omega t}}{-j\omega}\right]_{-2}^{2}$

$= 3\left(\dfrac{e^{-2j\omega} - e^{2j\omega}}{-j\omega}\right)$

$= 6\left(\dfrac{e^{2j\omega} - e^{-2j\omega}}{2j\omega}\right)$

$= \dfrac{6}{\omega}\sin 2\omega$

(b) We have $\mathscr{F}\{f(t)\} = F(\omega) = \dfrac{6\sin 2\omega}{\omega}$. Using the first shift theorem with

$a = -1$ we have

$$\mathscr{F}\{e^{-jt}f(t)\} = F(\omega + 1) = \dfrac{6}{\omega + 1}\sin 2(\omega + 1)$$

(c) $e^{-jt}f(t) = \begin{cases} 3e^{-jt} & -2 \le t \le 2 \\ 0 & \text{otherwise} \end{cases}$

So to evaluate its Fourier transform directly we must find:

$\mathscr{F}\{e^{-jt}f(t)\} = 3\displaystyle\int_{-2}^{2} e^{-jt}e^{-j\omega t}\, dt$

$= 3\displaystyle\int_{-2}^{2} e^{-(1+\omega)jt}\, dt$

$= 3\left[\dfrac{e^{-(1+\omega)jt}}{-j(1+\omega)}\right]_{-2}^{2}$

$= \dfrac{6}{1+\omega}\left(\dfrac{e^{2(1+\omega)j} - e^{-2(1+\omega)j}}{2j}\right)$

$= \dfrac{6}{1+\omega}\sin 2(1+\omega)$

as required. ▲

Example 13.27

Use the first shift theorem to find the function whose Fourier transform is

$\dfrac{1}{3 + j(\omega - 2)}$, given that $\mathscr{F}\{u(t)e^{-mt}\} = 1/(m + j\omega)$, $m > 0$.

Solution

From the given result we have:

$$\mathcal{F}\{u(t)e^{-3t}\} = \frac{1}{3 + j\omega} = F(\omega)$$

Now

$$\frac{1}{3 + j(\omega - 2)} = F(\omega - 2)$$

Therefore, from the first shift theorem with $a = 2$ we have:

$$\mathcal{F}\{e^{2jt}u(t)e^{-3t}\} = \frac{1}{3 + j(\omega - 2)}$$

Consequently the function whose Fourier transform is $\dfrac{1}{3 + j(\omega - 2)}$ is $u(t)e^{-(3-2j)t}$. ▲

Example 13.28

Find the Fourier transform of:

$$f(t) = \begin{cases} e^{-3t} & t \geq 0 \\ e^{3t} & t < 0 \end{cases}$$

Deduce the function whose Fourier transform is $G(\omega) = 6/(10 + 2\omega + \omega^2)$.

Solution

$$F(\omega) = \int_{-\infty}^{\infty} f(t)e^{-j\omega t}\, dt$$

$$= \int_{-\infty}^{0} e^{3t}e^{-j\omega t}\, dt + \int_{0}^{\infty} e^{-3t}e^{-j\omega t}\, dt$$

$$= \int_{-\infty}^{0} e^{(3-j\omega)t}\, dt + \int_{0}^{\infty} e^{-(3+j\omega)t}\, dt$$

$$= \left[\frac{e^{(3-j\omega)t}}{3 - j\omega}\right]_{-\infty}^{0} + \left[\frac{e^{-(3+j\omega)t}}{-(3+j\omega)}\right]_{0}^{\infty}$$

$$= \frac{1}{3 - j\omega} + \frac{1}{3 + j\omega}$$

$$= \frac{6}{9 + \omega^2}$$

Now

$$G(\omega) = \frac{6}{10 + 2\omega + \omega^2} = \frac{6}{(\omega + 1)^2 + 9} = F(\omega + 1)$$

Then, using the first shift theorem $F(\omega + 1)$ will be $\mathscr{F}\{e^{-jt}f(t)\}$, that is, the required function is

$$g(t) = \begin{cases} e^{(-3-j)t} & t \geq 0 \\ e^{(3-j)t} & t < 0 \end{cases} \quad \blacktriangle$$

13.11.3 *Second shift theorem*

■ If $F(\omega)$ is the Fourier transform of $f(t)$ then:

$$\mathscr{F}\{f(t - \alpha)\} = e^{-j\alpha\omega}F(\omega)$$

Example 13.29

Given that when $f(t) = \begin{cases} 1 & |t| \leq 1 \\ 0 & |t| > 1 \end{cases}$, $F(\omega) = \dfrac{2\sin\omega}{\omega}$, apply the second shift theorem to find the Fourier transform of:

$$g(t) = \begin{cases} 1 & 1 \leq t \leq 3 \\ 0 & \text{otherwise} \end{cases}$$

Verify your result directly.

Solution

The function $g(t)$ is depicted in Figure 13.30. Clearly $g(t)$ is the function $f(t)$ translated 2 units to the right, that is, $g(t) = f(t - 2)$. Now $F(\omega) = \dfrac{2\sin\omega}{\omega}$ is the Fourier transform of $f(t)$. Therefore, by the second shift theorem

$$\mathscr{F}\{g(t)\} = \mathscr{F}\{f(t - 2)\} = e^{-2j\omega}F(\omega) = \frac{2e^{-2j\omega}\sin\omega}{\omega}$$

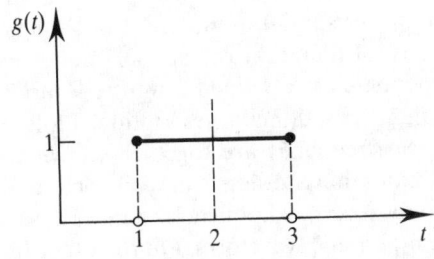

Figure 13.30 The function $g(t) = \begin{cases} 1 & 1 \leq t \leq 3 \\ 0 & \text{otherwise} \end{cases}$

To verify this result directly we must evaluate:

$$\mathcal{F}\{g(t)\} = \int_{-\infty}^{\infty} g(t)e^{-j\omega t}\,dt = \int_{1}^{3} e^{-j\omega t}\,dt = \left[\frac{e^{-j\omega t}}{-j\omega}\right]_{1}^{3}$$

$$= \frac{e^{-3j\omega} - e^{-j\omega}}{-j\omega} = e^{-2j\omega}\left(\frac{e^{-j\omega} - e^{j\omega}}{-j\omega}\right)$$

$$= \frac{2e^{-2j\omega}\sin\omega}{\omega}$$

as required. ▲

EXERCISES 13.7

1. Prove the first shift theorem.

2. Find the Fourier transform of:

$$f(t) = \begin{cases} 1 - t^2 & |t| \leq 1 \\ 0 & |t| > 1 \end{cases}$$

Use the first shift theorem to deduce the Fourier transforms of:

(a) $g(t) = \begin{cases} e^{3jt}(1 - t^2) & |t| \leq 1 \\ 0 & |t| > 1 \end{cases}$

(b) $h(t) = \begin{cases} e^{-t}(1 - t^2) & |t| \leq 1 \\ 0 & |t| > 1 \end{cases}$

3. Find the inverse Fourier transforms of:

(a) $\dfrac{1}{(\omega + 7)j + 1}$

(b) $\dfrac{2}{1 + 2(\omega - 1)j}$

4. Prove the second shift theorem.

5. Given $\mathcal{F}\{u(t)e^{-t}\} = 1/(1 + j\omega)$, use the second shift theorem to find:

$$\mathcal{F}\{u(t + 4)e^{-(t+4)}\}$$

Verify your result by direct integration.

6. Find, using the second shift theorem,

$$\mathcal{F}^{-1}\left\{6e^{-4j\omega}\frac{\sin 2\omega}{\omega}\right\}$$

13.12 Spectra

In the Fourier analysis of periodic waveforms we stated that although a waveform physically exists in the time (or spatial) domain it can be regarded as comprising components with a variety of temporal (or spatial) frequencies. The amplitude and phase of these components are obtained from the Fourier coefficients a_n and b_n. This is known as a frequency domain description. Plots of amplitude versus frequency and phase versus frequency are together known as the **spectrum** of a waveform. Periodic functions have **discrete** or **line spectra**, that is, the spectra assume non-zero values only at certain frequencies. Only a discrete set of frequencies is required to synthesize a periodic waveform. On the other hand when analysing non-periodic phenomena via Fourier transform techniques we find that, in general, a continuous range of frequencies is required. Instead of discrete spectra we have

continuous spectra. The modulus of the Fourier transform, $|F(\omega)|$, gives the spectrum amplitude while its argument $\arg(F(\omega))$ describes the spectrum phase.

Example 13.30

In Example 13.24 the Fourier transform of:

$$f(t) = \begin{cases} 1 & |t| \leq 1 \\ 0 & |t| > 1 \end{cases}$$

was found to be $F(\omega) = (2 \sin \omega)/\omega$. Sketch the spectrum of $f(t)$.

Solution

$F(\omega)$ is purely real. The spectrum of $f(t)$ is depicted by plotting $|F(\omega)|$ against ω as illustrated in Figure 13.31. Note that $\lim_{\omega \to 0} (\sin \omega)/\omega = 1$. (See Miscellaneous exercises in Chapter 9.) ▲

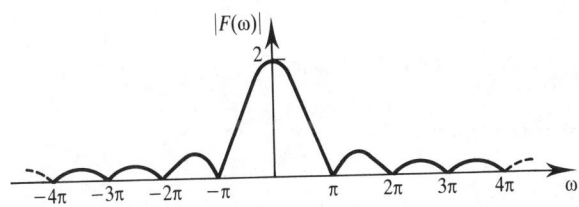

Figure 13.31   Spectrum of $f(t) = \begin{cases} 1 & |t| \leq 1 \\ 0 & |t| > 1 \end{cases}$.

Example 13.31 **Amplitude modulation**

Amplitude modulation is a technique that allows audio signals to be transmitted as electromagnetic radio waves. The maximum frequency of audio signals is typically 10 kHz. If these signals were to be transmitted directly then it would be necessary to use a very large antenna. This can be seen by calculating the wavelength of an electromagnetic wave of frequency 10 kHz using the formula

$$c = f\lambda$$

Here c is the velocity of an electromagnetic wave in a vacuum (3×10^8 m s^{-1}), f is the frequency of the wave and λ is its wavelength, and hence $\lambda = c/f = 30\,000$ m. It can be shown that an antenna must have dimensions of at least one quarter of the wavelength of the signal being transmitted if it is to be reasonably efficient. Clearly a very large antenna would be needed to transmit a 10 kHz signal directly. The solution is to have a **carrier signal** of a much higher frequency than the audio signal which is usually termed the **modulation signal**. This allows the antenna to be a reasonable size as a higher frequency signal has a lower wavelength. The arrangement for mixing the two signals is shown in Figure 13.32.

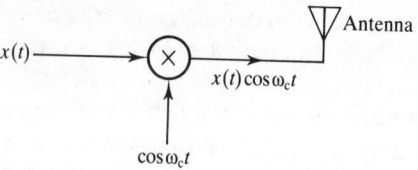

Figure 13.32 Amplitude modulation.

Let us now derive an expression for the frequency spectrum of an amplitude modulated signal given by:

$$\phi(t) = x(t) \cos \omega_c t$$

where ω_c is the angular frequency of the carrier signal and $x(t)$ is the modulation signal. Now $\phi(t) = x(t) \cos \omega_c t$ can be written as

$$\phi(t) = x(t) \frac{e^{j\omega_c t} + e^{-j\omega_c t}}{2}$$

Taking the Fourier transform and using the first shift theorem yields:

$$\mathscr{F}\{\phi(t)\} = \Phi(\omega) = \mathscr{F}\left\{ \frac{x(t)(e^{j\omega_c t} + e^{-j\omega_c t})}{2} \right\}$$

$$= \mathscr{F}\left\{ \frac{e^{j\omega_c t} x(t)}{2} \right\} + \mathscr{F}\left\{ \frac{e^{-j\omega_c t} x(t)}{2} \right\}$$

$$= \tfrac{1}{2}(X(\omega - \omega_c) + X(\omega + \omega_c))$$

where $X(\omega) = \mathscr{F}\{x(t)\}$, the frequency spectrum of the modulation signal.

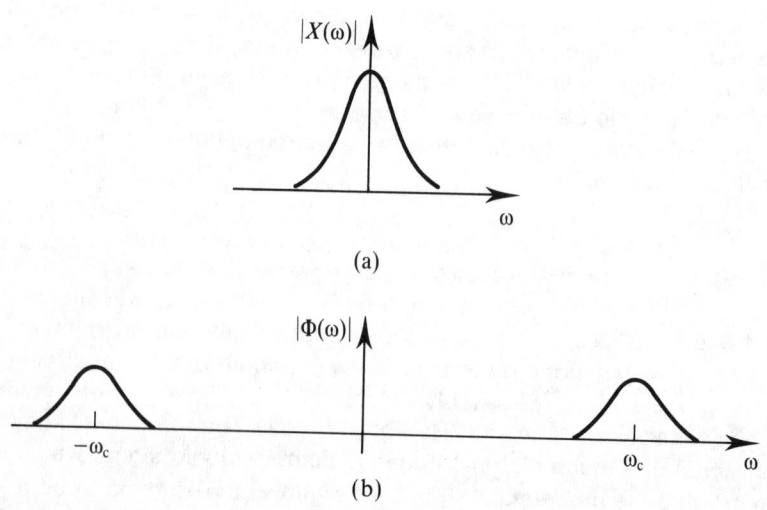

(a)

(b)

Figure 13.33 Amplitude modulation. (a) Spectrum of the modulation signal; (b) spectrum of the amplitude modulated signal.

Let us consider the case where the frequency spectrum, $|X(\omega)|$, has the profile shown in Figure 13.33(a). The frequency spectrum of the amplitude modulated signal, $|\Phi(\omega)|$, is shown in Figure 13.33(b). All of the frequencies of the amplitude modulated signal are much higher than the frequencies of the modulation signal thus allowing a much smaller antenna to be used to transmit the signal. This method of amplitude modulation is known as **suppressed carrier amplitude modulation** because the carrier signal is modulated to its full depth and so the spectrum of the amplitude modulated signal has no identifiable carrier component. ▲

EXERCISE 13.8

Show that the Fourier transform of the pulse

$$f(t) = \begin{cases} t + 1 & -1 < t < 0 \\ 1 - t & 0 < t < 1 \end{cases}$$

can be written as

$$F(\omega) = \frac{2}{\omega^2}(1 - \cos \omega)$$

Plot a graph of the spectrum of $f(t)$ for $-2\pi \le \omega \le 2\pi$. Write down an integral expression for the pulse which would result if the signal $f(t)$ were passed through a filter which eliminates all angular frequencies greater than 2π.

13.13 The $t–\omega$ duality principle

We have, from the definition of the Fourier integral,

$$f(t) = \frac{1}{2\pi} \int_{-\infty}^{\infty} F(\omega)e^{j\omega t}\, d\omega \tag{13.20}$$

where

$$F(\omega) = \int_{-\infty}^{\infty} f(t)e^{-j\omega t}\, dt \tag{13.21}$$

is the Fourier transform of $f(t)$. In Equation (13.20), ω is a dummy variable so, for example, Equation (13.20) could be equivalently written as

$$f(t) = \frac{1}{2\pi} \int_{-\infty}^{\infty} F(z)e^{jzt}\, dz \tag{13.22}$$

Then, from Equation (13.22), replacing t by $-\omega$ we find:

$$f(-\omega) = \frac{1}{2\pi} \int_{-\infty}^{\infty} F(z)e^{-j\omega z}\, dz = \frac{1}{2\pi} \int_{-\infty}^{\infty} F(t)e^{-j\omega t}\, dt$$

which we recognize as $\dfrac{1}{2\pi}$ times the Fourier transform of $F(t)$.

We have the following result:

■ If $F(\omega)$ is the Fourier transform of $f(t)$ then

$$f(-\omega) \text{ is } \frac{1}{2\pi} \times \text{(the Fourier transform of } F(t))$$

---

which is known as the $t - \omega$ duality principle.

We have seen in Example 13.24 that if

$$f(t) = \begin{cases} 1 & |t| \leq 1 \\ 0 & |t| > 1 \end{cases}$$

then $F(\omega) = \dfrac{2 \sin \omega}{\omega}$. This is depicted in Figure 13.34. From the duality principle we can immediately deduce that:

$$\mathcal{F}\left\{ \frac{2 \sin t}{t} \right\} = 2\pi f(-\omega) = 2\pi f(\omega)$$

since f is an even function (Figure 13.35). Unfortunately it is very difficult to verify this result in most cases because while one of the integrals is relatively straightforward to evaluate, the other is usually very difficult. However, we can use the result to derive a number of new Fourier transforms.

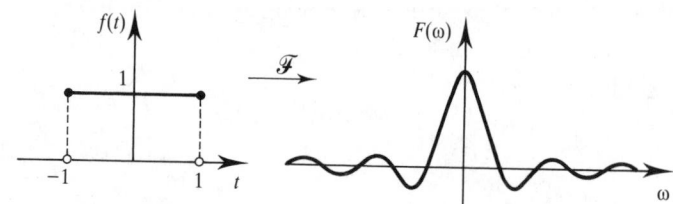

Figure 13.34   Illustrating the $t-\omega$ duality principle.

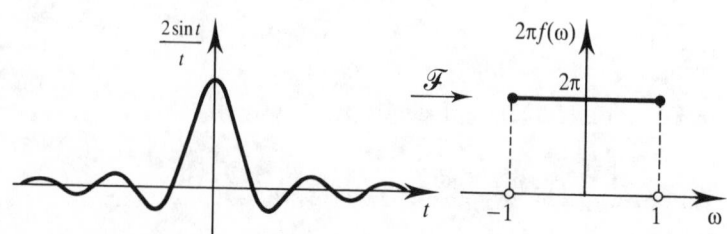

Figure 13.35   Illustrating the $t-\omega$ duality principle.

Example 13.32

Given that the Fourier transform of $u(t)e^{-t}$ is $1/(1+j\omega)$ use the duality principle to deduce the transform of $1/(1+jt)$.

Solution

We know $F(\omega) = 1/(1+j\omega)$ is the Fourier transform of $f(t) = u(t)e^{-t}$. Therefore $2\pi(u(-\omega)e^{\omega})$ is the Fourier transform of $1/(1+jt)$. ▲

13.14 Fourier transforms of some special functions

We saw in Section 13.12 that the Fourier transform tells us the frequency content of a signal. If we were to find the Fourier transform of a signal composed of only one frequency component, for example, $f(t) = \sin t$, we would hope that the exercise of finding the Fourier transform would result in a spectrum containing that single frequency.

Unfortunately if we try to find the Fourier transform of say $f(t) = \sin t$ problems arise since the integral:

$$\int_{-\infty}^{\infty} \sin t\, e^{-j\omega t}\, dt$$

cannot be evaluated in the usual sense because $\sin t$ oscillates indefinitely as $|t| \to \infty$. In particular, Condition (2) of Section 13.10 fails since $\int_{-\infty}^{\infty} |\sin t|\, dt$ diverges. There are many other functions which give rise to similar difficulties, for instance, the unit step function, polynomials and so on. All these functions fail to have a Fourier transform in its usual sense. However, by making use of the δ function it is possible to make progress even with functions like these.

13.14.1 *The Fourier transform of $\delta(t-a)$*

Example 13.33

Use the properties of the δ function to deduce its Fourier transform.

Solution

By definition:

$$\mathcal{F}\{\delta(t-a)\} = \int_{-\infty}^{\infty} \delta(t-a)e^{-j\omega t}\, dt$$

Next, recall the following property of the δ function.

$$\int_{-\infty}^{\infty} f(t)\delta(t-a)\, dt = f(a) \tag{13.23}$$

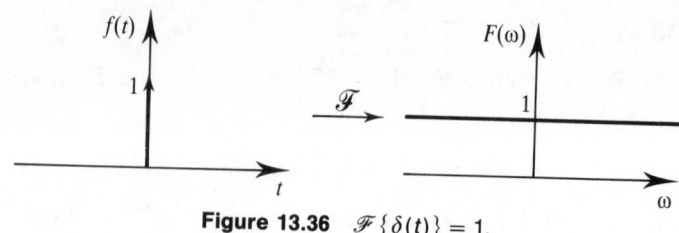

Figure 13.36 $\mathscr{F}\{\delta(t)\} = 1$.

for any reasonably well-behaved function $f(t)$. Using Equation (13.23) with $f(t) = e^{-j\omega t}$ we have

$$\mathscr{F}\{\delta(t-a)\} = \int_{-\infty}^{\infty} e^{-j\omega t}\,\delta(t-a)\,dt = e^{-j\omega a}$$

In particular, if $a = 0$ we have $\mathscr{F}\{\delta(t)\} = 1$. This result is depicted in Figure 13.36. ▲

Example 13.34

Apply the t–ω duality principle to the previous result. Interpret the result physically.

Solution

We have $f(t) = \delta(t)$ and $F(\omega) = 1$. The duality principle tells us that:

$$f(-\omega) = \delta(-\omega) \quad \text{which equals} \quad \frac{1}{2\pi}\mathscr{F}\{1\}$$

that is,

$$\mathscr{F}\{1\} = 2\pi\delta(\omega)$$

(since $\delta(-\omega) = \delta(\omega)$). This is illustrated in Figure 13.37. Physically $F(t) = 1$ can be regarded as a d.c. waveform. This result confirms that a d.c. signal has only one frequency component, namely zero. ▲

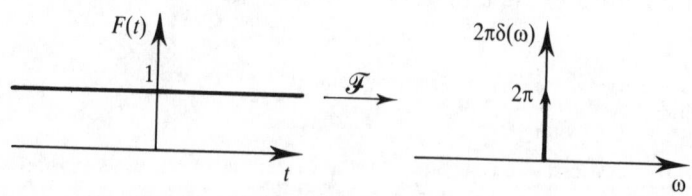

Figure 13.37 $\mathscr{F}\{1\} = 2\pi\delta(\omega)$.

Example 13.35

Given that $\mathscr{F}\{\delta(t-a)\} = e^{-j\omega a}$ find $\mathscr{F}\{e^{-jta}\}$.

Solution

We have $f(t) = \delta(t - a)$, $F(\omega) = e^{-j\omega a}$. Applying the $t\!-\!\omega$ duality principle we find

$$f(-\omega) = \delta(-\omega - a) = \frac{1}{2\pi}\mathscr{F}\{e^{-jta}\}$$

Therefore

$$\mathscr{F}\{e^{-jta}\} = 2\pi\delta(-\omega - a)$$
$$= 2\pi\delta(-(\omega + a))$$
$$= 2\pi\delta(\omega + a)$$

since $\delta(\omega)$ is an even function. ▲

13.14.2 *Fourier transforms of some periodic functions*

From Example 13.35 we have $\mathscr{F}\{e^{-jta}\} = 2\pi\delta(\omega + a)$ and also, replacing a by $-a$, $\mathscr{F}\{e^{jta}\} = 2\pi\delta(\omega - a)$. Adding these two expressions we find:

$$\mathscr{F}\{e^{-jta}\} + \mathscr{F}\{e^{jta}\} = 2\pi(\delta(\omega + a) + \delta(\omega - a))$$

Recalling the linearity properties of $\mathscr{F}$ we can write

$$\mathscr{F}\{e^{-jta} + e^{jta}\} = 2\pi(\delta(\omega + a) + \delta(\omega - a))$$

and using Euler's relations we find:

$$\mathscr{F}\{\cos at\} = \pi(\delta(\omega + a) + \delta(\omega - a))$$

We see that the spectrum of $\cos at$ consists of single lines at $\omega = \pm a$ corresponding to a single frequency component (Figure 13.38).

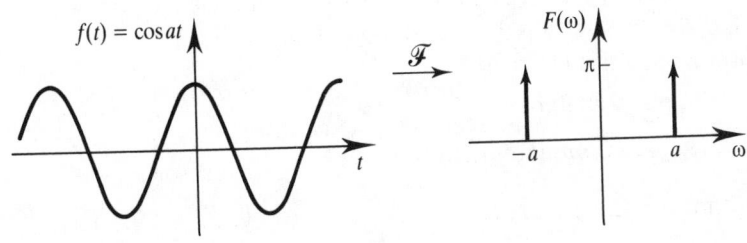

Figure 13.38   The spectrum of $\cos at$.

Example 13.36

Find $\mathscr{F}\{\sin at\}$.

Solution

Subtracting the previous expressions for $\mathscr{F}\{e^{jta}\}$ and $\mathscr{F}\{e^{-jta}\}$ and using Euler's relations we find:

$$\mathscr{F}\{e^{jta}\} - \mathscr{F}\{e^{-jta}\} = 2\pi(\delta(\omega - a) - \delta(\omega + a))$$

that is,

$$\mathscr{F}\left\{\frac{e^{jta} - e^{-jta}}{2j}\right\} = \frac{\pi}{j}(\delta(\omega - a) - \delta(\omega + a))$$

so that

$$\mathscr{F}\{\sin at\} = \frac{\pi}{j}(\delta(\omega - a) - \delta(\omega + a)) \quad \blacktriangle$$

13.15 The relationship between the Fourier transform and the Laplace transform

We have already noted (Section 13.10) the similarity between the Laplace transform and the Fourier transform. Let us now look at this a little more closely. We have:

$$\mathscr{F}\{f(t)\} = \int_{-\infty}^{\infty} f(t)e^{-j\omega t}\,dt \quad \text{and} \quad \mathscr{L}\{f(t)\} = \int_{0}^{\infty} f(t)e^{-st}\,dt$$

In the definition of the Laplace transform, the parameter s is complex and we may write $s = \sigma + j\omega$, so that:

$$\mathscr{L}\{f(t)\} = \int_{0}^{\infty} f(t)e^{-\sigma t}e^{-j\omega t}\,dt$$

Thus an additional factor, $e^{-\sigma t}$, appears in the integrand of the Laplace transform. For $\sigma > 0$ this represents an exponentially decaying factor, the presence of which means that the integral exists for a wider variety of functions than the corresponding Fourier integral.

Example 13.37

Find, if possible,

(a) the Laplace transform

(b) the Fourier transform

of $f(t) = u(t)e^{3t}$. Comment upon the result.

Solutions

(a) Either by integration, or from Table 11.1, we find:

$$\mathscr{L}\{u(t)e^{3t}\} = \frac{1}{s - 3} \quad \text{provided } s > 3$$

(b) $\mathscr{F}\{u(t)e^{3t}\} = \displaystyle\int_{0}^{\infty} e^{3t}e^{-j\omega t}\,dt = \int_{0}^{\infty} e^{(3-j\omega)t}\,dt = \left[\frac{e^{(3-j\omega)t}}{3 - j\omega}\right]_{0}^{\infty}$

Now, as $t \to \infty$, $e^{3t} \to \infty$, so that the integral fails to exist. Clearly, $u(t)e^{3t}$ has a Laplace transform but no Fourier transform. ▲

Suppose $f(t)$ is defined to be 0 for $t < 0$. Then its Fourier transform becomes:

$$\mathscr{F}\{f(t)\} = \int_0^\infty f(t)e^{-j\omega t}\, dt$$

and its Laplace transform is:

$$\mathscr{L}\{f(t)\} = \int_0^\infty f(t)e^{-st}\, dt$$

By replacing s by $j\omega$ in the Laplace transform we obtain the Fourier transform of $f(t)$ if it exists. Care must be taken here since we have seen that the Fourier transform may not exist for a function that nevertheless has a Laplace transform.

Example 13.38

Find the Laplace transforms of:

(a) $u(t)e^{-2t}$

(b) $u(t)e^{2t}$

Let $s = j\omega$ and comment upon the result.

Solutions

(a) $\mathscr{L}\{u(t)e^{-2t}\} = 1/(s + 2)$.

(b) $\mathscr{L}\{u(t)e^{2t}\} = 1/(s - 2)$.

Replacing s by $j\omega$ in (a) gives $1/(j\omega + 2)$. Similarly, replacing s by $j\omega$ in (b) gives $1/(j\omega - 2)$. Now

$$\mathscr{F}\{u(t)e^{-2t}\} = \frac{1}{j\omega + 2}$$

so that replacing s by $j\omega$ in the Laplace transform results in the Fourier transform. However, $\mathscr{F}\{u(t)e^{2t}\}$ does not exist and even though we can let $s = j\omega$ in the Laplace transform and obtain $1/(j\omega - 2)$, we cannot interpret this as a Fourier transform. ▲

The Fourier transform does possess certain advantages over the Laplace transform. While the Laplace transform can only be applied to functions which are zero for $t < 0$, the Fourier transform is applicable to functions with domain $-\infty < t < \infty$. In some applications where, for example, t represents not time but a spatial variable, it is often necessary to work with negative values.

The inverse Fourier transform is given by:

$$\mathscr{F}^{-1}\{F(\omega)\} = \frac{1}{2\pi} \int_{-\infty}^\infty F(\omega)e^{j\omega t}\, d\omega \tag{13.24}$$

The corresponding inverse Laplace transform requires advanced techniques in the theory of complex variables which are beyond the scope of this book. The existence of Equation (13.24) is not quite as advantageous as it may seem because it is often difficult to perform the required integration analytically.

Miscellaneous exercises

1. Find the half-range Fourier sine series representation of $f(t) = t \sin t$, $0 \leq t \leq \pi$.

2. Find the half-range sine series representation of $f(t) = \cos 2t$, $0 \leq t \leq \pi$.

3. Find (a) the half-range sine series, and (b) the half-range cosine series representation of the function defined in the interval $[0, \tau]$ by:

$$f(t) = \begin{cases} \dfrac{4t}{\tau} & 0 \leq t \leq \dfrac{\tau}{4} \\ \dfrac{4}{3}\left(1 - \dfrac{t}{\tau}\right) & \dfrac{\tau}{4} \leq t \leq \tau \end{cases}$$

4. Find the Fourier series representation of the function with period T defined by

$$f(t) = \begin{cases} V\,(\text{constant}) & |t| < T/6 \\ 0 & T/6 \leq |t| \leq T/2 \end{cases}$$

5. The output from a half-wave rectifier is given by:

$$i(t) = \begin{cases} I \sin \omega t & 0 < t < T/2 \\ 0 & T/2 < t < T \end{cases}$$

and is periodic with period $T = 2\pi/\omega$. Find its Fourier series representation.

6. Find the complex Fourier series representation of the function with period $T = 0.02$ defined by:

$$v(t) = \begin{cases} V\,(\text{constant}) & 0 \leq t < 0.01 \\ 0 & 0.01 \leq t < 0.02 \end{cases}$$

7. Find the Fourier series representation of the function with period 8 given by:

$$f(t) = \begin{cases} 2 - t & 0 < t < 4 \\ t - 6 & 4 < t < 8 \end{cases}$$

8. The r.m.s. voltage, $v_{\text{r.m.s.}}$, of a periodic waveform, $v(t)$, with period T, is given by

$$v_{\text{r.m.s.}} = \sqrt{\frac{1}{T} \int_0^T (v(t))^2 \, dt}$$

If $v(t)$ has Fourier coefficients a_n and b_n show, using Parseval's theorem, that

$$v_{\text{r.m.s.}} = \sqrt{\frac{1}{4} a_0^2 + \frac{1}{2} \sum_{n=1}^{\infty} (a_n^2 + b_n^2)}$$

9. If $f(t)$ has Fourier series:

$$f(t) = \frac{a_0}{2} + \sum_{n=1}^{\infty} \left(a_n \cos \frac{2n\pi t}{T} \right.$$

$$\left. + b_n \sin \frac{2n\pi t}{T} \right)$$

prove Parseval's theorem. *Hint:* multiply both sides by $f(t)$ to obtain:

$$(f(t))^2 = \frac{a_0 f(t)}{2}$$

$$+ \sum_{n=1}^{\infty} \left(a_n f(t) \cos \frac{2n\pi t}{T} + b_n f(t) \sin \frac{2n\pi t}{T} \right)$$

and integrate both sides over the interval $[0, T]$ using Equations (13.3)–(13.5).

10. Find the Fourier transforms of:

(a) $f(t) = \begin{cases} 1 - t^2 & |t| < 1 \\ 0 & \text{otherwise} \end{cases}$

(b) $f(t) = \begin{cases} \sin t & |t| < \pi \\ 0 & \text{otherwise} \end{cases}$

(c) $f(t) = \begin{cases} 1 & 0 < t < \tau \\ 0 & \text{otherwise} \end{cases}$

(d) $f(t) = \begin{cases} e^{-\alpha t} & t > 0 \\ -e^{\alpha t} & t < 0 \end{cases} \quad \alpha > 0$

11. Find the Fourier integral representations of

(a) $f(t) = \begin{cases} 3t & |t| < 2 \\ 0 & \text{otherwise} \end{cases}$

(b) $f(t) = \begin{cases} 0 & t < 0 \\ 6 & 0 < t < 2 \\ 0 & t > 2 \end{cases}$

Functions of several variables 14

14.1 Introduction

We have already discussed differentiation of functions of one variable. However, many functions depend upon two or more variables and we need to be able to calculate rates of change of such functions as these variables change. This is achieved by allowing one variable to change at a time, holding the others fixed. Differentiation under these conditions is called **partial differentiation**.

In engineering there are many functions which depend upon more than one variable. For example, the voltage on a transmission line depends upon position along the transmission line as well as time. The height of a liquid in a tank depends upon the flow rates into and out of the tank. We shall examine some more examples in this chapter.

14.2 Functions of more than one variable

In Chapter 7 we saw how to differentiate $y(x)$ with respect to x. Many standard derivatives were listed and some techniques explained. Since y is a function of x we call y the dependent variable and x the independent variable. The function y depends upon the one variable x. Consider the following example.

The area, A, of a circle, depends only upon the radius, r, and is given by

$$A(r) = \pi r^2$$

The rate of change of area w.r.t. the radius is $dA/dr = 2\pi r$. In practice functions often depend upon more than one variable. For example, the volume, V, of a cylinder depends upon the radius, r, and the height, h, and is given by

$$V = \pi r^2 h$$

V is the dependent variable; r and h are independent variables. V is a function of the two independent variables, r and h. We write $V = V(r, h)$.

Example 14.1 Electrical potential inside a cathode ray tube

Within a cathode ray tube the electrical potential, V, will vary with spatial position and time. Given Cartesian coordinates x, y and z, we can write:

$$V = V(x, y, z, t)$$

to show this dependence. Note that V is a function of four independent variables. ▲

Example 14.2 Power dissipated in a variable resistor

The power, P, dissipated in a variable resistor depends upon the instantaneous voltage across the resistor, v, and the resistance, r. It is given by:

$$P = \frac{v^2}{r}$$

Hence we may write $P = P(v, r)$ to show this dependence. The power is a function of two independent variables. ▲

As another example of a function of more than one variable consider a three-dimensional surface such as that shown in Figure 14.1. The height, z, of the surface above the x–y plane depends upon the x and y coordinates, that is, $z = z(x, y)$. If we are given values of x and y, then $z(x, y)$ can be evaluated, that is, the height of the surface above the point (x, y) can be found. The dependent variable, z, is a function of the independent variables x and y. Some important features are shown. The value of z at a **maximum point** is greater than the values of z at nearby points. Point A is such a point. As you move away from A the value of z decreases. A **minimum point** is similarly defined. At a minimum point, the value of z is smaller than the z value at nearby points. This is illustrated by point B. Point C illustrates an interesting point known as a **saddle point**. At a saddle point, z increases in one direction, axis D in the figure, but decreases in

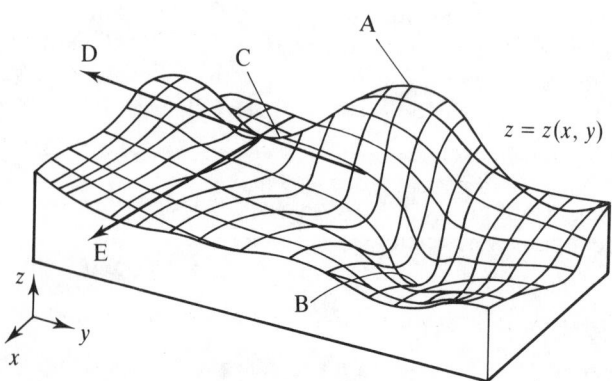

Figure 14.1 The height of the surface above the x–y plane is z.

the direction of axis E. These axes are at right angles to one another. The term 'saddle' is descriptive as a horse-saddle has the same shape.

14.3 Partial derivatives

Consider

$$z = z(x, y)$$

that is, z is a function of the independent variables x and y. We can differentiate z either with respect to x, or with respect to y. We need symbols to distinguish between these two cases. When finding the derivative w.r.t. x, the other independent variable, y, is held constant and only x changes. Similarly when differentiating w.r.t. y, the variable x is held constant. We write $\partial z/\partial x$ to denote differentiation of z w.r.t. x. It is called the first partial derivative of z w.r.t. x. Similarly, the first partial derivative of z w.r.t. y is denoted $\partial z/\partial y$. Referring to the surface $z(x, y)$, $\partial z/\partial x$ gives the rate of change of z moving only in the x direction, and hence y is held fixed.

■ If $z = z(x, y)$, then the **first partial derivatives** of z are

$$\frac{\partial z}{\partial x} \quad \text{and} \quad \frac{\partial z}{\partial y}$$

If we wish to evaluate a partial derivative, say $\partial z/\partial x$, at a particular point (x_0, y_0), we indicate this by

$$\frac{\partial z}{\partial x}(x_0, y_0) \quad \text{or} \quad \frac{\partial z}{\partial x}\bigg|_{(x_0, y_0)}$$

just as we did for functions of one variable.

Example 14.3

Given $z(x, y) = x^2y + \sin x + x \cos y$ find $\partial z/\partial x$ and $\partial z/\partial y$.

Solution

To find $\partial z/\partial x$ we differentiate z w.r.t. x, treating y as a constant. Note that since y is a constant then so is $\cos y$.

$$\frac{\partial z}{\partial x} = 2xy + \cos x + \cos y$$

In finding $\partial z/\partial y$, x and hence x^2 and $\sin x$ are held fixed, thus

$$\frac{\partial z}{\partial y} = x^2 - x \sin y \quad ▲$$

Example 14.4

Given $z(x, y) = 3e^x - 2e^y + x^2y^3$

(a) find $z(1, 1)$

(b) find $\partial z/\partial x$ and $\partial z/\partial y$ when $x = y = 1$.

Solutions

(a) $z(1, 1) = 3e^1 - 2e^1 + 1 = 3.718$

(b) $\dfrac{\partial z}{\partial x} = 3e^x + 2xy^3$

$$\frac{\partial z}{\partial y} = -2e^y + 3x^2y^2$$

When $x = y = 1$, then

$$\frac{\partial z}{\partial x} = 3e + 2 = 10.15$$

$$\frac{\partial z}{\partial y} = -2e + 3 = -2.44$$

At the point $(1, 1, 3.718)$ on the surface, the height of the surface above the x–y plane is increasing in the x direction, and decreasing in the y direction. Note that we could also write:

$$\left.\frac{\partial z}{\partial x}\right|_{(1, 1)} = 10.15 \quad \text{and} \quad \frac{\partial z}{\partial y}(1, 1) = -2.44$$

both sorts of notation being in common use.   ▲

Example 14.5 Eddy current losses

Eddy currents are circulating currents that arise in iron cores of electrical equipment as a result of an alternating current magnetic field. They lead to energy losses given by:

$$P_e = k_e f^2 B_{max}^2$$

where

P_e = eddy current losses (W per unit mass)

B_{max} = maximum value of the magnetic field wave (T),

f = frequency of the magnetic field wave (Hz),

k_e = a constant that depends upon factors such as the lamination thickness of the iron core.

Calculate $\partial P_e/\partial f$, and $\partial P_e/\partial B_{max}$.

Solution

$$\frac{\partial P_e}{\partial f} = 2k_e f B_{max}^2 \quad \text{and} \quad \frac{\partial P_e}{\partial B_{max}} = 2k_e f^2 B_{max} \quad \blacktriangle$$

Example 14.6

If $V(x, y) = \sin(xy)$, find $\partial V/\partial x$ and $\partial V/\partial y$.

Solution

To find $\partial V/\partial x$ we treat y as a constant. Recalling that $\dfrac{d}{dx}(\sin kx) = k \cos kx$

we find $\partial V/\partial x = y \cos(xy)$. To find $\partial V/\partial y$ we treat x as a constant. Thus $\partial V/\partial y = x \cos(xy)$. $\blacktriangle$

Example 14.7

Find the first partial derivatives of z where

(a) $z(x, y) = yxe^x$
(b) $z(x, y) = x^2 \sin(xy)$

Solutions

(a) To find $\partial z/\partial x$ we must treat y as a constant. However, differentiation of the factor xe^x will require use of the product rule. We find:

$$\frac{\partial z}{\partial x} = y\frac{\partial}{\partial x}(xe^x)$$
$$= y((1)e^x + xe^x)$$
$$= ye^x(x + 1)$$

To find $\partial z/\partial y$ the variable x is held constant and so

$$\frac{\partial z}{\partial y} = xe^x$$

(b) Observe the product term in the variable x which means we shall need to use the product rule. We find:

$$\frac{\partial z}{\partial x} = 2x \sin(xy) + x^2(y \cos(xy)) = 2x \sin(xy) + x^2 y \cos(xy)$$

To find $\partial z/\partial y$ we treat x as a constant and find

$$\frac{\partial z}{\partial y} = x^2(x \cos(xy)) = x^3 \cos(xy) \quad \blacktriangle$$

EXERCISES 14.1

1. Find the first partial derivatives of z where

   (a) $z = 2x + 3y - xy$

   (b) $z = x/y + x^2$

   (c) $z = \sqrt{x^2 + y^2}$

   (d) $z = 2x^2\sqrt{y} + 3\cos(xy)$

2. Find the first partial derivatives of y where

   (a) $y = 3\sin(2x + 3t)$

   (b) $y = 4\cos(4x - 6t)$

   (c) $y = 6x^3\sqrt{z} + 4\sqrt{x}\,z^4$

3. Given

   $$f(r, h) = 2r^2h - \sqrt{r}\,h$$

   evaluate the first partial derivatives of f when $r = 2$, $h = 1$.

14.4 Higher order derivatives

Just as functions of one variable have second and higher derivatives, so do functions of several variables. Consider

$$z = z(x, y)$$

The first partial derivatives of z are $\partial z/\partial x$ and $\partial z/\partial y$. The second partial derivatives are found by differentiating the first derivatives. We can differentiate either first partial derivative w.r.t. x or y to obtain various second partial derivatives:

differentiating $\dfrac{\partial z}{\partial x}$ w.r.t. x produces $\dfrac{\partial}{\partial x}\left(\dfrac{\partial z}{\partial x}\right) = \dfrac{\partial^2 z}{\partial x^2}$

differentiating $\dfrac{\partial z}{\partial x}$ w.r.t. y produces $\dfrac{\partial}{\partial y}\left(\dfrac{\partial z}{\partial x}\right) = \dfrac{\partial^2 z}{\partial y \partial x}$

differentiating $\dfrac{\partial z}{\partial y}$ w.r.t. x produces $\dfrac{\partial}{\partial x}\left(\dfrac{\partial z}{\partial y}\right) = \dfrac{\partial^2 z}{\partial x \partial y}$

differentiating $\dfrac{\partial z}{\partial y}$ w.r.t. y produces $\dfrac{\partial}{\partial y}\left(\dfrac{\partial z}{\partial y}\right) = \dfrac{\partial^2 z}{\partial y^2}$

For most common functions, the mixed derivatives, $\dfrac{\partial^2 z}{\partial y \partial x}$ and $\dfrac{\partial^2 z}{\partial x \partial y}$ are equal.

Example 14.8

Given

$$z(x, y) = 3xy^3 - 2xy + \sin x$$

find all second partial derivatives of z.

Solution

$$\frac{\partial z}{\partial x} = 3y^3 - 2y + \cos x \qquad \frac{\partial z}{\partial y} = 9xy^2 - 2x$$

$$\frac{\partial^2 z}{\partial x^2} = \frac{\partial}{\partial x}\left(\frac{\partial z}{\partial x}\right) = -\sin x \qquad \frac{\partial^2 z}{\partial y^2} = \frac{\partial}{\partial y}\left(\frac{\partial z}{\partial y}\right) = 18xy$$

$$\frac{\partial^2 z}{\partial x \partial y} = \frac{\partial}{\partial x}\left(\frac{\partial z}{\partial y}\right) = 9y^2 - 2 \qquad \frac{\partial^2 z}{\partial y \partial x} = \frac{\partial}{\partial y}\left(\frac{\partial z}{\partial x}\right) = 9y^2 - 2$$

Note that $\dfrac{\partial^2 z}{\partial x \partial y} = \dfrac{\partial^2 z}{\partial y \partial x}$.

Example 14.9

Given

$$H(x, t) = 3x^2 + t^2 + e^{xt}$$

verify that $\dfrac{\partial^2 H}{\partial x \partial t} = \dfrac{\partial^2 H}{\partial t \partial x}$.

Solution

$$\frac{\partial H}{\partial x} = 6x + te^{xt} \qquad \frac{\partial H}{\partial t} = 2t + xe^{xt}$$

$$\frac{\partial^2 H}{\partial t \partial x} = \frac{\partial}{\partial t}\left(\frac{\partial H}{\partial x}\right) = \frac{\partial}{\partial t}(6x + te^{xt}) = e^{xt} + txe^{xt}$$

$$\frac{\partial^2 H}{\partial x \partial t} = \frac{\partial}{\partial x}\left(\frac{\partial H}{\partial t}\right) = \frac{\partial}{\partial x}(2t + xe^{xt}) = e^{xt} + xte^{xt} \quad \blacktriangle$$

Third, fourth and higher derivatives are found in a similar way to finding second derivatives. The third derivatives are found by differentiating the second derivatives and so on.

Example 14.10

Find all third derivatives of $f(r, s) = \sin(2r) - 3r^4 s^2$.

Solution

$$\frac{\partial f}{\partial r} = 2\cos(2r) - 12r^3 s^2 \qquad \frac{\partial f}{\partial s} = -6r^4 s$$

$$\frac{\partial^2 f}{\partial r^2} = -4\sin(2r) - 36r^2 s^2 \qquad \frac{\partial^2 f}{\partial s^2} = -6r^4$$

$$\frac{\partial^2 f}{\partial r \partial s} = -24r^3 s$$

The third derivatives are $\dfrac{\partial^3 f}{\partial r^3}$, $\dfrac{\partial^3 f}{\partial r^2 \partial s}$, $\dfrac{\partial^3 f}{\partial r \partial s^2}$ and $\dfrac{\partial^3 f}{\partial s^3}$, and these are found by differentiating the second derivatives.

$$\frac{\partial^3 f}{\partial r^3} = \frac{\partial}{\partial r}\left(\frac{\partial^2 f}{\partial r^2}\right) = -8\cos(2r) - 72rs^2$$

$$\frac{\partial^3 f}{\partial r^2 \partial s} = \frac{\partial}{\partial r}\left(\frac{\partial^2 f}{\partial r \partial s}\right) = -72r^2 s$$

$$\frac{\partial^3 f}{\partial r \partial s^2} = \frac{\partial}{\partial r}\left(\frac{\partial^2 f}{\partial s^2}\right) = -24r^3$$

$$\frac{\partial^3 f}{\partial s^3} = \frac{\partial}{\partial s}\left(\frac{\partial^2 f}{\partial s^2}\right) = 0$$

Note that the mixed derivatives can be calculated in a variety of ways.

$$\frac{\partial^3 f}{\partial r^2 \partial s} = \frac{\partial}{\partial r}\left(\frac{\partial^2 f}{\partial r \partial s}\right) = \frac{\partial}{\partial s}\left(\frac{\partial^2 f}{\partial r^2}\right)$$

$$\frac{\partial^3 f}{\partial r \partial s^2} = \frac{\partial}{\partial r}\left(\frac{\partial^2 f}{\partial s^2}\right) = \frac{\partial}{\partial s}\left(\frac{\partial^2 f}{\partial r \partial s}\right) \qquad \blacktriangle$$

EXERCISES 14.2

1. Find $\partial z/\partial x$ and $\partial z/\partial y$ where $z(x, y)$ is given by

   (a) $x + 2xy$

   (b) $x/y + y/x$

   (c) $x \sin y$

   (d) $3e^{x+y}$

   (e) e^{xy}

   (f) $\sin(2xy)$

2. Calculate all second derivatives of v where $v(h, r) = r^2\sqrt{h}$.

3. Find all first, second and third derivatives of z where $z(x, y) = x^2/(y + 1)$.

14.5 Partial differential equations

Partial differential equations (p.d.e.) occur in many areas of engineering. If a variable depends upon two or more independent variables, then it is likely this dependence will be described by a p.d.e. The independent variables are often time, t, and space coordinates x, y, z.

One example is the wave equation. The displacement, u, of the wave depends upon time and position. Under certain assumptions, the motion of a wave travelling in one direction is given by:

$$\frac{\partial^2 u}{\partial t^2} = c^2 \frac{\partial^2 u}{\partial x^2}$$

where c is the speed of the wave. Under prescribed conditions the subsequent motion of the wave can be calculated as a function of position and time.

Equally important is Laplace's equation. This equation is used extensively in electrostatics. Under certain conditions the electrostatic potential in a region is described by a function $\phi(x, y)$ which satisfies Laplace's equation,

$$\frac{\partial^2 \phi}{\partial x^2} + \frac{\partial^2 \phi}{\partial y^2} = 0$$

This equation is so important that a whole area of mathematics, called potential theory, is devoted to the study of its solutions.

The transmission equation is another important p.d.e. The potential, u, in a transmission cable with leakage satisfies a p.d.e. of the form:

$$\frac{\partial^2 u}{\partial x^2} = A\frac{\partial^2 u}{\partial t^2} + B\frac{\partial u}{\partial t} + Cu$$

where A, B and C are constants relating to the physical properties of the cable.

The analytical and numerical solution of partial differential equations is a large field of mathematics and is beyond the scope of this book.

14.6 Scalar and vector fields

Imagine a large room filled with air. At any point, P, we can measure the temperature, ϕ, say. The temperature will depend upon whereabouts in the room we take the measurement. Perhaps, close to a radiator the temperature will be higher than near to an open window. Clearly the temperature ϕ is a function of the position of the point. If we label the point by its Cartesian coordinates, (x, y, z), then ϕ will be a function of x, y and z, that is,

$$\phi = \phi(x, y, z)$$

Additionally, ϕ may be a function of time but for now we will leave this additional complication aside. Since temperature is a scalar what we have done is define a scalar at each point $P(x, y, z)$ in a region. This is an example of a **scalar field**. The electrostatic potential in a region is another important scalar field.

Alternatively, suppose we consider the motion of a large body of fluid. At each point, fluid will be moving with a certain speed in a certain direction, that is, each small fluid element has a particular velocity, $\mathbf{v}$, depending upon whereabouts in the fluid it is. Since velocity is a vector what we have done is define a vector at each point $P(x, y, z)$. We now have a vector function of x, y and z, known as a **vector field**. Let us write:

$$\mathbf{v} = (v_x, v_y, v_z)$$

so that v_x, v_y and v_z are the $\mathbf{i}$, $\mathbf{j}$ and $\mathbf{k}$ components respectively of $\mathbf{v}$, that is,

$$\mathbf{v} = v_x\mathbf{i} + v_y\mathbf{j} + v_z\mathbf{k}$$

We note that v_x, v_y and v_z will each be functions of x, y and z. An electromagnetic force field is another important example of a vector field since at each point in a region a vector is defined which gives the magnitude and direction of the local electromagnetic field.

In Section 14.3 we described how to partially differentiate a scalar function of several variables. Let us now consider how we can partially differentiate vectors.

14.7 Partial differentiation of vectors

Consider the vector function $\mathbf{v} = v_x\mathbf{i} + v_y\mathbf{j} + v_z\mathbf{k}$, where each component v_x, v_y and v_z is a function of x, y and z. We can partially differentiate the vector with respect to x as follows:

$$\frac{\partial \mathbf{v}}{\partial x} = \frac{\partial v_x}{\partial x}\mathbf{i} + \frac{\partial v_y}{\partial x}\mathbf{j} + \frac{\partial v_z}{\partial x}\mathbf{k}$$

Partial differentiation with respect to y and z is defined in a similar way as are higher derivatives.

Example 14.11

If $\mathbf{v} = 3x^2y\mathbf{i} + 2xyz\mathbf{j} - 3x^4y^2\mathbf{k}$, find $\dfrac{\partial \mathbf{v}}{\partial x}, \dfrac{\partial \mathbf{v}}{\partial y}, \dfrac{\partial \mathbf{v}}{\partial z}$. Further, find $\dfrac{\partial^2 \mathbf{v}}{\partial x^2}$ and $\dfrac{\partial^2 \mathbf{v}}{\partial x \partial z}$.

Solution

We find

$$\frac{\partial \mathbf{v}}{\partial x} = 6xy\mathbf{i} + 2yz\mathbf{j} - 12x^3y^2\mathbf{k} \qquad \frac{\partial \mathbf{v}}{\partial y} = 3x^2\mathbf{i} + 2xz\mathbf{j} - 6x^4y\mathbf{k}$$

$$\frac{\partial \mathbf{v}}{\partial z} = 2xy\mathbf{j} \qquad \frac{\partial^2 \mathbf{v}}{\partial x^2} = 6y\mathbf{i} - 36x^2y^2\mathbf{k}$$

$$\frac{\partial^2 \mathbf{v}}{\partial x \partial z} = 2y\mathbf{j} \quad \blacktriangle$$

EXERCISES 14.3

1. If $\mathbf{v} = 3xyz\mathbf{i} + (x^2 - y^2 + z^2)\mathbf{j} + (x + y^2)\mathbf{k}$

   find $\dfrac{\partial \mathbf{v}}{\partial x}, \dfrac{\partial \mathbf{v}}{\partial y}, \dfrac{\partial \mathbf{v}}{\partial z}, \dfrac{\partial^2 \mathbf{v}}{\partial x^2}, \dfrac{\partial^2 \mathbf{v}}{\partial y^2}$, and $\dfrac{\partial^2 \mathbf{v}}{\partial z^2}$.

2. If $\mathbf{v} = \sin(xyz)\mathbf{i} + ze^{xy}\mathbf{j} - 2xy\mathbf{k}$ find

   $\dfrac{\partial \mathbf{v}}{\partial x}, \dfrac{\partial \mathbf{v}}{\partial y}, \dfrac{\partial \mathbf{v}}{\partial z}$.

3. If $\mathbf{v} = x\mathbf{i} + x^2y\mathbf{j} - 3x^3\mathbf{k}$, and $\phi = xyz$ find

   $\phi\mathbf{v}, \dfrac{\partial}{\partial x}(\phi\mathbf{v}), \dfrac{\partial \phi}{\partial x}\dfrac{\partial \mathbf{v}}{\partial x}$. Deduce that

   $$\frac{\partial}{\partial x}(\phi\mathbf{v}) = \phi\frac{\partial \mathbf{v}}{\partial x} + \frac{\partial \phi}{\partial x}\mathbf{v}$$

4. If $\mathbf{v} = \ln(xy)\mathbf{i} + 2xy\cos z\mathbf{j} - x^4yz\mathbf{k}$, find

   $\dfrac{\partial \mathbf{v}}{\partial x}, \dfrac{\partial \mathbf{v}}{\partial y}, \dfrac{\partial \mathbf{v}}{\partial z}, \dfrac{\partial^2 \mathbf{v}}{\partial x^2}, \dfrac{\partial^2 \mathbf{v}}{\partial y^2}$, and $\dfrac{\partial^2 \mathbf{v}}{\partial z^2}$.

14.8 The gradient of a scalar

Given a scalar function of x, y, z:

$$\phi = \phi(x, y, z)$$

we can partially differentiate it with respect to each of its independent variables to find $\partial\phi/\partial x$, $\partial\phi/\partial y$ and $\partial\phi/\partial z$. If we do this, the vector:

$$\frac{\partial\phi}{\partial x}\mathbf{i} + \frac{\partial\phi}{\partial y}\mathbf{j} + \frac{\partial\phi}{\partial z}\mathbf{k}$$

turns out to be particularly important. We shall call this vector the **gradient** of ϕ and denote it by

$$\nabla\phi \qquad \text{or} \qquad \text{grad } \phi$$

An alternative form of writing $\nabla\phi$ is

$$\left(\frac{\partial\phi}{\partial x}, \frac{\partial\phi}{\partial y}, \frac{\partial\phi}{\partial z}\right)$$

■ $\quad$ grad $\phi = \nabla\phi = \dfrac{\partial\phi}{\partial x}\mathbf{i} + \dfrac{\partial\phi}{\partial y}\mathbf{j} + \dfrac{\partial\phi}{\partial z}\mathbf{k}$

The process of forming a gradient applies only to a scalar field and the result is always a vector field.

It is often useful to write $\nabla\phi$ in the form

$$\left(\frac{\partial}{\partial x}, \frac{\partial}{\partial y}, \frac{\partial}{\partial z}\right)\phi$$

where the 'vector-like' quantity in brackets is called a **vector operator** and is regarded as operating on the scalar ϕ. Thus the vector operator, ∇, is given by

$$\left(\frac{\partial}{\partial x}, \frac{\partial}{\partial y}, \frac{\partial}{\partial z}\right)$$

Example 14.12

If $\phi = \phi(x, y, z) = 4x^3y \sin z$, find $\nabla\phi$.

Solution

$$\phi(x, y, z) = 4x^3y \sin z$$

so that by partial differentation we obtain:

$$\frac{\partial\phi}{\partial x} = 12x^2y \sin z$$

$$\frac{\partial \phi}{\partial y} = 4x^3 \sin z$$

$$\frac{\partial \phi}{\partial z} = 4x^3 y \cos z$$

Therefore

$$\nabla\phi = 12x^2 y \sin z\mathbf{i} + 4x^3 \sin z\mathbf{j} + 4x^3 y \cos z\mathbf{k} \quad \blacktriangle$$

Example 14.13

If $\phi = x^3 y + xy^2 + 3y$ find $\nabla\phi$, $\nabla\phi$ at $(0, 0, 0)$ and $|\nabla\phi|$ at $(1, 1, 1)$.

Solution

If

$$\phi = x^3 y + xy^2 + 3y$$

then

$$\frac{\partial \phi}{\partial x} = 3x^2 y + y^2$$

$$\frac{\partial \phi}{\partial y} = x^3 + 2xy + 3$$

$$\frac{\partial \phi}{\partial z} = 0$$

so that

$$\nabla\phi = (3x^2 y + y^2)\mathbf{i} + (x^3 + 2xy + 3)\mathbf{j} + 0\mathbf{k}$$

At $(0, 0, 0)$ $\nabla\phi = 0\mathbf{i} + 3\mathbf{j} + 0\mathbf{k} = 3\mathbf{j}$. At $(1, 1, 1)$ $\nabla\phi = 4\mathbf{i} + 6\mathbf{j} + 0\mathbf{k}$ so that $|\nabla\phi|$ at $(1, 1, 1)$ equals $\sqrt{4^2 + 6^2} = \sqrt{52}$. $\quad \blacktriangle$

EXERCISES 14.4

1. If $\phi = x^2 - y^2 - 3xyz$, find $\nabla\phi$. Evaluate $\nabla\phi$ at the point $(0, 0, 0)$.

2. If $\phi = x^2 yz^3$ find $\nabla\phi$, $\nabla\phi$ at $(1, 2, 1)$ and $|\nabla\phi|$ at $(1, 2, 1)$.

3. If $\mathbf{v} = \nabla\phi$, find ϕ when $\mathbf{v} = (2x - 4y^2)\mathbf{i} - 8xy\mathbf{j}$.

14.9 Divergence of a vector

Given a vector field $\mathbf{v} = \mathbf{v}(x, y, z)$ let us consider what happens when we differentiate its individual components. If

$$\mathbf{v} = v_x\mathbf{i} + v_y\mathbf{j} + v_z\mathbf{k}$$

we can take each component in turn and partially differentiate it with respect to x, y and z, respectively, that is, we can evaluate

$$\frac{\partial v_x}{\partial x} \qquad \frac{\partial v_y}{\partial y} \qquad \frac{\partial v_z}{\partial z}$$

If we add the calculated quantities the result also turns out to be a very useful quantity known as the **divergence** of v, that is,

$$\text{divergence of } \mathbf{v} = \frac{\partial v_x}{\partial x} + \frac{\partial v_y}{\partial y} + \frac{\partial v_z}{\partial z}$$

This is usually abbreviated to div $\mathbf{v}$. Alternatively, the notation $\nabla \cdot \mathbf{v}$ is often used. If we use the vector operator notation introduced in the previous section we have

$$\nabla \cdot \mathbf{v} = \left(\frac{\partial}{\partial x}, \frac{\partial}{\partial y}, \frac{\partial}{\partial z} \right) \cdot \mathbf{v}$$

$$= \left(\frac{\partial}{\partial x}, \frac{\partial}{\partial y}, \frac{\partial}{\partial z} \right) \cdot (v_x, v_y, v_z)$$

Interpreting the $\cdot$ as a scalar product we find:

$$\nabla \cdot \mathbf{v} = \frac{\partial v_x}{\partial x} + \frac{\partial v_y}{\partial y} + \frac{\partial v_z}{\partial z}$$

as before, although this is not a scalar product in the usual sense because $(\partial/\partial x, \partial/\partial y, \partial/\partial z)$ is a vector operator. We note that the process of finding the divergence is always performed on a vector field and the result is always a scalar field.

■ $\quad$ div $\mathbf{v} = \nabla \cdot \mathbf{v} = \dfrac{\partial v_x}{\partial x} + \dfrac{\partial v_y}{\partial y} + \dfrac{\partial v_z}{\partial z}$

Example 14.14

If $\mathbf{v} = x^2 z \mathbf{i} + 2y^3 z^2 \mathbf{j} + xyz^2 \mathbf{k}$ find div $\mathbf{v}$.

Solution

Partially differentiating the first component of $\mathbf{v}$ with respect to x we find

$$\frac{\partial v_x}{\partial x} = 2xz$$

Similarly,

$$\frac{\partial v_y}{\partial y} = 6y^2 z^2 \quad \text{and} \quad \frac{\partial v_z}{\partial z} = 2xyz$$

Adding these results we find:

$$\text{div } \mathbf{v} = \nabla \cdot \mathbf{v} = 2xz + 6y^2 z^2 + 2xyz \quad \blacktriangle$$

■ A vector field whose divergence is zero for all values of x, y and z is said to be **solenoidal**

Example 14.15

Show that the vector field

$$\mathbf{v} = x \sin y \mathbf{i} + y \sin x \mathbf{j} - z(\sin x + \sin y)\mathbf{k}$$

is solenoidal.

Solution

$$\nabla \cdot \mathbf{v} = \sin y + \sin x - (\sin x + \sin y) = 0$$

and hence $\mathbf{v}$ is solenoidal.   ▲

EXERCISES 14.5

1. If $\mathbf{A} = 3yz\mathbf{i} + 2xy\mathbf{j} + xyz\mathbf{k}$ find $\nabla \cdot \mathbf{A}$.

2. Find the divergences of the following vector fields.

   (a) $\mathbf{v} = x^2\mathbf{i} + y^2\mathbf{j} + z^2\mathbf{k}$.

   (b) $\mathbf{v} = e^{xy}\mathbf{i} + 2z \sin(xy)\mathbf{j} + x^3z\mathbf{k}$.

3. If $\mathbf{E} = x\mathbf{i} + z^2\mathbf{j} - yz\mathbf{k}$ find

   (a) $\mathbf{E} \cdot \mathbf{i}$,

   (b) $\mathbf{E} \cdot \mathbf{j}$,

   (c) $\mathbf{E} \cdot \mathbf{k}$,

   (d) $\nabla \cdot \mathbf{E}$.

14.10 The curl of a vector

A third quantity of some importance is known as the **curl**. It is defined rather like a vector product.

■ curl $\mathbf{v} = \nabla \times \mathbf{v}$

$$= \left(\frac{\partial}{\partial x}, \frac{\partial}{\partial y}, \frac{\partial}{\partial z} \right) \times (v_x, v_y, v_z)$$

$$= \begin{vmatrix} \mathbf{i} & \mathbf{j} & \mathbf{k} \\ \dfrac{\partial}{\partial x} & \dfrac{\partial}{\partial y} & \dfrac{\partial}{\partial z} \\ v_x & v_y & v_z \end{vmatrix}$$

This determinant is evaluated in the usual way except that we must regard $\partial/\partial x$, $\partial/\partial y$ and $\partial/\partial z$ as operators not multipliers. Thus, for example,

$$\begin{vmatrix} \dfrac{\partial}{\partial x} & \dfrac{\partial}{\partial y} \\ v_x & v_y \end{vmatrix} \text{ means } \dfrac{\partial v_y}{\partial x} - \dfrac{\partial v_x}{\partial y}$$

It only makes sense to let the operator $\partial/\partial x$ precede v_y, so we do not write quantities like $v_y \, \partial/\partial x$. Explicitly we have

$$\text{curl } \mathbf{v} = \left(\frac{\partial v_z}{\partial y} - \frac{\partial v_y}{\partial z} \right) \mathbf{i} + \left(\frac{\partial v_x}{\partial z} - \frac{\partial v_z}{\partial x} \right) \mathbf{j} + \left(\frac{\partial v_y}{\partial x} - \frac{\partial v_x}{\partial y} \right) \mathbf{k}$$

Example 14.16

If $\mathbf{v} = x^2yz\mathbf{i} - 2xy\mathbf{j} + yz\mathbf{k}$ find $\nabla \times \mathbf{v}$.

Solution

$$\begin{aligned} \nabla \times \mathbf{v} &= \begin{vmatrix} \mathbf{i} & \mathbf{j} & \mathbf{k} \\ \dfrac{\partial}{\partial x} & \dfrac{\partial}{\partial y} & \dfrac{\partial}{\partial z} \\ x^2yz & -2xy & yz \end{vmatrix} \\ &= \left[\frac{\partial(yz)}{\partial y} - \frac{\partial(-2xy)}{\partial z} \right] \mathbf{i} - \left[\frac{\partial(yz)}{\partial x} - \frac{\partial(x^2yz)}{\partial z} \right] \mathbf{j} + \left[\frac{\partial(-2xy)}{\partial x} - \frac{\partial(x^2yz)}{\partial y} \right] \mathbf{k} \\ &= z\mathbf{i} + x^2y\mathbf{j} - (2y + x^2z)\mathbf{k} \quad \blacktriangle \end{aligned}$$

We note that the process of forming the curl is always performed on a vector field and the result is another vector field.

■ A vector field whose curl is zero for all values of x, y and z is said to be **irrotational**.

Example 14.17

Show that the vector field

$$\mathbf{F} = ye^{xy}\mathbf{i} + xe^{xy}\mathbf{j} + 0\mathbf{k}$$

is irrotational.

Solution

$$\nabla \times \mathbf{F} = \begin{vmatrix} \mathbf{i} & \mathbf{j} & \mathbf{k} \\ \dfrac{\partial}{\partial x} & \dfrac{\partial}{\partial y} & \dfrac{\partial}{\partial z} \\ ye^{xy} & xe^{xy} & 0 \end{vmatrix}$$

$$= \left(\frac{\partial}{\partial y} 0 - \frac{\partial}{\partial z} xe^{xy} \right) \mathbf{i} - \left(\frac{\partial}{\partial x} 0 - \frac{\partial}{\partial z} ye^{xy} \right) \mathbf{j} + \left(\frac{\partial}{\partial x} xe^{xy} - \frac{\partial}{\partial y} ye^{xy} \right) \mathbf{k}$$

$$= 0\mathbf{i} + 0\mathbf{j} + ((xye^{xy} + e^{xy}) - (yxe^{xy} + e^{xy}))\mathbf{k}$$

$$= \mathbf{0}$$

The field is therefore irrotational. ▲

It is important to be able to combine the three operators grad, div and curl in sensible ways. For instance, because the gradient of a scalar is a vector we can consider evaluating its divergence, that is,

$$\nabla \cdot (\nabla \phi) = \nabla \cdot \left(\frac{\partial \phi}{\partial x}, \frac{\partial \phi}{\partial y}, \frac{\partial \phi}{\partial z} \right)$$

$$= \left(\frac{\partial}{\partial x}, \frac{\partial}{\partial y}, \frac{\partial}{\partial z} \right) \cdot \left(\frac{\partial \phi}{\partial x}, \frac{\partial \phi}{\partial y}, \frac{\partial \phi}{\partial z} \right)$$

$$= \frac{\partial^2 \phi}{\partial x^2} + \frac{\partial^2 \phi}{\partial y^2} + \frac{\partial^2 \phi}{\partial z^2}$$

This last expression is very important and is often abbreviated to simply

$$\nabla^2 \phi$$

pronounced 'del-squared ϕ', and occurs in Laplace's equation $\nabla^2 \phi = 0$.

Example 14.18

If $\phi = 2x^2 - y^2 - z^2$, find $\nabla \phi$, $\nabla \cdot (\nabla \phi)$ and deduce that ϕ satisfies Laplace's equation.

Solution

$$\nabla \phi = 4x\mathbf{i} - 2y\mathbf{j} - 2z\mathbf{k}$$

$$\nabla \cdot (\nabla \phi) = 4 - 2 - 2 = 0$$

that is,

$$\nabla^2 \phi = 0$$

Hence ϕ satisfies Laplace's equation. ▲

Example 14.19

If $\phi(x, y, z)$ is an arbitrary differentiable scalar field, show that $\text{curl}(\text{grad}\phi) = \nabla \times (\nabla\phi)$ is always zero.

Solution

Given $\phi = \phi(x, y, z)$ we have, by definition,

$$\nabla\phi = \frac{\partial\phi}{\partial x}\mathbf{i} + \frac{\partial\phi}{\partial y}\mathbf{j} + \frac{\partial\phi}{\partial z}\mathbf{k}$$

Then

$$\text{curl}(\text{grad}\phi) = \nabla \times (\nabla\phi)$$

$$= \begin{vmatrix} \mathbf{i} & \mathbf{j} & \mathbf{k} \\ \dfrac{\partial}{\partial x} & \dfrac{\partial}{\partial y} & \dfrac{\partial}{\partial z} \\ \dfrac{\partial\phi}{\partial x} & \dfrac{\partial\phi}{\partial y} & \dfrac{\partial\phi}{\partial z} \end{vmatrix}$$

$$= \left(\frac{\partial}{\partial y}\left(\frac{\partial\phi}{\partial z}\right) - \frac{\partial}{\partial z}\left(\frac{\partial\phi}{\partial y}\right)\right)\mathbf{i} - \left(\frac{\partial}{\partial x}\left(\frac{\partial\phi}{\partial z}\right) - \frac{\partial}{\partial z}\left(\frac{\partial\phi}{\partial x}\right)\right)\mathbf{j}$$

$$+ \left(\frac{\partial}{\partial x}\left(\frac{\partial\phi}{\partial y}\right) - \frac{\partial}{\partial y}\left(\frac{\partial\phi}{\partial x}\right)\right)\mathbf{k}$$

Now, since $\dfrac{\partial}{\partial x}\left(\dfrac{\partial\phi}{\partial y}\right) = \dfrac{\partial}{\partial y}\left(\dfrac{\partial\phi}{\partial x}\right)$ and similar results hold for the other mixed partial derivatives, it follows that

$$\nabla \times (\nabla\phi) = \mathbf{0}$$

for any ϕ whatsoever. ▲

■ For an arbitrary differentiable scalar field ϕ

$$\nabla \times (\nabla\phi) = \mathbf{0}$$

14.10.1 *Maxwell's equations*

Vector calculus provides a useful mechanism for expressing the fundamental laws of electromagnetism in a concise manner. These laws can be summarized by means of four equations, known as Maxwell's equations. Much of electromagnetism is concerned with solving Maxwell's equations for different boundary conditions.

Equation 1

$$\text{div } \mathbf{D} = \rho$$

where $\mathbf{D}$ = electric flux density, and ρ = charge density. This equation is a general form of Gauss's theorem which states that the total electric flux flowing out of a closed surface is equal to the electric charge enclosed by that surface.

Equation 2

$$\text{div } \mathbf{B} = 0$$

where $\mathbf{B}$ is the magnetic flux density.

This equation arises from the observation that all magnetic poles occur in pairs and therefore magnetic field lines are continuous, that is, there are no free magnetic poles. In contrast, electric field lines originate on positive charges and terminate on negative charges and so a net positive charge in a region leads to an outflow of electric flux.

Equation 3

$$\text{curl } \mathbf{E} = -\frac{\partial \mathbf{B}}{\partial t}$$

where $\mathbf{E}$ is the electric field strength. This equation is a statement of Faraday's law. A time varying magnetic field produces a space varying electric field.

Equation 4

$$\text{curl } \mathbf{H} = \mathbf{J} + \frac{\partial \mathbf{D}}{\partial t}$$

where $\mathbf{H}$ is the magnetic field strength and $\mathbf{J}$ is the free current density. This equation states that a time varying electric field gives rise to a space varying magnetic field.

The derivation of these equations is beyond the scope of this text but can be found in many books on electromagnetism. The power of these equations lies in their generality. The conciseness with which the main laws of electromagnetism can be expressed is a tribute to the utility of vector calculus.

EXERCISES 14.6

1. Find the curl of the vector field
   $\mathbf{v} = x\mathbf{i} - 3xy\mathbf{j} + 4z\mathbf{k}$.

2. If $\mathbf{v} = 3x\mathbf{i} - 2y^2z\mathbf{j} + 3xyz\mathbf{k}$ find $\nabla \times \mathbf{v}$.

3. If $\mathbf{A}$ is an arbitrary differentiable vector field show that the divergence of the curl of $\mathbf{A}$ is always 0.

4. If $\phi = 2x^2y - xz^3$ show that $\nabla^2\phi = 4y - 6xz$.

5. If $\mathbf{v} = xy\mathbf{i} - yz\mathbf{j} + (y + 2z)\mathbf{k}$ find $\text{curl}(\text{curl}(\mathbf{v}))$.

6. If $\phi = xyz$ and $\mathbf{v} = 3x^2\mathbf{i} + 2y^3\mathbf{j} + xy\mathbf{k}$ find $\nabla\phi$, $\nabla \cdot \mathbf{v}$, and $\nabla \cdot (\phi\mathbf{v})$. Show that $\nabla \cdot (\phi\mathbf{v}) = (\nabla\phi) \cdot \mathbf{v} + \phi\nabla \cdot \mathbf{v}$.

Miscellaneous exercises

1. Find all the first and second partial derivatives of

   (a) $z = 3x^4 - 3y^3 + 4xy$

   (b) $z = e^{x^2 - y^2}$

   (c) $z = z(x, y, t) = e^{-t} \cos(xy)$

2. If $\phi = 1/\sqrt{x^2 + y^2 + z^2}$, show that

   $$\frac{\partial^2 \phi}{\partial x^2} + \frac{\partial^2 \phi}{\partial y^2} + \frac{\partial^2 \phi}{\partial z^2} = 0$$

3. Functions satisfying Laplace's equation are called **harmonic functions**. Show that the following functions are harmonic.

   (a) $z = x^4 - 6x^2y^2 + y^4$

   (b) $z = 4x^3y - 4xy^3$

4. If $\mathbf{A} = x\mathbf{i} + y\mathbf{j} + z\mathbf{k}$ and $\mathbf{B} = \cos x\mathbf{i} - \sin x\mathbf{j}$, find

   (a) $\mathbf{A} \times \mathbf{B}$

   (b) $\nabla \cdot (\mathbf{A} \times \mathbf{B})$

   (c) $\nabla \times \mathbf{A}$

   (d) $\nabla \times \mathbf{B}$

   Verify that
   $\nabla \cdot (\mathbf{A} \times \mathbf{B}) = \mathbf{B} \cdot (\nabla \times \mathbf{A}) - \mathbf{A} \cdot (\nabla \times \mathbf{B})$.

5. For arbitrary differentiable scalar fields ϕ and ψ show that $\nabla(\phi\psi) = \psi\nabla\phi + \phi\nabla\psi$.

6. If $\psi = x^2 y$ and $\mathbf{a} = x\mathbf{i} + y\mathbf{j} + z\mathbf{k}$, find $\nabla\psi$, $\nabla \times \mathbf{a}$, $\nabla \times (\psi\mathbf{a})$. Show that
   $\nabla \times (\psi\mathbf{a}) = \psi\nabla \times \mathbf{a} + (\nabla\psi) \times \mathbf{a}$.

7. A scalar field ϕ is a function of x, z and t only. Vectors $\mathbf{E}$ and $\mathbf{H}$ are defined by:

   $$\mathbf{E} = \frac{1}{\varepsilon}\left(\frac{\partial\phi}{\partial z}\mathbf{i} - \frac{\partial\phi}{\partial x}\mathbf{k}\right) \qquad \mathbf{H} = -\frac{\partial\phi}{\partial t}\mathbf{j}$$

   where ε is a constant.

   (a) Show that $\nabla \cdot \mathbf{E} = 0$

   (b) Show that $\nabla \cdot \mathbf{H} = 0$

   Given that $\nabla \times \mathbf{E} = -\mu\,\partial\mathbf{H}/\partial t$, where μ is a constant, show that ϕ satisfies the partial differential equation

   $$\frac{\partial^2\phi}{\partial x^2} + \frac{\partial^2\phi}{\partial z^2} = \mu\varepsilon\frac{\partial^2\phi}{\partial t^2}$$

Probability and statistics 15

15.1 Introduction

Probability theory is applicable to several areas of engineering. One example is reliability engineering which is concerned with analysing the likelihood that an engineering system will fail. For most systems calculating the exact time of failure is not feasible but it is often possible to obtain a good estimate of whether or not a system will fail in a certain time interval. This is useful information to have for any engineering system, but it is vital if the failure of the system results in the possibility of injury or loss of life. Examples include the failure of high voltage switchgear so that the casing becomes live, or electrical equipment producing a spark while being used underground in a mine.

Probability theory is also used extensively in production engineering; particularly in the field of quality control. No manufacturing process produces components of exactly the same quality each time. There is always some variation in quality and probability theory allows this variation to be quantified. This enables some predictability to be introduced into the activity of manufacturing and gives an engineer the confidence to say components of a certain quality can be supplied to a customer.

A final example is in the field of communication engineering. Communication channels are subject to noise which is random in nature and so is most successfully modelled using probability theory. We will examine some of these concepts in more detail in this chapter.

15.2 Introducing probability

Consider a machine which manufactures electronic components. These must meet a certain specification. The quality control department regularly samples the components. Suppose, on average, 92 out of 100 components meet the specification. Imagine that a component is selected at random and let A be the outcome that a component meets the specification; let B be the outcome that a component does not meet the specification. Then we say the **probability** of A occurring is $92/100 = 0.92$ and the probability of B occurring is $8/100 = 0.08$. The probability is thus a measure of the likelihood of the occurrence of a particular outcome. We write:

$P(A) =$ probability of A occurring $= 0.92$

$P(B) =$ probability of B occurring $= 0.08$

We note that the sum of the probabilities of all possible outcomes is 1. The process of selecting a component is called a **trial**. The possible outcomes are also called **events**. In this example there are only two possible events; A and B. We can depict this situation using a Venn diagram as shown in Figure 15.1. The set of all possible outcomes is called the **sample space** and is represented by the universal set $\mathbb{E}$. The set A represents the event that a component meets the specification. Clearly the complement of A, denoted $\bar{A}$, represents the event that a component fails to meet the specification, that is, B.

Let E be an event. To define $P(E)$ in a formal way consider a trial repeated a very large number of times, n. The number of times the event E occurs, m say, is counted. Then $P(E) = m/n$. As a consequence, $0 \leq P(E) \leq 1$, that is the probability of any event occurring is a number between 0 and 1.

■ $0 \leq P(E) \leq 1$

If $P(E) = 0$ then the event will never happen; E is impossible. If $P(E) = 1$ then E will definitely happen; E is a certainty. If $P(E) > 0.5$, then E is more likely to happen than not.

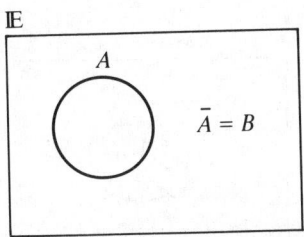

Figure 15.1 A: a component meets the specification. $\bar{A} = B$: a component fails to meet the specification.

Sometimes we know the probability of an event occurring from knowledge of the physical situation. For example, consider tossing a fair coin and let H be the event: the coin lands with the head facing up. Then clearly $P(H) = 1/2$. Similarly, if we roll a fair die and E is the event that a 6 is obtained, then $P(E) = 1/6$.

Suppose there are two possible outcomes, E_1 and E_2, of a trial. Then:

$$P(E_1) + P(E_2) = 1$$

If a trial has n possible outcomes, $E_1, E_2, \ldots, E_n$, then

$$P(E_1) + P(E_2) + \cdots + P(E_n) = 1$$

The sum of the probabilities of all possible events is 1, representing the total probability.

15.2.1 *Compound event*

A die is rolled and events E_1 and E_2 are

E_1: a 1, 2, 3 or 4 is obtained

E_2: an even score is obtained

Here the sample space is $\{1, 2, 3, 4, 5, 6\}$ because this is the set of possible outcomes. The event 'E_1 occurs and E_2 occurs' is called a **compound event**. This compound event occurs only when both E_1 and E_2 occur at the same time. In set theory notation we denote this by $E_1 \cap E_2$. Clearly $E_1 \cap E_2$ is the event of scoring a 2 or a 4. Hence $E_1 \cap E_2$ can occur in two out of the six possible ways. Therefore:

$$P(E_1 \cap E_2) = \tfrac{2}{6} = \tfrac{1}{3}$$

On a Venn diagram this intersection would be depicted as shown in Figure 15.2.

The event 'E_1 or E_2 occurs' is also called a compound event and is denoted $E_1 \cup E_2$. It occurs when a 1, 2, 3, 4 or a 6 is obtained. Hence $P(E_1 \cup E_2) = \tfrac{5}{6}$. On the Venn diagram this compound event is represented by the union of E_1 and E_2.

■ If events A and B both occur then this compound event is denoted $A \cap B$. If either event A or event B occurs then this compound event is denoted $A \cup B$.

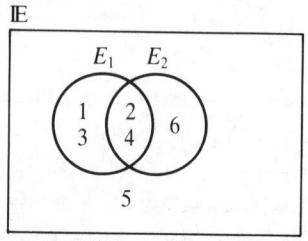

Figure 15.2 $E_1 \cap E_2$ corresponds to the compound event E_1 occurs and E_2 occurs.

Example 15.1 **Power supply to a computer**

Computers used in the control of life critical systems often have two separate power supplies. If one power supply fails then the other takes over. Let E_1 be the event that power supply 1 fails, and let E_2 be the event that power supply 2 fails. For the power supply to the computer to fail completely the compound event E_1 and E_2 must occur, that is, $E_1 \cap E_2$ occurs. ▲

EXERCISES 15.1

1. A fair die is rolled. The events $E_1, \ldots, E_5$ are defined as follows:

    E_1: an even number is obtained

    E_2: an odd number is obtained

    E_3: a score of less than 2 is obtained

    E_4: a 3 is obtained

    E_5: a score of more than 3 is obtained

    Find

    (a) $P(E_1)$, $P(E_2)$, $P(E_3)$, $P(E_4)$, $P(E_5)$

    (b) $P(E_1 \cap E_3)$

    (c) $P(E_2 \cap E_5)$

    (d) $P(E_2 \cap E_3)$

    (e) $P(E_3 \cap E_5)$

2. A trial can have three outcomes, E_1, E_2 and E_3. E_1 and E_2 are equally likely to occur. E_3 is three times more likely to occur than E_1. Find $P(E_1)$, $P(E_2)$ and $P(E_3)$.

3. A component is made by machines A and B. Machine A makes 70% of the components and machine B makes the rest. In all 90% of the components are acceptable, regardless of which machine made them. Find the probability that a component selected at random is:

    (a) unacceptable

    (b) acceptable and is made by machine A

    (c) unacceptable and is made by machine B.

15.3 Mutually exclusive events: the addition law of probability

Consider a machine which manufactures car components. Suppose each component falls into one of four categories:

    top quality

    standard

    substandard

    reject

After many samples have been taken and tested, it is found that under certain specific conditions the probability a component falls into a category is as shown in Table 15.1. The four categories cover all possibilities and so the probabilities must sum to 1. If 100 samples are taken, then on average 18 will be top quality, 65 of standard quality, 12 substandard and 5 will be rejected.

Table 15.1 The probability of a car component falling into one of four categories.

Category	Probability
top quality	0.18
standard	0.65
substandard	0.12
reject	0.05

Example 15.2

Using the data in Table 15.1 calculate the probability that a component selected at random is either standard or top quality.

Solution

On average 18 out of 100 components are top quality and 65 out of 100 are standard quality. So 83 out of 100 are either top quality or standard quality. Hence the probability a component is either top quality or standard quality is 0.83. The solution may be expressed more formally as follows. Let A be the event that a component is top quality. Let B be the event that a component is standard quality.

$$P(A) = 0.18 \qquad P(B) = 0.65$$

Then,

$$P(A \cup B) = 0.18 + 0.65 = 0.83$$

Note that in this example

$$P(A \cup B) = P(A) + P(B) \quad \blacktriangle$$

In Example 15.2 the events A and B could not possibly occur together. A component is either top quality or standard quality but cannot be both. We say A and B are **mutually exclusive** because the occurrence of one excludes the occurrence of the other. The result applies more generally.

■ If the occurrence of either of events E_i or E_j excludes the occurrence of the other, then E_i and E_j are said to be mutually exclusive events.

If E_i and E_j are mutually exclusive we denote this by:

$$E_i \cap E_j = \varnothing$$

where $\varnothing$ is the **impossible event**. On a Venn diagram E_i and E_j are shown as disjoint sets (see Figure 15.3). Suppose that $E_1, E_2, \ldots, E_n$ are n events and that

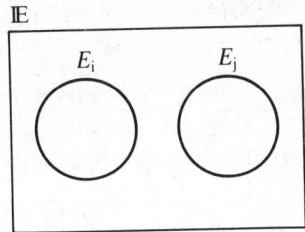

Figure 15.3 E_i and E_j are mutually exclusive events and so are depicted as disjoint sets.

in a single trial only one of these events can occur. The occurrence of any event, E_i, excludes the occurrence of all other events. Such events are mutually exclusive.

■ For mutually exclusive events the addition law of probability applies:

$$P(E_1 \text{ or } E_2 \text{ or} \ldots \text{or } E_n) = P(E_1 \cup E_2 \cup \cdots \cup E_n)$$
$$= P(E_1) + P(E_2) + \cdots + P(E_n)$$

Example 15.3

The lifespans of 5000 electrical components are measured to assess their reliability. The lifespan (L) is recorded and the results are shown in Table 15.2. Find the probability that a randomly selected component will last

(a) more than 3 years

(b) between 3 and 5 years

(c) less than 4 years

Table 15.2 The lifespans of 5000 electrical components.

Lifespan of component (yrs)	Number
$L > 5$	500
$4 < L \leq 5$	2250
$3 < L \leq 4$	1850
$L \leq 3$	400

Solutions

We define events A, B, C and D.

A: the component lasts more than 5 years

B: the component lasts between 4 and 5 years

C: the component lasts between 3 and 4 years

D: the component lasts 3 years or less

$P(A) = 500/5000 = 0.1$, $P(B) = 2250/5000 = 0.45$, $P(C) = 1850/5000 = 0.37$, $P(D) = 400/5000 = 0.08$. The events A, B, C and D are clearly mutually exclusive and so the addition law may be applied.

(a) $P(\text{component lasts more than 3 years}) = P(A \cup B \cup C)$

$$= P(A) + P(B) + P(C)$$

$$= 0.1 + 0.45 + 0.37$$

$$= 0.92$$

There is a 92% chance a component will last for more than 3 years.

(b) $P(\text{component lasts between 3 and 5 years}) = P(B \cup C)$

$$= P(B) + P(C)$$

$$= 0.45 + 0.37$$

$$= 0.82$$

(c) $P(\text{component lasts less than 4 years}) = P(C \cup D)$

$$= P(C) + P(D)$$

$$= 0.37 + 0.08$$

$$= 0.45 \quad \blacktriangle$$

EXERCISES 15.2

1. A component is classified as one of top quality, standard quality or substandard, with respective probabilities of 0.07, 0.85 and 0.08. Find the probability that a component is

(a) either top quality or standard quality

(b) not top quality.

2. The lifespan (L) of each of 2000 valves is measured and given in Table 15.3. Calculate the probability that the lifespan of a valve is

(a) more than 800 hours

(b) less than 600 hours

(c) between 400 and 800 hours

Table 15.3 The lifespans of 2000 valves.

Lifespan (hours)	Number
$L \geq 1000$	119
$800 \leq L < 1000$	520
$600 \leq L < 800$	931
$400 \leq L < 600$	230
$L < 400$	200

15.4 Complementary events

Consider an electronic circuit. Clearly, either the circuit works correctly or it does not work correctly. Let A be the event that the circuit works correctly, and let B be the event that the circuit does not work correctly. Either A or B must happen, and one excludes the other. The events A and B are said to be **complementary**. The Venn diagram corresponding to this situation is identical to that in Figure 15.1 where the set A corresponds to the event A, and the event B is represented by $\bar{A}$.

■ Two events, A and B, are complementary if they are mutually exclusive and in a single trial either A or B must happen.

Hence, if A and B are complementary then:

$$P(A) + P(B) = 1$$

and

$$P(A \cap B) = 0$$

It is usual to denote complementary events as A and $\bar{A}$. For example,

A: the component is top quality
$\bar{A}$: the component is not top quality
B: $n > 6$
$\bar{B}$: $n \leq 6$
C: the circuit has failed to meet the specification
$\bar{C}$: the circuit has met the specification

Recall from Example 15.3 that D was the event: the component lasts three years or less, and this had probability $P(D) = 0.08$. We could say:

D: the component lasts three years or less
$\bar{D}$: the component lasts more than three years

D and $\bar{D}$ are complementary events and so:

$$P(D) + P(\bar{D}) = 1$$
$$P(\bar{D}) = 1 - P(D) = 1 - 0.08 = 0.92$$

Example 15.4

Three transistors are tested. The probability that none of them works is 0.03. What is the probability that at least one transistor works?

Solution

We define the events A and $\bar{A}$:

A: all three transistors fail

$\bar{A}$: not all three transistors fail.

A and $\bar{A}$ are complementary events and so:

$$P(A) = 0.03 \qquad P(\bar{A}) = 1 - P(A) = 0.97$$

To say that not all three transistors fail is the same as saying that one or more transistors work. So we could say:

$\bar{A}$: at least one transistor works

Hence the probability that at least one transistor works is 0.97.   ▲

EXERCISES 15.3

1. Which pairs of events are complementary?

    A: the component is reliable

    B: there is only one component

    C: there are less than two components

    D: more than two components are reliable

    E: the component is unreliable

    F: there is more than one component

    G: most of the components are unreliable

2. Events A, $\bar{B}$, C and D are defined.

    A: the lifespan is 90 days or less

    $\bar{B}$: the machine is reliable

    C: all components have been tested

    D: at least three components from the sample are unreliable

State the events $\bar{A}$, B, $\bar{C}$ and $\bar{D}$.

15.5 Concepts from communication theory

Communication engineers find it useful to quantify information for the purposes of analysis. In order to do so a very restricted view of information is used. Information is seen in terms of knowledge of an event occurring. A highly improbable event occurring constitutes more information than an almost certain event occurring. This correlates to some extent with human experience as people tend to be much more interested in hearing about unlikely events. The **information**, I, associated with an event, is defined by

$$I = -\log p \qquad 0 < p \le 1$$

where p = probability of an event occurring. Notice that $p = 0$ is excluded from the domain of the function as the logarithm is not defined at 0. In practice this is not a problem because an event with zero probability never occurs. Often logarithms to the base 2 are used when calculating information as in many cases information arrives in the form of a 'bit stream' consisting of a series of binary numbers. For this case I has units of bits and is given by

$$I = -\log_2 p$$

A formula for evaluating logarithms to the base 2 is given in Chapter 1.

Example 15.5

Suppose that a computer generates a binary stream of data and that 1s and 0s occur with equal probability, that is, $P(0) = P(1) = 0.5$. Calculate the information per binary digit generated.

Solution

Here, $p = 0.5$ whether the binary digit is 0 or 1, so

$$I = -\log_2(0.5) = \frac{-\log_{10}(0.5)}{\log_{10} 2} = 1 \text{ bit} \quad \blacktriangle$$

Example 15.6

Suppose that a system generates a stream of upper case alphabetic characters and that the probability of a character occurring is the same for all characters. Calculate

(a) the information associated with the character G occurring

(b) the information associated with any single character.

Solutions

(a) $P(\text{G occurring}) = 1/26$,
$$I = -\log_2(1/26) = -\log_{10}(1/26)/\log_{10} 2 = 4.70 \text{ bits}$$

(b) All characters are equally likely to occur and so:

$$I = -\log_2\left(\frac{1}{26}\right) = \frac{-\log_{10}(1/26)}{\log_{10} 2} = 4.70 \text{ bits} \quad \blacktriangle$$

Often a series of events may occur that do not have the same probability. For example, if a stream of alphabetic characters is being generated then it is likely that some characters will occur more frequently than others and so have a higher probability associated with them. For this situation it becomes convenient to introduce the concept of average information. Given a source producing a set of events:

$$E_1, E_2, E_3, \ldots, E_n$$

with probabilities

$$p_1, p_2, p_3, \ldots, p_n$$

then for a long series of events the **average information** per event is given by:

$$H = -\sum_{i=1}^{i=n} p_i \log_2 p_i \text{ bits}$$

H is also termed the **entropy**.

Example 15.7

A source produces messages consisting of three characters, A, B and C. The probabilities of each of these characters occurring is $P(A) = 0.2$, $P(B) = 0.5$, $P(C) = 0.3$. Calculate the entropy of the signal.

Solution

$$H = -0.2 \log_2(0.2) - 0.5 \log_2(0.5) - 0.3 \log_2(0.3) = 1.49 \text{ bits} \quad \blacktriangle$$

Example 15.8

A source generates binary digits 0, 1, with probabilities $P(0) = 0.3$ and $P(1) = 0.7$. Calculate the entropy of the signal.

Solution

$$H = -0.3 \log_2(0.3) - 0.7 \log_2(0.7) = 0.881 \text{ bits} \quad \blacktriangle$$

Note that in Example 15.8, on average, each binary digit only carries 0.881 bits of information. In fact the maximum average amount of information that can be carried by a binary digit occurs when $P(0) = P(1) = 0.5$, as seen in Example 15.5. For this case $H = 1$. When the probabilities are not the same then one way of viewing the reduction in H is to think of the likely event being given too much of the signalling time given its lower information content. It is interesting to explore the two limiting cases, that is, (a) $P(0) = 0$, $P(1) = 1$; (b) $P(0) = 1$, $P(1) = 0$. In both cases it can be shown that $H = 0$. However, on examination this is reasonable because a continuous stream of 1s does not relay any useful information to the recipient and neither does a continuous stream of 0s.

The fact that some streams of symbols do not contain as much information as other streams of the same symbols leads to the concept of **redundancy**. This allows the efficiency with which information is being sent to be quantified and is defined as:

$$\text{redundancy} = \frac{\text{maximum entropy} - \text{actual entropy}}{\text{maximum entropy}}$$

A low value of redundancy corresponds to efficient transmission of information.

Example 15.9

Consider the source of binary digits examined in Examples 15.5 and 15.8. The maximum entropy for a binary stream is 1 bit per binary digit. Calculate the redundancy in each case.

Solution

For Example 15.5

$$\text{redundancy} = \frac{1 - 1}{1} = 0$$

For Example 15.8

$$\text{redundancy} = \frac{1 - 0.881}{1} = 0.119 \quad \blacktriangle$$

Example 15.10

A stream of data consists of four characters A, B, C, D with probabilities 0.1, 0.3, 0.2, 0.4, respectively. Calculate the redundancy.

Solution

It can be shown that the maximum entropy, H_{max}, corresponds to the situation in which the probability of each symbol is the same, that is 0.25.

$$H_{max} = 4 \times (-0.25 \log_2(0.25)) = 2 \text{ bits}$$

The actual entropy, H_{act}, is given by

$$H_{act} = -(0.1 \log_2(0.1) + 0.3 \log_2(0.3) + 0.2 \log_2(0.2) + 0.4 \log_2(0.4))$$

$$= 1.846 \text{ bits}$$

$$\text{redundancy} = \frac{2 - 1.846}{2} = 0.0770 \quad \blacktriangle$$

In the examples we have examined so far we have used the bit as the unit of information because the most common form of digital signalling uses binary digits. When there are only two possible events it is possible to represent an event by a single binary digit. However, if there is a larger number of possible events then several binary digits are needed to represent a single event. When calculating values for information and entropy in these examples an assumption was made that each event was represented by binary sequences or **codes** of the same length. It is only possible to do this efficiently if the number of events is a power of two, that is, 2, 4, 8, 16,.... In practice this problem does not arise because it is more common to produce codes that have a small number of binary digits for likely events and a long number of binary digits for unlikely events. This allows the redundancy of a data stream to be reduced. The design of such codes is known as **coding theory**. One complication is that most streams of data are not transmitted with 100% accuracy as a result of the presence of noise within the communication channel. It is often necessary to build extra redundancy into a code in order to recover these errors.

EXERCISES 15.4

1. A source generates six characters, A, B, C, D, E, F with respective probabilities 0.05, 0.1, 0.25, 0.3, 0.15, 0.15. Calculate the average information per character and the redundancy.

2. A visual display unit has a resolution of 600 rows by 800 columns. Ten different grey levels are associated with each pixel and their probabilities are 0.05, 0.07, 0.09, 0.10, 0.11, 0.13, 0.12, 0.12, 0.11, 0.10. Calculate the average information content in each picture frame.

15.6 Conditional probability: the multiplication law

Suppose two machines, M and N, both manufacture components. Of the components made by machine M, 92% are of an acceptable standard and 8% are rejected. For machine N, only 80% are of an acceptable standard and 20% are rejected. Consider now the event E:

E: a component is of an acceptable standard

If all the components are manufactured by machine M then $P(E) = 0.92$. However, if all the components are manufactured by machine N then $P(E) = 0.8$. If half the components are manufactured by machine M and half by machine N then $P(E) = 0.86$. To see why this is so consider 1000 components. Of the half made by machine M, $92\% \times 500 = 460$ will be of an acceptable standard. Of the half made by machine N, $80\% \times 500 = 400$ will be acceptable. Hence 860 of the 1000 components will be acceptable and so $P(E) = 860/1000 = 0.86$. Clearly, there are distinct probabilities of the same event; the probability changes as the conditions change. This is intuitive and leads to the idea of **conditional probability**.

We introduce a notation for conditional probability. Define events A and B by

A: the component is manufactured by machine M

B: the component is manufactured by machine N

Then the probability a component is of an acceptable standard, given it is manufactured by machine M is written as $P(E|A)$. We read this as the conditional probability of E given A. Similarly $P(E|B)$ is the probability of E happening, given B has already happened.

$$P(E|A) = 0.92 \qquad P(E|B) = 0.8$$

To be pedantic, all probabilities are conditional since the conditions surrounding any event can change. However, for many situations there is tacitly assumed a definite set of conditions which is always satisfied. The probability of an event calculated under only these conditions is known as the **unconditional probability**. If further well defined conditions are attached, the probability is conditional.

Example 15.11

Machines M and N manufacture a component. The probability the component is of an acceptable standard is 0.95 when manufactured by machine M and 0.83 when manufactured by machine N. Machine M supplies 65% of components; machine N supplies 35%. A component is picked at random.

(a) What is the probability the component is of an acceptable standard?

(b) What is the probability that a component is of an acceptable standard and is made by machine M?

(c) What is the probability the component is of an acceptable standard given it is made by machine M?

(d) What is the probability the component was made by machine M?

(e) What is the probability the component was made by machine M given it is of an acceptable standard?

(f) The component is not of an acceptable standard. What is the probability it was made by machine N?

Solutions

We define the events:

A: the component is manufactured by machine M

B: the component is manufactured by machine N

C: the component is of an acceptable standard

(a) Consider 1000 components. Then 650 are manufactured by machine M, 350 by machine N. Of the 650 manufactured by machine M, 95% will be acceptable, that is, $650 \times 95/100 = 617.5$. Of the 350 manufactured by machine N, 83% will be acceptable, that is, $350 \times 83/100 = 290.5$. So on average in 1000 components, $617.5 + 290.5 = 908$ will be acceptable, that is, $P(C) = 0.908$.

(b) From part (a) we know that out of 1000 components 617.5 will be made by machine M and be of an acceptable standard. Hence
$P(A \cap C) = 617.5/1000 = 0.6175$.

(c) We require $P(C|A)$. The probability a component is acceptable given it is manufactured by machine M is

$$P(C|A) = 0.95$$

(d) Machine M makes 65% of components, that is, $P(A) = 0.65$.

(e) We require $P(A|C)$. Consider again the 1000 components. On average 908 are acceptable. Of these 908 acceptable components, 617.5 are manufactured by machine M and 290.5 by machine N. We are told the component is acceptable and so we must restrict attention to the 908 acceptable components. So, out of 908 acceptable components, 617.5 are made by machine M, that is, $P(A|C) = 617.5/908 = 0.68$.

(f) We require $P(B|\bar{C})$. Consider 1000 components. Machine M manufactures 650 components of which 617.5 are acceptable and hence 32.5 are unacceptable. Machine M manufactures 350 components of which 290.5 are acceptable and 59.5 are unacceptable. There are 92 unacceptable components of which 59.5 were made by machine N.

Probability of the component being made by machine N given it is unacceptable is

$$P(B|\bar{C}) = \frac{59.5}{92} = 0.647$$

that is, almost 65% of unacceptable components are manufactured by machine N. ▲

15.6.1 *The multiplication law*

Consider events A and B for which $A \cap B \neq \varnothing$ as shown in Figure 15.4. Suppose we know that event A has occurred and we seek the probability that B occurs, that is, $P(B|A)$. Knowing event A has occurred we can restrict our attention to the set A. Event B will occur if any outcome is in $A \cap B$. Hence,

$$P(B|A) = \frac{P(A \cap B)}{P(A)}$$

■ The multiplication law of probability states:

$P(A \cap B) = P(A)P(B|A)$

Since the compound event 'A and B' is identical to 'B and A' we may also say:

$P(A \cap B) = P(B \cap A) = P(B)P(A|B)$

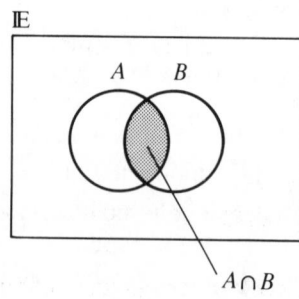

Figure 15.4   *A and B are not mutually exclusive.*

Consider Example 15.11(e). We require the probability that the component was manufactured by machine M, given it is acceptable. This is $P(A|C)$. Now

$$P(A \cap C) = P(C \cap A) = P(C)P(A|C)$$

$$P(A|C) = \frac{P(A \cap C)}{P(C)}$$

Now $P(A \cap C)$ is the probability the component is manufactured by machine M and is acceptable. This is known to be 0.6175, using Example 15.11(b). Also from Example 15.11(a) we see $P(C) = 0.908$. Hence,

$$P(A|C) = \frac{0.6175}{0.908} = 0.68$$

Example 15.12

A manufacturer studies the reliability of a certain component so that suitable guarantees can be given: 83% of components remain reliable for at least 5 years; 92% remain reliable for at least 3 years. What is the probability that a component which has remained reliable for 3 years will remain reliable for 5 years?

Solution

We define the events:

    A: a component remains reliable for at least 3 years

    B: a component remains reliable for at least 5 years

Then $P(A) = 0.92$, $P(B) = 0.83$. Note that these are unconditional probabilities. We require $P(B|A)$; a conditional probability.

$$P(A \cap B) = P(A)P(B|A)$$

$A \cap B$ is the compound event: a component remains reliable for at least 3 years and it remains reliable for at least 5 years. Clearly this is the same as the event B. So:

$$P(A \cap B) = P(B)$$

Hence

$$P(B) = P(A)P(B|A)$$

$$P(B|A) = \frac{P(B)}{P(A)} = \frac{0.83}{0.92} = 0.90$$

that is, 90% of the components which remain reliable for 3 years will remain reliable for at least 5 years. ▲

EXERCISES 15.5

1. A component is manufactured by machines 1 and 2. Machine 1 manufactures 72% of total production of the component. The percentage of components which are acceptable varies, depending upon which machine is used. For machine 1, 97% of components are acceptable and for machine 2, 92% are acceptable. A component is picked at random.

   (a) What is the probability it was manufactured by machine 1?

   (b) What is the probability it is not acceptable?

   (c) What is the probability that it is acceptable and made by machine 2?

   (d) If the component is acceptable what is the probability it was manufactured by machine 2?

   (e) If the component is not acceptable what is the probability it was manufactured by machine 1?

2. The lifespans (L) of 1500 components are measured and the information recorded in Table 15.4.

   (a) What is the probability that a component which is still working after 800 hours will last for at least 900 hours?

   (b) What is the probability that a component which is still working after 900 hours will continue to last for at least 1000 hours?

Table 15.4   The lifespans of 1500 components.

Lifespan (hours)	Number of components
$L \geq 1000$	210
$900 \leq L < 1000$	820
$800 \leq L < 900$	240
$700 \leq L < 800$	200
$L < 700$	30

15.7 Independent events

■ Two events are independent if the occurrence of either event does not influence the probability of the other event occurring.

Example 15.13

Machine 1 manufactures an electronic chip, A, of which 90% are acceptable. Machine 2 manufactures an electronic chip, B, of which 83% are acceptable. Two chips are picked at random; one of each kind. Find the probability that they are both acceptable.

Solution

The events E_1 and E_2 are defined.

E_1: chip A is acceptable

E_2: chip B is acceptable

$P(E_1) = 0.9 \qquad P(E_2) = 0.83$

A single trial consists of choosing two chips at random. We require the probability that the compound event, $E_1 \cap E_2$, is true. Using the multiplication law we have

$$P(E_1 \cap E_2) = P(E_1)P(E_2|E_1) = 0.9P(E_2|E_1)$$

$P(E_2|E_1)$ is the probability of E_2 happening given E_1 has happened. However, machine 1 and machine 2 are independent, so the probability of chip B being acceptable is in no way influenced by the acceptability of chip A. The events E_1 and E_2 are independent.

$$P(E_2|E_1) = P(E_2) = 0.83$$

Therefore

$$P(E_1 \cap E_2) = P(E_1)P(E_2) = (0.9)(0.83) = 0.75 \quad \blacktriangle$$

■ For independent events E_1 and E_2:

(1) $P(E_1|E_2) = P(E_1), \qquad P(E_2|E_1) = P(E_2)$
(2) $P(E_1 \cap E_2) = P(E_1)P(E_2)$

The concept of independence may be applied to more than two events. Three or more events are independent if every pair of events is independent. If $E_1, E_2, \ldots, E_n$ are n independent events then:

$$P(E_i|E_j) = P(E_i) \qquad \text{for any } i \text{ and } j, \, i \neq j$$

and

$$P(E_i \cap E_j) = P(E_i)P(E_j) \qquad i \neq j$$

A compound event may comprise several independent events; the multiplication law is extended in an obvious way.

$$P(E_1 \cap E_2 \cap E_3) = P(E_1)P(E_2)P(E_3)$$

$$P(E_1 \cap E_2 \cap E_3 \cap E_4) = P(E_1)P(E_2)P(E_3)P(E_4)$$

and so on.

Example 15.14

Machines 1, 2 and 3 manufacture resistors A, B and C, respectively. The probabilities of their respective acceptabilities are 0.9, 0.93 and 0.81. One of each resistor is selected at random.

(a) Find the probability that they are all acceptable.
(b) Find the probability that at least one resistor is acceptable.

Solutions

Define events E_1, E_2 and E_3 by:

E_1: resistor A is acceptable $P(E_1) = 0.9$

E_2: resistor B is acceptable $P(E_2) = 0.93$

E_3: resistor C is acceptable $P(E_3) = 0.81$

E_1, E_2 and E_3 are independent events.

(a) $P(E_1 \cap E_2 \cap E_3) = P(E_1)P(E_2)P(E_3) = (0.9)(0.93)(0.81) = 0.68$

(b) Let E_4 and E_5 be the events:

E_4: at least one resistor is acceptable

E_5: no resistor is acceptable

E_4 and E_5 are complementary events and so

$$P(E_4) + P(E_5) = 1$$

E_5 may be expressed as:

E_5: resistor A is not acceptable and resistor B is not acceptable and resistor C is not acceptable

that is, $\overline{E_1} \cap \overline{E_2} \cap \overline{E_3}$

$$P(E_5) = P(\overline{E_1} \cap \overline{E_2} \cap \overline{E_3}) = P(\overline{E_1})P(\overline{E_2})P(\overline{E_3})$$

$$= (1 - 0.9)(1 - 0.93)(1 - 0.81)$$

$$= (0.1)(0.07)(0.19)$$

$$= 0.001\,33$$

$$P(E_4) = 1 - P(E_5) = 1 - 0.001\,33 = 0.998\,67 \quad \blacktriangle$$

EXERCISES 15.6

1. *A* and *B* are two independent events with $P(A) = 0.7$ and $P(B) = 0.4$. The compound event: A occurs, then A occurs, then B occurs, is denoted *AAB*, and other compound events are denoted in a similar way. Calculate the probability of the following compound events.

   (a) *AAB*

   (b) *BAB*

   (c) *AAAA*

2. Capacitors are manufactured by four machines, 1, 2, 3 and 4. The probability a capacitor is manufactured acceptably varies according to the machine. The probabilities are 0.94, 0.91, 0.97 and 0.94, respectively, for machines 1, 2, 3 and 4.

   (a) A capacitor is taken from each machine. What is the probability all four capacitors are acceptable?

   (b) Two capacitors are taken from machine 1 and two from machine 2. What is the probability all four capacitors are acceptable?

   (c) A capacitor is taken from each

machine. Calculate the probability that at least three capacitors are acceptable.

(d) A capacitor is taken from each machine. From this sample of four capacitors, one is taken at random.

    (i) What is the probability it is acceptable and made by machine 1?

(ii) What is the probability it is acceptable and made by machine 2?

(e) A capacitor is taken from each machine. From this sample of four capacitors, one is taken at random. What is the probability it is acceptable? *Hint*: use the results in (d).

15.8 Random variables

We have introduced the idea of probability. Now we discuss random variables and how they can be described. Quantities whose variation contains an element of chance are called **random variables**. Some common examples are listed:

(a) the heights of women

(b) the weights of packs of butter sold in a shop

(c) the number of cars passing a given spot in one minute

(d) the shoe-sizes of men

(e) the length of time a machine works without failing.

Clearly all these quantities vary. In (a) and (b) the quantity varies continuously, that is, it can assume any value in some range. For example, a woman could have any height say between 4'0" and 7'0"; a pack of butter could have any weight between say 240 g and 260 g. These are examples of **continuous variables**. The variable itself will be recorded only to a certain accuracy, which depends upon the measuring device and the use to which the data will be put. For example, the heights of women could be measured to the nearest inch and recorded as 62", 67", 65", and so on. Although this data takes only integer values, the variable being measured is continuous.

    In (c) and (d) the variables can assume only a limited number of values. The number of cars passing a spot in one minute will be a non-negative integer 0, 1, 2, 3,.... The shoe-size of a man could be 4, 4.5, 5, 5.5,..., 14.5, 15. Variables such as these, which can assume only particular values are called **discrete variables**.

EXERCISES 15.7

1. Is the length of time a machine works without failing a continuous or a discrete variable?

2. State whether the following variables are continuous or discrete.

(a) monthly earnings of an employee

(b) the time taken to run a marathon

(c) the mark obtained in an examination

(d) the weight lost while dieting for three weeks

15.9 Probability distributions – discrete variable

The range of values that a variable can take does not give sufficient information about the variable. We need to know which values are likely to occur often and which values will occur only infrequently. For example, suppose x is a discrete random variable which can take values 0, 1, 2, 3, 4, 5 and 6. We may ask questions such as 'Which value is most likely to occur?', 'Is a 6 more likely to occur than a 5?', and so on. We need information on the probability of each value occurring. Suppose that information is provided and is given in Table 15.5. If x is sampled 100 times then on average 0 will occur 10 times, 1 will occur 10 times, 2 will occur 15 times and so on. Table 15.5 is called a **probability distribution** for the random variable x. Note that the probabilities sum to 1; the table tells us how the total probability is distributed among the various possible values of the random variable. Table 15.5 may be represented in graphical form (see Figure 15.5).

Table 15.5   The probability of a discrete value occurring.

x	0	1	2	3	4	5	6
$P(x)$	0.1	0.1	0.15	0.3	0.2	0.1	0.05

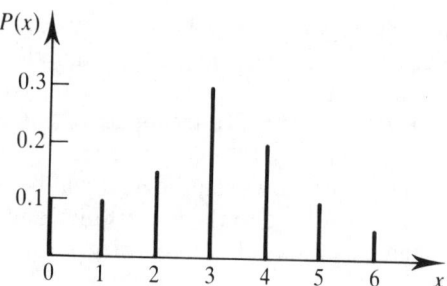

Figure 15.5   Plotted data of Table 15.5.

15.10 Probability density functions – continuous variable

Suppose x is a continuous random variable which can take any value on $[0, 1]$. It is impossible to list all possible values because of the continuous nature of the variable. There are infinitely many values on $[0, 1]$ so the probability of any one particular value occurring is zero. It is meaningful, however, to ask 'What is the probability of x falling in a sub-interval, $[a, b]$?'. Dividing $[0, 1]$ into sub-intervals and attaching probabilities to each sub-interval will result in a probability

Table 15.6 Probability that x lies in a given sub-interval.

x	$[0, 0.2)$	$[0.2, 0.4)$	$[0.4, 0.6)$	$[0.6, 0.8)$	$[0.8, 1.0]$
$P(x)$	0.1	0.25	0.35	0.2	0.1

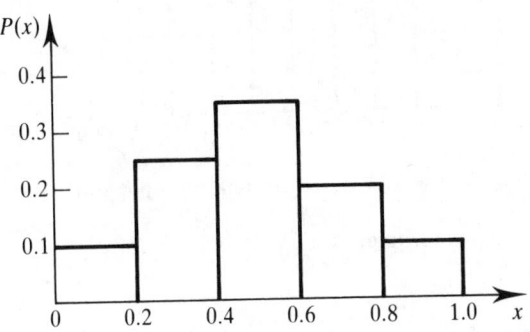

Figure 15.6 Plotted data of Table 15.6.

distribution. Table 15.6 gives an example. The probability that x will lie between 0.4 and 0.6 is 0.35, that is, $P(0.4 \leq x < 0.6) = 0.35$. Similarly,

$$P(0.2 \leq x < 0.4) = 0.25$$

Figure 15.6 shows the table in graphical form. By making the sub-intervals smaller a more refined distribution is obtained. Table 15.7 and Figure 15.7 illustrate this.

The probability that x lies in a particular interval is given by the sum of the heights of the rectangles on that interval. For example, the probability that x lies in $[0.5, 0.8]$ is $0.2 + 0.1 + 0.1 = 0.4$, that is, there is a probability of 0.4 that x lies somewhere between 0.5 and 0.8. Note that the sum of all the heights is 1, representing total probability.

Now consider a more general case in which we have n rectangles, each of width w, and heights $h_1, h_2, \ldots, h_n$. Noting that $h_1 + h_2 + \cdots + h_n = 1$, the total area is:

$$w \times (h_1 + h_2 + \cdots + h_n) = w$$

Consider some sub-interval $[a, b]$. We require the probability that x lies in this interval. Suppose the sum of the heights of the rectangles on this interval is H. Then the sum of the areas of the rectangles is wH.

Table 15.7 Refining the sub-intervals in Table 15.6.

x	$[0, 0.1)$	$[0.1, 0.2)$	$[0.2, 0.3)$	$[0.3, 0.4)$	$[0.4, 0.5)$
$P(x)$	0.03	0.07	0.1	0.15	0.15

x	$[0.5, 0.6)$	$[0.6, 0.7)$	$[0.7, 0.8)$	$[0.8, 0.9)$	$[0.9, 1.0]$
$P(x)$	0.2	0.1	0.1	0.07	0.03

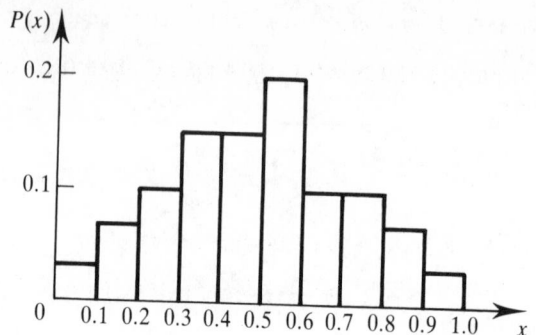

Figure 15.7   Plotted data of Table 15.7.

$P(a \leqq x \leqq b)$ = sum of heights of rectangles = H

We can write H as wH/w. Then

$$H = \frac{wH}{w}$$

$$= \frac{\text{sum of areas of rectangles on sub-interval } [a, b]}{\text{total area}}$$

The probability of x being in the sub-interval $[a, b]$ has been expressed in terms of area. If the probability varies continuously the probability distribution will be represented by a smooth curve, known as a **probability density function** (p.d.f.). This is denoted by $f(x)$. A p.d.f. is always scaled such that the total area under it is 1. This simplifies the calculation of probabilities. We consider the probability that $a \leqq x \leqq b$. Using the formula for probability in terms of area we see:

$$P(a \leqq x \leqq b) = \frac{\text{Area above } [a, b]}{\text{Total area}} = \frac{\text{Area above } [a, b]}{1}$$

Since $f(x)$ is a smooth curve, then:

■ $P(a \leqq x \leqq b)$ = Area above $[a, b]$ = $\displaystyle\int_a^b f(x) \, dx$

This is depicted in Figure 15.8.

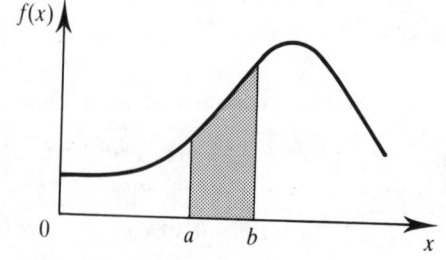

Figure 15.8   Shaded area represents $P(a \leqq x \leqq b)$.

Example 15.15

Suppose x is a continuous random variable taking any value on $[1, 4]$. Its p.d.f. $f(x)$, is given by:

$$f(x) = \frac{1}{2\sqrt{x}} \qquad 1 \le x \le 4$$

(a) Check that $f(x)$ is a suitable function for a p.d.f.
(b) What is the probability that (i) x lies on $[2, 3.5]$, (ii) $x \ge 2$, (iii) $x < 3$?

Solutions

(a) x can have any value on $[1, 4]$. For $f(x)$ to be a p.d.f., then the total area under it should equal 1, that is,

$$\int_1^4 f(x)\, dx = 1$$

$$\int_1^4 f(x)\, dx = \int_1^4 \frac{1}{2\sqrt{x}}\, dx = \left[\sqrt{x} \right]_1^4 = 1$$

Hence $f(x)$ is a suitable function for a p.d.f.

(b) (i) $\quad P(2 \le x \le 3.5) = \int_2^{3.5} f(x)\, dx = \left[\sqrt{x} \right]_2^{3.5} = 0.457$

(ii) $\quad P(x \ge 2) = \int_2^4 f(x)\, dx = [\sqrt{x}]_2^4 = 0.586$
(iii) $\quad P(x < 3) = \int_1^3 f(x)\, dx = [\sqrt{x}]_1^3 = 0.732$ ▲

Example 15.16

A random variable, z, has a p.d.f. $f(z)$ where

$$f(z) = e^{-z} \qquad 0 \le z < \infty$$

Calculate the probability that:

(a) $0 \le z \le 2$
(b) z is more than 1
(c) z is less than 0.5.

Solutions

Note that $\int_0^\infty e^{-z}\, dz = 1$ so that $f(z) = e^{-z}$ is suitable as a p.d.f.

(a) $P(0 \le z \le 2) = \int_0^2 e^{-z}\, dz = [-e^{-z}]_0^2 = 0.865$
(b) $P(z > 1) = \int_1^\infty e^{-z}\, dz = [-e^{-z}]_1^\infty = 0.368$
(c) $P(z < 0.5) = [-e^{-z}]_0^{0.5} = 0.393$ ▲

EXERCISES 15.8

1. $f(x) = kx^2$, k constant, $-1 \leq x \leq 1$. $f(x)$ is a p.d.f.

   (a) What is the value of k?
   (b) Calculate the probability that $x > 0.5$.
   (c) If $P(x > c) = 0.6$ then what is the value of c?

2. $f(x)$ is a p.d.f. for the random variable x, which can vary from 0 to 10. It is illustrated in Figure 15.9. What is the probability that x lies in $[2, 4]$?

3. A p.d.f. is given by:

   $$f(z) = 2e^{-2z} \qquad 0 \leq z < \infty$$

   (a) If 200 measurements of z are made,

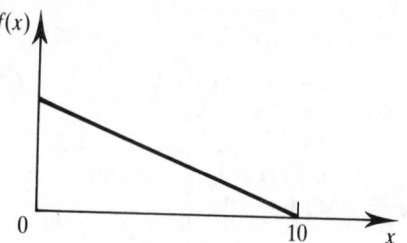

Figure 15.9 Probability density function for Exercise 15.8.2.

how many, on average, will be greater than 1?

   (b) If 50% of measurements are less than k, find k.

15.11 Mean value

If $\{x_1, x_2, x_3, \ldots, x_n\}$ is a set of n numbers, then the **mean value** of these numbers, denoted by $\bar{x}$ is

$$\blacksquare \quad \bar{x} = \frac{\text{sum of the numbers}}{n} = \frac{\Sigma x_i}{n}$$

$\bar{x}$ is sometimes called the **arithmetic mean**.

Example 15.17

Find the mean of -2.3, 0, 1, 0.7.

Solution

$$\bar{x} = \frac{-2.3 + 0 + 1 + 0.7}{4} = \frac{-0.6}{4} = -0.15 \quad \blacktriangle$$

The mean value is a single number which characterizes the set of numbers. It is useful in helping to make comparisons.

Example 15.18

A component is made by two machines, A and B. The lifespans of six components made by each machine are recorded in Table 15.8. Which is the preferred machine?

Table 15.8 The lifespans of six components made by machines A and B.

	Lifespan (hours)					
Machine A	92	86	61	70	58	65
Machine B	64	75	84	80	63	70

Solution

For components manufactured by machine A

$$\text{average lifespan} = \frac{432}{6} = 72$$

For components manufactured by machine B

$$\text{average lifespan} = \frac{436}{6} = 72.7$$

Machine B produces components with a higher average lifespan and so is the preferred machine. ▲

Example 15.19

A variable, x, can have values 2, 3, 4, 5 and 6. Many observations of x are made and denoted by x_i. They have corresponding frequencies, f_i. The results are as follows:

value (x_i)	2	3	4	5	6
frequency (f_i)	6	9	3	7	4

Calculate the mean of x.

Solution

The sum of all the measurements must be found. The value 2 occurs six times, contributing 12 to the total. Similarly, 3 occurs nine times contributing 27 to the total. Thus

$$\text{total} = 2(6) + 3(9) + 4(3) + 5(7) + 6(4) = 110$$

The number of measurements made is $6 + 9 + 3 + 7 + 4 = 29$. So,

$$\text{mean} = \bar{x} = \frac{110}{29} = 3.79 \quad ▲$$

Example 15.19 illustrates a general principle.

■ If values $x_1, x_2, \ldots, x_n$ occur with frequencies $f_1, f_2, \ldots, f_n$ then:

$$\bar{x} = \frac{\Sigma x_i f_i}{\Sigma f_i}$$

EXERCISES 15.9

1. Teaching staff are evaluated by students. A '0' is 'Terrible' and a '5' is 'Outstanding'. The 29 members of the department are evaluated and their scores recorded as follows:

score	0	1	2	3	4	5
number of staff	2	5	6	9	4	3

What is the mean score for the whole department?

2. Two samples of nails are taken. The first sample has 12 nails with a mean length of 2.7 cm, the second sample has 20 nails with a mean length of 2.61 cm. What is the mean length of all 32 nails?

15.12 Standard deviation

Although the mean indicates where the centre of a set of numbers lies, it gives no measure of the spread of the numbers. For example, $-1, 0, 1$ and $-10, 0, 10$ both have a mean of 0 but clearly the numbers in the second set are much more widely dispersed than those in the first. A commonly used measure of dispersion is the **standard deviation.**

Let $x_1, x_2, \ldots, x_n$ be n measurements with a mean $\bar{x}$. Then $x_i - \bar{x}$ is the amount by which x_i differs from the mean. The quantity $x_i - \bar{x}$ is called the **deviation** of x_i from the mean. Some of these deviations will be positive, some negative. The mean of these deviations is always zero (see Exercise 15.10.3) and so this is not helpful in measuring the dispersion of the numbers. To avoid positive and negative deviations summing to zero the squared deviation is taken, $(x_i - \bar{x})^2$. The **variance** is the mean of the squared deviations.

■ variance $= \dfrac{\Sigma(x_i - \bar{x})^2}{n}$

and

■ standard deviation $= \sqrt{\text{variance}}$

Standard deviation has the same units as the x_i.

Example 15.20

Calculate the standard deviation of

(a) $-1, 0, 1$

(b) $-10, 0, 10$

Solutions

(a) $x_1 = -1$, $x_2 = 0$, $x_3 = 1$. Clearly $\bar{x} = 0$.

$$x_1 - \bar{x} = -1 \qquad x_2 - \bar{x} = 0 \qquad x_3 - \bar{x} = 1$$

$$\text{variance} = \frac{(-1)^2 + 0^2 + 1^2}{3} = \frac{2}{3}$$

$$\text{standard deviation} = \sqrt{\frac{2}{3}} = 0.816$$

(b) $x_1 = -10$, $x_2 = 0$, $x_3 = 10$. Again $\bar{x} = 0$ and so $x_i - \bar{x} = x_i$, for $i = 1, 2, 3$.

$$\text{variance} = \frac{(-10)^2 + 0^2 + 10^2}{3} = \frac{200}{3}$$

$$\text{standard deviation} = \sqrt{\frac{200}{3}} = 8.165$$

As expected, the second set has a much higher standard deviation than the first. ▲

Example 15.21

Find the standard deviation of $-2, 7.2, 6.9, -10.4, 5.3$.

Solution

$$x_1 = -2, \ x_2 = 7.2, \ x_3 = 6.9, \ x_4 = -10.4, \ x_5 = 5.3, \ \bar{x} = 1.4$$

$$x_1 - \bar{x} = -3.4$$
$$x_2 - \bar{x} = 5.8$$
$$x_3 - \bar{x} = 5.5$$
$$x_4 - \bar{x} = -11.8$$
$$x_5 - \bar{x} = 3.9$$

$$\text{variance} = \frac{(-3.4)^2 + (5.8)^2 + (5.5)^2 + (-11.8)^2 + (3.9)^2}{5} = \frac{229.9}{5}$$

$$\text{standard deviation} = \sqrt{\frac{229.9}{5}} = 6.78 \quad ▲$$

Calculating $x_i - \bar{x}, i = 1, 2, \ldots, n$ is tedious for large n and so a more tractable form of the standard deviation is sought. Firstly observe that $\Sigma x_i = n\bar{x}$ and $\Sigma \bar{x}^2 = n\bar{x}^2$, since $\bar{x}^2$ is a constant. Now:

$$\Sigma(x_i - \bar{x})^2 = \Sigma x_i^2 - \Sigma 2x_i\bar{x} + \Sigma\bar{x}^2$$
$$= \Sigma x_i^2 - 2\bar{x}\Sigma x_i + n\bar{x}^2$$
$$= \Sigma x_i^2 - 2\bar{x}(n\bar{x}) + n\bar{x}^2$$
$$= \Sigma x_i^2 - n\bar{x}^2$$

Hence,

$$\blacksquare \quad \text{variance} = \frac{\Sigma x_i^2 - n\bar{x}^2}{n}$$

and

$$\text{standard deviation} = \sqrt{\frac{\Sigma x_i^2 - n\bar{x}^2}{n}}$$

Using these formulae it is not necessary to calculate $x_i - \bar{x}$.

Example 15.22

Repeat Example 15.21 using the newly derived formulae.

Solution

$$\Sigma x_i^2 = (-2)^2 + (7.2)^2 + (6.9)^2 + (-10.4)^2 + (5.3)^2 = 239.7$$

$$\text{standard deviation} = \sqrt{\frac{239.7 - 5(1.4)^2}{5}} = 6.78 \quad \blacktriangle$$

EXERCISES 15.10

1. Calculate the means and standard deviations of:

   (a) 1, 2, 3, 4, 5

   (b) 2.1, 2.3, 2.7, 2.6.

2. A set of measurements $\{x_1, x_2, x_3, \ldots, x_n\}$ has a mean of $\bar{x}$ and a standard deviation of s. What are the mean and standard deviation of the set $\{kx_1, kx_2, kx_3, \ldots, kx_n\}$ where k is a constant?

3. The mean of the numbers $\{x_1, x_2, x_3, \ldots, x_n\}$ is $\bar{x}$. Show that the sum of the deviations about the mean is 0, that is, show $\Sigma_{i=1}^{n}(x_i - \bar{x}) = 0$.

15.13 Expected value of a random variable

In Sections 15.11 and 15.12 we showed how to calculate the mean and standard deviation of a given set of numbers. No reference was made to probability distributions or p.d.f.s. Suppose now that we have knowledge of the probability distribution of a discrete random variable or the p.d.f. of a continuous random variable. The mean value of the random variable can still be found. Under these circumstances the mean value is known as the **expected value** or **expectation**.

15.13.1 *Expected value of a discrete random variable*

Suppose for definiteness that x is a discrete random variable with probability distribution as given in Table 15.9. In 100 trials x will have a value of zero 10 times on average, a value of one 20 times on average and so on. The mean value, that is, the expected value, is therefore

$$\text{expected value} = \frac{0(10) + 1(20) + 2(40) + 3(15) + 4(15)}{100} = 2.05$$

We could have arranged the calculation as follows:

$$\text{expected value} = 0(\tfrac{10}{100}) + 1(\tfrac{20}{100}) + 2(\tfrac{40}{100}) + 3(\tfrac{15}{100}) + 4(\tfrac{15}{100})$$
$$= 0(0.1) + 1(0.2) + 2(0.4) + 3(0.15) + 4(0.15)$$

Each term is of the form (value) × (probability). Thus,

$$\text{expected value} = \sum_{i=1}^{i=5} x_i P(x_i)$$

The symbol, μ, is used to denote the expected value of a random variable.

■ If a discrete random variable can take on values

$$x_1, x_2, \ldots, x_n$$

with probabilities $P(x_1), P(x_2), \ldots, P(x_n)$, then

$$\text{expected value of } x = \mu = \sum_{i=1}^{i=n} x_i P(x_i)$$

Table 15.9 Probability distribution for a discrete random variable x.

x	0	1	2	3	4
$P(x)$	0.1	0.2	0.4	0.15	0.15

Example 15.23

A random variable, *y*, has a known probability distribution given by:

y	2	4	6	8	10
P(y)	0.17	0.23	0.2	0.3	0.1

Find the expected value of *y*.

Solution

μ = expected value = $2(0.17) + 4(0.23) + 6(0.2) + 8(0.3) + 10(0.1) = 5.86$ ▲

15.13.2 *Expected value of a continuous random variable*

Suppose a continuous random variable, x, has p.d.f. $f(x)$, $a \leq x \leq b$. The probability that x lies in a very small interval, $[x, x + \delta x]$, is:

$$\int_{x}^{x+\delta x} f(t)\,dt$$

Since the interval is very small, f will vary only slightly across the interval. Hence the probability is approximately $f(x)\,\delta x$, see Figure 15.10. The contribution to the expected value as a result of this interval is:

(value) × (probability)

that is, $xf(x)\,\delta x$. Summing all such terms yields:

■ expected value = $\mu = \displaystyle\int_{a}^{b} xf(x)\,dx$

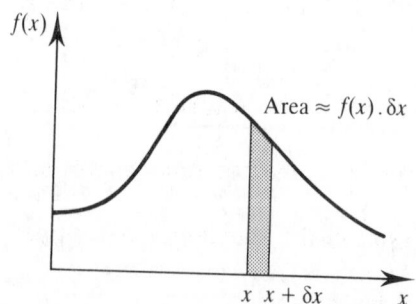

Figure 15.10 The shaded area represents the probability that *x* lies in the small interval [*x*, *x* + *δx*].

Example 15.24

A random variable has p.d.f. given by:

$$f(x) = \frac{1}{2\sqrt{x}} \qquad 1 \le x \le 4$$

Calculate the expected value of x.

Solution

$$\mu = \int_1^4 x \frac{1}{2\sqrt{x}} \, dx = \int_1^4 \frac{\sqrt{x}}{2} \, dx$$

$$= \left[\frac{x^{3/2}}{3}\right]_1^4 = \frac{7}{3}$$

So, if several values of x are measured, the mean of these values will be near to 7/3. As more and more values are measured the mean will get nearer and nearer to 7/3. ▲

EXERCISES 15.11

1. Calculate the expected value of the discrete random variable, h, whose probability distribution is:

h	1	1.5	1.7	2.1	3.2
P(h)	0.32	0.24	0.17	0.15	0.12

2. Is the expected value of a discrete random variable necessarily one of its possible values?

3. A random variable, z, has p.d.f. $f(z) = e^{-z}, 0 \le z < \infty$. Calculate the expected value of z.

4. A random variable, x, has p.d.f. $f(x)$ given by:

$$f(x) = \frac{5}{4x^2} \qquad 1 \le x \le 5$$

(a) Calculate the expected value of x.

(b) Ten values of x are measured. They are:

1.9, 2.9, 2.8, 2.1, 3.2, 3.4, 2.7, 2.3, 2.8, 2.7

Calculate the mean of the observations and comment on your findings.

15.14 Standard deviation of a random variable

15.14.1 *Standard deviation of a discrete random variable*

Recall from Section 15.12 that the standard deviation of a set of numbers, $\{x_1, x_2, \ldots, x_n\}$, is given by

$$\text{standard deviation} = \sqrt{\frac{\Sigma(x_i - \bar{x})^2}{n}}$$

Now suppose that x is a discrete random variable which can have values $x_1, x_2, x_3, \ldots, x_n$ with respective probabilities of $p_1, p_2, p_3, \ldots, p_n$, that is, we have

x	x_1	x_2	x_3	$\cdots$	x_n
$P(x)$	p_1	p_2	p_3	$\cdots$	p_n

Let the expected value of x be μ. Then the square of the deviation from the expected value has an identical probability distribution.

value:	$(x_1 - \mu)^2$	$(x_2 - \mu)^2$	$\cdots$	$(x_n - \mu)^2$
probability:	p_1	p_2	$\cdots$	p_n

The expected value of the mean squared deviation is the variance. The symbol σ^2 is used to denote the variance of a random variable.

■ variance $= \sigma^2 = \displaystyle\sum_{1}^{n} p_i(x_i - \mu)^2$

As before the standard deviation is the square root of the variance.

■ standard deviation $= \sigma = \sqrt{\sum p_i(x_i - \mu)^2}$

Example 15.25

A discrete random variable has probability distribution:

x	1	2	3	4	5
$P(x)$	0.12	0.15	0.23	0.3	0.2

Calculate:

(a) the expected value
(b) the standard deviation

Solutions

(a) $\mu = \Sigma x_i p_i = 1(0.12) + 2(0.15) + 3(0.23) + 4(0.3) + 5(0.2) = 3.31$

(b) $\sigma^2 = \Sigma p_i(x_i - \mu)^2$

$= 0.12(1 - 3.31)^2 + 0.15(2 - 3.31)^2 + 0.23(3 - 3.31)^2$

$\quad + 0.3(4 - 3.31)^2 + 0.2(5 - 3.31)^2$

$= 1.6339$

Standard deviation $= \sigma = \sqrt{1.6339} = 1.278$ ▲

15.14.2 *Standard deviation of a continuous random variable*

We simply state the formula for the standard deviation of a continuous random variable. It is analogous to the formula for the standard deviation of a discrete variable. Let x be a continuous random variable with p.d.f. $f(x)$, $a \leq x \leq b$. Then

$$\blacksquare \quad \sigma = \sqrt{\int_a^b (x - \mu)^2 f(x)\, dx}$$

Example 15.26

A random variable, x, has p.d.f. $f(x)$ given by:

$$f(x) = 1 \qquad 0 \leq x \leq 1$$

Calculate the standard deviation of x.

Solution

The expected value, μ, is found.

$$\mu = \int_0^1 xf(x)\, dx = \left[\frac{x^2}{2}\right]_0^1 = \frac{1}{2}$$

The variance can now be found.

$$\text{variance} = \sigma^2 = \int_0^1 (x - \tfrac{1}{2})^2 1\, dx$$

$$= \int_0^1 x^2 - x + \tfrac{1}{4}\, dx$$

$$= \left[\frac{x^3}{3} - \frac{x^2}{2} + \frac{x}{4}\right]_0^1 = \frac{1}{3} - \frac{1}{2} + \frac{1}{4} = \frac{1}{12}$$

Hence

$$\sigma = \sqrt{\tfrac{1}{12}} = 0.29$$

The standard deviation of x is 0.29. ▲

EXERCISES 15.12

1. A continuous random variable has p.d.f.

$$f(x) = \begin{cases} x + 1 & -1 \leq x \leq 0 \\ -x + 1 & 0 < x \leq 1 \end{cases}$$

(a) Calculate the expected value of x.
(b) Calculate the standard deviation of x.

2. A discrete random variable, w, has a known probability distribution.

w	-1	-0.5	0	0.5	1
$P(w)$	0.1	0.17	0.4	0.21	0.12

Calculate the standard deviation of w.

15.15 Permutations and combinations

Both permutations and combinations are used extensively in the calculation of probabilities.

15.15.1 *Permutations*

The following problem introduces permutations.

Example 15.27 Linking process control computers

A primary and a secondary route must be chosen from available routes A, B and C, for the linking of two process control computers in order to provide redundancy in case one fails. In how many ways can the choices be made?

Solution

The various possibilities are listed.

Primary route	Secondary route
A	B
B	A
A	C
C	A
B	C
C	B

There are six ways in which the choices can be made. Alternatively we could argue as follows. Suppose the primary route is chosen first. There are three choices: any one of A, B or C. The secondary route is then chosen from the two remaining routes, giving two possible choices. Together there are $3 \times 2 = 6$ ways of choosing a primary route and a secondary route. ▲

In Example 15.27 the choice AB is distinct from the choice BA, that is, the order is important. Choosing two routes from three and arranging them in order is an example of a **permutation**. More generally,

■ a permutation of n distinct objects taken r at a time is an arrangement of r of the n objects.

In forming permutations, the order of the objects is important. If three letters are chosen from the alphabet the permutation XYZ is distinct from the permutation ZXY. We pose the question 'How many permutations are there of n objects taken *r* at a time?' The following example will help to establish a formula for the number of permutations.

Example 15.28

Calculate the number of permutations there are of:

(a) 4 distinct objects taken 2 at a time
(b) 5 distinct objects taken 3 at a time
(c) 7 distinct objects taken 4 at a time

Solutions

(a) Listing all possible permutations is not feasible when the numbers involved are large. In choosing the first object, four choices are possible. In choosing the second object, three choices are possible. There are thus $4 \times 3 = 12$ permutations of 4 objects taken 2 at a time. Note that 12 may be written as:

$$12 = 4 \times 3 = \frac{4!}{2!} = \frac{4!}{(4-2)!}$$

(b) There are 5 objects available for the first choice, 4 for the second choice and 3 for the third choice. Hence there are $5 \times 4 \times 3 = 60$ permutations. Again note that:

$$60 = \frac{5!}{2!} = \frac{5!}{(5-3)!}$$

(c) There are 7 objects available for the first choice, 6 for the second, 5 for the third and 4 for the fourth. The number of permutations is $7 \times 6 \times 5 \times 4 = 840$. This may be written as $7!/(7-4)!$. ▲

The example illustrates the following general rule.

■ The number of permutations of *n* distinct objects taken *r* at a time, written $P(n, r)$, is

$$P(n, r) = \frac{n!}{(n-r)!}$$

Example 15.29

Find the number of permutations of:

(a) 10 distinct objects taken 6 at a time
(b) 15 distinct objects taken 2 at a time
(c) 6 distinct objects taken 6 at a time.

Solutions

(a) $P(10, 6) = 10!/(10 - 6)! = 10!/4! = 151\,200$

(b) $P(15, 2) = 15!/(15 - 2)! = 15!/13! = 210$

(c) $P(6, 6) = 6!/(6 - 6)! = 6!/0! = 720$

$P(6, 6)$ is simply the number of ways of arranging all six of the objects. ▲

Note that

$$P(n, n) = n!$$

This is the number of ways of arranging n given objects.

15.15.2 *Combinations*

Closely related to, but nevertheless distinct from permutations are **combinations**.

■ A combination is a selection of r distinct objects from n objects.

In making a selection the order is unimportant. For example, given the letters A, B and C, AB and BA are the same combination but different permutations. As with permutations we develop an expression for the number of combinations of n objects taken r at a time. Examples 15.30 and 15.31 help with this development.

Example 15.30

There are three routes, A, B and C joining two computers. In how many ways can two routes be chosen from A, B and C?

Solution

The possible combinations (selections) can be listed as:

AB, BC, AC

that is, there are three possible ways of making the selection. We can also use our knowledge of permutations to calculate the number of combinations. There are $P(3, 2)$ ways of arranging the three routes taken two at a time.

$$P(3, 2) = \frac{3!}{(3 - 2)!} = 6$$

Each combination of two routes can be arranged in $P(2, 2)$ ways, that is, each combination gives rise to two permutations. For example the combination AB could be arranged as AB or BA giving two permutations. Thus, the number of combinations is half the number of permutations. There are $6/2 = 3$ combinations. ▲

Example 15.31

Calculate the number of combinations of

(a) 6 distinct objects taken 4 at a time

(b) 10 distinct objects taken 6 at a time

Solutions

(a) Consider one combination of 4 objects. These 4 objects can be arranged in 4! ways, that is, each combination gives rise to 4! permutations. The number of permutations of 6 objects taken 4 at a time is:

$$P(6, 4) = \frac{6!}{2!}$$

Hence the number of combinations is

$$\frac{P(6, 4)}{4!} = \frac{6!}{2!4!} = 15$$

There are 15 combinations of 6 objects taken 4 at a time.

(b) Each combination, comprising 6 objects, gives rise to 6! permutations. The number of permutations of 10 objects taken 6 at a time is:

$$P(10,6) = \frac{10!}{4!}$$

Hence the number of combinations $= \dfrac{P(10, 6)}{6!} = \dfrac{10!}{4!6!} = 210$ ▲

We write $\binom{n}{r}$ to denote the number of combinations of n objects taken r at a time. A formula for $\binom{n}{r}$ is now developed.

Each combination of r objects gives rise to $r!$ permutations, but:

$$P(n, r) = \frac{n!}{(n - r)!}$$

■ The number of combinations of n distinct objects taken r at a time is:

$$\binom{n}{r} = \frac{P(n, r)}{r!} = \frac{n!}{(n - r)!r!}$$

Example 15.32

Calculate the number of combinations of

(a) 6 distinct objects taken 5 at a time

(b) 9 distinct objects taken 9 at a time

(c) 25 distinct objects taken 5 at a time.

Solutions

(a) $\binom{6}{5} = \frac{6!}{1!5!} = 6$

(b) $\binom{9}{9} = \frac{9!}{0!9!} = 1$

(c) $\binom{25}{5} = \frac{25!}{5!20!} = 53\,130$ ▲

We can generalize the result of Example 15.32b and state that $\binom{n}{n} = 1$.

Example 15.33

There are k identical objects and n compartments ($n \geq k$). Each compartment can hold only 1 object. In how many different ways can the k objects be placed in the n compartments?

Solution

The order in which the objects are placed is unimportant since all the objects are identical. Placing the k objects is identical to selecting k of the n compartments (see Figure 15.11). But the number of ways of selecting k compartments from n is precisely $\binom{n}{k}$. Hence the k objects can be placed in the n compartments in $\binom{n}{k}$ different ways.   ▲

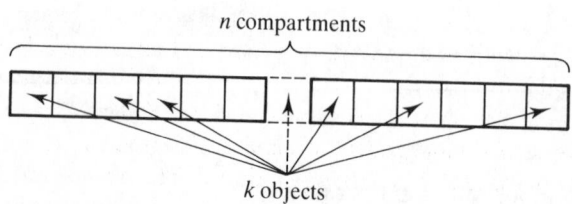

n compartments

k objects

Figure 15.11   Placing k objects in n compartments.

EXERCISES 15.13

1. Evaluate:

   (a) $P(8, 6)$ (b) $P(11, 7)$

   (c) $\begin{pmatrix} 12 \\ 9 \end{pmatrix}$ (d) $\begin{pmatrix} 15 \\ 12 \end{pmatrix}$

   (e) $\begin{pmatrix} 15 \\ 3 \end{pmatrix}$.

2. Write out explicitly:

   (a) $\begin{pmatrix} n \\ 0 \end{pmatrix}$ (b) $\begin{pmatrix} n \\ 1 \end{pmatrix}$

   (c) $\begin{pmatrix} n \\ 2 \end{pmatrix}$

3. The expansion of $(a + b)^n$ where n is a positive integer may be written with the help of combination notation.

$$(a + b)^n = a^n + \begin{pmatrix} n \\ 1 \end{pmatrix} a^{n-1}b$$

$$+ \begin{pmatrix} n \\ 2 \end{pmatrix} a^{n-2}b^2 + \cdots$$

$$+ \begin{pmatrix} n \\ n-1 \end{pmatrix} ab^{n-1} + \begin{pmatrix} n \\ n \end{pmatrix} b^n$$

$$= \sum_{k=0}^{n} \begin{pmatrix} n \\ k \end{pmatrix} a^{n-k}b^k$$

Expand

   (a) $(a + b)^4$

   (b) $(1 + x)^6$

   (c) $(p + q)^5$

4. Primary and secondary routes connecting two computers need to be chosen. Two primary routes are needed from eight which are suitable and three secondary routes must be chosen from four available. In how many ways can the routes be chosen?

5. A combination lock can be opened by dialling 3 correct letters followed by 3 correct digits. How many different possibilities are there for arranging the letters and digits? Is this more secure than a lock which has 7 digits? Is the word 'combination' being used correctly?

6. A nuclear power station is to be built on one of 20 possible sites. A team of engineers is commissioned to examine the sites and rank the three most favourable in order. In how many ways can this be done?

15.16 The binomial distribution

In a single trial or experiment a particular result may or may not be obtained. For example, in an examination, a student may pass or fail; when testing a component it may work or not work. The important point is that the two outcomes are complementary. Assuming that the probability of an outcome is fixed, such a trial is called a **Bernoulli trial** in honour of the mathematician, J. Bernoulli.

We address the following problem. In a single trial the outcome is either A or B, that is, $\bar{A}$. We refer to A as a **success** and B as a **failure**. If we know $P(A) = p$ then $P(B) = 1 - p$. If n independent trials are observed, what is the probability that A occurs k times, and B occurs $n - k$ times? The number of successful trials in n such experiments is a random variable; such a variable is said to have a **binomial distribution**. Let us consider a particular problem.

Example 15.34

A machine makes components. The probability that a component is acceptable is 0.9.

(a) If three components are sampled find the probability that the first is acceptable, the second is acceptable and the third is not acceptable.

(b) If three components are sampled what is the probability that exactly two are acceptable?

Solutions

Let the events A and B be defined thus:

A: the component is acceptable, $P(A) = 0.9$

B: the component is not acceptable, $P(B) = 0.1$

(a) We denote by AAB the compound event: the first is acceptable, the second is acceptable, the third is not acceptable. Since the three events are independent the multiplication law in Section 15.7 gives

$$P(AAB) = P(A)P(A)P(B) = (0.9)(0.9)(0.1) = (0.9)^2(0.1) = 0.081$$

(b) We are interested in the compound event in which two components are acceptable and one is not acceptable. We denote by AAB the compound event: the first is acceptable, the second is acceptable, the third is not acceptable. Compound events ABA and BAA have obvious interpretations.

   If exactly two components are acceptable then either AAB or ABA or BAA occurs. These compound events are mutually exclusive and we can therefore use the addition law (see Section 15.3). Hence

P(exactly two acceptable components)
$= P(AAB) + P(ABA) + P(BAA)$

From (a) $P(AAB) = 0.081$ and by similar reasoning
$P(ABA) = P(BAA) = 0.081$. Hence,

P(exactly two acceptable components) $= 3(0.081) = 0.243$ ▲

15.16.1 *Probability of k successes from n trials*

Let us now return to the general problem posed earlier. We define the compound event C.

   C: A occurs k times and B occurs $n - k$ times.

The k occurrences of event A can be distributed amongst the n trials in $\binom{n}{k}$ different ways (see Example 15.33). The probability of a particular distribution of k occurrences of A and $n - k$ occurrences of B is $p^k(1 - p)^{n-k}$. Since there are $\binom{n}{k}$ distinct distributions possible then:

■ $P(C) = \binom{n}{k}p^k(1 - p)^{n-k}$

Example 15.35

The probability a component is acceptable is 0.93. Ten components are picked at random. What is the probability that:

(a) at least 9 are acceptable

(b) at most 3 are acceptable?

Solutions

(a) $P(\text{exactly 9 components are acceptable}) = \binom{10}{9}(0.93)^9(0.07) = 0.364$

       $P(\text{exactly 10 components are acceptable}) = \binom{10}{10}(0.93)^{10} = 0.484$

Hence,

       $P(\text{at least 9 components are acceptable}) = 0.364 + 0.484 = 0.848$

(b) We require the probability that 0 are acceptable, 1 is acceptable, 2 are acceptable and 3 are acceptable.

    $P(0 \text{ are acceptable}) = \binom{10}{0}(0.93)^0(0.07)^{10} = 2.825 \times 10^{-12}$

    $P(1 \text{ is acceptable}) \;\;= \binom{10}{1}(0.93)^1(0.07)^9 = 3.753 \times 10^{-10}$

    $P(2 \text{ are acceptable}) = \binom{10}{2}(0.93)^2(0.07)^8 = 2.244 \times 10^{-8}$

    $P(3 \text{ are acceptable}) = \binom{10}{3}(0.93)^3(0.07)^7 = 7.949 \times 10^{-7}$

Hence,

    $P(\text{at most 3 are acceptable}) = 2.825 \times 10^{-12} + 3.753 \times 10^{-10}$

$$+ 2.244 \times 10^{-8} + 7.949 \times 10^{-7}$$

$$= 8.18 \times 10^{-7}$$

that is, the probability that at most three components are acceptable is almost zero; it is virtually impossible. ▲

15.16.2 *Mean and standard deviation of a binomial distribution*

Let the probability of success in a single trial be p and let the number of trials be n. The number of successes in n trials is a discrete random variable, x, with a binomial distribution. Then x can have any value from $\{0, 1, 2, 3, \ldots, n\}$, although clearly some values are more likely to occur than others. The expected value of x can be shown to be np. Thus, if many values of x are recorded, the mean of these will approach np.

■ Expected value of the binomial distribution $= np$

The standard deviation of the binomial distribution can also be found. This is given by:

■ Standard deviation of the binomial distribution $= \sqrt{np(1-p)}$

15.16.3 *Most likely number of successes*

When conducting a series of trials it is sometimes desirable to know the most likely outcome. For example, what is the most likely number of acceptable components in a sample of five tested.

Example 15.36

The probability a component is acceptable is 0.8. Five components are picked at random. What is the most likely number of acceptable components?

Solution

$$P(\text{no acceptable components}) = \binom{5}{0}(0.8)^0(0.2)^5 = 3.2 \times 10^{-4}$$

$$P(1 \text{ acceptable component}) = \binom{5}{1}(0.8)^1(0.2)^4 = 6.4 \times 10^{-3}$$

$$P(2 \text{ acceptable components}) = \binom{5}{2}(0.8)^2(0.2)^3 = 0.0512$$

$$P(3 \text{ acceptable components}) = \binom{5}{3}(0.8)^3(0.2)^2 = 0.2048$$

$$P(4 \text{ acceptable components}) = \binom{5}{4}(0.8)^4(0.2) = 0.4096$$

$$P(5 \text{ acceptable components}) = \binom{5}{5}(0.8)^5(0.2)^0 = 0.3277$$

The most likely number of acceptable components is four. ▲

Example 15.36 illustrates an important general result. Suppose we conduct n Bernoulli trials and wish to find the most likely number of successes. If $p =$ probability of success on a single trial, $i =$ most likely number of successes in n trials then:

$$p(n+1) - 1 < i < p(n+1)$$

In Example 15.36, $p = 0.8$, $n = 5$ and so

$$(0.8)(6) - 1 < i < (0.8)(6)$$

$$3.8 < i < 4.8$$

Since i is an integer, then $i = 4$.

EXERCISES 15.14

1. The probability a component is acceptable is 0.8. Four components are sampled. What is the probability that

   (a) exactly one is acceptable

   (b) exactly two are acceptable?

2. A machine requires all seven of its micro-chips to operate correctly in order to be acceptable. The probability a micro-chip is operating correctly is 0.99.

   (a) What is the probability the machine is acceptable?

   (b) What is the probability that 6 of the 7 chips are operating correctly?

   (c) The machine is re-designed so that the original 7 chips are replaced by 4 new chips. The probability a new chip operates correctly is 0.98. Is the new design more or less reliable than the original?

3. The probability a machine has a lifespan of more than 5 years is 0.8. Ten machines are chosen at random. What is the probability that

   (a) 8 machines have a lifespan of more than 5 years

   (b) all machines have a lifespan of more than 5 years

   (c) at least 8 machines have a lifespan of more than 5 years

   (d) no more than 2 machines have a lifespan of less than 5 years?

4. The probability a valve remains reliable for more than 10 years is 0.75. Eight valves are sampled. What is the most likely number of valves to remain reliable for more than 10 years?

15.17 The Poisson distribution

The Poisson distribution models the number of occurrences of an event in a given interval. Consider the number of emergency calls received by a service engineer in 1 day. We may know from experience that the number of calls is usually 3 or 4 per day, but occasionally it will be only 1 or 2, or even 0, and on some days it may be 6 or 7, or even more. This example suggests a need for assigning a probability to the number of occurrences of an event during a given time period. The Poisson distribution serves this purpose.

The number of occurrences of an event, E, in a given time period is a discrete random variable which we denote by X. We wish to find the probability that $X = 0$, $X = 1$, $X = 2$, $X = 3$, and so on. Suppose the occurrence of E in any time interval is not affected by its occurrence in any preceding time interval. For example, a car is not more, or less, likely to pass a given spot in the next ten seconds because a car passed (or did not pass) the spot in the previous ten seconds, that is, the occurrences are independent.

Let λ be the expected (mean) value of X, the number of occurrences during the time period. If X is measured for many time periods the average value of X will be λ. Under the given conditions X follows a Poisson distribution. The probability that X has a value r is given by:

■ $P(X = r) = \dfrac{e^{-\lambda}\lambda^{r}}{r!} \qquad r \in \mathbb{N}$

■ The expected value and variance of the Poisson distribution are both equal to λ.

Example 15.37

Records show that on average three emergency calls per day are received by a service engineer. What is the probability that on a particular day

(a) 3

(b) 2

(c) 4

calls will be received?

Solution

The number of calls received follows a Poisson distribution. The average number of calls is three per day, that is, $\lambda = 3$.

(a) $P(X = 3) = e^{-3}3^{3}/3! = 0.224$

(b) $P(X = 2) = e^{-3}3^{2}/2! = 0.224$

(c) $P(X = 4) = e^{-3}3^{4}/4! = 0.168$

The engineer will receive three calls on approximately 22 days in 100, he will receive two calls on approximately 22 days in 100 and will receive four calls on approximately 17 days in 100. ▲

Example 15.38

A workshop has several machines. During a typical month two machines will break down. What are the probabilities that in a month

(a) 0

(b) 1

(c) more than 2

will break down?

Solutions

λ = average number of machines that break down = 2
X = number of machines broken down

(a) $P(X = 0) = e^{-2}2^0/0! = 0.135$

(b) $P(X = 1) = e^{-2}2^1/1! = 0.271$

(c) $P(X > 2) = 1 - P(X = 0) - P(X = 1) - P(X = 2)$

$$= 1 - e^{-2} - 2e^{-2} - 2e^{-2} = 0.323 \quad \blacktriangle$$

15.17.1 *Poisson approximation to the binomial*

The Poisson and binomial distributions are related. Consider a binomial distribution, in which n trials take place and the probability of success is p. If n increases and p decreases such that np is constant, the resulting binomial distribution can be approximated by a Poisson distribution with $\lambda = np$. Recall that np is the expected value of the binomial distribution, and λ is the expected value of the Poisson distribution.

To illustrate the above point, Table 15.10 lists the probabilities for binomial and Poisson distributions with $n = 15$, $p = 0.05$ and hence $\lambda = 15(0.05) = 0.75$. The remaining probabilities are all almost 0.

As n increases and p decreases with np remaining constant, agreement between the two distributions becomes closer.

Table 15.10 The probabilities for binomial and Poisson distributions.

	Binomial (n, p)	Poisson (λ)
	$P(X = r)$; $n = 15$, $p = 0.05$	$P(X = r)$; $\lambda = 0.75$
$r = 0$	0.463 29	0.472 37
$r = 1$	0.365 76	0.354 27
$r = 2$	0.134 75	0.132 85
$r = 3$	0.030 73	0.033 21
$r = 4$	0.004 85	0.006 23
$r = 5$	0.000 56	0.000 90
$r = 6$	0.000 05	0.000 07
$r = 7$	0.000 00	0.000 01

Example 15.39

A workforce comprises 250 people. The probability a person is absent on any one day is 0.02. Find the probability that on a day

(a) 3

(b) 7

people are absent.

Solution

This problem may be treated either as a sequence of Bernoulli trials or as a Poisson process.

Bernoulli trials

The probabilities follow a binomial distribution.

     E: a person is absent.

     $n =$ number of trials $= 250$

     $p =$ probability that E occurs in a single trial $= 0.02$

     $X =$ number of occurrences of event E

(a) $P(X = 3) = \begin{pmatrix} 250 \\ 3 \end{pmatrix}(0.02)^3(0.98)^{247} = 0.140$

(b) $P(X = 7) = \begin{pmatrix} 250 \\ 7 \end{pmatrix}(0.02)^7(0.98)^{243} = 0.105$

Poisson process

Since n is large and p is small the Poisson distribution will be a good approximation to the binomial distribution

     $\lambda = np = 5$

(a) $P(X = 3) = e^{-5}(5)^3/3! = 0.140$

(b) $P(X = 7) = e^{-5}(5)^7/7! = 0.104$ ▲

EXERCISES 15.15

1. A computer network has several hundred computers. During an eight-hour period, there are on average seven computers not functioning. Find the probability that during an eight-hour period

  (a) 9

  (b) 5

do not function.

2. A workforce has on average two people absent through illness on any given day. Find the probability that on a typical day

  (a) 2

  (b) at least 3

  (c) less than 4

people are absent.

3. A machine manufactures 300 micro-chips per hour. The probability an individual chip is faulty is 0.01. Calculate the probability that:

   (a) 2

   (b) 4

   (c) more than 3

   faulty chips are manufactured in a particular hour. Use both the binomial and Poisson approximations and compare the resulting probabilities.

4. The probability of a disk drive failure in any week is 0.007. A computer service company maintains 900 disk drives. Use the Poisson distribution to calculate the probability of

   (a) 7

   (b) more than 7

   disk drive failures in a week.

5. The probability an employee fails to come to work is 0.017. A large engineering firm employs 650 people. What is the probability that on a particular day

   (a) 9

   (b) 10

   people are away from work?

15.18 The uniform distribution

Suppose the probability of an event occurring remains constant across a given interval of interest. Figure 15.12 illustrates such a p.d.f. $f(t)$. The area under $f(t)$ must equal 1 and so if the interval is of length T, the height of the rectangle is $1/T$. The p.d.f. is given by

$$f(t) = \begin{cases} 1/T & 0 < t < T \\ 0 & \text{otherwise} \end{cases}$$

Such a distribution is said to be **uniform**. The probability an event occurs in an interval $[a, b]$, $0 \le a < b \le T$, is then $(b - a)/T$. We shall make extensive use of this distribution in Section 15.21 when we deal with reliability engineering.

15.19 The exponential distribution

Suppose a random variable has a Poisson distribution, for example, the random variable could be the number of customers arriving at a service point, the number of

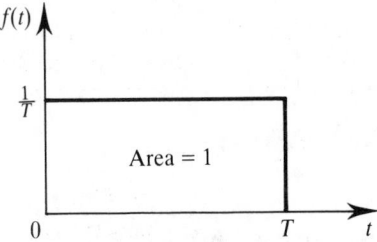

Figure 15.12 The uniform p.d.f.

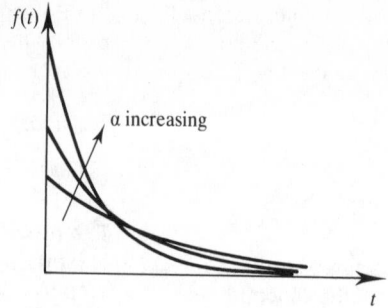

Figure 15.13   The exponential p.d.f. for various values of α.

telephone calls received at a switchboard, the number of machines breaking down in a week. Then the time between events happening is a random variable which follows an **exponential distribution**. Note that whereas the number of events is a discrete variable, the time between events is a continuous variable.

Let t be the time between events happening.

■   The exponential p.d.f. $f(t)$ is given by:

$$f(t) = \begin{cases} \alpha e^{-\alpha t} & t \geq 0, \\ 0 & \text{otherwise} \end{cases}$$

where $\alpha > 0$.

Figure 15.13 shows $f(t)$ for various values of α. The expected value of the distribution is given, by definition, as:

$$\text{expected value} = \mu = \int_0^\infty t\alpha e^{-\alpha t}\, dt = \frac{1}{\alpha}$$

For example, if

$$f(t) = 3e^{-3t} \qquad t \text{ in seconds} \qquad t \geq 0$$

then the mean time between events is $\frac{1}{3}$s, that is, on average there are three events happening per second.

15.20 The normal distribution

The normal probability density function, commonly called the **normal distribution**, is one of the most important and widely used. It is used to calculate the probable values of continuous variables, for example, weight, length, density, error

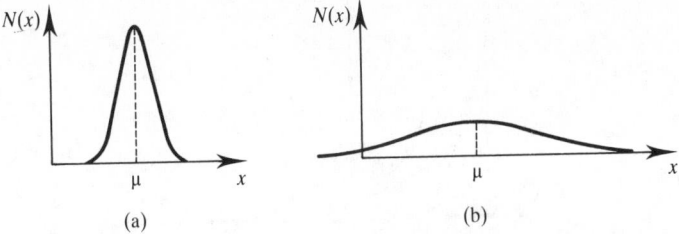

Figure 15.14 Two typical normal curves.

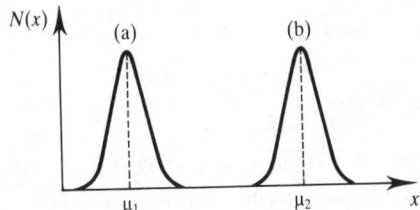

Figure 15.15 Two normal curves with the same standard deviation but different means.

measurement. Probabilities calculated using the normal distribution have been shown to reflect accurately those which would be found using actual data.

Let x be a continuous random variable with a normal distribution, $N(x)$. Then

$$\blacksquare \quad N(x) = \frac{1}{\sigma\sqrt{2\pi}} e^{-(x-\mu)^2/2\sigma^2}$$

where μ = expected (mean) value of x, σ = standard deviation of x. Figure 15.14 shows two typical normal curves. All normal distributions are bell-shaped and symmetrical about μ.

In Figure 15.14(a) the values of x are grouped very closely to the mean. Such a distribution has a low standard deviation. Conversely, in Figure 15.14(b) the values of the variable are spread widely about the mean and so the distribution has a high standard deviation.

Figure 15.15 shows two normal distributions. They have the same standard deviation but different means. The mean of the distribution in Figure 15.15(a) is μ_1 while the mean of that in Figure 15.15(b) is μ_2. Note that the domain of $N(x)$ is $(-\infty, \infty)$, that is, the domain is all real numbers. As for all distribution curves, the total area under the curve is 1.

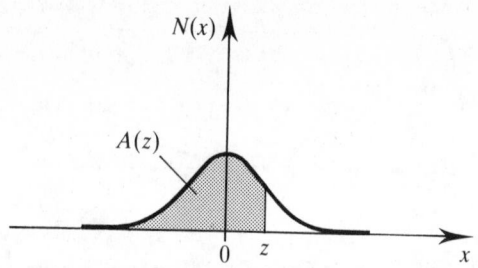

Figure 15.16 $A(z) = P(x < z) = \int_{-\infty}^{z} N(x)\, dx.$

15.20.1 *The standard normal*

A normal distribution is determined uniquely by specifying the mean and standard deviation. The probability that x lies in the interval $[a, b]$ is:

$$P(a \le x \le b) = \int_{a}^{b} N(x)\, dx .$$

The mathematical form of the normal distribution makes analytic integration impossible, so all probabilities must be computed numerically. As these numerical values would change every time the value of μ or σ was altered some standardization is required. To this end we introduce the **standard normal**. The standard normal has a mean of 0 and a standard deviation of 1.

Consider the probability that the random variable, x, has a value less than z. For convenience we call this $A(z)$.

$$A(z) = P(x < z) = \int_{-\infty}^{z} N(x)\, dx$$

Figure 15.16 illustrates $A(z)$. Values of $A(z)$ have been computed numerically and tabulated. They are given in Table 15.11. Using the table and the symmetrical property of the distribution, probabilities can be calculated.

Example 15.40

The continuous random variable x has a standard normal distribution. Calculate the probability that:

(a) $x < 1.2$

(b) $x > 1.2$

(c) $x > -1.2$

(d) $x < -1.2$

Solutions

(a) From Table 15.11

$$P(x < 1.2) = 0.8849$$

This is depicted in Figure 15.17.

Table 15.11 Cumulative normal probabilities.

z	$A(z)$	z	$A(z)$	z	$A(z)$	z	$A(z)$
0.00	0.500 000 0	0.51	0.694 974 3	1.02	0.846 135 8	1.53	0.936 991 6
0.01	0.503 989 4	0.52	0.698 468 2	1.03	0.848 495 0	1.54	0.938 219 8
0.02	0.507 978 3	0.53	0.701 944 0	1.04	0.850 830 0	1.55	0.939 429 2
0.03	0.511 966 5	0.54	0.705 401 5	1.05	0.853 140 9	1.56	0.940 620 1
0.04	0.515 953 4	0.55	0.708 840 3	1.06	0.855 427 7	1.57	0.941 792 4
0.05	0.519 938 8	0.56	0.712 260 3	1.07	0.857 690 3	1.58	0.942 946 6
0.06	0.523 922 2	0.57	0.715 661 2	1.08	0.859 928 9	1.59	0.944 082 6
0.07	0.527 903 2	0.58	0.719 042 7	1.09	0.862 143 4	1.60	0.945 200 7
0.08	0.531 881 4	0.59	0.722 404 7	1.10	0.864 333 9	1.61	0.946 301 1
0.09	0.535 856 4	0.60	0.725 746 9	1.11	0.866 500 5	1.62	0.947 383 9
0.10	0.539 827 8	0.61	0.729 069 1	1.12	0.868 643 1	1.63	0.948 449 3
0.11	0.543 795 3	0.62	0.732 371 1	1.13	0.870 761 9	1.64	0.949 497 4
0.12	0.547 758 4	0.63	0.735 652 7	1.14	0.872 856 8	1.65	0.950 528 5
0.13	0.551 716 8	0.64	0.738 913 7	1.15	0.874 928 1	1.66	0.951 542 8
0.14	0.555 670 0	0.65	0.742 153 9	1.16	0.876 975 6	1.67	0.952 540 3
0.15	0.559 617 7	0.66	0.745 373 1	1.17	0.878 999 5	1.68	0.953 521 3
0.16	0.563 559 5	0.67	0.748 571 1	1.18	0.880 999 9	1.69	0.954 486 0
0.17	0.567 494 9	0.68	0.751 747 8	1.19	0.882 976 8	1.70	0.955 434 5
0.18	0.571 423 7	0.69	0.754 902 9	1.20	0.884 930 3	1.71	0.956 367 1
0.19	0.575 345 4	0.70	0.758 036 3	1.21	0.886 860 6	1.72	0.957 283 8
0.20	0.579 259 7	0.71	0.761 114 9	1.22	0.888 767 6	1.73	0.958 184 9
0.21	0.583 166 2	0.72	0.764 237 5	1.23	0.890 651 4	1.74	0.959 070 5
0.22	0.587 060 4	0.73	0.767 304 9	1.24	0.892 512 3	1.75	0.959 940 8
0.23	0.590 954 1	0.74	0.770 350 0	1.25	0.894 350 2	1.76	0.960 796 1
0.24	0.594 834 9	0.75	0.773 372 6	1.26	0.896 165 3	1.77	0.961 636 4
0.25	0.598 706 3	0.76	0.776 372 7	1.27	0.897 957 7	1.78	0.962 462 0
0.26	0.602 568 1	0.77	0.779 350 1	1.28	0.899 727 4	1.79	0.963 273 0
0.27	0.606 419 9	0.78	0.782 304 6	1.29	0.901 474 7	1.80	0.964 069 7
0.28	0.610 261 2	0.79	0.785 236 1	1.30	0.903 199 5	1.81	0.964 852 1
0.29	0.614 091 9	0.80	0.788 144 6	1.31	0.904 902 1	1.82	0.965 620 5
0.30	0.617 911 4	0.81	0.791 029 9	1.32	0.906 582 5	1.83	0.966 375 0
0.31	0.621 719 5	0.82	0.793 891 9	1.33	0.908 240 9	1.84	0.967 115 9
0.32	0.625 515 8	0.83	0.796 730 6	1.34	0.909 877 3	1.85	0.967 843 2
0.33	0.629 300 0	0.84	0.799 545 8	1.35	0.911 492 0	1.86	0.968 557 2
0.34	0.633 071 7	0.85	0.802 337 5	1.36	0.913 085 0	1.87	0.969 258 1
0.35	0.636 830 7	0.86	0.805 105 5	1.37	0.914 656 5	1.88	0.969 946 0
0.36	0.640 576 4	0.87	0.807 849 8	1.38	0.916 206 7	1.89	0.970 621 0
0.37	0.644 308 8	0.88	0.810 570 3	1.39	0.917 735 6	1.90	0.971 283 4
0.38	0.648 027 3	0.89	0.813 267 1	1.40	0.919 243 3	1.91	0.971 933 4
0.39	0.651 731 7	0.90	0.815 939 9	1.41	0.920 730 2	1.92	0.972 571 1
0.40	0.655 421 7	0.91	0.818 588 7	1.42	0.922 196 2	1.93	0.973 196 6
0.41	0.659 097 0	0.92	0.821 213 6	1.43	0.923 641 5	1.94	0.973 810 2
0.42	0.662 757 3	0.93	0.823 814 5	1.44	0.925 066 3	1.95	0.974 411 9
0.43	0.666 402 2	0.94	0.826 391 2	1.45	0.926 470 7	1.96	0.975 002 1
0.44	0.670 031 4	0.95	0.828 943 9	1.46	0.927 855 0	1.97	0.975 580 8
0.45	0.673 644 8	0.96	0.831 472 4	1.47	0.929 219 1	1.98	0.976 148 2
0.46	0.677 241 9	0.97	0.833 976 8	1.48	0.930 563 4	1.99	0.976 704 5
0.47	0.680 822 5	0.98	0.836 456 9	1.49	0.931 887 9	2.00	0.977 249 9
0.48	0.684 386 3	0.99	0.838 912 9	1.50	0.933 192 8	2.01	0.977 784 4
0.49	0.687 933 1	1.00	0.841 344 7	1.51	0.934 478 3	2.02	0.978 308 3
0.50	0.691 462 5	1.01	0.843 752 4	1.52	0.935 744 5	2.03	0.978 821 7

Table 15.11 *Continued.*

z	$A(z)$	z	$A(z)$	z	$A(z)$	z	$A(z)$
2.04	0.979 324 8	2.22	0.986 790 6	2.40	0.991 802 5	2.58	0.995 060 0
2.05	0.979 817 8	2.23	0.987 126 3	2.41	0.992 023 7	2.59	0.995 201 2
2.06	0.980 300 7	2.24	0.987 454 5	2.42	0.992 239 7	2.60	0.995 338 3
2.07	0.980 773 8	2.25	0.987 775 5	2.43	0.992 450 6	2.70	0.996 533 0
2.08	0.981 237 2	2.26	0.988 089 4	2.44	0.992 656 4	2.80	0.997 444 9
2.09	0.981 691 1	2.27	0.988 396 2	2.45	0.992 857 2	2.90	0.998 134 2
2.10	0.982 135 6	2.28	0.988 696 2	2.46	0.993 053 1	3.00	0.998 650 1
2.11	0.982 570 8	2.29	0.988 989 3	2.47	0.993 244 3	3.20	0.999 312 9
2.12	0.982 997 0	2.30	0.989 275 9	2.48	0.993 430 9	3.40	0.999 663 1
2.13	0.983 414 2	2.31	0.989 555 9	2.49	0.993 612 8	3.60	0.999 840 9
2.14	0.983 822 6	2.32	0.989 829 6	2.50	0.993 790 3	3.80	0.999 927 7
2.15	0.984 222 4	2.33	0.990 096 9	2.51	0.993 963 4	4.00	0.999 968 3
2.16	0.984 613 7	2.34	0.990 358 1	2.52	0.994 132 3	4.50	0.999 996 6
2.17	0.984 996 6	2.35	0.990 613 3	2.53	0.994 296 9	5.00	0.999 999 7
2.18	0.985 371 3	2.36	0.990 862 5	2.54	0.994 457 4	5.50	0.999 999 9
2.19	0.985 737 9	2.37	0.991 106 0	2.55	0.994 613 9		
2.20	0.986 096 6	2.38	0.991 343 7	2.56	0.994 766 4		
2.21	0.986 447 4	2.39	0.991 575 8	2.57	0.994 915 1		

Source: Statistics Vol. Probability Inference and Decision by Hays, W. L. and Winkler, R. L. (Holt Rienhart Winston, 1970).

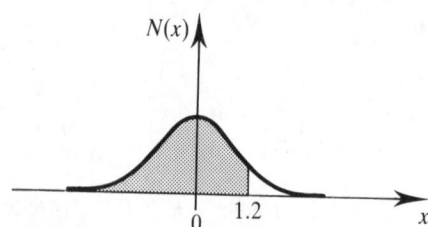

Figure 15.17 $P(x < 1.2) = 0.8849$.

(b) $P(x > 1.2) = 1 - 0.8849 = 0.1151$

This is shown in Figure 15.18.

(c) By symmetry $P(x > -1.2)$ is identical to $P(x < 1.2)$ (see Figure 15.19). So

$P(x > -1.2) = 0.8849$

(d) Using part (c) we find:

$P(x < -1.2) = 1 - P(x > -1.2) = 0.1151$

(see Figure 15.20). ▲

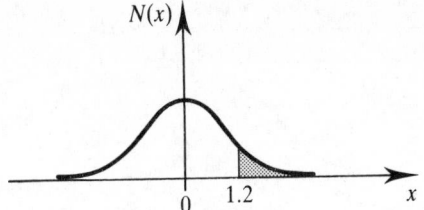

Figure 15.18 $P(x > 1.2) = 0.1151$.

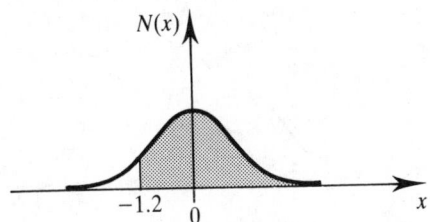

Figure 15.19 $P(x > -1.2) = 0.8849$.

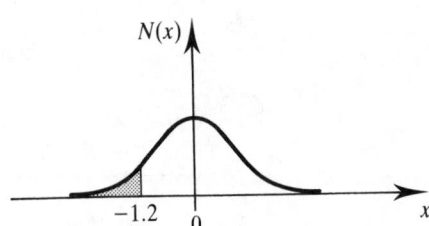

Figure 15.20 $P(x < -1.2) = 0.1151$.

Example 15.41

The continuous random variable v has a standard normal distribution. Calculate the probability that

(a) $0 < v < 1$

(b) $-1 < v < 1$

(c) $-0.5 \leq v \leq 2$

Solutions

(a) Figure 15.21 shows the area (probability) required.

$$P(v < 1) = 0.8413 \quad \text{using Table 15.11}$$

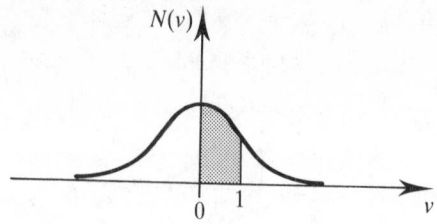

Figure 15.21 $P(0 < v < 1) = 0.3413$.

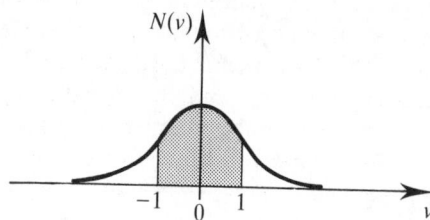

Figure 15.22 $P(-1 < v < 1) = 0.6826$.

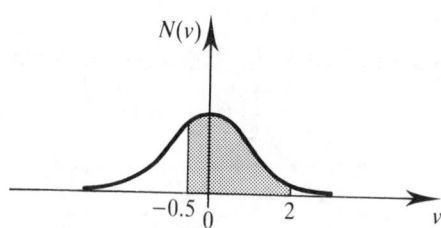

Figure 15.23 $P(-0.5 \leq v \leq 2) = 0.6687$.

$P(v < 0) = 0.5$ using symmetry

$P(0 < v < 1) = 0.8413 - 0.5 = 0.3413$

(b) Figure 15.22 shows the area (probability) required.

$P(-1 < v < 1) = 2 \times P(0 < v < 1)$ using symmetry

$= 2 \times 0.3413 = 0.6826$

This tells us that 68.3% of the values of *v* are within one standard deviation of the mean.

(c) Figure 15.23 shows the area (probability) required.

$P(v \leq 2) = 0.9772$

$$P(v \leq -0.5) = P(v > 0.5) = 1 - P(v < 0.5)$$
$$= 1 - 0.6915 = 0.3085$$
$$P(-0.5 \leq v \leq 2) = 0.9772 - 0.3085 = 0.6687$$

Note that whether inequalities defining v are strict or not, is of no consequence in calculating the probabilities. ▲

15.20.2 *Non-standard normal*

Table 15.11 allows us to calculate probabilities for a random variable with a standard normal distribution. This section shows us how to use the same table when the variable has a non-standard distribution. A non-standard normal has a mean value other than 0 and/or a standard deviation other than 1. The non-standard normal is changed into a standard normal by application of a simple rule. Suppose the non-standard distribution has a mean μ and a standard deviation σ. Then all non-standard values are transformed to standard values using:

Non-standard → Standard

$$X \rightarrow \frac{X - \mu}{\sigma}$$

Example 15.42

A random variable, h, has a normal distribution with mean 7 and standard deviation 2. Calculate the probability that

(a) $h > 9$
(b) $h < 6$
(c) $5 < h < 8$

Solutions

(a) Applying the transformation gives:

$$9 \rightarrow \frac{9 - 7}{2} = 1$$

So $h > 9$ has the same probability as $x > 1$, where x is a random variable with a standard normal distribution

$$P(h > 9) = P(x > 1) = 1 - P(x < 1) = 1 - 0.8413 = 0.1587$$

(b) Applying the transformation gives

$$6 \rightarrow \frac{6 - 7}{2} = -0.5$$

So $h < 6$ has the same probability as $x < -0.5$.

$$P(x < -0.5) = P(x > 0.5) = 1 - P(x < 0.5) = 0.3085$$

(c) Applying the transformation to 5 and 8 gives:

$$5 \to \frac{5-7}{2} = -1 \qquad 8 \to \frac{8-7}{2} = 0.5$$

and so we require $P(-1 < x < 0.5)$. Therefore

$$P(x < 0.5) = 0.6915 \qquad P(x < -1) = 0.1587$$

and then

$$P(-1 < x < 0.5) = 0.6915 - 0.1587 = 0.5328 \quad \blacktriangle$$

EXERCISES 15.16

1. A random variable, x, has a standard normal distribution. Calculate the probability that x lies in the following intervals.

   (a) (0.25, 0.75)

   (b) (−0.3, 0.1)

   (c) within 1.5 standard deviations of the mean

   (d) more than two standard deviations from the mean

   (e) (−1.7, −0.2)

2. The scores from IQ tests have a mean of 100 and a standard deviation of 15. What should a person score in order to be described as in the top 10% of the population?

3. A machine produces car pistons. The diameter of the pistons follows a normal distribution, mean 6.04 cm with a standard deviation of 0.02 cm. The piston is acceptable if its diameter is in the range 6.010 cm to 6.055 cm. What percentage of pistons is acceptable?

4. The random variable, x, has a normal distribution. How many standard deviations above the mean must the point P be placed if the tail-end is to represent

   (a) 10%

   (b) 5%

   (c) 1%

   of the total area? (See Figure 15.24).

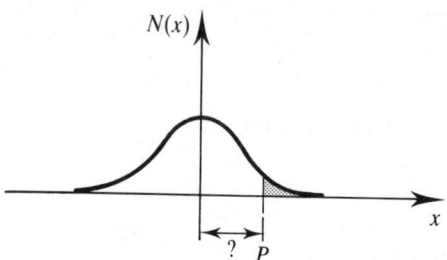

Figure 15.24 Graph for Exercise 15.16.4.

5. Consider Figure 15.25. The two tail-ends have equal area. How many standard deviations from the mean must A and B be placed if the tail-ends are:

   (a) 10%

   (b) 5%

   (c) 1%

   of the total area?

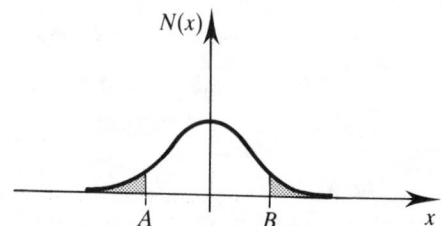

Figure 15.25 Graph for Exercise 15.16.5.

15.21 Reliability engineering

Reliability engineering is an important area of study. Unreliable products lead to human frustration, financial loss and in the case of life-critical systems, can lead to death. As the complexity of engineering systems has increased, mathematical methods of assessing reliability have grown in importance. Probability theory forms a central part of the design of highly reliable systems. For most items the failure rate changes with time. A common pattern is exhibited by the appropriately named 'bath tub' curve, illustrated in Figure 15.26.

Car part failures are quite well modelled by this distribution. For example, a crankshaft may fail quite quickly as a result of a manufacturing defect. If it does not, then there is usually a long period during which the likelihood of failure is low. After many years the probability of failure increases.

Consider the probability of an item failing over a total time period T. This period, T, is the time during which the item is functioning. The time taken to repair the item is not considered in this calculation but is considered a little later on. Suppose the probability of failure is evenly distributed over this period, that is, the probability of failure is modelled by the uniform distribution (see Section 15.18). It is important to note that this is a fairly simplistic assumption. If N = number of failures of the item over a time period T, it is possible to define a mean failure rate, k, by

$$k = \frac{N}{T}$$

For example, if an item fails 10 times in a period of 5 years we define the mean failure rate to be $k = 10/5 = 2$, that is, 2 failures per year. Because of the uniform distribution of the failures across the time period, the quantity k is constant. To illustrate this consider the previous example with a period of 10 years. During this time the item will fail 20 times and so

$k = \frac{20}{10} = 2$ failures per year

as found earlier.

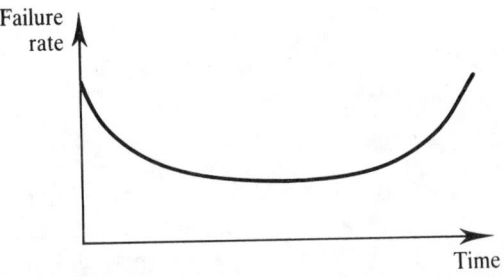

Figure 15.26 The 'bath tub' curve.

Another useful term is the **mean time between failures** (MTBF), which is given by

$$MTBF = \frac{1}{k}$$

The term MTBF is only used for items that are repairable.

Let the interval T be divided into n small sub-intervals each of length δT, that is, $n\,\delta T = T$. Suppose each sub-interval is so small that only one failure can occur during it. Note that since repair time has been neglected it is always possible to have a failure in a sub-interval. So the N failures which occur during time T, occur in N distinct sub-intervals. The probability of failure occurring in a particular sub-interval, P_f, is then given by

$$P_f = \frac{\text{number of sub-intervals in which failure occurs}}{\text{total number of sub-intervals}}$$
$$= \frac{N}{n}$$

In each sub-interval, δT, the probability of an item not failing, P_{nf}, is

$$P_{nf} = 1 - P_f = 1 - \frac{N}{n} = \frac{n-N}{n}$$

As the item progresses through each of the successive sub-intervals δT it can be thought of as undergoing a series of trials, the result of which is failure or non-failure. Therefore, the probability of the item not failing as it passes through all of the n sub-intervals can be obtained by multiplying each of the sub-interval probabilities of not failing, that is

$$\text{probability of not failing in time } T = \left(\frac{n-N}{n}\right)^n$$

As the sub-interval δT becomes small, that is, $\delta T \to 0$, the number of sub-intervals, n, becomes large, that is, $n \to \infty$. Hence we need to consider the quantity:

$$\lim_{n \to \infty} \left(\frac{n-N}{n}\right)^n$$

This limit can be shown to be e^{-N} (see Appendix VI). Hence:

$$\text{probability of not failing in time } T = e^{-N}$$

Finally, the probability of an item failing one or more times – items can be repaired and fail again – in the time interval T is given by:

$$P(T) = 1 - e^{-N} = 1 - e^{-kT}$$

$P(T)$ is the probability of at least one failure in the time period T. As a consequence of the constancy of k, this formula can be used to calculate the probability of an item failing one or more times in an arbitrary time period, t, in which case:

$$P(t) = 1 - e^{-kt}$$

$$(15.1)$$

It is important to stress that this formula only applies if the probability of failure is evenly distributed.

Example 15.43

A factory process line makes use of twelve controllers to maintain process variables at their correct values; all 12 controllers need to be working in order for the process line to be operational. Records show that each controller fails, on average, once every six months. Calculate the probability of the process line being stopped as a result of a controller failure within a time period of one month.

Solution

The mean failure rate of each controller is 1/6 breakdowns per month. Given that there are 12 controllers the overall failure rate for controllers is $12/6 = 2$ breakdowns per month. Using Equation (15.1) with $k = 2$ and $t = 1$ gives: the probability, $P(t)$, of the process line being stopped because of a controller failure within a period of one month as:

$$P(t) = 1 - e^{-kt} = 1 - e^{-2} = 0.865 \quad \blacktriangle$$

So far we have ignored the 'downtime' associated with an item waiting to be repaired after it has failed. This can be a significant factor with many engineering systems. The simplest possibility is that an item is out of action for a fixed period of time T_r while it is being repaired. If there are N failures during a period T then the total downtime is NT_r. The time the item is available, T_a, is given by:

$$T_a = T - NT_r$$

A useful quantity is the **fractional dead time**, D, which is the ratio of the mean time the item is in the dead state to the total time. In this case:

$$D = \frac{NT_r}{T} \tag{15.2}$$

Another useful quantity is the **availability**, A, which is the ratio of the mean time in the working state to the total time. For the present model:

$$A = \frac{T - NT_r}{T} = 1 - D$$

The failure rate model developed previously was based on the time the item was working rather than the total time. When the repair time is included k becomes:

$$k = \frac{N}{T_a}$$

and so

$$N = kT_a = k(T - NT_r)$$

$$N + kNT_r = kT$$

and therefore

$$N = \frac{kT}{1 + kT_r}$$

So using Equation (15.2)

$$D = \frac{kT}{1 + kT_r} \frac{T_r}{T} = \frac{kT_r}{1 + kT_r}$$

Also

$$A = 1 - D = 1 - \frac{kT_r}{1 + kT_r} = \frac{1}{1 + kT_r} \tag{15.3}$$

For more complicated repair characteristics the equations for D and A are correspondingly more complicated.

Example 15.44

The electrical supply to a large factory has a mean time between failures of 350 hours. When the supply fails it takes 3 hours to repair the failure and restore the supply. Calculate the average availability of the electrical supply to the factory.

Solution

Failure rate of the supply is k where $k = 1/350$ failures per hour. Using Equation (15.3) with $T_r = 3$ hours gives:

$$A = \frac{1}{1 + 3/350} = 0.992 \qquad \text{that is, } 99.2\%$$

The supply is up and running for 99.2% of the time. ▲

So far we have only examined systems in which failure was caused by one or more components each with the same failure rate or MTBF. A more common situation is one in which the different components of a system have different degrees of reliability. It is still useful to be able to calculate the overall reliability of the system although the analysis is more complicated. In order to do so it is necessary to define the term reliability. From Equation (15.1) we know that $P(t)$ defines the probability of one or more failures during a time period, t. Therefore the probability of no failures is given by $1 - P(t)$. The quantity $1 - P(t)$ is called the **reliability** of the system during a time period t, and is denoted $R(t)$, that is,

$$R(t) = 1 - P(t) = e^{-kt} \tag{15.4}$$

$R(t)$ can be interpreted as the probability a component works properly during a period t. We now examine the reliability of two simple system configurations.

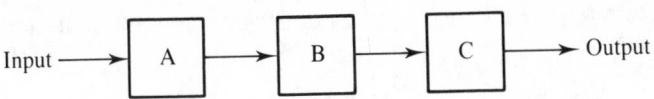

Figure 15.27   Series system.

15.21.1 *Series system*

A **series system** is one in which all the components of a system must operate satisfactorily if the system is to function correctly. Consider a system consisting of three components, shown in Figure 15.27. The reliability of the system is the product of the reliabilities of the individual components, that is,

$$R = R_A R_B R_C \tag{15.5}$$

This formula is a direct consequence of the fact that the failure of any one of the components is an independent event. So the probability of the system not failing, that is, its reliability, is the product of the probabilities of each of the components not failing. (See Section 15.7 for independent events.)

Example 15.45

A radio system consists of a power supply, a transmitter/amplifier and an antenna. During a 1000 hour period the reliability of the various components is as follows

$$R_{ps} = 0.95$$
$$R_{ta} = 0.93$$
$$R_a = 0.99.$$

Calculate the overall reliability of the radio system.

Solution

$$R = R_{ps} R_{ta} R_a = 0.95 \times 0.93 \times 0.99 = 0.87$$

that is, there is an 87% chance that the radio will not fail during a 1000 hour period. ▲

   Using Equations (15.4) and (15.5) it is possible to generate a formula for the reliability of a system consisting of n series components.

$$R = e^{-k_1 t} e^{-k_2 t} \ldots e^{-k_n t} = e^{-(k_1 + k_2 + \ldots + k_n)t}$$

where $k_1, k_2, \ldots, k_n$ are the mean failure rates of the n components,.

Example 15.46

A satellite link is being used to transmit television pictures from America to England. The components of this system are the television studio in New York, the transmitter ground station, the satellite and the receiver ground station in London. The MTBF of each of the components is as follows:

$$MTBF_{ts} = 1000 \text{ hours}$$
$$MTBF_{tgs} = 2000 \text{ hours}$$
$$MTBF_{s} = 500\,000 \text{ hours}$$
$$MTBF_{rgs} = 5000 \text{ hours}$$

Calculate the overall reliability of the system during a 28 day period and a yearly period of 365 days.

Solution

First we calculate the mean failure rate of each of the components.

$$k_{ts} = \frac{1}{1000} = 1 \times 10^{-3} \text{ failures per hour}$$

$$k_{tgs} = \frac{1}{2000} = 5 \times 10^{-4} \text{ failures per hour}$$

$$k_{s} = \frac{1}{500\,000} = 2 \times 10^{-6} \text{ failures per hour}$$

$$k_{rgs} = \frac{1}{5000} = 2 \times 10^{-4} \text{ failures per hour}$$

So the overall reliability of the system is:

$$R = e^{-(k_{ts} + k_{tgs} + k_s + k_{rgs})t} = e^{-(1 \times 10^{-3} + 5 \times 10^{-4} + 2 \times 10^{-6} + 2 \times 10^{-4})t}$$

$$= e^{-1.702 \times 10^{-3}t}$$

For $t = 28 \times 24$ hours, $R = 0.319$.
For $t = 365 \times 24$ hours, $R = 3.35 \times 10^{-7}$. Clearly during a yearly period the system is almost certain to fail at least once. ▲

15.21.2 *Parallel system*

A **parallel system** is one in which several components are in parallel and all of them must fail for the system to fail. The case of three components is shown in Figure 15.28. The probability of all three components failing in a time period, t, is the product of the individual probabilities of each component failing, that is, $(1 - R_A)(1 - R_B)(1 - R_C)$. So the overall system reliability is:

$$R = 1 - (1 - R_A)(1 - R_B)(1 - R_C)$$

This formula can be generalized to the case of n components in parallel quite easily.

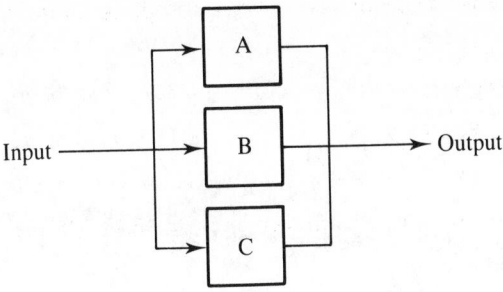

Figure 15.28 Parallel system.

Example 15.47

A process control computer is supplied by two identical power supplies. If one fails then the other takes over. The MTBF of the power supplies is 1000 hours. Calculate the reliability of the power supply to the computer during a 28 day period. Compare this with the reliability if there is no standby power supply.

Solution

$$k = \frac{1}{1000} = 0.001 \text{ failures per hour}$$

$$R_{ps} = e^{-kt} = e^{-0.001 \times 24 \times 28} = 0.511$$

Therefore the overall reliability of the power supply to the computer is:

$$R = 1 - (1 - 0.511)(1 - 0.511) = 0.761$$

Clearly the existence of a standby power supply does improve the reliability of the power supply to the computer. In practice, the figure would be much higher than this because once a power supply failed the standby would take over and maintenance engineers would quickly repair the failed power supply. Therefore the period of time on which the computer was relying on one power supply would be very small. ▲

We have only examined the simplest possible reliability models. In practice, reliability engineering can often be very complicated. Some models can cater for maintenance strategies of the sort we touched on in the previous example. The effect of non-uniform failure rates can also be catered for. Also, many systems consist of a mixture of series and parallel subsystems of the type we have discussed as well as more complicated failure modes than we have examined.

EXERCISES 15.17

1. A computer contains 20 circuit boards each with a MTBF of three months. Calculate the probability that the computer will suffer a circuit board failure during a monthly period.

2. Three pumps are required to feed water to a boiler in a power station in order for it to be fully operational. The MTBF of each of the pumps is ten days. Calculate the probability that the boiler will not be fully operational during a monthly period if there are only three pumps available.

3. A process control computer has a power supply with a MTBF of 600 hours. If the power supply fails then it takes two hours to replace it. Calculate the average availability of the computer power supply.

4. A radar station has a MTBF of 1000 hours. The average repair time is ten hours. Calculate the average availability of the radar station.

5. A process line to manufacture bread consists of four main stages: mixing of ingredients, cooking, separation and finishing, packaging. Each of the stages has a MTBF of

$$MTBF_m = 30 \text{ hours}$$

$$MTBF_c = 10 \text{ hours}$$

$$MTBF_{sf} = 20 \text{ hours}$$

$$MTBF_p = 15 \text{ hours}.$$

Calculate the probability that the process line will be stopped during an eight-hour shift.

6. A remote water pumping station has two pumps. Only one pump is needed in normal operation; the other acts as a standby. Access to this station is difficult and so if a pump fails it is not easy to repair it immediately. Given that the MTBF of the pumps is 3000 hours, calculate the overall reliability of the pumping station during a 28 day period in which the maintenance engineers are unable to repair a pump which fails. Calculate the improvement in reliability that would be obtained if a second standby pump was installed in the pumping station.

7. Repeat the calculation carried out in Exercise 15.17.6 with the following information. The pumps, which are different to those in Exercise 15.17.6, can be considered to consist of two components: a motor and the pump itself. The MTBF of the motors is 4000 hours and that of the pumps is 6000 hours.

Miscellaneous exercises

1. Find the mean and standard deviation of
   8, 6.9, 7.2, 8.4, 9.6, 10.3, 7.4, 9.0

2. Three machines, A, B and C, manufacture a component. Machine A manufactures 35% of the components, machine B manufactures 40% of the components and machine C makes the rest. A component is either acceptable or not acceptable; 7% of components made by machine A are not acceptable, 12% of components made by machine B are not acceptable and 2% of those made by machine C are also not acceptable.

   (a) Find the probability that a component is made either by machine A or machine B.

   (b) Two components are picked at random. What is the probability that they are both made by machine B?

   (c) Three components are picked at random. What is the probability they are each made by a different machine?

   (d) A component is picked at random. What is the probability it is not acceptable?

(e) A component is picked at random. It is not acceptable. What is the probability it was made by machine B?

(f) A component is picked at random and is acceptable. What is the probability it was made by either machine A or machine B?

3. A continuous random variable, x, has a p.d.f. $f(x)$, given by

$$f(x) = 3x^2 \qquad 0 \le x \le 1$$

(a) Find $P(0 \le x \le 0.5)$

(b) Find $P(x > 0.3)$

(c) Find $P(x < 0.6)$

(d) Find the expected value of x.

(e) Find the standard deviation of x.

4. A discrete random variable, y, has a probability distribution given by

y	−0.25	0.25	0.75	1.25	1.75
P(y)	0.25	0.20	0.10	0.15	0.30

(a) Find the expected value of y.

(b) Find the standard deviation of y.

Appendices

Appendix I

Numerical solution of non-linear equations

It is often necessary to solve equations of the form $f(x) = 0$. For example,

$$f(x) = x^3 - 3x^2 + 7 = 0 \qquad f(x) = \ln x - \frac{1}{x} = 0$$

To 'solve' means to find values of x which satisfy the given equation. These values are known as **roots**. Equations where the unknown quantity, x, occurs only to the first power are called **linear equations**. Otherwise an equation is **non-linear**. A simple, but rough and ready way of finding a root of an equation $f(x) = 0$ is to sketch a graph of $y = f(x)$ as shown in Figure I.1.

The roots are those values of x where the graph cuts or touches the x axis and so a sketch will provide estimates of roots. Furthermore we note from Figure I.1 that the values of $f(x)$ on opposite sides of the root will usually have different signs. Therefore we can locate solutions of $f(x) = 0$ by looking for changes in sign of $f(x)$. Generally there is no analytical way of solving the equation $f(x) = 0$ and so we must resort to approximate or numerical methods of solution. An **iterative** technique produces a sequence of approximate solutions which may converge to the desired root. Iterative techniques can fail in that the sequence produced can diverge. Whether or not this happens depends upon the equation being solved and the availability of a good estimate at the start. One such method – the Newton–Raphson technique – has been described in Section 7.11. An alternative method is that of **simple iteration**. While convergence is generally not as rapid as that produced by the Newton–Raphson technique, the method is particularly simple to apply. It requires that the equation being solved be rewritten in the form $x = g(x)$. An estimate, x_0 of the root is made, perhaps from sketching a graph, and this value is substituted into the right-hand side of $x = g(x)$. This yields another estimate, x_1. The process is repeated until, hopefully, the sequence

663

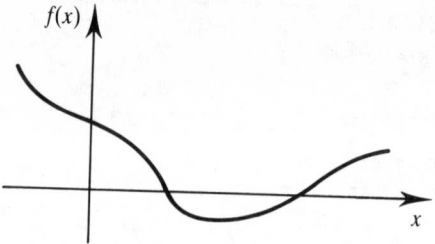

Figure I.1   A root of $f(x) = 0$ occurs where the graph touches or cuts the x axis.

of numbers produced converges to a root. Formally we express this as:

$$x_{n+1} = g(x_n)$$

Unfortunately, the method of simple iteration does not always work because the sequence x_n diverges. Whether or not the method works depends upon the particular rearrangement of the equation and also the initial estimate x_0. It is particularly simple to program on a microcomputer. A check would be built into the program to test whether or not successive estimates have converged.

Example I

Solve $f(x) = e^{-x} - x = 0$ by simple iteration.

Solution

Rearrange the equation to obtain $x = e^{-x}$. Suppose we estimate $x_0 = 0$. Then $x_1 = e^{-x_0} = e^{-0} = 1$. Similarly $x_2 = e^{-x_1} = e^{-1} = 0.368$. The process is continued. In general we have:

$$x_{n+1} = e^{-x_n}$$

The calculation is shown in Table I.1. The sequences converges (eventually) to 0.567 (three decimal places) so that $x = 0.567$ is a root of $e^{-x} - x = 0$.   ▲

Table I.1
Iterative
solution to
Example I.1.

n	x_n
0	0
1	1
2	0.368
3	0.692
4	0.501
5	0.606
6	0.546
⋮	⋮
	0.567

Finally, if the equation to be solved involves trigonometric functions, angles must be measured in radians not degrees.

Appendix II

Laws of indices

It is often necessary to multiply a number by itself repeatedly, for example $a \times a \times a \times \cdots \times a$. If the number a is multiplied by itself k times we write the result as a^k. The number k is called a **power** or **index** while the number a is often referred to as a **base**. The following laws hold:

$$a^m \times a^n = a^{m+n} \qquad \frac{a^m}{a^n} = a^{m-n} \qquad (a^m)^n = a^{mn} \qquad a^{-m} = \frac{1}{a^m}$$

Furthermore

$$a^0 = 1 \qquad \text{that is, any number raised to the power 0 is 1}$$

and

$$a^1 = a \qquad \text{that is, any number raised to the power 1 is itself}$$

Roots of numbers can also be expressed using indices:

$$\sqrt[m]{a} = a^{1/m}$$

so that, for example, we can write $\sqrt[2]{a} = \sqrt{a} = a^{1/2}$.

Appendix III

Radian measure

Angles can be measured in units of either radians or degrees. A complete revolution is defined as 360 degrees ($360°$) or 2π radians. It is easy to use this fact to convert between the two measures. We have

$$360° = 2\pi \text{ radians}$$

$$1° = \frac{2\pi}{360} = \frac{\pi}{180} \text{ radians}$$

$$1 \text{ radian} = \frac{180}{\pi} \text{ degrees} \approx 57.3°$$

Your calculator should be able to work with angles measured in both radians and degrees. Usually the MODE button allows you to select the appropriate measure. Remember that in all work involving calculus angles must be measured in radians.

Appendix IV

Factorials

The number $5 \times 4 \times 3 \times 2 \times 1$ is referred to as '5 factorial' and is written as 5!.
Generally if $n \in \mathbb{N}^+$, $n! = n(n-1)(n-2)\ldots 3.2.1$. We define $0! = 1$. You will
probably find that your calculator is able to compute factorials. Frequently
expressions involving factorials need to be simplified. For example,

$$\frac{n!}{(n-1)!} = \frac{n(n-1)(n-2)\ldots 3.2.1}{(n-1)(n-2)\ldots 3.2.1}$$

$$= n$$

Appendix V

A sequence as a sum of weighted impulses

Suppose we have a continuous function of t, say $f(t)$ as shown in Figure V.1. If
we sample this function at intervals T we obtain a sampled sequence $f[k]$, $k \in \mathbb{N}$,
that is,

$$f[0], f[1], f[2]\ldots$$

This is depicted in Figure V.2. The sequence is a function of the discrete

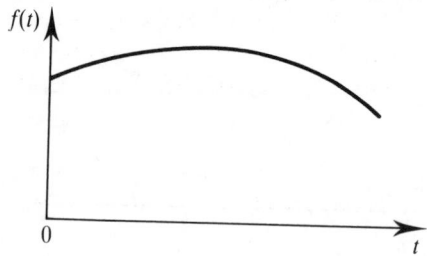

Figure V.1   A continuous function $f(t)$.

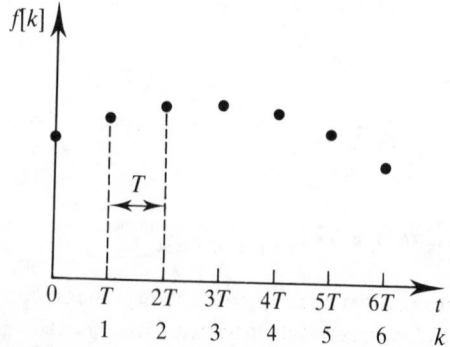

Figure V.2   The sampled sequence $f[k]$.

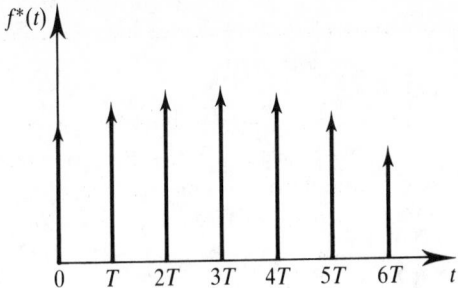

Figure V.3 Representation of f[k] as a sum of impulses.

variable k. Instead of working with the sequence directly it is often convenient to express it in the form:

$$f*(t) = \;= \sum_{k=0}^{\infty} f[k]\,\delta(t - kT) \tag{V.1}$$

where $f*$ denotes that the original function $f(t)$ has been sampled. Observe that Equation (V.1) takes the form of a function of the continuous variable t. One of the reasons for writing the sequence in this form is that transform techniques (for example, the Laplace transform) can be applied. The representation given by Equation (V.1) is justified as follows.

The δ function, $\delta(t - kT)$, is zero everywhere except when $t = kT$, $k \in \mathbb{N}$, so that $f*(t)$ is zero except when $t = kT$. It is evident that Equation (V.1) lends itself to situations in which events of interest are happening at discrete points in time. Suppose we now select such a point, for example, $t = mT$, $m \in \mathbb{N}$. Then

$$f*(mT) = \sum_{k=0}^{\infty} f[k]\,\delta(mT - kT)$$

The δ function is zero everywhere except where its argument is zero, that is, where $k = m$, hence

$$f*(mT) = f[m]\,\delta(0)$$

This represents an impulse of strength $f[m]$ occurring when $t = mT$. In general,

$$f*(t) = \sum_{k=0}^{\infty} f[k]\,\delta(t - kT)$$

represents a sum of impulses each of strength $f[k]$ occurring at $t = kT$ as shown in Figure V.3. It is in this sense that Equation (V.1) is used as a valid representation of the sequence $f[k]$.

Appendix VI

The binomial expansion of $\left(\dfrac{n - N}{n}\right)^n$

Consider the binomial expansion of $\left(\dfrac{n - N}{n}\right)^n$

$$\left(\frac{n - N}{n}\right)^n = \left(1 - \frac{N}{n}\right)^n$$

$$= 1 + n\left(-\frac{N}{n}\right) + \frac{n(n - 1)}{2!}\left(-\frac{N}{n}\right)^2$$

$$+ \frac{n(n - 1)(n - 2)}{3!}\left(-\frac{N}{n}\right)^3 + \cdots$$

$$= 1 - N + \left(\frac{n - 1}{n}\right)\frac{N^2}{2!} - \left(\frac{n - 1}{n}\right)\left(\frac{n - 2}{n}\right)\frac{N^3}{3!} + \cdots$$

$$= 1 - N + \left(1 - \frac{1}{n}\right)\frac{N^2}{2!} - \left(1 - \frac{1}{n}\right)\left(1 - \frac{2}{n}\right)\frac{N^3}{3!} + \cdots$$

Suppose now we let $n \to \infty$. We find,

$$\lim_{n \to \infty} \left(\frac{n - N}{n}\right)^n = 1 - N + \frac{N^2}{2!} - \frac{N^3}{3!} + \cdots$$

But this is the power series expansion of e^{-N} (see Section 3.5). We conclude that:

$$\lim_{n \to \infty} \left(\frac{n - N}{n}\right)^n = e^{-N}$$

In particular, note that if $N = -1$

$$\lim_{n \to \infty} \left(\frac{n + 1}{n}\right)^n = e$$

Appendix VII

The Greek alphabet

A	α	alpha	I	ι	iota	P	ρ	rho
B	β	beta	K	κ	kappa	Σ	σ	sigma
Γ	γ	gamma	Λ	λ	lambda	T	τ	tau
Δ	δ	delta	M	μ	mu	Y	υ	upsilon
E	ε	epsilon	N	ν	nu	Φ	ϕ	phi
Z	ζ	zeta	Ξ	ξ	xi	X	χ	chi
H	η	eta	O	o	omicron	Ψ	ψ	psi
Θ	θ	theta	Π	π	pi	Ω	ω	omega

Appendix VIII

SI units and prefixes

Throughout this book SI units have been used. Below is a list of these units together with their symbols.

Quantity	SI unit	Symbol	Quantity	SI unit	Symbol
length	metre	m	force	newton	N
mass	kilogram	kg	power	watt	W
time	second	s	electric charge	coulomb	C
frequency	hertz	Hz	potential difference	volt	V
electric current	ampere	A	resistance	ohm	Ω
temperature	kelvin	K	capacitance	farad	F
energy	joule	J	inductance	henry	H

	Prefix	Symbol
10^{12}	tera	T
10^{9}	giga	G
10^{6}	mega	M
10^{3}	kilo	k
10^{2}	hecto	h
10^{1}	deca	da
10^{-1}	deci	d
10^{-2}	centi	c
10^{-3}	milli	m
10^{-6}	micro	μ
10^{-9}	nano	n
10^{-12}	pico	p

Solutions

Chapter 1

EXERCISES 1.1

1. (a) Square the input and then multiply by 2; domain $(-\infty, \infty)$, range $[0, \infty)$.

   (b) Square the input, then subtract 1; domain $[0, \infty)$, range $[-1, \infty)$.

   (c) Multiply input by 3 and subtract 4; domain $[0, \infty)$, range $[-4, \infty)$.

   (d) Cube the input; domain $(-\infty, \infty)$, range $(-\infty, \infty)$.

   (e) Multiply input by 0.5, and then add 2; domain $[-2, 10]$, range $[1, 7]$.

   (f) Multiply input by 3 and then subtract 2; domain $[3, 8]$, range $[7, 22]$.

2. (a) 19 (b) -11 (c) $5x + 4$ (d) $5x + 9$ (e) $15x + 4$
   (f) $5x^2 + 4$

3. (a) -4 (b) 16 (c) 41 (d) $5x^2 - 4$ (e) $20t^2 - 20t + 1$

4. 3183 ohms

5. (a) $f^{-1}(x) = x - 4$ (b) $g^{-1}(t) = (t-1)/3$
   (c) $y^{-1}(x) = x^{1/3}$ (d) $h^{-1}(t) = 3t + 8$

6. (a) $9t^2 + 12t + 5$ (b) $1/t^2 + 1$ (c) $3/t + 2$ (d) $1/(t^2 + 1)$
   (e) $9/t^2 + 12/t + 5$

7. (a) $6t + 3$ (b) $(t-3)/2$ (c) $t/3$ (d) $(t-3)/6$ (e) $(t-3)/6$

8. Piecewise continuous, discontinuity at $t = 2$.

9. Discontinuities at $t = 0, 2$.

EXERCISES 1.2

1. (a) 3 (b) 5 (c) 4 (d) 2 (e) 2 (f) 3

3. (a) 34/3 (b) 12 (c) 2, 3 (d) 3, -4 (e) 1.372, -4.372 (f) 5
   (g) 1, 2, 3 (h) 1, -2, 3 (i) 3, 4, -6

4. (a) -4, 2 (b) -0.742, 6.742 (c) -5.162, 1.162
   (d) -0.681, 14.681 (e) -9.933, 4.933

5. (a) -0.36 (b) -1.99, 1.36 (c) -1

EXERCISES 1.3

1. (a) $x = 3$, $f = 2$ (b) $s = -1$, $g = 1$ (c) $h = 0$ (d) $x = 0$, $y = 1$
   (e) $x = 1$, $x = 2$, $r = 0$ (f) $z = -1$, $z = 6$, $y = 0$
   (g) $s = 1$, $s = -1$, $G = 0$

2. $x = -2/3$, $l = 2x/3 - 7/9$

3. $x = -3/2$, $p = 2x + 1/2$

4. $y = x$

5. (a) $x = 3$ (b) $s = -1$ (d) $x = 0$ (e) $x = 1$, $x = 2$
   (f) $z = -1$, $z = 6$ (g) $s = 1$, $s = -1$

EXERCISES 1.4

1. (a) $(2x-1)/(x+2)(x-3)$ (b) $\dfrac{(5x^2 + 19x + 6)}{2x(x+2)(x+3)}$

   (c) $(9x + 4)/2x^2(x+4)$ (d) $\dfrac{(2x^2 + x + 4)}{(x+2)^2(x-3)}$

   (e) $\dfrac{(19x^3 + 25x^2 + 53x - 9)}{2x(x^2 + 3)(7x - 1)}$

2. (a) $1/(x+1) + 3/(2x-3)$ (b) $7/(x+3) - 4/(x+2)$
   (c) $1/(s+1) + 2/(s+1)^2$ (d) $x - x/(x^2 + 1)$ (e) $1 - 1/(s^2 + 1)$

   (f) $\dfrac{2s+1}{s^2+s+1} - \dfrac{1}{s-4}$

   (g) $11/9(x-2) + 4/3(x-2)^2 - 2/9(x+1)$
   (h) $2s + 1 - 3/(s^2 + s - 1)$ (i) $x + 1 + 1/(x+1) + 1/(x+2)$

EXERCISES 1.5

2. (a) $e^{-x/3}$ (b) e^{2t-5} (c) $1 + e^x$ (d) e^{-14} (e) e^{8t}

4. (a) 10^{-6} (b) 10^{-5} (c) 3.3×10^{-6} (d) 5.6×10^{-6}

EXERCISES 1.6

2. (a) $2 \ln t$ or $\ln t^2$ (b) $16 \log t$ (c) $\ln(y^7/3)$ (d) $\ln 2$
   (e) $\log(9x^{3/2}/2)$

3. (a) 1.0148 (b) −0.4130 (c) −0.2877 (d) −1.3863 (e) 1.3117
(f) 2.0553 (g) 0.5485

4. (a) 20 dB (b) 80 dB (c) 64.1 dB (d) 30.46 dB

5. Preamplifier gain = 26.02 dB   main amplifier gain = 17.50 dB
total gain = 43.52 dB

6. (a) 3 (b) 3.9069 (c) 1.4110 (d) 1.7356 (e) 1.5079 (f) 1.8515

EXERCISES 1.7

1. (a) cosec 2x   (b) −1

EXERCISES 1.8

2. (a) 0.8481 (b) 0.8957 (c) 1.2490

EXERCISES 1.9

1. (a) $2\pi/7$ (b) π (c) 2π (d) $2\pi/3$ (e) π (f) $3\pi/2$

2. (2/3) sin($\pi t + k$) or (2/3) cos($\pi t + k$)

3. (a) Phase angle = 3, time displacement = 3 (b) −3, −3/2
(c) 0.2, 0.4 (d) −2, −2 (e) 4/5, 4/3 (f) −0.858, −0.286
(g) π, 1/2 (h) −3, −3/5π (i) 2, 6/π (j) −3π, −3π

4. (a) (i) $\sqrt{41}$ sin($t+0.675$) (ii) $\sqrt{41}$ sin($t−5.608$)
(iii) $\sqrt{41}$ sin($t+5.387$) (iv) $\sqrt{41}$ cos($t−0.896$)
(b) (i) $\sqrt{8}$ sin($3t+3\pi/4$) (ii) $\sqrt{8}$ sin($3t−5\pi/4$)
(iii) $\sqrt{8}$ cos($3t+\pi/4$) (iv) $\sqrt{8}$ cos($3t−\pi/4$)
(c) (i) $\sqrt{52}$ sin($2t+5.300$) (ii) $\sqrt{52}$ sin($2t−0.983$)
(iii) $\sqrt{52}$ cos($2t+3.730$) (iv) $\sqrt{52}$ cos($2t−2.554$)
(d) (i) $\sqrt{10}$ sin($5t+4.391$) (ii) $\sqrt{10}$ sin($5t−1.893$)
(iii) $\sqrt{10}$ cos($5t+2.820$) (iv) $\sqrt{10}$ cos($5t−3.463$)

EXERCISES 1.10

1. $(a+b)\cosh x + (a−b)\sinh x$

2. $((a+b)/2)e^x + ((a−b)/2)e^{−x}$

EXERCISES 1.11

1. (a) $x>4$ or $x<−4$ (b) $−2 \leqq y \leqq 4$
(c) $t \geqq −3$ or $t \leqq −9$ (d) $−3 < t < 3$
(e) no value of t satisfies this

MISCELLANEOUS EXERCISES 1

1. (a) multiply input by 7, then subtract 2
(b) square the input, then subtract 2
(c) calculate e^x where x is the input, multiply by 3 and then add 4
(d) take the cosine of the input, multiply by 3 and then add 2
(e) multiply by 2, calculate exponential, subtract 1 and then divide
by 2
(f) multiply by 100π, add $\pi/2$, take sine and multiply by 6
(g) cube input, add to twice the input, and add 5

2. (a) $y^{−1}(x) = x/2$ (b) $f^{−1}(t) = (t+3)/8$
(c) $f^{−1}(x) = 0.5 \ln x$

3. (a) sin λ (b) sin($t−\lambda$) (c) sin($t+\lambda$)

4. $g(\lambda) = \ln(\lambda^2 + 1)$, $g(t−\lambda) = \ln((t−\lambda)^2 + 1)$

5. (a) $\dfrac{(19x^2 + 22x + 4)}{2x(x+2)(2x+1)}$

(b) $\dfrac{(3s^3 + 12s^2 + 22s + 12)}{2s^2(4s+3)}$

(c) $\dfrac{2(3s^4 + 30s^3 + 120s^2 + 202s + 120)}{s^2(s+2)(s+3)(s+4)}$

6. (a) $\sqrt{13}$ sin($\omega t + 0.5880$), amplitude = $\sqrt{13}$, phase angle = 0.5880
(b) 0.765 sin($\omega t + 1.1781$), amplitude = 0.765, phase angle = 1.1781
(c) 5.596 sin($\omega t + 2.10$), amplitude = 5.596, phase angle = 2.10
(d) 1.581 sin($\omega t + 0.4636$), amplitude = 1.581, phase angle = 0.4636
(e) −3 sin ωt, amplitude = 3, phase angle = π

7. (a) ln 6 (b) ln t^7 (c) e^8 (d) x (e) $e^{2x} + 2 + e^{−2x}$

8. (a) 5, −7 (b) 0, 1, −6 (c) ±1.817 (d) −3.689, 1.689
(e) 0.5, 2.5

Chapter 2

EXERCISES 2.1

1. $A \cap B = \{3, 5\}$, $A \cup B = \{1, 3, 4, 5, 6, 7, 9, 11\}$

2. (a) $\{0, 1, 2, 3, 4, 5, 6, 7, 8, 9\}$ (b) $\{0, 3, 6, 9\}$

5. (a) function (b) not a function since 0 is mapped to two elements
(c) not a function since 2 is not mapped to anything

6. r is not a function since 1 is not mapped to anything

7. (a) $\{x: x \in \mathbb{R}\}$ (b) $\{t: t \in \mathbb{R} \text{ and } t \geqq 2\}$
(c) $\{x: x \in \mathbb{R}, -1 \leqq x \leqq 7\}$ (d) $\{x: x \in \mathbb{R}, x \geqq −3\}$

(e) $\{t: t\in\mathbb{R}, t \geq 184\}$ (f) $\{t: t\in\mathbb{R}, -59 \leq t \leq 337\}$
(g) $\{t: t\in\mathbb{R}, -1 \leq t \leq 1\}$ (h) $\{t: t\in\mathbb{R}, 0 \leq t \leq 3\}$
(i) $\{x: x\in\mathbb{R}, x \geq 1\}$

EXERCISES 2.2

3. (a) $X = (A\cdot B)\cdot(C + D + E)$
(b) $X = (A\cdot\bar{B} + \bar{A}\cdot B) + (C\cdot D) + (E + F)$
4. (a) $A + B\cdot C$ (b) $A + B + C + D$
(c) $(A\cdot B) + (A\cdot\bar{B}\cdot D)$. This can be further simplified to $A\cdot(B + D)$.

MISCELLANEOUS EXERCISES 2

1. (a) $S = \{-7, -6, -5, -4, -3, 3, 4, 5, 6, 7\}$ (b) $S = \{3, 4, 5, 6, 7\}$
(c) $S = \{3\}$
2. $A\cap B = \{16, 17, 18, 19, 20\}$
$A\cup B = \{m: m\in\mathbb{Z}, m \geq -10\}$
3. (a) $\{x: x\in\mathbb{R}, e^{-6} \leq x \leq e^{15}\}$ (b) $\{x: x\in\mathbb{R}, 0 \leq x \leq 2\}$
(c) $\{x: x\in\mathbb{R}, x \geq 2\}$ (d) $\{x: x\in\mathbb{R}, x \neq 1\}$
(e) $\{x: x\in\mathbb{R}, -6 \leq x \leq 6\}$ (f) $\{x: x\in\mathbb{R}, 0 \leq x \leq 3\}$
(g) $\{x: x\in\mathbb{R}, x \geq 1\}$ (h) $\{x: x\in\mathbb{R}, x \geq \sinh 2\}$
4. (a) $\bar{A} + \bar{B}\cdot C$ (b) $A\cdot B + A\cdot C + B\cdot\bar{C}$ (c) $A\cdot(B + C)$

Chapter 3

EXERCISES 3.1

2. cos 2, cos 3, cos 4, cos 5, cos 6, cos 7
3. 0, 3/2, 4, 15/2, 12
4. (a) 1, 1/2, 1/4, 1/8, 1/16   (b) 2, 1, −1, −5, −13
5. 11, 21, 31, 41
6. 1, 1, 2, 6, 15
7. 5, 11, 26, 59, 137
8. (a) 10th term = 35, 19th term = 62
(b) 10th term = −19, 19th term = −46
9. (a) $-2b/3$ (b) $b(4 - k)/3$ (c) −45/16
10. 1/2, 1/64
11. 1536, 1 572 864
12. only (c) must be true
13. ±128
14. 3
15. 0

16. (a) limit does not exist   (b) 1/2   (c) 1
17. 16
18. (a) limit does not exist   (b) limit does not exist   (c) 0
(d) 3/5   (e) 0

EXERCISES 3.2

1. 15 060
2. (a) 650   (b) −490   3. 23.5
4. 41
5. 1.494, $S_\infty = 3/2$
6. 19 946
7. 650
8. 3/4
9. $\pm 1/\sqrt{2}$
10. $\sum_1^\infty (-1)^{n+1}/n$
11. $1 + 1/z + 1/z^2 + 1/z^3 + 1/z^4 + 1/z^5$
13. (a) converges   (b) diverges
14. series diverges
15. (a) converges absolutely   (b) converges absolutely
(c) diverges   (d) conditionally convergent
16. converges

EXERCISES 3.3

1. $a^8 + 8a^7b + 28a^6b^2 + 56a^5b^3 + 70a^4b^4 + 56a^3b^5 + 28a^2b^6 + 8ab^7 + b^8$
2. $16x^4 + 96x^3y + 216x^2y^2 + 216xy^3 + 81y^4$
3. $a^5 - 10a^4b + 40a^3b^2 - 80a^2b^3 + 80ab^4 - 32b^5$
4. $729 - 2916x + 4860x^2 - 4320x^3$
5. $1 + 5x + 45x^2/4 + 15x^3$
6. $1 + x - x^2/2 + x^3/2 - 5x^4/8$ valid for $-1/2 < x < 1/2$
7. $1 - 2x + 5x^2/2 - 5x^3/2 + 35x^4/16$ valid for $-2 < x < 2$
8. $1 - 1/2x + 3/8x^2 - 5/16x^3$
9. (a) $1 + 4x^2 + 6x^4 + 4x^6 + x^8$ (b) $1 + 4/x^2 + 6/x^4 + 4/x^6 + 1/x^8$
10. (a) $1 + 1/2x - 1/8x^2 + 1/16x^3$ valid for $|x| > 1$
(b) $x^{-1/2}(1 + x/2 - x^2/8 + x^3/16)$ valid for $|x| < 1$
11. (a) $1 - 4x^2 + 10x^4 - 20x^6$ valid for $|x| < 1$
(b) $x^{-8}(1 - 4/x^2 + 10/x^4 - 20/x^6 \cdots)$ valid for $|x| > 1$

EXERCISES 3.4

1.

x	e^x	Sum to 4 terms
0	1	1
0.1	1.1052	1.1052
0.5	1.6487	1.6458
1	2.7183	2.6667

Values are in close agreement when x is small

2. (a) $1 - 2x^2 + 2x^4/3 + \cdots$ (b) $1 - x^2/8 + x^4/384 - \cdots$
3. e
4. $x + x^2 + x^3/3$
5. (a) $1 - x + x^2/2! - x^3/3! + \cdots$
   (b) $\cosh x = 1 + x^2/2! + x^4/4! + \cdots$, $\sinh x = x + x^3/3! + x^5/5! + \cdots$

MISCELLANEOUS EXERCISES 3

1. (a) $1, -1, 1, -1, 1$ (b) $1, -1/3, 1/5, -1/7, 1/9$
2. (a) $8k - 7$ $k \geqq 1$ (b) $(-1)^k$ $k \geqq 1$
4. $1 + 5x + 15x^2 + 30x^3$
5. $(1 - x + 2x^2/3 - 10x^3/27 + \cdots)/27$
7. 0
8. 27/8
9. 0
11. $0, 2, 3, 3, -1, -7$
12. (a) 0 (b) no limit (c) no limit (d) a (e) no limit
13. (a) $-40/9$ (b) 19
14. 2, 3/4

Chapter 4

EXERCISES 4.1
1. (a) $\overrightarrow{DB}$ (b) $\overrightarrow{DE}$
2. $\mathbf{q} - \mathbf{p}, \mathbf{q} + \mathbf{p}, \mathbf{q} + \mathbf{r} + \mathbf{p}$

EXERCISES 4.2
1. $\overrightarrow{PQ} = -6\mathbf{i} + 3\mathbf{j}$ $|\overrightarrow{PQ}| = 6.71$
2. $3\mathbf{i} + 4\mathbf{j}, \mathbf{i} - 5\mathbf{j}, -2\mathbf{i} - 9\mathbf{j}$ unit vector: $\dfrac{-1}{\sqrt{85}}(2\mathbf{i} + 9\mathbf{j})$

3. $\dfrac{1}{\sqrt{26}}(4\mathbf{i} - \mathbf{j} + 3\mathbf{k}), \dfrac{1}{3}(-2\mathbf{i} + 2\mathbf{j} - \mathbf{k}), \dfrac{1}{\sqrt{61}}(-6\mathbf{i} + 3\mathbf{j} - 4\mathbf{k})$
4. $8\mathbf{i} - 9\mathbf{j}, 2\mathbf{i} + 2\mathbf{j}, -6\mathbf{i} + 11\mathbf{j}, -21\mathbf{i} + 40\mathbf{j}$
5. $\dfrac{1}{\sqrt{50}}(\mathbf{i} + 7\mathbf{j})$
6. $\overrightarrow{PQ} = -3\mathbf{i} + \mathbf{j} + \mathbf{k}$, distance from P to Q = 3.32, $\dfrac{1}{4}(13\mathbf{i} + 5\mathbf{j} + 13\mathbf{k})$
7. P and Q
8. 21.66, 10.44, 5.39, 0.81, 21.5$\mathbf{i}$ + 16.8$\mathbf{j}$ + 5.6$\mathbf{k}$

EXERCISES 4.3
1. $-22, -22, 58, 20$
2. 19
3. 18.4°
4. 7
5. 14, 16, 26
6. $2|\mathbf{q}|^2, 7|\mathbf{q}|^2, -2|\mathbf{q}|^2$
7. $\sqrt{3}, 3, -1, 101.1°$
8. $3, 7, -8, 112.4°$
10. $c(\mathbf{i} + 2\mathbf{j})$, c constant

EXERCISES 4.4
1. (a) $5\mathbf{i} + 7\mathbf{j} + 3\mathbf{k}$ (b) $-5\mathbf{i} - 7\mathbf{j} - 3\mathbf{k}$
2. $-10\mathbf{i} - 5\mathbf{j} + 10\mathbf{k}$
3. (a) $-\mathbf{i} - 3\mathbf{j} + 2\mathbf{k}$ (b) $\mathbf{i} - \mathbf{j}$
4. $\sqrt{89}, \sqrt{26}, 48.01, 0.9980$
5. (a) $\mathbf{q} \times \mathbf{p}$ (b) $2\mathbf{q} \times \mathbf{p}$
6. $\dfrac{1}{\sqrt{6}}(\mathbf{i} - \mathbf{j} - 2\mathbf{k})$, 0.775
7. $\dfrac{1}{\sqrt{27}}(\mathbf{i} - 5\mathbf{j} + \mathbf{k})$
8. $-\mathbf{i} - 46\mathbf{j} + 17\mathbf{k}$

EXERCISES 4.5
1. $2, \sqrt{15}, 6, \sqrt{7}$
2. **a** and **e**, **b** and **d**

MISCELLANEOUS EXERCISES 4

1. (a) 24, $-7\mathbf{i} - 32\mathbf{j} + 17\mathbf{k}$ (b) 18, **0**
2. **0**
3. $\sqrt{54}$, $\sqrt{66}$, 57, 0.9548
4. $\sqrt{41}$, $\sqrt{19}$, $\sqrt{154}$, 0.4446
5. $\dfrac{1}{\sqrt{139}}$ **a**, $\dfrac{1}{\sqrt{20}}$ **b**, $\dfrac{1}{\sqrt{2296}}(-12\mathbf{i} - 6\mathbf{j} - 46\mathbf{k})$
6. $5\mathbf{i} + 7\mathbf{j} - 4\mathbf{k}$, 22
7. $(\mathbf{a} \times \mathbf{b}) \times \mathbf{c} = 4\mathbf{i} - 12\mathbf{j} - 16\mathbf{k}$ $\mathbf{a} \cdot \mathbf{c} = 8$ $\mathbf{b} \cdot \mathbf{c} = 4$
   $\mathbf{a} \times (\mathbf{b} \times \mathbf{c}) = -2\mathbf{i} - 22\mathbf{j} - 8\mathbf{k}$
9. $\dfrac{1}{\sqrt{50}}(\mathbf{i} + 7\mathbf{j})$, $\dfrac{1}{\sqrt{5}}(\mathbf{i} + 2\mathbf{j})$, $18.4°$
10. $\dfrac{\mu q}{56\pi}(\mathbf{i} + 11\mathbf{j} + 7\mathbf{k})$

Chapter 5

EXERCISES 5.1

1. (a) 15 (b) -19 (c) $\begin{pmatrix} 47 \\ 12 \end{pmatrix}$ (d) $\begin{pmatrix} -7 & 18 \\ -14 & 10 \end{pmatrix}$ (e) $\begin{pmatrix} 1 & 0 \\ 0 & 1 \end{pmatrix}$

2. (a) $A + D$ does not exist, $\begin{pmatrix} -8 & 0 \\ -3 & 0 \end{pmatrix}$, $D - E$ does not exist

   (b) $\begin{pmatrix} 3 \\ 10 \end{pmatrix}$, BA does not exist, $\begin{pmatrix} -4 & -3 \\ 12 & 16 \end{pmatrix}$, $\begin{pmatrix} -7 & 5 \\ -21 & 19 \end{pmatrix}$, EB does not exist,

   DA does not exist, DB does not exist, $\begin{pmatrix} 6 & 4 & 2 \\ 3 & 2 & 1 \end{pmatrix}$,

   BE does not exist, $\begin{pmatrix} 3 & 5 & 3 \\ 10 & 17 & 8 \end{pmatrix}$

3. (c) $\begin{pmatrix} -49 & 7 \\ 0 & 28 \end{pmatrix}$, $(-9, -6, -3)$, $\begin{pmatrix} 2k & 3k & 4k \\ k & 2k & -k \end{pmatrix}$ (c) $\begin{pmatrix} 0 & 0 \\ 1 & 2 \end{pmatrix}$, $\begin{pmatrix} 0 & 0 \\ 1 & 2 \end{pmatrix}$, $\begin{pmatrix} -3 \\ 2 \end{pmatrix}$

   (a) $\begin{pmatrix} 1 & 2 \\ 0 & 0 \end{pmatrix}$, $\begin{pmatrix} 2 \\ -3 \end{pmatrix}$, (b) $\begin{pmatrix} 0 & 0 \\ 1 & 2 \end{pmatrix}$

4. $\begin{pmatrix} 7 & 17 & 23 \\ -1 & 10 & 19 \\ 24 & 10 & 4 \end{pmatrix}$, $\begin{pmatrix} 8 & 16 & 17 \\ 17 & 4 & 8 \\ 23 & 8 & 9 \end{pmatrix}$

5. $\begin{pmatrix} 18 & 14 \\ 7 & 11 \end{pmatrix}$, $\begin{pmatrix} 86 & 78 \\ 39 & 47 \end{pmatrix}$

6. $\begin{pmatrix} -10 & 16 \\ -20 & 18 \end{pmatrix}$, $\begin{pmatrix} 0 & 10 \\ -14 & 8 \end{pmatrix}$, $\begin{pmatrix} 3 & 4 \\ -6 & 9 \end{pmatrix}$, $\begin{pmatrix} -15 & 48 \\ -72 & 57 \end{pmatrix}$

7. (a) $\begin{pmatrix} 1 & -2 \\ 1 & 1 \end{pmatrix}$, $\begin{pmatrix} 5 \\ -2 \\ 1 \end{pmatrix}$ (b)

8. $\begin{pmatrix} 46 & 29 \\ 29 & 78 \end{pmatrix}$, $\begin{pmatrix} 8 & 37 \\ 36 & 13 \\ 19 & -6 \end{pmatrix}$

9. (a) $\begin{pmatrix} 10 \\ 31 \\ 18 \end{pmatrix}$ (b) $\begin{pmatrix} 8 \\ 36 \\ 19 \end{pmatrix}$ (c) $\begin{pmatrix} 11 \\ 28 \end{pmatrix}$ (d) $\begin{pmatrix} 11 \\ 28 \end{pmatrix}$ (e) $\begin{pmatrix} 31 \\ 33 \\ 38 \\ 18 \end{pmatrix}$

EXERCISES 5.2

1. $\begin{pmatrix} x \\ y \\ z \end{pmatrix}$

3. $\begin{pmatrix} 2 & 4 & -1 \\ 1 & 2 & 3 \\ 3 & 1 & 2 \end{pmatrix}$, $\begin{pmatrix} 1 & 0 & 3 \\ -7 & 2 & 4 \\ 0 & 5 & 5 \end{pmatrix}$, $\begin{pmatrix} 11 & 0 & 20 \\ 7 & -20 & 15 \\ 5 & 21 & 25 \end{pmatrix}$

   $\begin{pmatrix} 11 & 7 & 5 \\ 0 & -20 & 21 \\ 20 & 15 & 25 \end{pmatrix}$

4. diagonal matrix

EXERCISES 5.3

1. $\dfrac{1}{64} \begin{pmatrix} 8 & -6 \\ 4 & 5 \end{pmatrix}$

$AB = \begin{pmatrix} 1 & 4 \\ -3 & 9 \end{pmatrix}$, $(AB)^{-1} = \frac{1}{21}\begin{pmatrix} 9 & -4 \\ 3 & 1 \end{pmatrix}$, $B^{-1} = \frac{1}{7}\begin{pmatrix} 3 & -1 \\ 1 & 2 \end{pmatrix}$,

$A^{-1} = \frac{1}{3}\begin{pmatrix} 3 & -1 \\ 0 & 1 \end{pmatrix}$

3. $\begin{pmatrix} \cos\omega t & \sin\omega t & 0 \\ -\sin\omega t & \cos\omega t & 0 \\ 0 & 0 & 1 \end{pmatrix}$

4. $-132, -132$

5. $\begin{pmatrix} ae+bg & af+bh \\ ce+dg & cf+dh \end{pmatrix}$, $ad-bc$, $eh-fg$, $(ad-bc)(eh-fg)$

6. $\begin{pmatrix} -2 & 1 \\ 3/2 & -1/2 \end{pmatrix}$ $a=-2, b=5$

EXERCISES 5.4

1. 20, 33, 39
2. 1
3. 55, 504
4. $-164, -164$ Note $|A|=|A^T|$
5. (a) $x=y=z=1$ (b) $x=-1, y=2, z=3$
6. (a) -133 (b) $47, -39, 22$ (c) -133
7. $5i - 19j - 87k$

EXERCISES 5.5

1. (a) $|A|=-43$

adj$(A) = \begin{pmatrix} -2 & -5 & -7 \\ 7 & -4 & 3 \\ -18 & -2 & 23 \end{pmatrix}$, $A^{-1} = \frac{1}{43}\begin{pmatrix} 2 & 5 & 7 \\ -7 & 4 & -3 \\ 18 & 2 & -23 \end{pmatrix}$

(b) $|A|=25$

adj$(A) = \begin{pmatrix} 2 & 7 & 1 \\ 19 & -21 & -3 \\ 15 & -10 & -5 \end{pmatrix}$, $A^{-1} = \frac{1}{25}\begin{pmatrix} 2 & 7 & 1 \\ 19 & -21 & -3 \\ 15 & -10 & -5 \end{pmatrix}$

(c) $|A|=0$ adj$(A) = \begin{pmatrix} -16 & 7 & -1 \\ 32 & -14 & 2 \\ 16 & -7 & 1 \end{pmatrix}$ A^{-1} does not exist

2. $|P|=230$ adj$(P) = \begin{pmatrix} 71 & 52 & 55 \\ 52 & 64 & 50 \\ 55 & 50 & 75 \end{pmatrix}$ $P^{-1} = \frac{1}{230}\begin{pmatrix} 71 & 52 & 55 \\ 52 & 64 & 50 \\ 55 & 50 & 75 \end{pmatrix}$

EXERCISES 5.6

1. (a) $x=1, y=2$ (b) $x=2, y=3$ (c) $x=-1, y=2$
(d) $x=2, y=0, z=4$ (e) $x=1, y=2, z=3$ (f) $x=y=z=0$

EXERCISES 5.7

1. (a) $x=7, y=-6$ (b) $x=3, y=-5, z=2$
(c) $x=1-\mu, y=2\mu, z=\mu$ (d) $x=-3, y=1, z=4$
(e) inconsistent
2. $x=2, y=1, z=4$
3. (a) $\frac{1}{5}\begin{pmatrix} 2 & -1 \\ -3 & 4 \end{pmatrix}$ (b) $\frac{1}{15}\begin{pmatrix} 5 & -5 & 5 \\ -4 & 13 & -16 \\ 3 & -6 & 12 \end{pmatrix}$ (c) $\frac{1}{24}\begin{pmatrix} -9 & 6 & -3 \\ -37 & 22 & 1 \\ 11 & -2 & 1 \end{pmatrix}$

EXERCISES 5.8

1. (a) Jacobi

$x_1 = -0.2500, y_1 = 0, z_1 = 0.2500$
$x_2 = -0.3125, y_2 = -0.0417, z_2 = 0.3125$
$x_3 = -0.3177, y_3 = -0.0521, z_3 = 0.3490$

Gauss–Seidel

$x_1 = -0.2500, y_1 = 0.0417, z_1 = 0.2917$
$x_2 = -0.3333, y_2 = -0.0417, z_2 = 0.3542$
$x_3 = -0.3281, y_3 = -0.0634, z_3 = 0.3637$

(b) Jacobi

$x_1 = 0.8000, y_1 = 1, z_1 = 1.5000$
$x_2 = 0.9000, y_2 = 1.5750, z_2 = 2.4000$
$x_3 = 0.9650, y_3 = 1.8250, z_3 = 2.7375$

Gauss–Seidel

$x_1 = 0.8000, y_1 = 1.2000, z_1 = 2.5000$
$x_2 = 1.0600, y_2 = 1.8900, z_2 = 2.9750$
$x_3 = 1.0170, y_3 = 1.9980, z_3 = 3.0075$

(c) Jacobi

$x_1 = 4.2500, \; y_1 = 3, \; z_1 = 0.2000$
$x_2 = 3.4500, \; y_2 = 1.6500, \; z_2 = -0.9000$
$x_3 = 4.0625, \; y_3 = 1.5500, \; z_3 = -0.8500$

Gauss–Seidel

$x_1 = 4.2500, \; y_1 = 1.5833, \; z_1 = -1.1833$
$x_2 = 4.1500, \; y_2 = 1.2222, \; z_2 = -1.2156$
$x_3 = 4.2484, \; y_3 = 1.1787, \; z_3 = -1.2636$

MISCELLANEOUS EXERCISES 5

1. 1

2. $A^{-1} = \dfrac{1}{6}\begin{pmatrix} 0 & 3 & 0 \\ 0 & 3 & -6 \\ 2 & -2 & 4 \end{pmatrix}$ $\quad A^2 = \begin{pmatrix} 7 & -3 & 0 \\ 0 & 4 & 6 \\ -2 & 2 & 3 \end{pmatrix}$

3. (a) -8 (b) $\begin{pmatrix} 0 & -8 & 0 \\ 0 & -4 & 4 \\ -2 & 3 & 1 \end{pmatrix}$ (c) $\dfrac{1}{8}\begin{pmatrix} 0 & 8 & 0 \\ 0 & 4 & -4 \\ 2 & -3 & -1 \end{pmatrix}$

4. $\dfrac{1}{21}\begin{pmatrix} -7 & 6 & -10 \\ -14 & 3 & -5 \\ 7 & 0 & 7 \end{pmatrix}$ $\quad x = 1, \; y = -1, \; z = 2$

5. $x = 2\mu - \lambda, \; y = 1 + 2\lambda - 2\mu, \; z = \lambda, \; w = \mu$

6. $x = 0.082, \; y = 0.184, \; z = 0.082$

7. $x = 1.462, \; y = -0.308, \; z = -0.385$

Chapter 6

EXERCISES 6.1

1. (a) $\pm j$ (b) $\pm 2j$ (c) $\pm \sqrt{7/3}\,j$ (d) $-1/2 \pm (\sqrt{3}/2)j$ (e) $1 \pm \sqrt{3}j$
(f) $-3/2 \pm (\sqrt{7}/2)j$ (g) $-3/4 \pm (\sqrt{15}/4)j$ (h) $-3/2 \pm (\sqrt{7}/2)j$

2. $2, \; 5/6 \pm (\sqrt{47}/6)j$

3. (a) $1/2 - (1/2)j$ (b) $2j$ (c) $2/5 - (4/5)j$ (d) $1/2 + (1/2)j$
(e) $(23 - 11j)/26$

4. (a) $1 + j$ (b) $3 + 2j$ (c) $6 + 12j$ (d) $47 + 32j$ (e) $2 - (1/2)j$

5. $x^2 - 2x + 10 = 0$

6. $x = 2, \; y = 1; \; x = -2, \; y = -1$

7. (a) $-62/85, \; -61/85$ (b) $1, 0$ (c) $0, \; -1$ (d) $0, \; 1/4$

8. (a) $x = 3$ (b) $x = \pm j$ (c) $x = -1/2 \pm (\sqrt{3}/2)j$

9. $s = -1 \pm 2j$

EXERCISES 6.2

1. $|z_1| = \sqrt{13}, \; \arg(z_1) = -0.5880$ $\quad |z_2| = 1, \; \arg(z_2) = -\pi/2$
$|z_3| = 1, \; \arg(z_3) = \pi$ $\quad |z_4| = \sqrt{20}, \; \arg(z_4) = -2.0344$
$|z_5| = 3, \; \arg(z_5) = 0$

2. (a) $\sqrt{10}\,\underline{/\,-0.3218}$ (b) $2\,\underline{/\,0}$ (c) $1\,\underline{/\,-\pi/2}$ (d) $13\,\underline{/\,1.9656}$

3. (a) $2, \; 5\pi/6$ (b) $4\sqrt{2}, \; \pi/4$
$z_1 z_2 = 8\sqrt{2}\,\underline{/\,13\pi/12}$ $\quad z_1/z_2 = \sqrt{2}/4\,\underline{/\,7\pi/12}$

4. $1 + j, \; \sqrt{3} + j, \; \sqrt{3} - j$

6. $-j/\omega C$

7. $z_1 z_2 = 12 \,(\cos 110° + j \sin 110°)$
$z_1/z_2 = (4/3)\,(\cos 30° - j \sin 30°)$

8. 4

EXERCISES 6.3

1. (a) $3, \; \pi/4$ (b) $2, \; -\pi/6$

2. (a) $2.5, \; 4.3301$ (b) $-0.5, \; 0.8660$

3. $6e^{(\pi/6)j}, \; \mathrm{Re}(z) = 5.1962, \; \mathrm{Im}(z) = 3$

4. real part: $e^{\sigma T} \cos \omega T$, imaginary part: $e^{\sigma T} \sin \omega T$

5. ej

6. $\sqrt{2}\, e^{(-3\pi/4)j}$

EXERCISES 6.4

1. $\cos 9\theta + j \sin 9\theta, \; \cos(\theta/2) + j \sin(\theta/2)$

2. (a) $\cos 7\theta + j \sin 7\theta$ (b) $\cos 10\theta + j \sin 10\theta$

3. (a) $1\,\underline{/\,\pi/3 + 2n\pi/3}$ $\quad n = 0, 1, 2$
(b) $2^{1/8}\,\underline{/\,\pi/16 + n\pi/2}$ $\quad n = 0, 1, 2, 3$

4. $8^{1/6}\,\underline{/\,\pi/12 + 2n\pi/3}$ $\quad n = 0, 1, 2$

6. $\sqrt{2}\, e^{(-\pi/4)j}, \; (1/\sqrt{2})\, e^{(5\pi/12)j}$

7. $\sqrt{5}\,\underline{/\,\pi/4 + n\pi/2}$ $\quad n = 0, 1, 2, 3$

8. $1\,\underline{/\,\pi/10 + 2n\pi/5}$ $\quad n = 0, 1, 2, 3, 4$

11. $16 \cos^5 \theta - 20 \cos^3 \theta + 5 \cos \theta$

12. $16 \sin^5 \theta - 20 \sin^3 \theta + 5 \sin \theta$

MISCELLANEOUS EXERCISES 6

2. $9/41 + (40/41)j$, $2/13 − (3/13)j$, $4/13$, $\dfrac{x}{(x^2 + y^2)} + \dfrac{y}{(x^2 + y^2)}j$

3. $|-j| = 1$, $\arg(-j) = -\pi/2$; $|-3| = 3$, $\arg(-3) = \pi$; $|1 + j| = \sqrt{2}$, $\arg(1 + j) = \pi/4$, $|\cos\theta + j\sin\theta| = 1$, $\arg(\cos\theta + j\sin\theta) = \theta$

5. (a) $\cos 6\theta + j\sin 6\theta$ (b) $\cos 3\theta − j\sin 3\theta$
   (c) $\cos(\theta − \phi) + j\sin(\theta − \phi)$

7. $e^{j\omega t} − e^{-j\omega t} = 2j\sin\omega t$

8. $1 + \cos 2\omega t + j\sin 2\omega t$

10. $1.5831 + 0.4607\,j$

11. $2^{1/4}\,\angle\,\pi/6 + n\pi/2$ $n = 0, 1, 2, 3$

12. $[s − (−3 + 2j)][s − (−3 − 2j)]$

13. $2[s − (−2 + (\sqrt{6}/2)j)][s − (−2 − (\sqrt{6}/2)j)]$

Chapter 7

EXERCISES 7.1

1. (a) 1 (b) 2 (c) not defined (d) 1 (e) 2
2. (b) $t = 4$ (c) (i) 9 (ii) not defined (iii) 11

EXERCISES 7.2

1. (a) 24 (b) −12 (c) 6
2. −0.25

EXERCISES 7.3

1. (a) 4 (b) 2 (c) 5 (d) 8
2. y is linear in x, that is, $y = ax + b$
3. (a) always negative (b) always positive (c) always positive
   (d) always negative
4. (a) $2x$ (b) $−2x + 2$
5. $4x$, 12, −8, 4, 0
6. $4 − 2t$, 0

EXERCISES 7.4

1. (a) $t = 1$ (b) $t = n\pi$ (c) derivative exists everywhere
   (d) 0 (e) 0 (f) 0

EXERCISES 7.5

1. (a) $2t$ (b) $9t^8$ (c) $−3t^{-4}$ (d) 1 (e) $−t^{-2}$ (f) $−2t^{-3}$ (g) $3e^{3t}$
   (h) $−3e^{-3t}$ (i) $−5e^{-5t}$ (j) $0.5t^{-1/2}$ (k) $2\cos(2t + 3)$
   (l) $\sin(4 − t)$ (m) $0.5\sec^2(t/2 + 1)$
   (n) $−3\operatorname{cosec}(3t + 7)\cot(3t + 7)$
   (o) $\operatorname{cosec}^2(1 − t)$ (p) $2\sec(2t − \pi)\tan(2t − \pi)$
   (q) $1/\sqrt{1 − (t + \pi)^2}$ (r) 0 (s) $−2/(1 + (−2t − 1)^2)$
   (t) $−4/\sqrt{1 − (4t − 3)^2}$ (u) $6\operatorname{sech}^2 6t$ (v) $2\sinh(2t + 5)$
   (w) $0.5\cosh((t + 3)/2)$ (x) $\operatorname{sech}(−t)\tanh(−t)$
   (y) $−(2/3)\operatorname{cosech}^2(2t/3 − 1/2)$ (z) $1/\sqrt{(t + 3)^2 − 1}$

2. (a) $−0.5x^{-3/2}$ (b) $2e^{2x/3}/3$ (c) $−0.5e^{-x/2}$ (d) $1/x$
   (e) $−(2/3)\operatorname{cosec}((2x − 1)/3)\cot((2x − 1)/3)$ (f) $\pi/\{1 + (\pi x + 3)^2\}$
   (g) $2\operatorname{sech}^2(2x + 1)$ (h) $−3/\sqrt{9x^2 + 1}$ (i) $−\omega\operatorname{cosec}^2(\omega x + \pi)$
   (j) $−5\operatorname{cosec}(5x + 3)\cot(5x + 3)$ (k) $−3\sin 3x$ (l) $3\sec 3x\tan 3x$
   (m) $2\sec^2(2x + \pi)$ (n) $−0.5\operatorname{cosech}((x − 1)/2)\coth((x − 1)/2)$
   (o) $2/7\{1 − ((2x + 3)/7)^2\}$

EXERCISES 7.6

1. (a) $12x^2 − 10x$ (b) $15\cos 5t + 8e^{4t}$ (c) $4\cos 4t − 6\sin 2t − 1$
   (d) $3\sec^2 3z$ (e) $6e^{3t} − 8\cos 2t$ (f) $−3/t^4 − 2.5\sin 5t$
   (g) $2w^2 + 2e^{4w}$ (h) $0.5x^{-1/2} + 1/2x$

2. (a) 2.8244 (b) 0 (c) −0.3012 (d) 5 (e) 33.5194 (f) 14.195

3. (a) $\omega e^{\omega t}$ (b) $−\omega e^{-\omega t}$

4. (a) $6/\sqrt{1 − 4t^2} + 15/\sqrt{1 − 9t^2}$ (b) $1/2[1 + (t + 2)^2] − 4/\sqrt{t(1 − t)}$
   (c) $6\cosh(3t − 1) − 2\sinh((t − 3)/2)$
   (d) $−2\operatorname{cosech} 4t\coth 4t − 9\operatorname{sech} 6t\tanh 6t$
   (e) $1/\sqrt{((t + 1)/2)^2 + 1} + 3/2\sqrt{((1 − t)/2)^2) − 1}$
   (f) $6/\{1 − (2t + 3)^2\} − 6/\{1 − (3t + 2)^2\}$
5. (a) $t^2 − 5t + 4$ (b) 1, 4
6. $y = 40x − 53$
7. $\pi/4, 3\pi/4$
8. (a) 1 (b) −1.5183

EXERCISES 7.7

1. (a) $\cos^2 x − \sin^2 x$ (b) $(1/t)\tan t + \ln t\sec^2 t$ (c) $e^{2t}(2t^3 + 3t^2 + 2)$
   (d) $e^x\sqrt{x} + 1/2\sqrt{x})$ (e) $e^t(2\cos^2 t + \sin t\cos t − 1)$
   (f) $3[2\cosh 2t\cosh 3t + 3\sinh 2t\sinh 3t]$

(g) $(\sin^3 t + \sin^2 t)/(\cos^2 t) + 2\sin t + 1$

(h) $4[\cosh(t+1)\cosh(1-t) - \sinh(t+1)\sinh(1-t)]$

2. (a) $-\csc^2 x$ (b) $\{\ln t \sec^2 t - (\tan t)/t\}/(\ln t)^2$

(c) $e^{2t}(2t^3 - 3t^2 + 2)/(t^3 + 1)^2$ (e) $\dfrac{e^x(-x^2 + x) + 2x + 1}{(e^x + 1)^2}$

(d) $\dfrac{-3x^4 - 4x^3 + 27x^2 + 6x + 2}{(x^3 + 1)^2}$

(f) $\dfrac{2\cosh 2t \cosh 3t - 3\sinh 2t \sinh 3t}{(\cosh 3t)^2}$

(g) $(-e^{3t} - 2e^{2t} + e^t)/(e^{2t} + 1)^2$

3. (a) $300t^2(t^3 + 1)^{99}$ (b) $9\sin^2(3t + 2)\cos(3t + 2)$ (c) $2x/(x^2 + 1)$

(d) $(2t + 1)^{-1/2}$ (e) $-3\sin(2x - 1)/\sqrt{\cos(2x - 1)}$ (f) $-(t + 1)^{-2}$

(g) $an(at + b)^{n-1}$

4. 0, 2

5. (a) -0.5403 (b) 0 (c) 109.2 (d) 2 (e) -2.0742

6. $y = 2x + 1$

EXERCISES 7.8

1. (a) $(1 + 9x^2)/(4y - 1)$ (b) $\dfrac{(4x\sqrt{x} - 1)\sqrt{y}}{(1 - 6y^2\sqrt{y})\sqrt{x}}$

(c) $\dfrac{(2/3)(e^x\sqrt{2x + 3y - 1})}{3x^2 + 3y^2 - 2y^4}$ (d) $y(1/2(1 + x) + 1 - 2/x)$

(f) $\dfrac{\cos(x + y)}{1 - \cos(x + y)}$ (g) $\dfrac{2(x^2 + y^2 - x)}{3x^2 + 3y^2 + 2y}$

(e) $\dfrac{2xy(4y^2 - 3)}{e^{(x/2 - 2y)}x(x + 4)}$

(h) $\dfrac{e^{(x/2 - 2y)}x(x + 4)}{2(1 + 2y)}$

2. (a) $3t/2$ (b) $e^t\cos t$ (c) $(t/(1 + t))^2$ (d) $-3/2\sin 2t$

(e) $-2e^{2t}/3$ (f) $(e^t - e^{-t})/(e^t + e^{-t})$

3. (a) $z(4/t - 6/(1 - t) + 4/(2 + t))$ (b) $y(6x/(1 + x^2) + 7 - 6/(2 + x))$

(c) $x(3/(1 + t) + 4/(2 + t) + 5/(3 + t))$

(d) $y(4\cot t - 8t/(2 - t^2) - 6e^t/(1 + e^t))$

4. -0.1387

5. (a) 192 (b) -2 (c) $-1/3$ (d) $\dfrac{dx}{dy} = -3t^4$; when $t = 2$, $\dfrac{dx}{dy} = -48$

6. $y = (-4x + 6)/5$, $y = (4x + 24)/5$

EXERCISES 7.9

1. (a) 1.02 (b) 1.85 (c) 7.15 (d) 1.76 (e) 0.64

EXERCISES 7.10

1. (a) $2t + 1$, 2

(b) $6t^2 - 2t$, $12t - 2$ (c) $2\cos 2t$, $-4\sin 2t$ (d) $k\cos kt$, $-k^2 \sin kt$

(e) $6e^{3t} - 2t$, $18e^{3t} - 2$ (f) $1/(t + 1)^2$, $-2/(t + 1)^3$

(g) $-2\sin(t/2)$, $-\cos(t/2)$ (h) $e^t(t + 1)$, $e^t(t + 2)$

(i) $4\cosh 4t$, $16\sinh 4t$ (j) $2\sin t \cos t$, $2\cos 2t$

2. $-1/2$

3. (a) $6t + 1$ (b) 6

4. 7/2

5. (a) concave up (b) concave up (c) concave down

6. concave up on $(0, \infty)$, concave down on $(-\infty, 0)$

7. (a) -3.08 (b) $-1/2$ (c) 3

EXERCISES 7.11

1. (a) $(-1, -4)$ minimum (b) $(-0.5, 4.25)$ maximum

(c) $(0, 10)$ maximum, $(1, 59/6)$ minimum, $(1/2, 119/12)$ point of inflexion

(d) $(4, -131/3)$ minimum, $(-5, 77.83)$ maximum, $(-1/2, 17.08)$ point of inflexion

(e) $(0, 0)$ point of inflexion (f) $(0, 0)$ minimum

(g) $(0, 0)$ maximum, $(1, -1)$ minimum, $(-1, -1)$ minimum,

$(1/\sqrt{3}, -5/9)$, $(-1/\sqrt{3}, -5/9)$ points of inflexion

(h) $(1, 2)$ minimum, $(-1, -2)$ maximum

(i) $(1, -2/3)$ minimum, $(-1, 2/3)$ maximum, $(0, 0)$, $(1/\sqrt{2}, -7/12\sqrt{2})$,

$(-1/\sqrt{2}, 7/12\sqrt{2})$, are also point of inflexion

(j) $(0, 0)$ point of inflexion

EXERCISES 7.12

1. (a) $\cos t\mathbf{i} - \sin t\mathbf{j}$ (b) $-\sin t\mathbf{i} - \cos t\mathbf{j}$ (c) 1

MISCELLANEOUS EXERCISES 7

1. 1

2. $\cos t$

3. (a) $\sqrt{R^2 + \omega^2 L^2 - 2L/C + 1/(\omega^2 C^2)}$

(b) $\dfrac{\omega L^2 - 1/\omega^3 C^2}{\sqrt{R^2 + \omega^2 L^2 - 2L/C + 1/(\omega^2 C^2)}}$

(c) $\omega = 1/\sqrt{LC}$ produces a minimum value of Z

4. $-R/(E - iR)$

5. -0.3784

7. (a) $-2e^{-t} - 3\sin(t/2)$ (b) $-6(-t+2)^5$ (c) $2\cos 4t$
   (d) $-2x/(x^2+1)^2$
9. First iteration: 1.859, second iteration: 1.857
10. (a) (2, 61) maximum, (5, 34) minimum, (7/2, 95/2) point of inflexion
    (b) $(1/\sqrt{3}, -2/3\sqrt{3})$ minimum, $(-1/\sqrt{3}, 2/3\sqrt{3})$ maximum,
    (0, 0) point of inflexion
11. $x^x(\ln x + 1)$
12. (a) $2t\mathbf{i} + \mathbf{k}$ (b) $2\mathbf{j}$ (c) -3 (d) $-4t\mathbf{i} + 2t\mathbf{j} + (6t^2 + 2)\mathbf{k}$

Chapter 8

EXERCISES 8.1

1. (a) $3x + x^2/2 + \ln|x| + c$ (b) $e^{2x}/2 + e^{-2x}/2 + c$
   (c) $-(2\cos 3x)/3 + (\sin 3x)/3 + c$
   (d) $0.5 \ln|\sec(2t+\pi) + \tan(2t+\pi)| + 2\ln|\sin(t/2 - \pi)| + c$
   (e) $2\ln|\sec(t/2)| + (1/3)\ln|\cosec(3t - \pi) - \cot(3t - \pi)| + c$
   (f) $-\cos x + x^2/6 - e^{-x} + c$ (g) $(1/3)\ln|\sec 3x + \tan 3x| + c$
   (h) $t^3/3 + 2t - 1/t + c$ (i) $-e^{-2x}/6 + c$
   (j) $(1/4)\ln|\sec(4t-3)| + 2\cos(-t-1) + c$ (k) $x + (2/3)\ln|\sin 3x| + c$
   (l) $-2\cos(t/2) - 6\sin(t/2) + c$ (m) $t^3/3 - 2t^2 + 4t + c$
   (n) $-3e^{-t} + 2e^{-t/2} + c$ (o) $7x - x^7 - e^{-x} + c$
   (p) $k^2t + kt^2 + t^3/3 + c$ (q) $-k\cos t - (\sin kt)/k + c$
   (r) $(1/5)\tan^{-1}(t/5) + c$ (s) $\sin^{-1}(t/5) + c$
   (t) $6\tan^{-1} x + x/6 + x^3/18 + c$
2. speed: $t + t^2/4 + c$, distance: $t^2/2 + t^3/12 + ct + d$
3. (a) $(\cosh ax)/a + c$ (b) $(\sinh ax)/a + c$
   (c) $(3/2)\cosh 2x + (1/4)\sinh 4x + c$
4. (a) $(1/2)\tan^{-1}(x/2) + c$ (b) $(1/2\sqrt{2})\tan^{-1}(x/\sqrt{2}) + c$
   (c) $(3/\sqrt{2})\tan^{-1}(\sqrt{2x}) + c$ (d) $\sin^{-1}(x/3) + c$
   (e) $2\sin^{-1}(x/2) + c$
   (f) $(-7/\sqrt{3})\sin^{-1}(\sqrt{3}x/\sqrt{2}) + c$
5. $3\ln|x| + x + c$
6. (a) $2t - 2e^{-t/2} + c$ (b) $-(1/2)e^{-t/2}$
8. (a) $-t^{-2} - \ln|t| + c$ (b) $(-4/3)e^{-3t} - e^{-t} + c$
   (c) $(1/4)\ln|\sin 4x| + c$ (d) $(1/6)\ln|\cosec 3x - \cot 3x| + c$
   (e) $x + \tan^{-1} x + c$ (f) $t + c$

EXERCISES 8.2

1. (a) $1/2$ (b) $1/2$ (c) 1

2. (a) $-65/12$ (b) 2.275 (c) -0.639 (d) -0.9255 (e) 1.2313
   (f) 1.1175
3. $1/6$
4. $4/3$
5. (a) 1.0839 (b) 0.6468 (c) 0 (d) 2.3504
6. $4.7756, 2.3130$
7. 0.4142
8. 39
9. 2.3026
10. (a) -3.6202 (b) 3.1945 (c) 0 (d) 5.4158

EXERCISES 8.3

1. (a) 2.8819×10^9 (b) 2322 (c) -1.2×10^{36} (d) $(2/9)(3t+1)^{3/2} + c$
   (e) $(9y - 2)^{18}/162 + c$ (f) 9.092×10^{-5} (g) $(-1/3)\cos(t^3) + c$
   (h) $e^{x^3 + 1}/3 + c$ (i) 0.5009 (j) $2/3$ (k) $(2/3)(\sin t)^{3/2} + c$
2. (a) $(x^2/2)\ln|x| - x^2/4 + c$ (b) $t\sin t + \cos t + c$ (c) 4.2281
   (d) $x\ln|x| - x + c$ (e) $(-t^2\cos 2t)/2 + (t\sin 2t)/2 + (\cos 2t)/4 + c$
3. (a) 1.8767 (b) $\ln|t+1| - 2(t+1)^{-1} + c$
   (c) $\ln|t| + (1/2)\ln|t^2 + 1| + 3\tan^{-1} t + c$
   (d) $5\ln|t-2| - 2\ln|2t-1| + c$
4. (a) $(1/3)\ln|t^3 + 1| + c$ (b) $3/8$ (c) 0.2657
   (d) $(-1/2)\ln|x-2| - 2\ln|x-3| + (5/2)\ln|x-4| + c$
   (e) $\ln|e^x + 1| + c$
5. (a) 1.5778 (b) 317.3 (c) $\pi(1+\omega)$ (d) 1.0149 (e) 1.2105
   (f) 0.9730
6. 3.8805

EXERCISES 8.4

1. (a) average $= 2$, r.m.s. $= 2.0817$ (b) average $= 0$, r.m.s. $= 0.7071$
   (c) average $= 1.5556$, r.m.s. $= 1.5811$
   (d) average $= (1 - \cos\omega\pi)/\pi\omega$, r.m.s. $= \sqrt{\dfrac{1}{2}\left(1 - \dfrac{\sin 2\omega\pi}{2\omega\pi}\right)}$
   (e) average $= 1.1752$, r.m.s. $= 1.3466$
   (f) average $= 0.9589$, r.m.s. $= 0.9595$
   (g) average $= 5$, r.m.s. $= 5.2915$
   (h) average $= 0.5493$, r.m.s. $= 0.5774$
   (i) average $= 2.1752$, r.m.s. $= 2.2724$
   (j) average $= (1 - \cos\omega + \sin\omega)/\omega$, r.m.s. $= \sqrt{1 + \dfrac{\sin^2\omega}{\omega}}$

EXERCISES 8.5

1. (a) 1 (b) $1/k$ (c) does not exist (d) 1 (e) does not exist
2. (a) 1 (b) e^4 (c) e^{-3} (d) 36 (e) 1 (f) 1 (g) e^{-ka} (h) 1

EXERCISES 8.6

1. (a) 6 (b) 5.75 (c) 8.5417 (d) 15.333
2. (a) 6 (b) 15.5 (c) 14 (d) 33.5 (e) 6.75
3. (a) 4 (b) 2 (c) 10 (d) 2 (e) $(e^{4k} - e^{3k})/k$

EXERCISES 8.7

1. (a) 0.3139 (b) 0.5467
2. (a) 2.7955 (b) 15.1164
3. (a) trapezium rule: 1.5900, Simpson's rule: 1.5681
   (b) trapezium rule: 0.2464, Simpson's rule: 0.2460

EXERCISES 8.8

1. $6\mathbf{i} + 1.3484\mathbf{k}$
2. (a) $\mathbf{i} - 3\mathbf{j} + \mathbf{k}$ (b) $2\mathbf{i} - 6\mathbf{j} + 2\mathbf{k}$
3. (a) $0.333\mathbf{i} + 0.6321\mathbf{j} + 0.5\mathbf{k}$ (b) $6.333\mathbf{i} + 0.0855\mathbf{j} + 2.5\mathbf{k}$
   (c) $21\mathbf{i} + 0.3496\mathbf{j} + 7.5\mathbf{k}$
4. no

MISCELLANEOUS EXERCISES 8

1. $-(L/R)\ln|E - iR| + c$
2. $(L/R)\,e^{Rt/L}(t - L/R) + c$
3. 24
4. (a) average $= 0$, r.m.s. $= \sqrt{(A^2 + B^2)/2}$
   (b) average $= 2B/\pi$, r.m.s. $= \sqrt{(A^2 + B^2)/2}$
5. (a) $-(-2t + 0.1)^5/10 + c$ (b) $-(1 + x)\cos x + \sin x + c$
   (c) 0.7957 (d) 2.9205 (e) $\ln|t| - \ln|t + 1| + 1/(t + 1) + c$
   (f) 0.3066
6. (a) 1.8111 (b) 1.8101
7. (a) 1 (b) 5/2 (c) 2.85 (d) 3 (e) 0.6
8. (a) does not exist (b) 1.0039 (c) does not exist
9. $(1/2)\mathbf{i} - \mathbf{j} + (3/2)\mathbf{k}$
10. (a) 25/3 (b) 4 (c) 23/3 (d) 46/3

Chapter 9

EXERCISES 9.1

1. $t^2/2 - 4t + 19/2$
2. (a) 0, 2, -2 (b) $-x^3/3 + 2x^2 - 3x + 10/3$ (c) 2.0573
3. (a) $1 - t^2/2! + t^4/4! - t^6/6! + t^8/8! - \cdots$
   (b) $t - t^2/2 + t^3/3 - t^4/4 + t^5/5 - \cdots$
4. (a) (i) x^2 (ii) x^2 (b) $x + x^2 - x^3/3! + x^5/5! - x^7/7! + \cdots$

MISCELLANEOUS EXERCISES 9

1. (a) $-t + 7$ (b) 4.1, 3.8 (c) 4.105, 3.82
2. $6x^2 - 12x + 8$
3. (a) et (b) $-2.97t + 3.11$ (c) $-0.25t + 3.11$
4. $\ln 2 + x/2 - x^2/8 + x^3/24 - x^4/64 + \cdots$
5. (a) 4 (b) $y'' + xy'' - 2y' = 2x$ (c) 4 (d) $1 + 2x + 2x^2 + 2x^3/3$
   (e) 2.5833
6. (a) k (b) 0 (c) not defined
7. $(4a + 3b + 2c + d)t - 3a - 2b - c + e$

Chapter 10

EXERCISES 10.1

5. $y(x) = 3e^{2x}$, $y(x) = e^{2x}$
6. (a) y is the dependent variable; x is the independent variable; first order, first degree, linear
   (b) y is the dependent variable; x is the independent variable; second order, first degree, non-linear
   (c) x is the dependent variable; t is the independent variable; third order, first degree, non-linear

EXERCISES 10.2

1. $x = e^t + 2e^{2t}$
2. $y = xe^x$
3. particular solution: $\sin \omega t/\omega$
   general solution: $(A + B)\cos \omega t + (A - B)\mathbf{j}\sin \omega t$

EXERCISES 10.3

1. $t\ln|t| - t + c$, $t\ln|t| - t + 2$
2. (a) $kx^2/2 + c$ (b) Ae^{-kx} (c) $-1/(x + c)$ (d) $y^2 = 2(c - \cos x)$
   (e) $y^2 = x^2 + 4x + c$ (f) $x/(A - 2\ln|x|)$ (g) $x^6/6 = t^5/5 + c$

3. (a) $x = Ae^{t^2/2}$ (b) $y^2 = x^2 + c$ (c) $x = \sin^{-1}(kt)$
   (d) $x = (1 + At^2)/(1 - At^2)$
4. $2 + Ae^{t^2/2}$, $2 + 3e^{t^2/2}$

EXERCISES 10.4

1. $-2t - 1 + ce^{2t}$, $-2t - 1 + 5e^{2(t-1)}$
2. $2x + 3 + ce^{-x}$
3. $1 - e^{-t^2/2}$
4. $L((R/L)\sin\omega t - \omega\cos\omega t)/(R^2 + L^2\omega^2) + ce^{-Rt/L}$
5. $x^4/5 + c/x$
6. $3t/2 + c/t$
7. $1/2 - (3/2)e^{-t^2}$
8. $(e^t + c)/t^3$

EXERCISES 10.5

1. (a) $Ae^x + Be^{2x}$ (b) $Ae^x + Bxe^x$ (c) $A\sin 3x + B\cos 3x$
   (d) $A + Be^{2x}$
2. $Lk^2 + Rk + 1/C = 0$ $i(t) = Ae^{k_1 t} + Be^{k_2 t}$

where

$$k_1, k_2 = \frac{-R \pm \sqrt{\dfrac{R^2 C - 4L}{C}}}{2L}$$

3. $e^{-x/2}(A\cos\sqrt{3}x/2 + B\sin\sqrt{3}x/2)$

EXERCISES 10.6

1. $Ae^{3t} + Be^{-t} - 2$
2. $(5/2)e^{3t}$
3. $-2te^{-t}$
4. $Ae^{2x} + Be^{-x} - 3$
5. $Ae^{-2x} + Be^{-x} + (3/2)\sin 2x - (1/2)\cos 2x$,
   $(3/2)\,e^{-2x} + (3/2)\sin 2x - (1/2)\cos 2x$
6. x
7. (a) $Ae^t + Be^{5t} + 3/5$ (b) $Ae^t + Bte^t + t^2 e^t/2$
8. $Ae^{-11270t} + Be^{-88730t}$
9. $e^{-4t}[A\sin 3t + B\cos 3t] + (14\sin 3t + 18\cos 3t)/13$

EXERCISES 10.7

$y'' + 3y' + 2y = 0$ $y_1(t) = Ae^{-2t} + Be^{-t}$

$y_2(t) = -(2/3)Ae^{-2t} - (B/2)e^{-t}$ $\begin{pmatrix} y_1' \\ y_2' \end{pmatrix} = \begin{pmatrix} 2 & 6 \\ -2 & -5 \end{pmatrix} \begin{pmatrix} y_1 \\ y_2 \end{pmatrix}$

EXERCISES 10.8

1. Exact: $y = x\ln|x| - 0.1931x$

x_i	y_i (h = 0.5)	y_i (h = 0.25)	y (exact)
2.00	1.0000	1.0000	1.0000
2.25	—	1.3750	1.3901
2.50	1.7500	1.7778	1.8080
2.75	—	2.2056	2.2509
3.00	2.6000	2.6661	2.7165

2. Exact: $y = -x - 1 + e^x$

x_i	y_i (h = 0.25)	y (exact)
0	0.0000	0.0000
0.25	0.0000	0.0340
0.50	0.0625	0.1487

x_i	y_i (h = 0.1)	y (exact)
0	0.0000	0.0000
0.1	0.0000	0.0052
0.2	0.0100	0.0214
0.3	0.0310	0.0499
0.4	0.0641	0.0918
0.5	0.1105	0.1487

3. Exact:
$$v = \frac{\sin 100\pi t - \pi\cos 100\pi t + \pi e^{-100t}}{\pi^2 + 1}$$

t_i	v_i (h = 0.005)	v (exact)
0	0.0000	0.0000
0.005	0.0000	0.2673
0.010	0.5000	0.3954

t_i	v_i (h = 0.002)	v (exact)
0	0.0000	0.0000
0.002	0.0000	0.0569
0.004	0.1176	0.1919
0.006	0.2843	0.3354
0.008	0.4176	0.4178
0.010	0.4517	0.3954

EXERCISES 10.9

1.

x_i	y_i (h = 0.5)	y_i (h = 0.25)	y (exact)
2.00	1.0000	1.0000	1.0000
2.25	—	1.3889	1.3901
2.50	1.8000	1.8057	1.8080
2.75	—	2.2476	2.2509
3.00	2.7017	2.7123	2.7165

2.

x_i	y_i (h = 0.25)	y (exact)
0	0.0000	0.0000
0.25	0.0313	0.0340
0.50	0.1416	0.1487

3.

t_i	v_i (h = 0.005)	v (exact)
0	0.0000	0.0000
0.005	0.2500	0.2673
0.010	0.2813	0.3954

EXERCISES 10.10

1.

(a)

x_i	y_i
0	1.0000
0.2	1.3085
0.4	2.0640

(b)

x_i	y_i
0	1.0000
0.1	1.1230
0.2	1.3085
0.3	1.5958
0.4	2.0649

2. (a) Repeating Exercise 10.8, Question 1

x_i	y_i (h = 0.5)	y_i (h = 0.25)	y (exact)
2.00	1.0000	1.0000	1.0000
2.25	—	1.3900	1.3901
2.50	1.8078	1.8079	1.8080
2.75	—	2.2507	2.2509
3.00	2.7163	2.7164	2.7165

(b) Repeating Exercise 10.8, Question 2

x_i	y_i (h = 0.25)	y (exact)
0.00	0.0000	0.0000
0.25	0.0340	0.0340
0.50	0.1487	0.1487

x_i	y_i (h = 0.1)	y (exact)
0	0.0000	0.0000
0.1	0.0050	0.0052
0.2	0.0210	0.0214
0.3	0.0492	0.0499
0.4	0.0909	0.0918
0.5	0.1474	0.1487

t_i	v_i (h = 0.002)	v (exact)
0.002	0.0588	0.0569
0.004	0.1903	0.1919
0.006	0.3273	0.3354
0.008	0.4032	0.4178
0.010	0.3777	0.3954

x_i	y_i (h = 0.1)	y (exact)
0.0	0.0000	0.0000
0.1	0.0052	0.0052
0.2	0.0214	0.0214
0.3	0.0499	0.0499
0.4	0.0918	0.0918
0.5	0.1487	0.1487

t_i	v_i (h = 0.002)	v (exact)
0	0.0000	0.0000
0.002	0.0569	0.0569
0.004	0.1919	0.1919
0.006	0.3352	0.3354
0.008	0.4178	0.4178
0.010	0.3954	0.3954

(c) Repeating Exercise 10.8, Question 3

t_i	v_i (h = 0.005)	v (exact)
0	0.0000	0.0000
0.005	0.2675	0.2673
0.010	0.3959	0.3954

MISCELLANEOUS EXERCISES 10

1. (a) $x = Ae^{2t}$ (b) $x = 3 \ln|1 + t| + c$ (c) $y = 1/(A − \sin x)$
2. $y = 2x + 3$
3. $x = t + c/t$
4. $y = Ae^{-4x} + x/4 − 1/16$
5. $x = e^{2t}(4 \cos t + \sin t)/17 + ce^{-2t}$
6. $y = A \cos 4x + B \sin 4x + x^2/16 − 1/128$
7. $y = (39e^x + 11e^{-4x})/25 + xe^x/5$
8. 4.7937
9. 0.7535, −0.7535
10. $A \sin 4t + B \cos 4t$, 12
11. $\sin t$, $x(t) = (\cos 2t − (1/2) \cos 4t + c)/(4 \sin t)$

Chapter 11

EXERCISES 11.1

1. (a) $6/(s^2 + 36)$ (b) $s/(s^2 + 16)$ (c) $6/(9s^2 + 4)$
(d) $9s/(9s^2 + 16)$ (e) $24/s^5$ (f) $120/s^6$ (g) $1/(s + 3)$
(h) $1/(s − 3)$ (i) $1/(s + 4)$ (j) $(s^2 − 9)/(s^2 + 9)^2$
(k) $2s/(s^2 + 1)^2$ (l) $3/[(s + 1)^2 + 9]$ (m) $(s + 5)/[(s + 5)^2 + 49]$

EXERCISES 11.2

1. (a) $6/s^3 − 4/s$ (b) $8/(s^2 + 16) + 11/s − 1/s^2$
(c) $2/s − 2/s^3 + 48/s^5$ (d) $3/(s − 2) + 4/(s^2 + 1)$
(e) $1/(s^2 + 9) − 16s/(4s^2 + 1)$ (f) $72/(s − 5)^5 + 1/s^2$
(g) $(3s + 2)/(s^2 − 4)$ (h) $(4s + 7)/[(s + 1)^2 + 9]$
2. (a) $(3s^2 + 6s + 2)/(s^2 + 3s + 3)$ (b) $(3s^2 − 18s + 26)/(s^2 − 5s + 7)$
(c) $(12s^2 + 12s − 1)/(4s^2 + 8s + 7)$
3. (a) $4s e^{-s}/(s^2 + 1)$ (b) $12s e^{-2s}/(s^2 + 1)$ (c) $2s e^{-4s}/(s^2 + 1)$

EXERCISES 11.3

1. (a) $sY − 3$ (b) $s^2Y − 3s − 1$ (c) $3s^2Y − sY − 9s$
(d) $(s^2 + 2s + 3)Y − 3s − 7$ (e) $(3s^2 − s + 2)Y − 9s$
(f) $(−4s^2 + 5s − 3)Y + 12s − 11$
2. (a) $(3s − 2)F − 6$ (b) $(3s^2 − s + 1)F − 6s − 7$
(c) $s^3F − 2s^2 − 3s + 1$ (d) $(2s^3 − s^2 + 4s − 2)F − 4s^2 − 4s − 3$
3. $1/(s + 1)^2$

EXERCISES 11.4

1. (a) $3/2$ (b) $4 − t^2/2$ (c) $30t$ (d) $e^{-2t}/3$ (e) $3 \cos 3t − (7/3) \sin 3t$
(f) $δ(t) − 2e^{4t}$ (g) $e^{-4t} \cos t$ (h) $5e^{-4t} \sin t$
(i) $6e^{-4t} \cos t − 7e^{-4t} \sin t$ (j) $e^{-t} \cos \sqrt{6}t − (1/\sqrt{6}) e^{-t} \sin \sqrt{6}t$
(k) $t e^{-t/2}/2$

EXERCISES 11.5

1. (a) $2e^t + e^{-2t}$ (b) $e^{-t} + 3e^{3t/2}/2$ (c) $2δ(t) + e^{-2t}$
(d) $\{4 + e^{-t} \cos 2t + (11/2)e^{-t} \sin 2t\}/5$ (e) $e^{-t}(2 − 2\cos t − \sin t)$
(f) $e^{-2t}(1 + 2t)$ (g) $e^{-2t}(3 \cosh \sqrt{3}t − (8/\sqrt{3}) \sinh \sqrt{3}t) − \cos t$
(h) $e^t(\cos \sqrt{2}t + \sqrt{2} \sin \sqrt{2}t) + 2e^{-2t}$

EXERCISES 11.6

1. (a) $(3/2 + 11j/4)/(s − a) + (3/2 − 11j/4)/(s − b)$
where $a = −3 + 2j, b = −3 − 2j$; $e^{-3t}[3 \cos 2t − (11/2) \sin 2t]$
(b) $(1 − 3j/2)/(s − a) + (1 + 3j/2)/(s − b)$
where $a = 1 + j, b = 1 − j$, $e^t(2 \cos t + 3 \sin t)$
(c) $(2 + 8j/3)/(s − a) + (2 − 8j/3)/(s − b) + 2$
where $a = 1 + 3j, b = 1 − 3j$, $e^t(4 \cos 3t − (16/3) \sin 3t) + 2δ(t)$
(d) $1 + (3/2 − j/2\sqrt{2})/(s − a) + (3/2 + j/2\sqrt{2})/(s − b)$
where $a = 1 + \sqrt{2}j, b = 1 − \sqrt{2}j$
$δ(t) + e^t(3 \cos \sqrt{2}t + (1/\sqrt{2}) \sin \sqrt{2}t)$
(e) $(−1 + 5j/4)/(s − a) + (−1 − 5j/4)/(s − b)$
where $a = 1 + 2j, b = 1 − 2j$, $e^t(−2 \cos 2t − (5/2) \sin 2t)$

EXERCISES 11.7

1. (a) $e^{-t} − e^{-2t}$ (b) $t^2/3 − 2t/9 + 2/27 − 2e^{-3t}/27$
2. (a) $t^2/2$ (b) $t^4/12$ (c) $−t − 1 + e^t$ (d) $t − \sin t$
3. (a) $e^t − 1$ (b) $(e^t − e^{-3t/2})/5$ (c) $(1 − e^{-3t/2})/3$
4. (a) $t − 1 + e^{-t}$ (b) $(e^{2t} − e^{-3t})/5$ (c) $(\sin t − t \cos t)/2$

EXERCISES 11.8

1. (a) $3 \sin t + 2t$ (b) $-3 \cos 2t + 2e^{-t}$ (c) $t + 6e^{-2t}$
   (d) $e^{-2t}/4 + e^{2t}/4 - (1/2) \cos 2t + 1$ (e) $x = t^2 + 3$, $y = 6 - t$
   (f) $x(t) = 6e^t/5 + 2e^{-t} + 13e^{6t}/17$, $y(t) = 6e^{-3t/2}/5 - 2e^{-t} + 9e^t/5$

2. $-2te^{-t} + 8e^{-t}/7 + 13e^{6t}/17$

EXERCISES 11.9

(a) $1/(s^2 + 1)$ (b) $1/(2s^2 + s - 1)$ (c) $1/(s^2 + s - 2)$ (d) $1/(s^2 - 4)$

EXERCISES 11.10

1. (a) $3/s + 1$ (b) $-6/s + 4$ (c) $3e^{-2s}/s + e^{-2s}$ (d) $e^{3s}/s - e^{4s}$
   (e) $e^{-4s}/2s + 3e^{-4s}$
2. (a) $2u(t) - \delta(t)$ (b) $2u(t)/3 + \delta(t)/2$ (c) $3u(t) - 2\delta(t)$
   (d) $4\delta(t) - 3u(t)$
3. (b) $(1 - e^{-s})/s^2(1 + e^{-s})$
4. (b) $2(s + e^{-2.5s} - e^{-1.5s})/s^2(1 - e^{-2.5s})$
5. $(30 \cos 2t + 20 \sin 2t)e^{-t}/13 - 30e^{-4t}/13$

MISCELLANEOUS EXERCISES 11

1. (a) $(s + 3)/(s^2 + 4s + 6)$ (b) $3(s - 1)/(s^2 - 4s + 6)$
   (c) $2(2s^2 + 5s + 5)/(s + 1)(s^2 + 2s + 3)$ (d) $e^{-s}(s + 1)/(s^2 + 2)$
   (e) $4e^{-3s}(s + 1)/(s^2 + 2)$ (f) $e^{-2(s+2)}(s + 3)/(s^2 + 4s + 6)$
2. (a) $12s/(s^2 + 9)^2$ (b) $-3(s^2 - 4)/2(s^2 + 4)^2$ (c) $1/(s + 1)$
   (d) $e^{-(s+1)}/(s + 1)$ (e) $3e^{-s}(2/s^3 + 2/s^2 + 1/s)$ (f) $e^{-2(s+1)}$
3. (a) $e^{-t} + e^{-2t}$ (b) $2e^{-3t} + 2e^{3t}$ (c) $2(2 \sin 2t + \cos 4t)$
   (d) $u(t - 2)(t - 2)^3$ (e) $u(t - 1)t$
4. (a) $-11e^{-3t}/36 + 7e^t/4 - 2t/3 - 4/9$
   (b) $e^t[-(1/5) \cos 2t + (13/20) \sin 2t] + (1/5) \cos t - (1/10) \sin t$
   (c) $x = t$, $y = t/2$ (d) $x = e^t - 1$, $y = 2e^t$
6. (b) $v_0(t) = 5 - 5e^{-100t}$ (c) $100e^{-100t}$ (d) $500te^{-100t}$
7. $(1 - e^{-as})/s(1 + e^{-as})$
8. (c) $3e^{-3t/2}/2 - e^{-t}/2$ (d) $-e^{-t} + 4e^{-2t} - 3e^{-3t}$

Chapter 12

EXERCISE 12.1

(i) first order; linear; inhomogeneous (ii) second order; non-linear
(iii) second order; linear; inhomogeneous (iv) second order; non-linear
(v) first order; non-linear

EXERCISE 12.2

(a) $p[k + 1] - 3p[k + 2] = p[k - 1]$
(b) $R[n] - R[n - 1] - R[n - 2] = n + 1$ $R[0] = 1$, $R[1] = -2$
(c) $q[n] + (t - 2)q[t - 3] = 3q[t - 1]$ $q[1] = 0$, $q[2] = -2$
(d) $T[m + 2] + (m + 1)T[m] = (m + 2)^2$ $T[0] = T[1] = 1$
(e) $y[k - 2] - y[k] = (s[k - 1] - s[k + 1])/2$

EXERCISE 12.3

$a[n] = (I[n] + I[n - 1] + I[n - 2])/3$

EXERCISES 12.4

1. $s[n] = s[n - 1] + Tv[n]$

EXERCISES 12.5

1. 0, 7, -5, 14
2. $4/7$, $63/4$, $64/63$
3. (a) 5, $29/2$ (b) $1/3$, $7/18$ (c) $29/3$, $52/9$

EXERCISES 12.6

1. (a) $1/z(z - 1)$ (b) $4/z^5(z - 1)$ (c) $3z/(z - 1)^2$ (d) $ez/(ez - 1)$
   (e) $(z^2 + 2z + 3)/z^2$ (f) 3 (g) $2z/(z - 1)$
2. (a) $z(z - \cos 3)/(z^2 - 2z \cos 3 + 1)$ (b) $z/(z - e)$
   (c) $(z^2 - ze^{-2} \cos 1)/(z^2 - 2ze^{-2} \cos 1 + e^{-4})$
   (d) $(ze^4 \sin 2)/(z^2 - 2ze^4 \cos 2 + e^8)$ (e) $z/(z - 4)$ (f) $z/(z + 3)$
   (g) $z/(z^2 + 1)$ (h) $z^2/(z^2 + 1)$
3. (a) $(-4)^k$ (b) $(1/2)^k$ (c) $(-1/3)^k$ (d) e^{3k} (e) $\sin(\pi k/2)$
4. $e^{1/z}$

EXERCISES 12.7

1. (a) $3z/(z - 4) + 7z(z + 1)/(z - 1)^3$
   (b) $(3ze^{-1} \sin 4)/(z^2 - 2ze^{-1} \cos 4 + e^{-2}) - z/(z - 1)^2$
2. (a) $T^2z(z + 1)/(z - 1)^3$ (b) $4Tz/(z - 1)^2$
   (c) $z \sin 2T/(z^2 - 2z \cos 2T + 1)$ (d) $1/z^3(z - 1)$ (e) $z/(z - e^{3T})$
3. $1/z^2(z - 1)^2$
6. (a) $ze^b/(ze^b - 1)^2$ (b) $ez \sin 1/(e^2z^2 - 2ez \cos 1 + 1)$
7. $4z/(z - 3)$, $12z/(z - 3)$
8. 0, 4, 8, 16, 32, $\ldots$, $4/(z - 2)$

EXERCISES 12.8

1. (a) $4(4^k)$ (b) $13(7/3)^k/30 - (-1)^k/10$
(c) $u[k-2]4(3)^{k-2} - \delta[k-2])/3$
   May be written as: $3^{k-2}u[k-2] + 3^{k-3}u[k-3]$
(d) $19k(3^k)/6 + 3u[k]/4 + 5(3^k)/4$

2. (a) $-2(2^k) + 2(3^k)$ (b) $e^{-k}k$ (c) $\delta[k] - 2\delta[k-1] + 4^k - 3^k$
(d) $((1/3)^k + (-1/3)^k)/2$ (e) $21.05u[k] - 19.05 (0.905)^k$

3. $2 - 5/z - 1/z^2 + 6/z^3$ $f[0] = 2, f[1] = -5, f[2] = -1, f[3] = 6,$
   $f[k] = 0 \quad k \geq 4$

4. $10(2^k - u[k])$

EXERCISES 12.9

1. (a) $x[k] = 3 - 2(3)^k$ (b) $2^k/3 + 5(1/2)^k/3$ (c) k
(d) $10(4^k) - 5(k4^k)$ (e) $(u[k] - (-1)^k)/2$
   May be expressed as $x[k] = 0 \quad k$ even
   $= 1 \quad k$ odd

2. $(2^{k-1} - 1)u[k-1]$
3. $(-1)^k - (-2)^k$

MISCELLANEOUS EXERCISES 12

1. (i) second order, n, x, linear, inhomogeneous
   (ii) second order, k, y, linear, inhomogeneous
   (iii) first order, z, y, non-linear
   (iv) second order, n, z, linear, homogeneous
   (v) third order, k, q, non-linear

2. $1.1547, 1.0503, 1.4260$

3. (a) $3(n-1)y[n] - y[n-2] = (n-1)^2$
(b) $z[k+1] + (3 + (k-1)/2)z[k-1] = \sqrt{k-1}z[k-2]$
(c) $x[3]x[n+1] - x[2]x[n] = (n+2)^2$

4. $u[k] - a^k$

5. (a) $12(2^k) - 12(3^k) + 5k \, 3^k$ (b) $[u[k] - (-5/3)^k/3]/2$

6. $1/z^l$

7. $v_o/v_i = z/(2z-1)$

Chapter 13

EXERCISES 13.1

1. (a) fundamental frequency is 50 Hz, amplitude 3
   second harmonic has frequency of 100 Hz, amplitude 4
   third harmonic has frequency of 150 Hz, amplitude 0.7

(b) fundamental frequency is 20/π Hz, amplitude 1
   second harmonic is missing
   third harmonic has frequency of 60/π Hz, amplitude 0.5
   fourth and fifth harmonics are missing
   sixth harmonic has frequency of 120/π Hz, amplitude 0.3

2. (a) $\sqrt{13}\cos(t + 0.983)$; amplitude = $\sqrt{13}$, phase = 0.983
(b) $3.24\cos(t + 4.867)$; amplitude = 3.24, phase = 4.867
(c) $3\cos 3t$; amplitude = 3, phase = 0
(d) $\sqrt{13}\cos(2t - 0.983)$; amplitude = $\sqrt{13}$, phase = -0.983

4. (a) $f(t) = 1 \quad 0 \leq t \leq 1$
        $= 0 \quad 1 < t < 2 \quad$ period = 2

(b) $f(t) = 2t \quad 0 \leq t \leq 1/2$
            $= 2 - 2t \quad 1/2 < t < 1 \quad$ period = 1

(c) $f(t) = 0 \quad 0 \leq t \leq 1/2$
            $= t/2 - 1/2 \quad 1 < t < 3 \quad$ period = 3

EXERCISES 13.2

1. (a) odd (b) neither (c) odd (d) even
2. (a) even (b) odd (c) even (d) odd (e) neither
3. (a) 0 (b) 0 (c) 0 (d) 0 (e) 1 (f) 0

EXERCISES 13.3

1. $1/2 + (2/\pi)\sin(\pi t/5) + (2/3\pi)\sin(3\pi t/5) + (2/5\pi)\sin(5\pi t/5)$
   $+ (2/7\pi)\sin(7\pi t/5)\ldots$

2. $\pi/4 - (2/\pi)\cos t - \sin t + (1/2)\sin 2t - (2/9\pi)\cos 3t - (1/3)\sin 3t\ldots$

3. $\pi^2/3 + 2\pi\sin t - 4\cos t - \pi\sin 2t + \cos 2t + (2\pi/3)\sin 3t$
   $- (4/9)\cos 3t\ldots$

4. $8\{2\sin t + (2/3)\sin 3t + (2/5)\sin 5t + \ldots\}/\pi$

5. $1/2 + 2\{(2/\pi)\cos \pi t - \sin \pi t - (1/2)\sin 2\pi t + (2/9\pi)\cos 3\pi t$
   $- (1/3)\sin 3\pi t + \ldots\}/\pi$

6. $\pi^2/6 + [(\pi^2 - 4)/\pi]\sin t - 2\cos t - (\pi/2)\sin 2t + (1/2)\cos 2t$
   $+ [(9\pi^2 - 4)/27\pi]\sin 3t - (2/9)\cos 3t + \ldots$

7. $2\sin t$

EXERCISES 13.4

1. $3\pi/2 + (6/\pi)\sum_1^\infty [(\cos n\pi - 1)/n^2]\cos nt$
2. $-2\sum_1^\infty (\cos n\pi/n)\sin nt$
3. $2/\pi - (2/\pi)\sum_2^\infty [(\cos n\pi + 1)/(n^2 - 1)]\cos nt$
4. $e - 1 + 2\sum_1^\infty [(e\cos n\pi - 1)/(n^2\pi^2 + 1)]\cos n\pi t$
5. $(4/\pi)\sum_1^\infty [\sin(n\pi t/2)/n]$

EXERCISES 13.5

1. (a) $\sum_{-\infty}^{\infty} (j/2n\pi)(\cos n\pi - 1) e^{jn\pi t/2}$
   (b) $(1/2) \sum_{-\infty}^{\infty} [(e^{-jn\pi+1} - e^{-1+jn\pi})/(1 - jn\pi)] e^{jn\pi t}$
   (c) $\sum_{-\infty}^{\infty} A(1 + e^{-jn\pi}) e^{jn\omega t}/2\pi(1 - n^2)$

EXERCISES 13.6

1. (a) $(\sin 3\omega)/2\omega$ (b) $(1 - \cos 2\omega)/\omega^2$ (c) $2\alpha/(\alpha^2 + \omega^2)$
   (d) $(1 + j\omega)/[(1 + j\omega)^2 + 1]$ (e) $\tau/(1 + j\omega\tau)$
2. (a) $1/(\alpha + j\omega)$ (b) $1/(s + \alpha)$
3. $(2 \sin 2(1 - \omega))/(1 - \omega)$

EXERCISES 13.7

2. $4(\cos \omega)/(-\omega^2) + 4(\sin \omega)/\omega^3$
   (a) $[-4 \cos(\omega - 3)]/(\omega - 3)^2 + [4 \sin(\omega - 3)]/(\omega - 3)^3$
   (b) $[-4 \cos(\omega - j)]/(\omega - j)^2 + [4 \sin(\omega - j)]/(\omega - j)^3$
3. (a) $u(t)e^{-t}e^{-7jt}$ (b) $e^{it}u(t)e^{-t/2}$
5. $e^{4j\omega}/(1 + j\omega)$
6. 3 for $2 \leq t \leq 6$, 0 otherwise

EXERCISES 13.8

$(1/2\pi) \int_{-2\pi}^{2\pi} 2(1 - \cos \omega) e^{j\omega t}/\omega^2 \, d\omega$

MISCELLANEOUS EXERCISES 13

1. $-\sum_1^{\infty} [4n(1 + \cos n\pi)/\pi(n + 1)^2(n - 1)^2] \sin(nt)$
2. $(2/\pi) \sum_1^{\infty} [n(1 - \cos n\pi)/(n^2 - 4)] \sin(nt)$
3. (a) $(32/3\pi^2) \sum_1^{\infty} [\sin(n\pi/4) \sin(n\pi t/\tau)]/n^2$
   (b) $1/2 + (8/\pi^2) \sum_1^{\infty} [(4 \cos(n\pi/4) - \cos n\pi - 3)/3n^2] \cos\left(\dfrac{n\pi t}{\tau}\right)$
4. $v/3 + (2v/\pi) \sum_1^{\infty} [\sin(n\pi/3)\cos(2n\pi t/T)]/n$
5. $1/\pi + (1/2) \sin \omega t - (1/\pi) \sum_2^{\infty} [(\cos n\pi + 1)/(n^2 - 1)] \cos n\omega t$
6. $(v/2\pi) \sum_{-\infty}^{\infty} j[(\cos n\pi - 1)/n] e^{100n\pi jt}$
7. $(8/\pi^2) \sum_1^{\infty} [(1 - \cos n\pi)/n^2] \cos(n\pi t/4)$
10. (a) $4(\sin \omega - \omega \cos \omega)/\omega^3$ (b) $(-2j \sin \omega\pi)/(1 - \omega^2)$
    (c) $(\sin \omega\tau/\omega) + j(\cos \omega\tau - 1)/\omega$ (d) $-2j\omega/(\alpha^2 + \omega^2)$
11. (a) $(1/2\pi) \int_{-\infty}^{\infty} (6/\omega^2)(2j\omega \cos 2\omega - j \sin 2\omega) e^{j\omega t} \, d\omega$
    (b) $(1/2\pi) \int_{-\infty}^{\infty} (6/\omega)[\sin 2\omega + (\cos 2\omega - 1)j] e^{j\omega t} \, d\omega$

Chapter 14

EXERCISES 14.1

1. (a) $\partial z/\partial x = 2 - y$ $\partial z/\partial y = 3 - x$
   (b) $\partial z/\partial x = 1/y + 2x$ $\partial z/\partial y = -x/y^2$
   (c) $\partial z/\partial x = x/\sqrt{x^2 + y^2}$ $\partial z/\partial y = y/\sqrt{x^2 + y^2}$
   (d) $\partial z/\partial x = 4x\sqrt{y} - 3y \sin(xy)$ $\partial z/\partial y = x^2/\sqrt{y} - 3x \sin(xy)$
2. (a) $\partial y/\partial x = 6 \cos(2x + 3t)$ $\partial y/\partial t = 9 \cos(2x + 3t)$
   (b) $\partial y/\partial x = -16 \sin(4x - 6t)$ $\partial y/\partial t = 24 \sin(4x - 6t)$
   (c) $\partial y/\partial x = 18x^2\sqrt{z} + 2z^4/\sqrt{x}$ $\partial y/\partial z = 16z^3\sqrt{x} + 3x^3\sqrt{z}$
3. $\partial f/\partial r|_{(2,1)} = 7.646$ $\partial f/\partial h|_{(2,1)} = 6.586$

EXERCISES 14.2

1. (a) $1 + 2y, 2x$ (b) $1/y - y/x^2, 1/x - x/y^2$ (c) $\sin y, x \cos y$
   (d) $3e^{x+y}, 3e^{x+y}$ (e) ye^{xy}, xe^{xy} (f) $2y \cos(2xy), 2x \cos(2xy)$
2. $\partial^2 v/\partial r^2 = 2h^{1/2}, \partial^2 v/\partial h^2 = -h^{-3/2}r^2/4, \partial^2 v/\partial h \, \partial r = \partial^2 v/\partial r \, \partial h = rh^{-1/2}$
3. $\partial z/\partial x = 2x/(y + 1), \partial^2 z/\partial x^2 = 2/(y + 1), \partial^3 z/\partial x^3 = 0$
   $\partial z/\partial y = -x^2/(y + 1)^2, \partial^2 z/\partial y^2 = 2x^2/(y + 1)^3, \partial^3 z/\partial y^3 = -6x^2/(y + 1)^4$
   $\partial^2 z/\partial x \, \partial y = -2x/(y + 1)^2, \partial^3 z/\partial y \, \partial x^2 = -2/(y + 1)^2,$
   $\partial^3 z/\partial x \, \partial y^2 = 4x/(y + 1)^3$

EXERCISES 14.3

1. $\partial v/\partial x = 3yzi + 2xj + k$ $\partial v/\partial y = 3xzi - 2yj + 2yk$
   $\partial v/\partial z = 3xyi + 2zj$ $\partial^2 v/\partial x^2 = 2j$ $\partial^2 v/\partial y^2 = -2j + 2k$
   $\partial^2 v/\partial z^2 = 2j$
2. $\partial v/\partial x = yz \cos(xyz)i + yze^{xy}j - 2yk$
   $\partial v/\partial y = xz \cos(xyz)i + xze^{xy}j - 2xk$
   $\partial v/\partial z = xy \cos(xyz)i + e^{xy}j$
3. $\phi v = x^2yzi + x^3y^2zj - 3x^4yzk$
   $\partial(\phi v)/\partial x = 2xyzi + 3x^2y^2zj - 12x^3yzk$
   $\partial \phi/\partial x = yz$
   $\partial v/\partial x = i + 2xyj - 9x^2k$
4. $\partial v/\partial x = 1/x + 2y \cos z j - 4x^3yzk$
   $\partial v/\partial y = 1/y + 2x \cos z j - x^2zk$
   $\partial v/\partial z = -2xy \sin z j - x^4yk$
   $\partial^2 v/\partial x^2 = -1/x^2 - 12x^2yzk$
   $\partial^2 v/\partial y^2 = -1/y^2$
   $\partial^2 v/\partial z^2 = -2xy \cos z j$

EXERCISES 14.4

1. $\nabla \phi = (2x - 3yz)i + (-2y - 3xz)j - 3xyk$ $\nabla \phi|_{(0,0,0)} = 0$
2. $\nabla \phi = 2xyz^3i + x^2z^3j + 3x^2yz^2k$ $\nabla \phi|_{(1,2,1)} = 4i + j + 6k$
   $|\nabla \phi||_{(1,2,1)} = \sqrt{53}$
3. $\phi = x^2 - 4y^2x + c$

EXERCISES 14.5

1. $\nabla \cdot \mathbf{A} = 2x + xy$

2. (a) $2x + 2y + 2z$ (b) $ye^{xy} + 2xz \cos(xy) + x^3$ (c) $-yz$ (d) $1 - y$

3. (a) x (b) z^2 (c) $-yz$ (d) $1 - y$

EXERCISES 14.6

1. $-3y\mathbf{k}$

2. $(3xz + 2y^2)\mathbf{i} - 3yz\mathbf{j}$

5. $\mathbf{j} - \mathbf{k}$

6. $\nabla\phi = yz\mathbf{i} + xz\mathbf{j} + xy\mathbf{k}$ $\nabla \cdot \mathbf{v} = 6x + 6y^2$
   $\nabla \cdot (\phi\mathbf{v}) = 9x^2yz + 8xy^3z + x^2y^2$

MISCELLANEOUS EXERCISES 14

1. (a) $\partial z/\partial x = 12x^3 + 4y$, $\partial z/\partial y = -9y^2 + 4x$, $\partial^2 z/\partial x^2 = 36x^2$,
   $\partial^2 z/\partial y^2 = -18y$, $\partial^2 z/\partial x\,\partial y = 4$

   (b) $\partial z/\partial x = 2xe^{x^2-y^2}$, $\partial z/\partial y = -2ye^{x^2-y^2}$, $\partial^2 z/\partial x^2 = (4x^2 + 2)e^{x^2-y^2}$,
   $\partial^2 z/\partial y^2 = (-2 + 4y^2)e^{x^2-y^2}$, $\partial^2 z/\partial x\,\partial y = -4xy\,e^{x^2-y^2}$,

   (c) $\partial z/\partial x = -ye^{-t}\sin(xy)$, $\partial z/\partial y = -xe^{-t}\sin(xy)$,
   $\partial z/\partial t = -e^{-t}\cos(xy)$, $\partial^2 z/\partial x^2 = -y^2e^{-t}\cos(xy)$,
   $\partial^2 z/\partial y^2 = -x^2e^{-t}\cos(xy)$, $\partial^2 z/\partial t^2 = e^{-t}\cos(xy)$
   $\partial^2 z/\partial x\,\partial y = -e^{-t}(xy\cos(xy) + \sin(xy))$, $\partial^2 z/\partial t\,\partial x = ye^{-t}\sin(xy)$
   $\partial^2 z/\partial t\,\partial y = xe^{-t}\sin(xy)$

4. (a) $z\sin x\mathbf{i} + z\cos x\mathbf{j} + (-y\cos x - x\sin x)\mathbf{k}$
   (b) $z\cos x$ (c) $\mathbf{0}$ (d) $-\cos x\mathbf{k}$

6. $\nabla\psi = 2xy\mathbf{i} + x^2\mathbf{j}$ $\nabla x\mathbf{a} = \mathbf{0}$ $\nabla x(\psi\mathbf{a}) = x^2z\mathbf{i} - 2xyz\mathbf{j} + (2xy^2 - x^3)\mathbf{k}$

Chapter 15

EXERCISES 15.1

1. (a) 1/2, 1/2, 1/6, 1/6, 1/2 (b) 0 (c) 1/6 (d) 1/6 (e) 0

2. 0.2, 0.2, 0.6

3. (a) 0.1 (b) 0.63 (c) 0.03

EXERCISES 15.2

1. (a) 0.92 (b) 0.93

2. (a) 0.3195 (b) 0.215 (c) 0.5805

EXERCISES 15.3

1. A and E, C and F

2. $\bar{A}$: the lifespan is more than 90 days
   B: the machine is unreliable
   $\bar{C}$: some components have not been tested
   $\bar{D}$: two or fewer components are unreliable

EXERCISES 15.4

1. 2.3905, 0.0752

2. 3.2790

EXERCISES 15.5

1. (a) 0.72 (b) 0.044 (c) 0.2576 (d) 0.2695 (e) 0.4909

2. (a) 0.8110 (b) 0.2039

EXERCISES 15.6

1. (a) 0.196 (b) 0.112 (c) 0.2401

2. (a) 0.7800 (b) 0.7317 (c) 0.9808 (d) (i) 0.235 (ii) 0.2275 (e) 0.94

EXERCISES 15.7

1. continuous

2. (a) discrete (b) continuous (c) discrete (d) continuous

EXERCISES 15.8

1. (a) 3/2 (b) 7/16 (c) -0.5848

2. 0.28

3. (a) 27 (b) 0.3466

EXERCISES 15.9

1. 2.586

2. 2.64

EXERCISES 15.10

1. (a) mean = 3, standard deviation = 1.414
   (b) mean = 2.425, standard deviation = 0.238

2. mean = $k\bar{x}$, standard deviation = ks

EXERCISES 15.11

1. 1.668

2. no

3. 1

4. (a) 2.012 (b) 2.68

EXERCISES 15.12

1. (a) 0 (b) 0.4082
2. 0.5598

EXERCISES 15.13

1. (a) 20 160 (b) 1 663 200 (c) 220 (d) 455 (e) 455
2. (a) 1 (b) n (c) $n(n − 1)/2$
3. (a) $a^4 + 4a^3b + 6a^2b^2 + 4ab^3 + b^4$
   (b) $1 + 6x + 15x^2 + 20x^3 + 15x^4 + 6x^5 + x^6$
   (c) $p^5 + 5p^4q + 10p^3q^2 + 10p^2q^3 + 5pq^4 + q^5$
4. 112
5. 17 576 000. Yes, it is more secure. Combination is not being used correctly: permutation lock would be better.
6. 6840

EXERCISES 15.14

1. (a) 0.0256 (b) 0.1536 (c) 0.9224. New design is less reliable
2. (a) 0.9321 (b) 0.0659 (c) 0.678 (d) 0.678
3. (a) 0.3020 (b) 0.1074 (c) 0.678 (d) 0.678
4. 6

EXERCISES 15.15

1. (a) 0.1014 (b) 0.1277
2. (a) 0.2707 (b) 0.3233 (c) 0.8571
3.

	Binomial	Poisson
(a)	0.2244	0.2240
(b)	0.1689	0.1680
(c)	0.353	0.353

4. (a) 0.1435 (b) 0.2983
5. (a) 0.1075 (b) 0.1188

EXERCISES 15.16

1. (a) 0.1747 (b) 0.1577 (c) 0.8664 (d) 0.0455 (e) 0.3762
2. 119
3. 71%
4. (a) 1.28 (b) 1.645 (c) 2.33
5. (a) 1.64 (b) 1.96 (c) 2.57

EXERCISES 15.17

1. 0.999
2. 0.9998
3. 0.9967
4. 0.9901
5. 0.865
6. 0.960, 0.992
7. 0.940, 0.985

MISCELLANEOUS EXERCISES 15

1. mean = 8.35, standard deviation = 1.1314
2. (a) 0.75 (b) 0.16 (c) 0.21 (d) 0.0775 (e) 0.619 (f) 0.734
3. (a) 0.125 (b) 0.973 (c) 0.216 (d) 0.75 (e) 0.194
4. (a) 0.775 (b) 0.799

Index

Emboldened page references show where a term has been defined in text.